CONCRETE SOLUTIONS

T0227543

PROCEEDINGS OF THE INTERNATIONAL CONFERENCE ON CONCRETE SOLUTIONS, PADUA, ITALY, 22–25 JUNE 2009

Concrete Solutions

Editor

Michael Grantham
Concrete Solutions, Margate, Kent, UK
Queen's University Belfast, Northern Ireland, UK

Carmelo Majorana & Valentina Salomoni
University of Padova, Italy

CRC Press
Taylor & Francis Group
Boca Raton London New York

CRC Press is an imprint of the
Taylor & Francis Group, an **informa** business

CRC Press
Taylor & Francis Group
6000 Broken Sound Parkway NW, Suite 300
Boca Raton, FL 33487-2742

First issued in paperback 2017

© 2009 Taylor & Francis Group, London, UK
CRC Press is an imprint of Taylor & Francis Group, an Informa business

Typeset by Charon Tec Ltd (A Macmillan Company), Chennai, India

No claim to original U.S. Government works

This book contains information obtained from authentic and highly regarded sources. Reasonable efforts have been made to publish reliable data and information, but the author and publisher cannot assume responsibility for the validity of all materials or the consequences of their use. The authors and publishers have attempted to trace the copyright holders of all material reproduced in this publication and apologize to copyright holders if permission to publish in this form has not been obtained. If any copyright material has not been acknowledged please write and let us know so we may rectify in any future reprint.

Except as permitted under U.S. Copyright Law, no part of this book may be reprinted, reproduced, transmitted, or utilized in any form by any electronic, mechanical, or other means, now known or hereafter invented, including photocopying, microfilming, and recording, or in any information storage or retrieval system, without written permission from the publishers.

For permission to photocopy or use material electronically from this work, please access www. copyright.com (http://www.copyright.com/) or contact the Copyright Clearance Center, Inc. (CCC), 222 Rosewood Drive, Danvers, MA 01923, 978-750-8400. CCC is a not-for-profit organization that provides licenses and registration for a variety of users. For organizations that have been granted a photocopy license by the CCC, a separate system of payment has been arranged.

Trademark Notice: Product or corporate names may be trademarks or registered trademarks, and are used only for identification and explanation without intent to infringe.

Published by: CRC Press/Balkema
                P.O. Box 447, 2300 AK Leiden, The Netherlands
                e-mail: Pub.NL@taylorandfrancis.com
                www.crcpress.com – www.taylorandfrancis.co.uk – www.balkema.nl

ISBN-13: 978-0-415-55082-6 (hbk)
ISBN-13: 978-1-138-11632-0 (pbk)

Visit the Taylor & Francis Web site at
http://www.taylorandfrancis.com

and the CRC Press Web site at
http://www.crcpress.com

Concrete Solutions – Grantham, Majorana & Salomoni (Eds)
© 2009 Taylor & Francis Group, London, ISBN 978-0-415-55082-6

Table of Contents

Patch repairs

Repair strategy

Repair with composites

Testing

Concrete Solutions – Grantham, Majorana & Salomoni (Eds)
© 2009 Taylor & Francis Group, London, ISBN 978-0-415-55082-6

Preface

The Concrete Solutions Conference series is a British, Italian, French joint venture comprising GR Technologie Ltd. of the UK, the University of Padova, Italy (Galileo's University!) and INSA Rennes, of France. The conference is co-sponsored by the ACI and supported by the UK's Concrete Society, the Institute of Concrete Technology, and the Corrosion Prevention Association.

The first two conferences attracted wide international audiences and presented the latest in cutting-edge technology approaches to the repair of deteriorated concrete structures. This conference has attracted participation from 34 Countries, which is no mean achievement! Again an excellent rage of diverse papers has been captured, with many useful case studies and plenty of useful research papers, too.

This book contains the Proceedings of the third conference, held in Padova, Italy, at the prestigious University of Padova. It presents the best in research, practical application, theory and strategy from this important international conference.

Papers are presented on electrochemical repair, patch repair, repair with composites, repair of fire damaged structures, repair strategy, general repair topics and testing and inspection.

Such sharing of international knowledge is the key to understanding of concrete problems, how to prevent them and how to repair them economically and practically. It is essential reading for researchers and engineers alike.

Concrete Solutions – Grantham, Majorana & Salomoni (Eds)
© 2009 Taylor & Francis Group, London, ISBN 978-0-415-55082-6

Acknowledgements

The Editors would like to acknowledge the help of the Technical Committee in organising and publicising the Concrete Solutions Conference.

Prof. Michael Grantham – Concrete Solutions and Queen's University Belfast.
Prof. Carmelo Majorana – University of Padova, Italy
Dr. Valentina Salomoni – University of Padova, Italy
Prof. Christophe Lanos – IUT Rennes, France
Dr. Raoul Jauberthie – INSA Rennes, France
Dr. Carmen Andrade – Instituto Eduardo Torroja – Spain
Dr. Ulrich Schneck – Citec GmbH – Dresden
Kevin Armstrong – States of Jersey, Public Services Dept.
Dr. Peter Robery – Halcrow Group, UK
Dr. John Broomfield – Corrosion Consultant – UK
Dr. Calin Mircea – INCERC National Institute for Building Research – Romania

The support of the sponsors, supporting organisations and exhibitors is gratefully acknowledged.

Electrochemical repair

Concrete Solutions – Grantham, Majorana & Salomoni (Eds)
© 2009 Taylor & Francis Group, London, ISBN 978-0-415-55082-6

Rapid electrokinetic nanoparticle transport in concrete

Henry E. Cardenas
Department of Mechanical and Nanosystems Engineering, Louisiana Tech University, Ruston, Louisiana, USA

Faisal Syed
Department of Electrical Engineering, Louisiana Tech University, Ruston, Louisiana, USA

ABSTRACT: Lithium bearing compounds have long been identified as exhibiting a capacity to suppress alkali silica reaction (ASR). Until now, the prospect of getting lithium into concrete was confined to the batching operation of new construction or to surface spray applications. The latter processes typically yielded very shallow penetration levels of less than 3 cm. In light of recent advances in electrokinetic nanoparticle processing, this study explored the possibility of rapidly driving lithium deep into concrete by coating it onto a 20 nm silica particle. Trials were conducted on 30-cm cylinders of relatively immature concrete. Treatments were carried out at current densities that were as much as an order of magnitude higher than the typical levels permitted for electrochemical processes applied to concrete such as cathodic protection or electrochemical chloride extraction. This work showed that lithium transport exceeded 30 cm and resultant cylinder strengths were found to increase 30% on average even at extremely high current densities. Scanning electron microscopic analysis indicated radical alteration of the microstructure. In light of these findings the paper further introduces the concept of electromutagenic processing as a next generation methodology for electrochemical repair.

1 INTRODUCTION

ASR is a chemical reaction between reactive silica found in certain aggregates and the hydroxides of alkali metals. Stanton first described the general features of alkali silica reaction (Stanton, T.E., 1940). This reaction results in formation of hygroscopically expansive gel. Due to expansion, stresses are induced on the surrounding hardened cement paste resulting in micro cracking and volumetric expansion (Nizar *et al.* 2004, Chiara, 1995, and Berube *et al.* 2004). Chemical admixtures such as lithium salts have been found to inhibit ASR (McCoy and Caldwell, 1951). Various lithium bearing compounds have been investigated for practical use as ASR suppressing agents (Ramyar *et al.* 2004, Hudec and Banahene, 1993, Stark, 1994, and Prezzi, *et al.* 1998). Getting lithium into hardened concrete is a continuing challenge that is limited by long processing time and poor penetration. The current study examines the use of silica nanoparticles as lithium carriers that can permit high speed electrokinetic transport.

The possibility of rapidly transporting nanoparticles into concrete by applying an electric potential had been explored by Cardenas and Struble (2006). A reduction in permeability was achieved by transporting silica nanoparticles in hardened cement paste. Recently lithium-coated nanoparticles have been used in electrokinetic treatment of concrete masonry units (Cardenas *et al.* 2008). Nanoparticle treatment of masonry units with 50 nm lithium-coated silica

particles resulted in an increase in strength of 316%. Electrokinetic transport of nanoscale alumina-coated silica particles (24 nm in size) resulted in controlling corrosion of reinforcement in concrete (Kupwade-Patil and Cardenas, 2008). In another study involving sulfate attack in concrete, specimens were exposed to sulfates after being treated with alumina-coated silica nanoparticles. The treated specimens were 33% higher in strength, 13% lower in porosity, and free of sulfates (Kupwade-Patil and Cardenas 2008).

Cracking due to ASR results in a reduction of strength and stiffness. This damage can effectively increase the porosity, making the structure more vulnerable to chemical attacks. In the case of civil structures such as locks and dams, ASR damage can cause failure of moving parts that are surrounded by expanding concrete (Charlwood and Solymar, 1995. Newell and Wagner, 1995, and Collins et al. 2004). ASR affected structures can also suffer from displacement, deformation, and loss of rigidity (Stephane, *et al.* 2006). Remediation often requires the costly removal of material or replacement.

At this point, the mechanism by which lithium salts inhibit expansion has yet to be unequivocally established (Hudec and Banahene, 1993, and Feng et al. 2005]. It is generally assumed that Li ions partially substitute for K^+ or Na^+ ions in the ASR gel. Sakaguchi and others confirmed no visible ASR by examining a pyrex glass with a lithium containing cement paste interface using an X-ray spectrometer (Sakaguchi et al. 1989, and Feng et al. 2005). The study

Table 1. Batch composition.

Material	Mass	
	(kg)	(lb)
Water	8.39	18.5
Gravel	42.41	93.5
Cement	16.56	36.5
Sand	25.63	56.5
NaOH	0.09	0.2

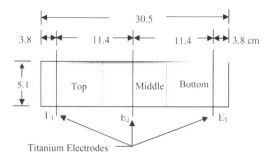

Figure 1. Schematic diagram of the specimen.

confirmed the presence of a lithium silicate at the surface of the aggregate particles. It was also reported that the lithium-silica reaction was more favorable than the competing sodium-silica or the potassium-silica reactions.

In general, Li delivery is accomplished by simple addition to the mix. For hardened concrete, surface spray application has yielded limited penetration. Electrokinetic transport to date has also exhibited limited penetration (Takao, et al. 2006). Within this region, the Li^+ treatment was successful in suppressing expansion due to ASR.

The current work was distinctive in exploring the use of lithium-coated silica nanoparticles for rapid (high current) delivery of lithium several centimeters into concrete. The silica particle interiors were intended to provide some degree of microstructural healing that was needed to compensate for the use of extremely high current densities. These elevated current densities were required to provide deep penetration in a reasonable amount of time. Perhaps the most unique feature of this work was the radical and unexpected change in microstructure and strength that can now be referred to as an electromutagenic process.

2 PROCEDURES

The mix design in this study consisted of type I portland cement, aggregate, and water in the ratio of 1.9:8.1:1. Table 1 shows the batch composition. A W/C ratio of 0.5 was used for all batching procedures. The gravel and sand aggregates were obtained from Chaparral Sand & Gravel, El-Paso, Texas. The aggregate exhibited a fineness modulus of 3.31. In order to encourage a possible alkali-silica reaction (ASR), NaOH pellets were added to the water as per ASTM C129. Each trial consisted of 4 treated specimens and 4 untreated control specimens.

Figure 1 illustrates a schematic side view of the 30 cm (12 inches) deep specimen used in this work. After the treatment, the specimens were sectioned into lengths of 10.2 cm (10 cmches) each as indicated by the dashed lines in Figure 1. The sections are denoted as top, middle and bottom. The top section is the one that had the treatment reservoir attached to it (as shown in Figure 2) and the Bottom one is the section that was immersed in NaOH, also shown in Figure 2. E1, E2 and E3 are titanium wire electrodes. These

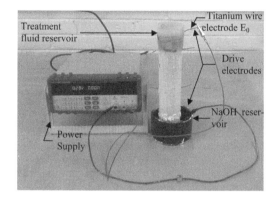

Figure 2. Nanoparticle treatment setup.

consisted of mixed metal oxide coated wire of diameter 0.15 cm (0.06 inches), manufactured by Corrpro Companies Inc, Medina, Ohio. The electrodes E1 and E3 are located 3.8 cm (1.5 inches) from each end and E2 is in the middle of the specimen (15.2 cm (6 inches) from either end). See Figure 1.

The specimens were cured in 1 M NaOH for a period of 10 days at approximately 20°C. The side of the specimens were coated with masonry sealant (Dry Lock Masonry sealant, United Gilsonite Labs, Jacksonville, IL) leaving 3.81 cm (1.5 inches) uncoated adjacent to each end. The sealant prevented the loss of moisture along the sides of each specimen. Figure 2 shows the specimen setup for treatment application. The treatment fluid reservoir was glued to the top portion of each specimen. Electrode E_0 (titanium wire) was inserted through the top reservoir, where it served as one of the drive electrodes. Another plastic container was used as a NaOH reservoir at the bottom section of each specimen.

A 50 ml portion of proprietary lithium coated silica suspension was placed into the top reservoir at the beginning of treatment. The particles ranged from 50 to 600 nm with a medium size of 140 nm. The bottom portion of the specimen was partly immersed in a one molar NaOH solution. Electrodes E_0 and E_3 (see Figure 1) were used as drive electrodes. Each treatment was conducted for 12 days.

Three treatment cases were studied during this work. Table 2 contains the treatment matrix. All the

Table 2. Treatment case matrix.

Case	Current Density (A/m^2)
Low current	3
Medium current	6
High current	10

treatments were applied under controlled current conditions. After setting the current at a selected current density, the voltage across each specimen was monitored daily. Due to the generally low availability of high voltage for on-site treatments the potentials applied across the specimens were restricted to 50 V. After a maximum allowed potential of 50 V was encountered, the treatment mode was shifted to control voltage.

Upon completion of the treatment, the treated and untreated control specimens were cut into three equal sections of 10.2 cm (4 inches) in length and labeled as top, middle and bottom. Each section was then tested for compressive strength. A sample of each section was later taken for SEM evaluation and porosity measurement. Another sample from each section was extracted to determine lithium content via atomic absorption.

3 RESULTS AND DISCUSSION

The impact of the treatment on strength and porosity is presented in Tables 3 and 4. In these figures each data point represents an average of four trials. For the low current density treatment (3 A/m^2) the top section of the treated specimens exhibited a 56% increase in strength followed by the bottom sections with an increase of 23%. See Table 3. The middle sections exhibited a 12% increase in strength. The specimens were then tested for any change in porosity due to treatment. These values are listed in Table 4. The top section exhibited an increase in porosity from 7.0% in the untreated controls to 7.5 % among the treated specimens. The middle sections exhibited a decrease from 7.0% in the controls to 6.8% in treated specimens. Decreases in porosity were also observed for the bottom sections, which were at 8% for the controls and 7.6% among the treated specimens. As a point of reference, the range of error in these measurements was ±0.5%. This indicates that the net porosity changes were not significant.

The highest average strength increase was observed in the top section of the specimen. This result appears to indicate that a relatively small portion of the particles penetrated beyond the top section. It is possible that the voltage applied to drive the particles was not high enough to provide a significant depth of penetration. Alternatively the particles were accumulating in the top section due to some physical or chemical constraint. At the other end of the specimen, there was an apparent deposit of efflorescence observed just above electrode E$_3$ (See Figure 3). This build up could be associated with the increased strength of the bottom

Table 3. Strength increases.*

Location of Sample	Current Density		
	Low (3 A/m^2)	Medium (6 A/m^2)	High (10 A/m^2)
Top	56%	50%	32%
Middle	12%	22%	33%
Bottom	23%	16%	36%
Average	30%	29%	34%
Error range	±9%	±9%	±9%

* Each value is an average of 3 section results.

Table 4. Porosity changes.*

Location of Sample	Current Density		
	Low (3 A/m^2)	Medium (6 A/m^2)	High (10 A/m^2)
Top Change	+0.5%	+0.1%	−1.4%
Middle Change	−0.2%	−0.2%	−1.2%
Bottom Change	−0.4%	−0.4%	−1.2%
Ave. Porosity	7.5%	7.0%	7.0%
Error range	±0.5%	±0.5%	±0.5%

* Each value is an average of 3 section results.

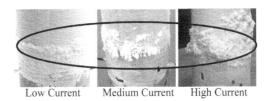

Low Current   Medium Current   High Current

Figure 3. Images showing efflorescence of material near the bottom of each specimen at electrode E3. The current density increases correlated with increasing material discharge.

section by more than what was observed in the middle section. The accumulation of particles at this electrode and the adjacent material may have resulted in a more densified microstructure in the bottom section as compared to the middle. The efflorescence of material shown in Figure 3 could be due to the accumulation of unreacted particles and other species near electrode E$_3$. This electrode was immersed in the NaOH reservoir. It can be speculated that due to this physical arrangement there was no visible efflorescence observed either 1.3 cm (1/2 inch) above or anywhere below the electrode. This is because the escaping material would have drifted out into the NaOH solution in these areas.

It is informative to analyze the amount of particles available to fill in the pores of the concrete. The average porosity of the untreated control was 7%. The volume of the entire cylinder specimen was 620 ml. 7% of the volume of this cylinder was the pore volume, which was 43 ml. The volume of the particles available in the treatment fluid was 6 ml.

From these values it is obvious that there were not enough particles to fill the pores completely. A decrease of 1.0% in porosity was the best that could be expected after the treatment. The measured change in porosity was in the range of 0.6%, which was lower than expected. This value was not much larger than the range of error ($\pm 0.5\%$) that was observed for these measurements. There are two likely reasons for this lower than expected porosity impact. Efflorescence of material out of the specimen may have driven nanoparticles out of the specimen. In addition to treatment species, the excreted material may have included ions such as sodium, potassium and calcium which originated within the concrete. Future work will examine the content of the efflorescent material. Based on these observations, the low particle volume dosage combined with gelling and efflorescence may have limited the porosity reduction that could be achieved from the nanoparticle treatment.

In the case of the medium current density treatment ($6\,A/m^2$), there was an average strength increase of 50% observed in the top section of the specimens (See Table 3). These specimens also exhibited no significant change in porosity (Table 4). A strength gain of 22% and 16% was observed in the middle and bottom sections of the specimens respectively. An average porosity decrease of 0.2% and 0.4% was observed in the middle and bottom sections respectively. Here again the range of error for the porosity measurement was approximately $\pm 0.5\%$, indicating that the change in porosity was not significant.

After the treatment the strength of the specimens from the top section exhibited a 50% increase from 1200 psi to 1800 psi. By comparison, the low current trial exhibited a 55% increase in strength from 900 psi to 1400 psi. The increase in strength of the middle section was 10% more when compared to that of specimens treated at low current density (12% low versus 22% medium). This may have come at the expense of the top section which gained 5% less strength in this trial as compared to the low current case.

The decrease in strength gain of the bottom section from 22% in the low current case to 16% in medium current case does not carry clear significance. The amount of material expelled from the medium current density specimens was at least 100% greater than compared to the low current case (See Figure 3). This is not surprising since twice the current would be expected to transport twice as much material.

The porosity and strength of the treated and untreated control specimens after treatment at high current density ($10\,A/m^2$) are also listed in Tables 3 and 4. An increase in strength of 32% was observed in the top section of the specimen. The middle and the bottom sections of the specimen exhibited a rise of 33% and 36% in strength respectively. An average porosity decrease of 1% was observed at each section of the specimen (top, middle and bottom sections).

The 32% strength benefit observed in the top section of the specimen, was 23% less than the low current case and 18% less than the medium current case. The decrease in strength benefit in the top section could possibly be due to the high voltage (50 V) developed across the specimen during treatment, causing damage to the microstructure in this region. An elevated electric field strength would increase the kinetic energy of the particles as a quadratic function of the velocity. In contrast, the higher voltage may also have permitted the extraction of some loosely bonded particles from the gel that had formed in the treatment suspension (at the top of the specimen). This would have provided more particles for strengthening. The voltage across each specimen was restricted to a maximum value of 50 V which is comparable to typical voltage limits for onsite electrochemical treatments.

The average strength increase observed in the middle and bottom sections of the high current specimens was 33% and 36% respectively. Due to the high current and voltage across the specimens a reduction in strength was expected throughout. The increase in strength in the middle and bottom sections could have been due to the porosity reduction provided by the nanoparticles. The nanoparticles may have filled micro or nano cracks developed as a result of the high current and voltage. Based on these observations, it is concluded that current densities significantly higher than $1\,A/m^2$ can be employed for treatments without degradation in strength when restorative nanoparticles are driven in during treatment.

As evident in Figure 3 the amount of material expelled at high current density was approximately twice the that observed during the medium current density treatment. The location from where the material was expelled remained unchanged at all three current levels. As observed in all the three treatment cases, increases in current density resulted in proportionally larger amounts of efflorescence.

Examination of Table 3 indicates that an increase in current density reduced the strength benefit of the top section and increased the strength benefit in the middle and bottom sections. It can also be noted that increasing current density tended to result in reduced error ranges. Another interesting point is that the high current density produced a relatively uniform distribution of strength benefit in all the sections.

Samples from the top, middle and bottom sections of each specimen were taken to examine lithium penetration. Atomic absorption was used to determine lithium content. Table 5 shows levels of lithium detected. Each data point is an average of three trials. The error range was 0.38 ppm.

The lithium content of the treatment liquid (lithium coated silica suspension) was found to be 10 ppm. Following treatment, the top section of each concrete specimen exhibited the highest amount of lithium. The low, medium and high current density treatments resulted in 2.5, 3.6 and 2.2 ppm of lithium respectively in the top section. In two cases, the expelled material was also examined for lithium. 4.9 ppm and 5.2 ppm of lithium was found in the expelled material from the medium and high current density treatments respectively.

Table 5. Lithium Content in parts per million (ppm)*

Location of Sample	Current Density		
	Low (3A/m^2)	Medium (6A/m^2)	High (10A/m^2)
Top	2.5	3.6	2.2
Middle	0.8%	0.4	1.0
Bottom	0.8%	0.4	1.0
Expelled mass	—	4.8	5.2

* Each value is an average of 3 section results.

As shown in Table 5, the medium and high current density treatments were successful in driving almost all the lithium content from the treatment liquid. The total lithium content found in the medium and high current density treatment cases correlates to a starting suspension content of 9.2 ppm and 9.4 ppm respectively (original lithium content was 10 ppm). It is possible that some of the lithium might have been retained in the treatment solution due to formation of gel during treatment. Additional lithium may also have escaped into the NaOH solution in which the bottom of each specimen was immersed.

In the low current density treatment, the average lithium content throughout the specimen was observed to be 1.4 ppm. This value did not include the lithium content of the expelled material. The middle and the bottom sections exhibited 0.4 ppm of lithium volume content relatively. 4.8 ppm of lithium was found in the material that was expelled from the sample. This relatively high amount of lithium content in the expelled material may have come at the expense of the middle and bottom sections which exhibited the lowest values. There was a decrease of lithium content in medium and bottom sections from 0.8 ppm for low current case to 0.4 ppm for the medium current case (which exhibited the lowest values).

In the high current density treatment, there was a decrease in the lithium content in the top section from 3.6 ppm in the medium current case, down to 2.2 ppm in the high current case. An average increase of 0.6 ppm was observed in the middle and bottom sections of the high current specimens as compared to the medium current treated specimens. A negligible increase from 4.9 to 5.2 ppm of lithium was detected in the expelled material for the medium and high current cases respectively.

The decrease in average strength gain of the top section of the high current specimen, from 50% for the medium current case to 32% appeared to coincide with shifting of lithium and probably nanoparticles from the top section towards the middle and bottom sections. The lithium coated silica particles being strongly driven from the top section deeper into the specimens could have damaged the microstructure in the top region resulting in lowered strength. Based on these observations, lithium and nanoparticle penetration was achieved to a 30 cm (12 inch) depth using EN treatment.

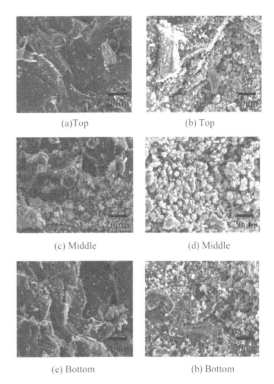

(a)Top (b) Top

(c) Middle (d) Middle

(e) Bottom (b) Bottom

Figure 4. (a), (c), and (e) are the SEM images from the top, middle and bottom sections of a specimen treated at low current density (3 A/m^2). (b), (d), and (f) are the SEM images of untreated control specimens from top, middle and bottom sections respectively.

As mentioned earlier the 7% porosity of a given specimen was equivalent to a pore volume of 43 ml. The amount of lithium stabilized nanoparticles available in 50 ml of treatment liquid was only 6 ml. This limited amount of treatment material could be a possible reason for insufficient volume of nanoparticles available to completely block the pores. Gelling of the solution and efflorescence observed due to treatment may have further limited the amount of lithium content found in specimen. Even with a low volume of delivered nanoparticles, a considerable increase of at least 29% in strength was achieved in all three treatment cases.

Fracture samples were extracted from the top, middle and bottom sections of each specimen for scanning electron microscopic (SEM) examination. Figure 4 contains images of the untreated control specimens and the treated specimens for the low current density case. At all three levels it is clear that the treatment dramatically altered the microstructure. Similar microstructural changes were observed for all three current density cases. In all cases, the top sections of the treated specimens tended to exhibit the most planar microstructure as compared to the untreated controls and as compared to the middle and bottom sections that received treatment. The top section also appeared to have the fewest micro cracks.

In the low current density treatment, the top section exhibited the maximum amount of lithium content of 2.5 ppm as compared to 0.8 ppm measured for the other two sections (see Table 5). The highly planar morphology observed in the SEM image of the top section also appears to coincide with the maximum strength increase of 56% in this section compared to 12% and 23% in the middle and bottom sections.

This work provided new insight on deeply penetrating lithium delivery, the use of extremely high processing currents, and the use of an electrochemical treatment to significantly alter the microstructure and strength of concrete. These radical alterations may possibly be associated with the *in situ* formation of a lithium silicate (Chung-Chemg et al. 2006, and Kamadaa et al. 2005). This possibility will be furthered examined in future work. At this juncture these findings appear to constitute the first recorded case of an electromutagenic process applied to concrete.

4 CONCLUSIONS

The electrokinetic treatment successfully achieved penetration of lithium and nanoparticles to a depth of 30 cm (12 inches). An increase in treatment current density resulted in decreasing strength benefit in the top sections of the treated specimens. At the same time there was an increase of strength benefit in the middle and bottom sections of treated specimens that correlated with the rise in current density. This work showed for the first time that high current densities ($>1 A/m^2$) can be employed for treatments without degradation in strength when restorative nanoparticles are utilized.

It was interesting to note that the highest applied current density appeared to provide the most even distribution of the strength benefit.

Each increase in current density resulted in ever larger amounts of efflorescence issuing from near the bottom electrode of each treated specimen. It appears that the combination of the relatively low dosage of nanoparticles and efflorescence resulted in virtually no reduction of porosity due to treatment.

An average strength benefit of ~30% was achieved in all the three cases even with an apparently small volume of nanoparticles being retained in the specimen. The radical change in microstructure combined with the increase in strength constitutes the first recorded case of an electromutagenic process applied to concrete.

REFERENCES

ASTM C1293, "Standard Test Method for Determination of Length Change of Concrete Due to Alkali-Silica Reaction", Philadelphia, pp. 10–17, February 2001.

Bérubé, M.A., Tremblay, C., Fournier, B., Thomas, M.D., Stokes, D.B., "Influence of Lithium-Based Products Proposed for Counteracting ASR on the Chemistry of Pore Solution and Cement Hydrates", Cement and Concrete Research, Vol. 34, No. 9, pp. 1645–1660, September 2004.

Cardenas, H., Lvov, Y., Kurukunda, A., "Electrokinetic Assembly of Polymeric and Pozzolanic Nanoparticle Phases Within Concrete", Proceedings, Society for the Advancement of Material and Process Engineering, Long Beach, CA, May 2008.

Cardenas, H., and Struble, L., "Electrokinetic Nanoparticle Treatment of Hardened Cement Paste for Reduction in Permeability", American Society of Civil Engineers Journal of Materials in Civil Engineering, Vol. 18, No. 4, July/August 2006.

Charlwood, R.G., Solymar, Z.V., "Long-Term Management of AAR-Affected Structures: An International Perspective", 2nd International Conference on Alkali-Aggregate Reaction in Hydroelectric Plants and Dams, Chattanooga, TN, p. 19, 1995.

Chiara, F.F., "Alkali Silica Reaction and High Performance Concrete", Building and Fire Research Laboratory, National Institute of Standards and Technology, August 1995.

Chung-Cherng, L., Pouyan Shen ; Chang H.M., YANG Y. J., "Composition Dependent Structure and Elasticity of Lithium Silicate Glasses: Effect of ZrO_2 Additive and the Combination of Alkali Silicate Glasses", Journal of the European Ceramic Society, Vol. 26, No.16, pp. 3613–3620, 2006.

Collins, C.L., Ideker, J.H., Willis, G.S., Kurtis, K.E., "Examination of the Effects of LiOH, LiCl, and LiNO3 on Alkali–Silica Reaction", Cement and Concrete Research, Vol. 34, No. 8, pp.1403–1415, August 2004.

Feng, X., Thomas, M.D.A., Brenner, T.W., Balcom, B.J., Folliard, K.J., "Studies on Lithium Salts to Mitigate ASR-induced Expansion in New Concrete: a Critical Review", Cement and Concrete Research, Vol. 35, No. 9, pp. 1789–1796, September 2005.

Hudec, P.P., Banahene, N.K., "Chemical Treatments and Additives for Controlling Alkali Reactivity", Cement and Concrete Composites, Vol.15, No. 1-2, pp. 21–26, 1993.

Kamadaa, K., Izawab, k., Yamashitab, S., Tsutsumib, Y., Enomotoa, N., Hojoa, J., Matsumoto, Y., "Crack-Free Lithium Ion Injection into Alkali Silicate Glass Using Li+ Conducting Microelectrodes", Solid State Ionics, Vol. 176, No. 11-12, pp. 1073-1078, March 2005.

Kupwade-Patil, K., Cardenas, H., "Corrosion Mitigation in Concrete Using Electrokinetic Injection of Reactive Composite Nanoparticles", Proceedings, Society for the Advancement of Material and Process Engineering, Long Beach, CA, May 2008.

Kupwade-Patil, K., Cardenas, H., "Composite Nanoparticle Treatments for Mitigation of Sulfate Attack in Concrete", Society for the Advancement of Material and Process Engineering, Long Beach, CA, May 2008.

McCoy, W.J.,Caldwell, A.G., "A New Approach to Inhibiting Alkali-Aggregate Expansion," Journal of the American Concrete Institute, 47, pp. 693–706, 1951.

Newell, V.A., Wagner, C.D., " A Review of the History of Alkali-Aggregate Reaction at the Three of the Tennessee Valley Authority dams", 2nd International Conference on Alkali-Aggregate Reaction in Hydroelectric Plants and Dams, Chattanooga, TN, p. 57, 1995.

Nizar, S., Marc-André, B., Fournier, B., Benoit, B., Benoit, D., "Evaluation of the Expansion Attained to date by Concrete Affected by Alkali–Silica Reaction. Part I: Experimental Study", Canadian Journal of Civil Engineering, Vol. 31, No. 5, pp. 826-845, 2004.

Prezzi, M., Monteiro, P.J.M., Sposito, G., "Alkali-Silica Reaction: Part 2: The Effect of Chemical Additives", ACI Materials Journal, Vol. 95, No. 1, pp. 3–10, 1998.

Ramyar, K., Çopuroğlu, O., Andiç, O., Fraaij, A.L.A.A., "Comparison of Alkali–Silica Reaction Products of Fly-Ash or Lithium-Salt Bearing Mortar under Long-Term Accelerated Curing", Cement and Concrete Research, Vol. 34, No. 7, pp. 1179–1183, 2004.

Sakaguchi, T., Takakura, M., Kitagawa, A., Hori, T., Tomozawa, F., Abe, M., "The Inhibitive Effect of Lithium Compounds on Alkali-Silica Reaction", Proceedings of 8th International Conference on Alkali-Aggregate Reaction, Elsevier Applied Science, Kyoto, p. 229, 1989.

Stanton, T.E., "Expansion of Concrete through Reaction between Cement and Aggregate," Proceedings, American Society of Civil Engineers, Vol. 66, Part 2, No. 10, pp.1781-1811, New York, 1940.

Stark, D., "Alkali-Silica Reaction in Concrete, in: Klieger, P., Lamond, J.F. (Eds), Significance of Tests and Properties of Concrete and Concrete Making Materials", STP 169C, ASTM Publications, Philadelphia, PA, pp. 365–371, 1994.

Stéphane, P., Alain, S., Bruno, C., Geneviève, T.F., Jean-Michel, T., Hélène, T.C., Eric, B., "Influence of Water on Alkali-Silica Reaction: Experimental Study and Numerical Simulations", Journal of Materials in Civil Engineering, Vol. 18, No. 4, pp. 588–596, 2006.

Takao, U., Yukihiro, Y., Keisuke, Y., Akira, N., "Influence of Electrochemical Treatment Condition on Electrophoresis of Lithium Ion and Asr Expansion of Concrete", Japan Cement Association Proceedings of Cement & Concrete, Vol 59, pp. 483-489, 2006.

Wagner, C.D., Newell, V.A., " Modification to Hiawassee Dam and Planned Modification to Fontana and Chickamauga Dams by the Tennessee Valley Authority to Manage Alkali-Aggregate Reaction", 2nd International Conference on Alkali-Aggregate Reaction in Hydroelectric Plants and Dams, Chattanooga, TN, p. 83, 1995.

Concrete Solutions – Grantham, Majorana & Salomoni (Eds)
© 2009 Taylor & Francis Group, London, ISBN 978-0-415-55082-6

Cathodic protection of reinforced concrete swimming pools

Paul Chess
C P International ApS, Denmark

John Drewett
Concrete Repairs Ltd, UK

ABSTRACT: Many swimming pools were built in the 1960's and 1970's as part of leisure facilities or schools. These typically comprise a cast insitu reinforced concrete base and walls with cast insitu promenade decks around the pool. These structures are now often in severe distress due to corrosion of the reinforcement in the walls and promenade decks. The use of cathodic protection to stop this degradation of the reinforcement, by the use of anodes in the pool water, has been developed to be the standard repair method in Denmark. The anodes are installed in boxes, recessed into the pool side walls of the swimming pool. The economic benefits of this form of repair are so impressive that it is now being used in the United Kingdom (UK).

This paper describes a recent installation of this type in the UK at Chiltern Pools Leisure Centre, Amersham, England. In this installation, the main swimming pool walls and base are protected using anodes in the pool water. The cathodic protection system is computer controlled. This gives an accurate continuous control of the output current to each part of the structure based on real time readings from reference electrodes. This gives a better and more even protection from corrosion, increases the life of the anodes and switches off the anodes in the water when bathers are in the pool. The computer control system has a modem and a telephone connection allowing remote monitoring and control of the system.

1   INTRODUCTION

In 1992, Cathodic Protection International ApS (CPI) supplied a cathodic protection system, using anodes in the pool water, for a swimming pool in Denmark. This was the first commercial use of this type of cathodic protection for a swimming pool structure in Denmark. Since then CPI has supplied cathodic protection for more than 60 swimming pools in Denmark, with at least 50 of them computer controlled. Other countries have discovered the benefits of cathodic protection of swimming pools, and CPI has supplied these systems to Norway, Sweden, Luxembourg and UK.

The cathodic protection of a swimming pool can consist of two different methods:

- The first method is only suitable for the pool structure and involves anodes placed in the water, which we call water anodes. These are normally recessed into the wall and fixed in plastic boxes sized to replace one or two tiles. The current is passed through a perforated lid to the water, from the water through the grout between the tiles and through the concrete to the reinforcement steel. A 25 m × 12 m swimming pool would normally require 6 – 12 water anodes. Water anodes are also applicable for concrete expansion tanks.
- The second and "conventional" cathodic protection method for protecting a swimming pool uses anodes

mounted from the outside (i.e. the dry side of the pool) in the corrosion affected area. Anode types used are internal anodes, conductive paint, titanium mesh, ribbon or others applied directly to the outside of the corrosion affected area. This method can be used instead of the water anodes, and is applied from the outside of the pool. In Denmark it is normal for internal anodes to be used to protect the parts of the structure that the water anodes cannot reach, such as support columns, promenade deck slabs and diving towers.

To control the performance of the cathodic protection system a number of reference electrodes are normally installed. In the water, zinc reference electrodes are often used. A 25 m × 12 m swimming pool would normally require 4 – 10 zinc reference electrodes. In the concrete, silver/silver chloride reference electrodes are generally used for absolute potentials and depolarisation and the cheaper mixed metal oxide reference electrodes are used for depolarisation measurements only. Often, cast in reference electrodes will be calibrated against an external reference electrode.

The use of computer control has been shown to be very advantageous for cathodic protection systems in swimming pools, especially for water anode systems. The current requirement can vary significantly in different areas of the pool, and at different times of the day. A computer control system can take readings from many reference electrodes into account and adjust the individual anode outputs to optimise performance on a real time basis.

2 STRUCTURAL PROBLEMS AT CHILTERN LEISURE CENTRE

Chiltern swimming pool was constructed in 1972 on behalf of the local council. The pool has since become a very popular facility in this town. The pool building comprises a steel portal framed building with insitu concrete pool basins and pool side walkway slabs. Access is made available to the dry side of the pool basins via a pool side service void.

The main pool basin and the pool building frame have experienced significant structural decay over their life, most notably in the form of spalled concrete due to expansive corrosive reactions occurring in the reinforcement and corrosion at the base of the steel stanchions. The majority of the areas of damage to the pool basin are located in areas of inadequate concrete cover to the steel, which was noted to be zero over significant areas. It is believed that the lack of cover was the result of poor quality control during the construction of the pool. This lack of quality control was confirmed by the large amounts of formwork timber left encased in the concrete, and the large number of steel formwork bolts and brackets left protruding from the concrete surfaces.

The precast concrete building frame would be expected to have been built with a higher level of quality control than the insitu components. Unfortunately, however, the building frame was not designed specifically for use in swimming pools with only 25mm of concrete cover having been provided to the steel (mild exposure conditions for a grade C30 concrete). In reality the exposure conditions within the pool side service void are very severe. BS EN206-1/BS8500 would now classify this as an XD3 Exposure Class, calling, for example, for a C32/40 concrete and a minimum 45mm cover.

In areas of the tile grout and the expansion and addition to the poor concrete construction quality, significant contraction joint sealants around the pool have failed. This has resulted in a steady flow of chlorinated water over the exposed concrete faces.

The resulting damage to the concrete basin and steel portal frame can be seen in Figures 1 and 2.

Chiltern District Council undertook a full structural appraisal of the pool building frame and the pool basins in the damaged state to assess the risks to the public and staff at the leisure facility. The levels of reinforcement were determined following inspections of the areas of spalled concrete. It was found that the reinforcement provided in the pool basin was the minimum area of reinforcement and was not required for the long term stability of the pool basin. The pool basin was shown to be of adequate strength and stiffness even allowing for a large loss in cross section due to the spalled concrete. The building frame was assessed with the assumption

Figure 1. Areas of spalled concrete on the main pool basin wall. The very low cover to the steel reinforcement can be seen.

Figure 2. Encrustation of the main pool basin wall. These deposits are normally signs of leakage.

that a plastic hinge had formed at the location of the severe spalling. The frame was also shown to be stable in its damaged state. Despite having shown that the structure was stable in the ultimate limit state it was concluded that concrete repairs should be undertaken to help prevent further decay and serviceability problems in the future.

Investigations were undertaken by CRL Surveys, a division of Concrete Repairs Ltd, into the causes of the degradation of the concrete. Half cell potential and covermeter surveys were undertaken and concrete samples were taken from a number of locations around the pool basin and the frame .The samples were tested for chloride content, carbonation, sulphate and concrete strength. It was shown that sulphate levels were well within the prescribed limits, the carbonation levels were low and that the concrete was of a reasonable quality. The chloride content results however showed a large variation across the samples tested with a number demonstrating very high levels of chlorides, which was consistent with the half cell survey.

It was concluded that the main factor affecting the decay of the structure was the high level of chlorides

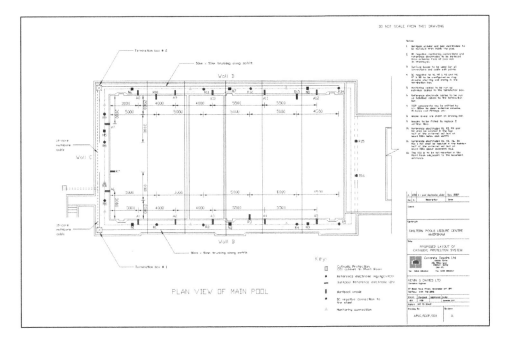

Figure 3.   Schematic layout for the ICCP system.

due to the penetration of swimming pool water into the concrete and the very low levels of cover to the concrete.

Given the importance of the facility it was considered vital that any proposals for remedial works provide a robust long term solution. Consideration was given to a number of options:-

- Patch Repairs to concrete – The concrete would be repaired over the damaged areas only. It is possible that in future years additional repairs would be required as the building continued to decay.
- Chloride Extraction – Consideration was given to the extraction of the chlorides from the concrete. It was considered that due to the high risk of future penetration of chlorides into the concrete that this would only provide a temporary solution and would need to be repeated in future years.
- Cathodic Protection – Given the low levels of concrete cover to the reinforcement and the constant risk of chloride penetration it was concluded that the only satisfactory long term solution for the repair of the pool basin and pool building was the patch repair of damaged concrete followed by the application of a cathodic protection system to the areas at risk. The cathodic protection system was specified to have a design life of 25 years minimum.

3   DESCRIPTION OF REHABILITATION WORKS

Concrete Repairs Ltd (CRL) were employed through competitive tender to carry out the design and installation of the impressed current cathodic protection (ICCP) system and associated works to Chiltern Leisure Centre Swimming Pool. CRL have designed and installed ICCP systems for 20 years including several swimming pool schemes.

CRL engaged the services of Kevin Davies, Corrosion Consultant, to complete the ICCP design, commission the installation and carry out the following 12 months monitoring. Dr John Broomfield acting for the client approved the final design prior to the site work proceeding.

The schematic layout for the ICCP system can be seen in Figure 3.

The contract was awarded on 8th October 2007 and after design approval, the work started on 14th December 2007, with completion on 14th January 2008.

Concrete repairs to the pool walls, beams, columns and promenade slab soffit were carried out using a general purpose, prebagged repair mortar. Pad repairs were used to increase cover to reinforcement on certain pool wall patches. Shuttered repairs using flowing prebagged concrete were used on columns that were very badly damaged. Around the pool the steel stanchions were beginning to deteriorate due to corrosion. These were repaired by grit-blasting and recoating and some were strengthened with the addition of a steel collar.

Once the repairs were complete, the anodes for the cathodic protection system were installed. These were CPI durAqua water anodes and were applied to the main pool walls.

Fourteen LD15 silver/silver chloride (Ag/AgCl) reference electrodes, supplied by Castle Electrodes Ltd., were embedded in 25 mm diameter x 120 mm deep holes adjacent to the reinforcement to monitor

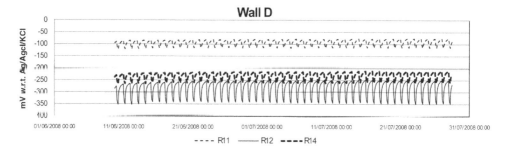

Figure 4. 'Instant Off' readings against the Ag/AgCl reference electrodes.

potentials for controlling the anode outputs and evaluating the system's performance.

The pool anodes were proprietary modules comprising a monolithic base and removable lid, both manufactured from XLPE. The twelve anode boxes were mounted in place, flush with the tiles on the internal face of the pool around its perimeter. The pool anodes work by using the pool water as the current distributing electrolyte. The current distributor, housed in the XLPE boxes consists of a 3 mm diameter mixed metal oxide (MMO) coated titanium wire welded to a 20 mm wide x 120 mm long length of MMO coated expanded titanium mesh.

An array of four zinc reference electrodes were fixed in similar but smaller XLPE boxes around the pool perimeter to monitor the performance of the system.

Connections were made to the anodes and reference electrodes outside the pool wall in trunking, running around the dry side pool perimeter. Care was taken to ensure that all wiring holes passing through the pool wall were back filled correctly using a water-tight epoxy mortar. DC negative connections were made to the pool wall reinforcement ensuring that a degree of redundancy was maintained. The perimeter trunking carried all DC positive anode wires, DC negative and reference electrode wires back to the transformer rectifier enclosure.

A wall mounted enclosure in the machinery room adjacent to the pool contains the computer controlled transformer rectifier units that provide DC current to each of the anode zones, and electronics that carry out 'instant off' potential measurements from the reference electrodes. The computer stores transformer rectifier outputs, 'instant off' potentials from the reference electrodes, monthly 24-hour depolarisation data and other data on flash memory. A dedicated phone line and modem allow all information on the system to be retrieved directly by CPI in Denmark or CRL in England. Any other computer containing a modem and with the correct software and password can be used to remotely monitor the system.

At Chiltern Leisure Centre Swimming Pool the installation of a cathodic protection system resulted in considerable savings in both pool closure time and construction costs compared to traditional refurbishment techniques, which in this case would have involved major structural repairs to remove chloride contaminated concrete.

The pool anode system was both easier and less costly to install than conventional internal anode or mesh and render overlay alternatives. As well as ease of installation, the pool anode system took far less time to install than traditional systems. This shorter installation period was a major advantage when the possible liquidated damages were taken into account. A further consideration was that the amount of work being carried out in the confined space around the dry side of the pool perimeter was kept to a minimum.

4 RESULTS AND EXPERIENCE

The ICCP system was commissioned in stages between 12 January and 15 May 2008 after final configuration of the software, timer and infra red interruption devices.

The first performance monitoring report was undertaken in July 2007. The total current demand was 572 mA at 4.5V DC. The graphical plots provided in Figures 4 & 5 show the steel potentials recorded as 'Instant Off' readings against the Ag/AgCl reference electrodes on the dry side of the pool wall and the zinc reference electrodes on the wet side of the pool wall. Note that the readings for the zinc reference electrodes have been converted to Ag/AgCl readings for clarity. As the wet side of the pool reacts more rapidly to the application of the ICCP the four zinc reference electrodes were shown on a separate graph.

From the graphs the effect of the approximate 5 hour daily application of the ICCP can be seen to polarize the steel. When the pool is in use, the current is switched off and the potentials decay steadily over the remaining 19 hour period.

The steel in the pool basin is becoming more negative in potential (particularly on reference electrode 16 – the most deeply buried), demonstrating that the steel in this region is becoming, in an electrochemical sense, more reactive or less passivated. The most probable reason for this behaviour is there is insufficient oxygen to reform the oxide layer on the steel. This can be explained by the cathodic reaction using up the oxygen. If sufficient oxygen cannot diffuse through the concrete to re-supply the cathode reaction then there

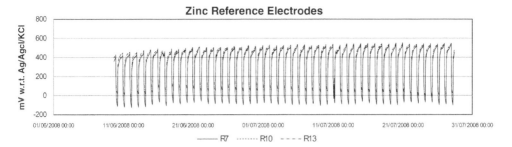

Figure 5. 'Instant Off' readings against the Zinc reference electrodes.

is an oxygen deficiency, which causes the potential to drop. This does not mean that corrosion is occurring as oxygen is required in the corrosion reaction, and if no oxygen is present no corrosion will occur.

Once every month, a 24-hour depolarisation test is also automatically carried out by the computer controlling the cathodic protection system. This test is showing more than 100mVdepolarisation on all the steel associated with the reference electrodes.

5 CONCLUSIONS

The swimming pool at Chiltern Leisure Centre was a severely corrosion damaged structure with only a limited structural life if temporary repairs had been used, or large financial and operational implications if total replacement had been recommended.

The use of cathodic protection as a repair technique is now becoming a standard technique in many parts of Europe, but for the UK there is some novelty in protecting the pool basin steel reinforcement using the pool water as electrolyte. The use of water as an electrolyte is common in the conventional cathodic protection industry.

The barriers for using water anodes in swimming pools have previously been the aesthetics and the necessity, for safety and vandalism reasons, to avoid contact between the bathers and the anodes. The installation of the anodes recessed into the walls has overcome these barriers. Additionally this way of installing the anodes is relatively cheap, fast and simple compared to applying anodes to the outside walls, and has less to go wrong. Using water anodes reduced both the closure time of the swimming pool and the work in confined spaces giving a considerable cost saving.

The results of the application show that cathodic protection is being effectively achieved, even after only a short time in operation.

Experience with cathodic protection of swimming pools over the past 15 years has shown that if installed correctly cathodic protection is able to arrest corrosion effectively. Water anodes have shown to be a cheap and durable way of protecting the basin effectively.

Concrete Solutions – Grantham, Majorana & Salomoni (Eds)
© 2009 Taylor & Francis Group, London, ISBN 978-0-415-55082-6

Active protection of FRP wrapped reinforced concrete structures against corrosion

Sangeeta Gadve
Department of Structural Engineering, Sardar Patel College of Engineering, Mumbai, India

Abhijit Mukherjee
Thapar university, Patiala, India

S.N. Malhotra
Department of Material Science and Metallurgical Engineering, Indian Institute of Technology Bombay; Mumbai, India

ABSTRACT: Large numbers of reinforced concrete (RC) structures that have been damaged due to corrosion of steel reinforcement are rehabilitated with fibre reinforced polymer (FRP) composites. This paper investigates active protection of the steel embedded in concrete that is treated with surface bonded carbon FRP. The electrically conductive carbon fibre is used as an anode while the reinforcing bars are used as a cathode. Concrete cylinder specimens with embedded steel bars are immersed in salt water and anodic current is passed through the reinforcement to initiate cracking in concrete due to accelerated corrosion of steel. Carbon FRP sheets have been adhesively bonded to the cylinders. The adhesive has been modified to impart electrical conductivity. Specimens were exposed to highly corrosive environment for a specified time. Pullout strength, mass loss and half cell potential of steel are reported as measures of performance of the samples. The proposed technique has been very effective in retarding the corrosion of steel.

1 INTRODUCTION

Reinforced concrete has been developed and applied extensively for a century. However, there are many instances of premature failure of reinforced concrete components due to corrosion of reinforcement. The economic implications of such damage are enormous. In a tropical country like India, where approximately 80% of the annual rainfall takes place in the two monsoon months, corrosion related problems are more alarming. India also has a very long coastline where marine weather prevails. Typically, a building in the coastal region requires major restoration work within fifteen years of its construction.

Recent developments in the field of FRP have resulted in highly efficient construction materials. FRPs are being used increasingly to rehabilitate corrosion affected structures. The efficiency of FRPs in enhancement of bending (Mukherjee and Rai 2008) and shear (Mukherjee and Joshi 2005) capacities of flexure elements and enhancement of confinement of concrete in compression elements (Mukherjee et al. 2004) is well established. An important spin-off from the FRP treatment of RC structures can be their resistance to corrosion. There have been various attempts of passive protection of steel reinforcement with surface bonded FRP, both in the laboratory and on site (Sen 2003). FRPs are unaffected by electro-mechanical/ electro-chemical deterioration. Therefore, unlike steel

reinforcement, they can be applied on the surface of concrete without any apprehension of environmental degradation. Arguably, FRP wraps prevent the increase in volume of RC members due to rusting by applying confinement pressure, thereby preventing dislodging of concrete cover.

The FRP wraps provide a barrier layer that should impede further corrosion of steel. FRP has been applied primarily to compensate the lost steel and to improve confinement of concrete. However, there is near unanimity that FRP wraps have slowed down the rate of corrosion, albeit in varying degrees.

The general procedure of laboratory experiments has been to accelerate corrosion in steel embedded in concrete and then applying the FRP to observe its effects on corrosion (Soudki et al. 2003 & 2006). Corrosion has been accelerated through application of impressed potential on the reinforcements. (Debaiky et al. 2002). The main indicators of performance may be mass loss of reinforcement, pull out strength, electric resistance, half-cell potentials and potential scans. The suitability of performance indicators depends on the method of imparting corrosion. It is noted that the benefit of FRP depends on several factors such as, the adhesive, fibre, method of application and environment.

So far researchers have used FRP as a passive barrier layer that would impede ingress of moisture and corrosive chemicals. One class of FRPs, namely carbon FRP (CFRP) is electrically conductive. Therefore, they may be used in active protection of RC structures. The authors are unaware of any previous investigation on active protection of the steel using FRP. This paper investigates the use of FRP wraps for active protection of steel in concrete. We briefly introduce active protection.

2 ACTIVE PROTECTION

The corrosion of steel reinforcing bars is an electrochemical process that requires a flow of electric current and several chemical reactions. The three essential components of a galvanic corrosion cell are anode, cathode and electrolyte. The idea of active protection is based on cathodic protection of corroding metals. The principle of active protection lies in connecting an external anode to the metal to be protected and passing an electric direct current to make all areas of the metal surface cathodic. The external anode may be a sacrificial galvanic anode, where the current is a result of the potential difference between the two metals, or it may be an impressed current anode, where the current is impressed from an external dc power source. In this paper, we investigate impressed current anode system. In the impressed current system an inert (zero or low dissolution) anode is employed and an external source of dc power is used to impress a current from an external anode onto the cathode (reinforcing steel) surface. There have been many alternative anode systems developed for active protection to RC structures such as highway bridge substructures, buildings, and marine structures (Wyatt 1993). A variety of anodes, such as titanium wire (Jiangyuan and Chung 1997) titanium strip and titanium mesh (Koleva 2007), zinc spray (Page and Sergi 1994) and conductive systems such as conductive concrete mortar (Bertolini et al. 2004) and carbon fibre reinforced overlay (Jiangyuan and Chung 1997) have been attempted by the researchers. The driving voltages for protection vary with the type of anode and environmental conditions. Typical operating current densities range between 0.2 and 2.0 mA/m² for cathodic prevention of new reinforced concrete structures, and between 2 and 20 mA/m² for cathodic prevention of existing salt contaminated structures (Daily, 2003). A cathodic current density of 5–20 mA/m² to the steel reinforcement reduces its corrosion rate to negligible values (Bertolini et al. 2004). Over-voltage or higher current densities for prolonged periods may lead to damage such as degradation of the steel-concrete bond, associated with softening of the cement matrix in contact with the metal (Locke et al. 1983; Rasheeduzzafar et al. 1993). There has been an enhanced risk of expansive alkali silica reaction in cathodic regions of concrete with siliceous aggregates. It has been reported that the risk is considerably reduced if the cathodic current density is uniformly and consistently maintained at a level lower than 20 mA/m² (Sergi and Page 1994). Although significant research has been done on active protection of RC the application of the technique has remained meager due to the requirement of special anodes that are often expensive.

In a previous investigation the authors have reported the efficacy of FRP wraps for passive protection of RC structures (Gudve et al. 2008). In addition to passive protection, the FRP materials that are electrically conductive can be designed to offer active protection as well. This has not been reported in literature so far (Mukherjee 2008). With the FRP wraps acting as anode no other anode would be necessary. Therefore, the cost of active protection can be brought down significantly. Present work investigates active protection using surface bonded carbon FRP sheets as anode.

3 EXPERIMENTAL PROGRAM

The experimental program was carried out in following steps:

- Casting cylindrical reinforced concrete specimens
- Inducing initial corrosion into the reinforced concrete specimens
- Wrapping the pre-corroded specimens with CFRP and GFRP sheets
- Subjecting the wrapped specimens to further corrosion by exposing them to salt mist while applying active protection
- Monitoring corrosion
- Carrying out destructive tests

3.1 *Preparation of test specimens*

In the present program, cylindrical specimens with an embedded steel bar (Fig. 1) were used. The height and the diameter of the specimen were 230 mm and 100 mm respectively, with an accuracy of ±1 mm. A standard reinforcing bar of 330 mm length and 12 mm nominal diameter of Fe 415 grade was used.

The bar was shot blasted to SA 2.5 surface and immediately dipped in oil. The white shining surface was maintained in the laboratory until the oil was removed and it was embedded in concrete. Before placing the bar in concrete a groove of 2 mm diameter was drilled on one end of the bar and a copper stud was fixed in the groove. The plug was used for electrical connections. Teflon tape was wound around the bar at two locations-bottom edge and at the interface with concrete top face. This served as a bond breaker and also the embedded length of the bar in concrete was maintained precisely at 152 mm.

The bar was placed in such a way that transverse clear cover was 45 mm and bottom cover was 51 mm.

Before casting of the test specimens, each reinforcing bar was weighed to 0.1 gm accuracy. The protruded part of the steel bar was coated with liquid epoxy resin for corrosion protection.

Ordinary Portland cement of nominal strength 43 MPa, fine aggregate (medium-sized natural/river sand) and crushed stone coarse aggregate with a maximum size of 20 mm was used in the concrete. The ratio of cement:sand:coarse aggregate was 1:2.16:2.44.

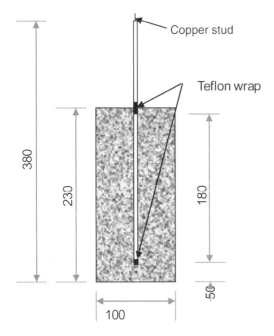

Figure 1. Cylindrical reinforced concrete specimen.

The water-cement ratio was 0.42 and aggregate-cement ratio of 4.6. The resulting strength of concrete was 40 MPa.

3.2 Inducing corrosion

The objective of inducing corrosion to the bar is to simulate corrosion damaged concrete. The commonly used methods of inducing corrosion in RC specimens are salt mist (Debaiky et al. 2002), Chloride diffusion (Masoud and Soudki 2006), alternate drying and wetting in salt water (Debaiky et al. 2002 and Soudki 2006) and impressing anodic current. (Lee et al. 2000)

In this investigation, the impressed current technique was used. The specimens were kept immersed in 3.5% NaCl solution for 24 hours to ensure full saturation. A stainless steel (SS) mesh rolled into a hollow, open cylinder was used as cathode.

The cathode and the specimen were placed in 3.5% sodium chloride (NaCl) solution. The level of NaCl solution was 3 cm below the top surface of the specimen to alleviate corrosion at the steel-concrete interface. The DC regulated power supplier used in the present study could supply 500 mA DC at 60 V. The reinforcing steel bar was connected to the positive terminal of the external DC source and the negative terminal was connected to the SS mesh. A 100 mA direct constant current was impressed between the reinforcing bar and the SS mesh. In this investigation a constant current was preferred because the aim of this investigation was to examine the active protection through maintenance of constant cathodic current. A total of 12 specimens were used in this experiment, divided into three groups. Constant current was impressed for 2 days, 4 days and 8 days into four specimens of each

Figure 2. Specimens after initial exposure.

group respectively. Every day anode to cathode voltage corresponding to constant current of 100 mA was monitored for all specimens. Half-Cell potential of the corroding reinforcement bar was also noted everyday with reference to a standard Ag/AgCl electrode. Using this method the concrete was found to develop cracks within two days. The cracks initiated at the surface of the cylinder and ran along the direction of the reinforcement on the sides of the cylinder (Fig. 2).The voltage between the reinforcing steel anode and SS mesh cathode decreased with time indicating that the resistance had gone down with the progression of the crack.

3.3 Wrapping of pre-corroded specimens

Two fibre materials are popular in the rehabilitation of structures in India: glass and carbon. Owing to the electrical conductivity of carbon fiber, only CFRP has been used for active protection. However, comparison has been made between the active and the passive protection offered by both glass and carbon fibre sheets. The fibres are applied in the form unidirectional sheets. Glass fibre sheets are thicker than the carbon fibre sheets. In this investigation, two often used commercially available unidirectional CFRP and GFRP sheets and compatible epoxy adhesive were used. The properties of the sheets are presented in Table 1.

The samples were air dried prior to the application of FRP wraps. Manufacturer's specification was followed in the application of the wraps. Out of the four specimens in a group one was kept unwrapped. One layer of CFRP and GFRP sheet was wrapped around the two test specimens respectively with the fibre along the circumferential direction of the cylinder. The entire length of the cylinder was covered.

A 25 mm overlap was provided at the ends of the sheets. The remaining specimen was used for active protection. This CFRP wrapped test specimen was additionally provided with adhesively bonded 25–30 mm wide vertically oriented carbon sheet (Fig. 3), so that uniform distribution of direct current throughout the specimen was possible for effective application

Table 1. Properties of Fibres used in the experiment.

Material	Thickness	Tensile Strength	Tensile Modulus	Ultimate Strain
	mm	GPa	GPa	
Carbon Sheet Net Fibre (CS)	0.13	3.79	230	0.015
Carbon Sheet Net Fibre (CS)	0.35	2.3	76	0.018
Adhesive	–	1.5	4.3	0.020

Figure 4. Active protection applied to FRP wrapped specimens exposed to salt mist.

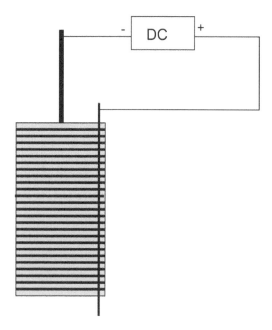

Figure 3. Schematic representation of actively protected specimen.

of active protection. The epoxy adhesive used was modified to be conductive.

3.4 Active protection

Since carbon is electrically conductive, an attempt was made to use this property in applying active protection to the reinforced concrete system without using any external anode. In this case, the carbon fibre sheets that were wrapped around the cylindrical reinforced concrete specimen themselves were used as anodes and the reinforcing steel bar as cathode. To achieve this, the wrapping system had to be modified in two ways.

One of the requirements of the system was to make its electric conductivity uniform. A ribbon of carbon fibres was stitched through the CFRP sheet in the perpendicular direction of the fibres. The ribbon was extended beyond the sheet by about 25 mm. It was used as the anode terminal for supplying electricity to the CFRP sheet. The ribbon pressed against the fibres of the sheet and thus ensured proper contact and uniform conductivity (Mukherjee 2008). The only non-conductive part in the system was the epoxy adhesive used to bond carbon sheets onto the concrete. In the present study, the epoxy was made conductive by mixing conductive particulates into the epoxy. An experiment was carried out to find out the optimum amount of the particulate to be mixed into the epoxy, such that epoxy became sufficiently conductive without losing the required consistency for proper coverage of the concrete surface. The workability and conductivity were studied by adding 2% to 20% conductive particulates by mass of epoxy (Mukherjee 2008).

An external DC power supply was used to impress the constant current for active protection. The positive terminal of the DC power supplier was connected to the protruding ribbon of the carbon sheet and the negative terminal was connected to the reinforcing bar, to be protected from corrosion. A constant current of 50 mA was impressed between carbon fibre, cathode and reinforced steel, anode. To simulate the practical condition of applying active protection to reinforced concrete structures in a corrosive environment, the specimens were kept in a salt mist chamber, with all necessary electrical connections for active protection for 60 days (Fig. 4).

3.5 Exposure of wrapped specimens

To simulate corrosion damaged structures, prior to the application of wrap, an initial exposure was applied. In practice, the FRP wraps are applied on structures that are corroded to varying degrees. Therefore, different exposure durations were chosen prior to the application of the wrap. Three exposure durations – 2, 4 and 8 days were applied. In two days the first crack appeared in all the samples. In 4 and 8 days the crack became wider and corrosion products oozed out in larger volumes. The control samples were not wrapped. GPP and CPP indicate passive protection with Glass and Carbon FRP wraps respectively. CAP indicates active protection that is active protection applied to the RC specimens using Carbon FRP wraps. Control-2, GPP-2, CPP-2 and CAP-2 represent samples with initial corrosion induced by exposing the samples to anodic current for 2 days. Similar sample identifications were used for the 4 day and 8 day samples too.

All specimens with varying degrees of initial corrosion, treated with surface bonded glass and carbon FRP sheets were then exposed to severe corrosive environment created in a salt mist chamber. The salt mist chamber was designed as per IS: 11864 (Fig. 4).

The salt mist test was carried out with 5% NaCl solution at 50°C. Injection of salt mist was carried out for 8 hours and then off for 16 hours, keeping the samples in the chamber. On switching off, the temperature inside the chamber would have subsequently come down to ambient temperature (~27°C). The total duration of exposure in the salt mist chamber under the specified conditions was 1500 hours i.e. 60 days.

3.6 Corrosion monitoring

Several parameters were monitored during the entire process. Some of these were non-destructive, such as half cell potential and cell voltages. These studies were carried out simultaneously on all specimens. Half cell potential was noted every day. Cell voltages were observed every day during induction of initial corrosion by impressing anodic current into the reinforcing bar embedded in concrete. Other parameters were destructive in nature, such as pullout and mass loss tests. These were carried out at the termination of the salt mist exposure. The results of the tests are discussed in the following section. Prior research suggested maintenance of half cell potential to threshold levels. 'Instant-off' monitoring potential was used for the specimens exposed to the active protection. A standard silver/silver chloride (Ag/AgCl) electrode was used as a reference electrode.

Takewaka (1993), suggests that if some corrosion had been taking place on steel in concrete, subsequent corrosion can be stopped by setting the rebar potential to less than −550 mV (sat.Ag/AgCl). An "instantaneous off" (IR free) steel potential (measured between 0.1 and 1 s) after switching off the dc source) more negative than −720 mV with respect to an Ag/AgCl/0.5M KCl electrode is recommended for prevention of steel from corrosion (Page and Sergi, 1994). A corrosion protection potential maintained at a level less negative than 900 mV with respect to an Ag/AgCl/0.5M KCl electrode is proposed for prestressing steel in the European Draft Standard (1996). In addition, it is specified that no "instantaneous off" (IR free) steel potential may be more negative than −1,100 mV (for reinforcing steel) or −900 mV (for prestressing steel) with respect to an Ag/AgCl/0.5M KCl electrode.

In the present study, the corrosion protection potential was obtained from a potentiodynamic scan by extrapolating the cathodic polarization curve. A tangent to the cathodic polarization curve was drawn.

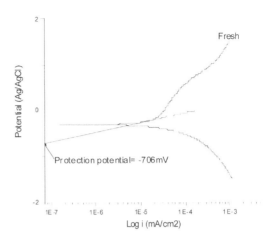

Figure 5.   Corrosion Protection Potential (mV).

Its intercept with the voltage axis was determined. That voltage was maintained for active protection. Fig. 5 shows an example. The maximum value of the voltage was −750 mV. The required voltage was maintained throughout the period of exposure. Every day, the instant-off potential of each cathodically protected specimen was observed and confirmed it was equal to or more negative than −750 mV with respect to a standard Ag/AgCl electrode.

4   RESULTS AND DISCUSSION

The results were obtained from two types of tests. The non destructive tests were carried out during the exposure of the samples to salt mist. After the complete exposure, destructive tests were carried out. After 60 days of exposure to salt mist, pullout tests were carried out on all the specimens. This was done by securing the cylinder in a universal testing machine (UTM) and attaching the grip on to the protruding portion of the reinforcing bar. The control specimens offered very little resistance to bar pull out. The pullout force was much larger in the case of passively protected wrapped samples. It may be recalled that there was considerable corrosion and consequent loss of metal at the top interface. As a result, the reinforcing bar of all specimens broke off at the interface (Fig. 6).

Therefore, the original specimens were cut down to a height of 117 mm, (Fig. 7) to expose the reinforcing bar sufficiently for holding it in the grip of the UTM. The pullout test was again performed on the cut specimens. The actively protected specimens did not exhibit excessive corrosion at the top interface. At the time of pull out test the bar did not break at the interface. This illustrates that active protection has been effective in nucleation of corrosion in the bars.

After the pullout test, the corroded bars were cleaned of all corrosion products as per ASTM G1-90 and weighed to determine their mass loss. Table 2 shows that the control samples had the highest mass loss.

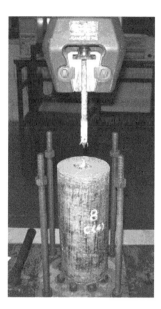

Figure 6. Bar fracture at the interface.

Figure 7. Cut specimen for pullout test.

The passively protected samples had a much lower loss of mass. However, the mass loss in the actively protected samples was the least. The mass loss in these samples was around 400% less than the corresponding CFRP wrapped samples in passive protection. This exhibits that active protection system has been very effective in avoiding the corrosion in bars.

The pullout strengths have also been reported in Table 2. It may be recalled that the bars in the CPP and GPP specimens had broken at the interface and the cylinders were cut to expose the bars for gripping. Therefore, the pullout strengths have been calculated based on different embedded lengths for those specimens. The pullout strengths of the control samples were the least. The passively protected samples had the highest strength. Surprisingly, although the actively protected samples had a much lower loss of bar mass, their pullout strengths were lower than that of the

Table 2. Pullout strength and mass loss.

Control-2	6.47	8.86
Control-4	4.29	5.87
Control 8	10.71	5.21
GPP-2	2.86	12.99
GPP-4	2.5	13.05
GPP-8	8.57	11.15
CPP-2	6.2	13
CPP-4	4.6	10.32
CPP-8	8.27	11.2
CAP-2	1.65	7.81
CAP-4	1.38	4.15
CAP-8	2.87	5.63

passively protected samples. The bars used in these experiments were the standard reinforcing bars that have helical ribs on their surface to enhance bond. A close examination of the actively protected bars revealed that although the bars had very little corrosion products on their surface the ribs in these bars were much less pronounced. This may be one of the reasons for the reduced bond strength. Bond degradation in actively protected samples has also been reported by Chang, 2001. The author postulates that sodium ions migrate to the interface during cathodic polarization of the rebar and result in softening of C-S-H gel. The bars exert bursting pressure on concrete due to corrosion. In the absence of the FRP wrap, tensile stresses develop in concrete that result in its cracking and spalling. The FRP wrapped samples resist the expansion pressure by developing a hoop stress. As a result, concrete goes into compression. This results in better grip of concrete on steel. In the case of actively protected samples, the corrosion products did not develop. Therefore, the compressive force in concrete was also absent. The ribs on the bars, on the other hand, were lost. This resulted in loss of bond strength. For the success of the active protection technique it is imperative that an optimum protection current density is achieved such that the loss of bond strength is avoided.

5 CONCLUDING REMARKS

A novel system for active protection of steel bars in concrete is reported. The electrical conductivity of the carbon FRP is utilized in creating an anode around the steel. Thus the proposed system eliminates any requirement of an external anode and has a very favorable impact on cost.

The efficacy of the system was established through a set of experiments and performance measurements, both non-destructive and destructive. It was observed that surface bonded FRP wrapping protects steel in concrete and reduces the rate of corrosion to a great extent. The protected samples exhibited lower mass loss and corrosion current. The actively protected system had the least loss of mass.

Although the actively protected system was superior in all other evaluations it reported lower pullout strength. This indicates poorer bond between concrete and steel. The bars that had helical ribs on the surface had considerably lost them in the case of active protection. A judicious selection of current density may reduce the loss of the ribs while offering the same level of protection. Prior literature has also reported a loss of bond due to active protection and has postulated softening of the interface as the responsible phenomenon. Research is being initiated to investigate this phenomenon and the results will be reported in future.

REFERENCES

Chang, J., J., (2001). "A study of the bond degradation of rebar due to cathodic protection current." *Cement and Concrete Research*, (32), 657–663.

Daily, S. F., (2003). "Understanding corrosion and cathodic protection of reinforced concrete structures." *Corrpro Companies, Inc.*

Debaiky, A., Green, M., and Hope, B., (2002). "Carbon fibre-reinforced polymer wraps for corrosion control and rehabilitation of reinforced concrete columns." *ACI Mater. J.*, 99(2), 142–152.

Gadve, S., Mukherjee, A., Malhotra, S., N., (2009). "Corrosion of Steel Reinforcements Embedded in FRP Wrapped Concrete," *J. Constr. Build. Mater.* (23), 153–161.

Hearn, N., and Aiello, J., (1998). "Effect of mechanical restraint on the rate of corrosion in concrete." *Can. J. Civ. Eng.*, 25(1), 81–86.

Jiangyuan, H., and Chung, D., D., L., (1997). "Cathodic Protection of Steel Reinforced Concrete Facilitated By using Carbon Fibres Reinforced Mortar or Concrete." *Cement and Concrete Research*, 27(5), 649–656.

Koleva, D., A., Wit, J., H., W., de, Breugel, K., van, Lodhi, Z., F., and Westing E., van, (2007). "Investigation of Corrosion and Cathodic Protection in Reinforced Concrete. I. Application of Electrochemical Techniques." *Journal of the Electrochemical Society*, 154 (4), 52–61.

Lee, C., Bonacci, J., Thomas, M., Khajenpour, S., and Hearn, N., (2000). "Accelerated Corrosion and Repair of Reinforced Concrete Columns Using CFRP Sheets." *Can. J. Civ. Eng.*, 27(5), 949–959.

Locke, C., E., Dehghanian, C., and Gibbs, L., (1983). "Effect of impressed current on bond strength between steel rebar and concrete." *Proc., NACE Corrosion 83, Paper 183*, National Association of Corrosion Engineers, Houston.

Luca, B., Fabio, B., Tommaso, P., Pietro, P., (2004). "Effectiveness of a conductive cementitious mortar anode for cathodic protection of steel in concrete." *Cement and Concrete Research*, (34), 681–694.

Masoud, S., Soudki, K., A., (2006). "Evaluation of corrosion activity in FRP repaired RC beams." *Cement and Concrete Composites*, (28), 969–977.

Masoud, S., Soudki, K., and Topper, T., (2002). "CFRP-strengthened and corroded RC beams under monotonic and fatigue loads." *J. Compos. Constr.*, 5(4), 228–236.

Mukherjee, A., (2000) "An improved method for protection of reinforced concrete structures from corrosion", Indian Patent Application, 803/DEL/2008.

Mukherjee, A., Rai, G., L., (2008) "Performance of Reinforced Concrete Beams Externally Prestressed with Fibre Composites." *J. Constr. Build. Mater.* (in press).

Mukherjee, A., Gadve S., (2008) "An improved method for protection of reinforced concrete structures from corrosion." Indian Patent Application, 803/DEL/2008.

Page, C. L. and Sergi, G., (2000). "Developments in Cathodic Protection applied to Reinforced Concrete." *J. Mater. Civ. Eng.*, 12(1), 8–15.

Rasheeduzzafar, A., M., G., and Al-Sulaimani, G., J., (1993). "Degradation of bond between reinforcing steel and concrete due to cathodic protection current." *ACI Materials J.*, 90(1), 8–15.

Sen, R., (2003). "Advances in the application of FRP for repairing corrosion damage." *Prog. Struct. Engg Mater.* 5, 99–113.

Sergi, G., and Page, C., L., (1994). "The effects of cathodic protection on alkali-silica reaction in reinforced concrete: Stage 2." *TRL Proj. Rep.62*, Transport Research Laboratory, Crowthorne, Berkshire, U.K.

Takewaka, K., (1993). "Cathodic Protection for Reinforced Concrete and Prestressed Concrete Structures." *Corrosion Science,* 35 (5–8), 1617–1626.

Wyatt, B., S., (1993). "Cathodic protection of steel in concrete." *Corrosion Science*, (35).

Concrete Solutions – Grantham, Majorana & Salomoni (Eds)
© 2009 Taylor & Francis Group, London, ISBN 978-0-415-55082-6

Electrochemical desalination of concrete and electrochemical soil remediation – Differences and similarities

L.M. Ottosen
Department of Civil Engineering, Technical University of Denmark, Lyngby, Denmark

M. Castellote
Institute of Construction Science "Eduardo Torroja", IETcc (CSIC), Madrid, Spain

ABSTRACT: Electrokinetic methods are developed for removal of pollutants from concrete and soil (chloride and heavy metals, respectively), but even though there are many topics of common interest to these methods there is no tradition for collaboration. In order to utilize the possible synergy from such collaboration a new RILEM committee "Electrokinetics in Civil and Environmental Engineering (EPE)" has been initiated. In the present paper some issues for beneficial collaboration are outlined such as: neutralizing pH changes at the electrodes, minimizing matrix changes, increasing mobility and electromigration of target ions, influence from the degree of interconnected pores on the overall removal process, and the electric field distribution at varying water content.

1   INTRODUCTION

When applying an electric field to a fine-porous, moist material, transport of ions, ionic species and water occurs. This means that transport of matter is possible even in dense materials where a pressure gradient is inefficient. Electrokinetics have been suggested to be utilized in civil and environmental engineering for various purposes, but the implementation of the methods has been relatively slow considering the very good results that have been obtained in the laboratory and pilot scale trials with e.g. chloride removal from concrete and removal of heavy metals from contaminated soils. One major reason for the slow implementation is that the fundamentals of the methods are not well understood, meaning that success or failure cannot always be sufficiently foreseen prior to an action.

Even though there are many fundamental topics of common interest to civil- and environmental engineering in the use of electrokinetics, there is no tradition for collaboration. However, this is anticipated to change and collaboration is expected, since a newly formed RILEM technical committee "EPE   Electrokinetics in Civil and Environmental Engineering" has started its work and this committee will work in the years 2008–2012.

The present paper has the aim of pointing out fundamental topics where knowledge transfer between the two engineering fields could improve the existing knowledge, as well as pointing out scientific areas of shared interest for further investigation. The paper is based on a literature survey and own experiences.

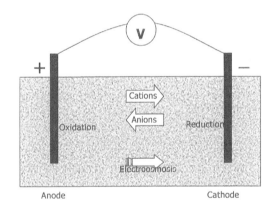

Figure 1   Electrochemical processes in a moist porous material under action of an applied electric DC field.

2   ELECTROCHEMICAL PROCESSES

When an electric field is applied to a pair of electrodes in a moist porous material various processes occur (see figure 1).

In the pores of the material, ions will flow, as ions carry the current. Cations will be transported towards the cathode and anions towards the anode. This movement of ions in the applied electric field is *electromigration*. At the electrodes, the electrode processes transfer the current carried by ions to current carried by electrons in the metallic electrodes. At the anode it will be *oxidation processes* and at the cathode *reduction processes*. Under some circumstances *electroosmosis* will occur, i.e. transport of water in the porous medium under the action of the applied electric field.

Owing to the electrode processes, the environment around the electrodes changes. Which electrode processes occur depend on the electrode material, the type and concentration of ions in the pore water and the applied potential. Electrolysis of water is often a major electrode process at both anode and cathode during the processes discussed in this paper. Anode process: $H_2O \rightarrow 2H^+ + \frac{1}{2}O_{2(g)} + 2e^-$ and cathode process: $2H_2O + 2e^- \rightarrow 2OH^- + H_2$ (g). Thus acidification occurs from the anode whereas the material becomes alkaline from the cathode.

3 ELECTROKINETICS IN CONCRETE AND SOIL

3.1 Some important material characteristics

Concrete is a porous material. When soil is remediated in-situ, the soil is compact and it can be regarded as a porous material. Most porous materials like soil and concrete carry a surface charge and have in common the presence of pores with unbalanced charges and corresponding unbound ions (Marry et al. 2003). The charged surfaces are counterbalanced by ions of opposite sign in the diffuse electric double layer. Ions in the solution with the same sign as the charged surface are co-ions and they are represented to a much lesser extent in the electric double layers than the counterions. Charge balance is always maintained throughout the system at all times, i.e. the porous media and electrolyte system must be electrically neutral.

Important similarities when electro-remediating concrete and soil:

- They contain solids, liquids and sometimes air
- They are in-homogenous materials
- Pore size range from nano-pores to macro-pores
- Cracks play an important role in contaminant transport into both materials
- Pollutants bind to internal surfaces - heavy metals to soil; chlorides to concrete

Some major differences are:

- Concrete has a high pH. Soil pH is most often neutral to slightly acidic
- Soils most often contain organic particles, concrete not

3.2 The electrode placement and neutralization of pH changes

In the methods of electrochemical desalination of concrete the reinforcement steel is used as cathode and the anode is placed externally at the outer surface of the concrete. The anode is preferably a titanium mesh embedded in an easily removable material which can buffer the produced acid. The material must further have good water retention properties, such as paper pulp (Tritthart, 1999) or blankets of cotton wool soaked

in e.g. $Ca(OH)_2$ solution (Page & Yu, 1995). Another possibility is to use an ion exchanger at the anode. The ion exchanger will adsorb the chloride and buffer the produced acid since the ion exchanger initially is filled with OH^- (Schneck et al. 2006). When concrete is contaminated by radioactive or heavy metals, remediation of it can also involve external electrodes (Castellote et al 2002). In the electrochemical soil remediation methods, both electrodes are externally placed. Most pilot and full scale actions have been carried out in-situ and the electrodes have been placed in electrode wells directly into the polluted soil (the wells were separated from the soil by a porous membrane) (Lageman, 1993). Inside the electrode wells there is circulation of electrolyte solutions. Most often an acid is used at the cathode to hinder alkalinisation of the soil because alkalinisation will cause the heavy metals to precipitate and be immobile for electromigration. Other possibilities for hindering the alkaline front have been developed e.g. separation of catholyte and soil by a cation exchange membrane (e.g. Ottosen et al. 2005) The acidic front developing from the anode, on the other hand, is often utilized for desorption and mobilization of the heavy metals. Thus, during soil remediation, the alkaline front is neutralized, whereas the acid is allowed to enter the soil. In the concrete it is exactly the opposite. However, since both methods do use neutralization systems, it may be beneficial to share experiences about the practical application.

4 ELECTROMIGRATION, POROSITY AND WATER CONTENT

The electric current can be carried by ions in the electric double layer and by ions dissolved in the brine, with rather different characteristics (Pengra & Wong 1996). Unlike in solutions, the ions in a porous material are not able to move directly to the opposite pole by the shortest route. Instead, they have to find their way along the tortuous pores and around the particles or air filled voids that block the direct path. Moreover, the ions can be transported only in continuous, moist pores, but not in closed ones and ions are only transported in the liquid phase. The electromigration rate of ions in the porous media depends on the pore volume, geometry and the water content. Further adsorption/desorption phenomena may influence the electromigration significantly as discussed in the next section.

Charged porous media are filled at least by counter-ions and often water and occasionally co-ions. The dynamics of counter-ions, co-ions and water molecules depend strongly on the water content of the medium. For very compacted media, the water content is low and the ions are slowed down. For less compacted media the dynamics of inserted water and counter-ions tend towards that in bulk solution (Marry et al. 2003). The electric resistance of e.g. concrete is influenced by the moisture content. The relationship is not asymptotic, in that after a certain moisture content the change in resistance with increased moisture content is insignificant (Saeem et al. 1996). In both soils and concretes the water content can vary significantly even over small distances. This means that the electric field strength can be expected to vary as well, because the water content influences the resistivity of the material. The electric field distribution in these porous materials with varying water content is of importance to the overall success of electrokinetic contaminant removal. If the current bypasses volumes with the least water these volumes are not remediated. Thus it is important to clarify at which water content and water distribution this problem occurs for proper management.

The electric field is strongest where the conductivity is highest (i.e. where most electromigration occurs), but where that is at pore level is not clarified in depth in these highly inhomogeneous pore systems. The relative effects on the flow pattern of surface charge, electric double layer thickness, ion concentration in pore solution outside the double layer, pore geometry, tortuosity pore size distribution, water content etc. need clarification in order to optimize the electrokinetic extraction methods. Such fundamental studies can be of benefit to both soil remediation and chloride removal from concrete.

5 MOBILITY OF CHLORIDE AND HEAVY METALS

It is only free ions in the pore water (in exchangeable sites or in the brine) which are directly mobile for electromigration. Thus the relation between free and bound contaminants is highly important and release of the bound fraction is necessary during the electrokinetic treatment in order to obtain a successful removal. Next to the electrokinetic transport processes, several other processes occur within the porous material during application of the electric field: pH changes, ion exchange, redox changes, complexation, adsorption/desorption, precipitation/dissolution, mineral degradation, and structural changes. These changes can limit the electrokinetic extraction of contaminants, but on the contrary intelligent utilization of these processes can also be used for optimization of the methods.

Chloride ions bind physically and chemically to the pore surfaces within the cement. This binding is not permanent though and chloride ions will be released again if the concentration of free ions drops, as there is a balance between free and bound ions (Wang et al 2000). In contrast, the distribution between bound and free heavy metals in the soil is only very little affected by the concentration of free heavy metals. Once bound to the solid surface, the heavy metals may even be very difficult to extract again. Strong acidification or use of complexing agents is most often necessary in order to mobilize the heavy metals for electromigration (Ottosen et al 2005). Thus for chloride removal from concrete, the kinetics for desorption of chloride as the free chloride ions are removed during the action is the major factor for mobilization. For

heavy metal polluted soil, sufficient desorption cannot be obtained only by removing the free heavy metal ions (such are not even present in older industrially polluted soils). Here the soil environment need to be changed to obtain desorption and most often acidification is used.

Increase or decrease of ions in the pore solution influence the transport number of the target ions, e.g. due to the increase in the OH^- concentration generated at the cathode, the transference number of chloride ions decreases during chloride extraction from concrete (Castellote et al. 2000). Likewise H^+ ions may be the major cation transporting current during soil remediation (Ottosen et al 2008). Optimization of transport number for target ions could be of shared interest to the two methods.

6 MATRIX CHANGES

Matrix changes occur under the action of the applied electric field and these include both physico-chemical and hydrological changes. Some solid phases are weathered and new ones can be formed. Increased fundamental understanding of the effects and side effects, when applying the electric field to a porous material, can probably lead to improvements of the known technologies and possibly to new applications.

The electrolysis of water influences pH around the electrodes as described. In the case of chloride removal from concrete it is important to hinder the acid in entering the concrete matrix since the acid will dissolve the cement and increase porosity. The alkaline front from the cathode on the other hand is not regarded as problematic since concrete itself has a high pH. The transport of OH^- into the concrete is even beneficial in the electrochemical method for re-alkalization, since in this method the major aim is to obtain a high pH level where the reinforcement corrosion is significantly lowered compared to in the carbonated concrete.

In soils the acidification from the anode which is used for mobilization of heavy metals results in increased weathering. The soil particles are attacked and weathered and it has been seen that Ca, Mn, Mg and Fe have been mobilized from the acidified parts of the soil and moved towards the cathode (Ottosen et al. 2001).

For concrete, the cement dissolution is highly unwanted because the increased porosity gives a vulnerable to the concrete and avoiding acidification is build into the methods. For soils, however, it is difficult to mobilize the heavy metals by acidification without changing the matrix. Use of complexing agents which attack the soil less has been suggested, e.g. where the heavy metals are mobilized at high pH. An example is a Cu polluted calcareous soil. Here addition of ammonia to the soil causes formation of $Cu[NH_3]^{2+}$ complexes and thus Cu can electromigrate even in the alkaline soil. The major reason for adding ammonia in this case was to avoid dissolution of carbonates which would be highly acid demanding and further current would be wasted for removal of Ca^{2+} instead of Cu^{2+} (Ottosen et al. 2005).

When the electric field passes through a pore, the chemistry of the pore is changed and the effect of this has only gained little attention, except from cases where strong pH changes are involved. But what does it mean to the solid surfaces when the chemical equilibrium in the pore is changed fast? Will it result in increased weathering of some porous material? The answer must be highly dependent on material and kinetics. Some research on the side effects on the microstructure of concrete after applying electrical fields have been carried out (Kuroi & Sueyoshi 1987, Castellote et al. 1999, Castellote et al 2002, Gerard, 1999). Even if the current density is low under application of the electric field, it will be unevenly distributed over the pores and where the current density is highest, there may be greater risk for material dissolution and the flow pattern must be investigated and modelled for prediction of the pattern in each case, so the necessary precautions can be taken.

7 CONCLUSIONS

Concrete and soil have many important characteristics in common in relation to electrokinetic treatment so many scientific and practical topics are of shared interest. At the electrodes neutralization of pH changes is necessary to both methods and though the neutralization is not at the same electrode, there are shared practical issues. Increased fundamental understanding of electrokinetics in in-homogenous matrices at pore level could be a benefit in optimization of both methods. This includes distribution of the electric field lines in porous materials with non-uniform distribution of water and matrix changes. Further increase of transport number for target ions is a possibility for optimization of both methods.

REFERENCES

Castellote, M., Andrade, C., Alonso, C. 1999 Changes in the concrete pore size distribution due to electrochemical chloride migration trials, *ACI Materials Journal*, 96-M39, 314–319.
Castellote, M. Andrade, C., Alonso, C. 2000 Electrochemical removal of chlorides. Modeling of the extraction, resulting profiles and determination of the efficient time of treatment. *Cem. & Concrete Res.* 30, 615–621
Castellote, M.; Andrade, C., Alonso C. 2002 Non destructive decontamination of concrete from radioactive metals and heavy species by application of electrical fields, *Environ. Sci. Technol.* 36(10); 2256–2261

Castellote, M., Andrade, C., Alonso, C., Turrillas, X., Kvick, Å., Terry, A., Vaughan, G. Campo, J. 2002 Synchrotron radiation diffraction study of the microstructure changes in cement paste due to accelerated leaching by application of electrical fields". *J. American Ceramic Society*, 85(2)
Gerard, B., Pijaudier-Cabot, G and Le Ballego, C. 1999 Calcium leaching of cement based materials: a chemomechanics application. Construction materials theory and application. Hans-Wolf Reinhart. Zum 60 Geburtstag. Ed. H von R. Eligehausen, ibiden-Verlag Stuttgart, 313–329
Kuroi, T. & Sueyoshi, T. 1987 Basic study on softening phenomenon of cement paste due to action of electric current" *CAJ review*, 164–167.
Lageman, R. 1993 Electroreclamation. Applications in The Netherlands. *Environ. Sci. Technol.* 27(13), 2648–2650
Marry, V.; Dufreche, J.F.; Jardat, M.; Meriguet, G.; Turq, P.; Grun, F. 2003 Dynamics and transport in charged porous media. *Colloids and Surfaces A: Physicochem. Eng. Aspects* 222, 147–153
Ottosen, L.M.; Christensen, I.V.; Rörig-Dalgaard, I.; Jensen, P.E.; Hansen H.K. 2008 Utilization of electromigration i civil and environmental engineering – processes, transport rates and matrix changes. *Journal of Environmental Science and Health* Part A, 43, 795–809
Ottosen L.M., Pedersen A.J., Ribeiro A.B. and Hansen H.K. 2005 Case study on the strategy and application of enhancement solutions to improve remediation of soils contaminated with Cu, Pb and Zn by means of electrodialysis. *Engineering Geology*, 77, 317–329
Ottosen, L.M.; Villumsen, A.; Hansen, H.K; Ribeiro, AB.; Jensen, P.E.; Pedersen, A.J. 2001 Electrochemical soil remediation – Accelerated soil weathering? EREM2001 3rd Symposium and Status Report on Electrokinetic Remediation. Chapter 5.
Page, C.L., Yu, S.W. 1995 Potential effects of electrochemical desalination of concrete on alkali-silica reaction. *Mag. of Concrete Res.*, 47, 23–31
Pengra, D.B., Wong, P.Z. 1996 Electrokinetic phenomena in porous media. *Disordered Materials and Interfaces*, 407, 3–14
Saeem, M., Shameem, M., Hussain, S.E., Maslehuddin, M. 1996 Effect of moisture, chloride and sulphate contamination on the electrical resistivity of Portland cement concrete. *Construction and Building Material*, 10(3), 209–214
Schneck, U., Grünzig, H.; Vonau, W.; Herrmann, S.; Berthold, F. 2006 Chloride measuring unit for the improvement of security and performance of electrochemical chloride extraction. Concrete solutions: Proceedings of the Second International Conference, 280–286
Tritthart, J. 1999 Ion transport in cement paste during electrochemical chloride removal. *Advances in Cement Research*, 11(4), 149–160
Wang, Y., Li, L.Y., Page, C.L. 2000 Efficiency investigation of chloride removal from concrete by using an electrochemical method, In: Hughes, B.P. et al (eds.), Proc. 10th Concrete Communication Conf., Birmingham, BCA Publication 98.003, Crowthorne, 2000, 223–235

Concrete Solutions – Grantham, Majorana & Salomoni (Eds)
© 2009 Taylor & Francis Group, London, ISBN 978-0-415-55082-6

Chloride transport in masonry (brick and cement mortar) in isolated and combined systems

I. Rörig-Dalgaard

Department of Civil Engineering, Technical University of Denmark, Denmark

ABSTRACT: Deterioration of building materials (clay bricks, concrete, etc.) can have several reasons, one being the presence of water and salts. The presence of salts is still an unsolved problem. A new method for repair could be the use of electrochemical methods. Previously, transport in isolated systems of brick or concrete has been studied. In the present work, chloride and sodium transport in bricks and cement mortars were studied in both isolated brick systems and combined systems with brick and an intervening mortar joint.

In the study no significant differences in chloride and sodium transport velocities were seen. Additional ion accumulation in specific areas (neither brick nor mortar) was not measured and consequently no significant differences between electromigration in isolated and combined systems were seen.

1 INTRODUCTION

Rendered or plastered structures suffer from peeling when exposed to damaging building salts, figure 1. Plaster or render peeling from traditional structures results in a poor appearance, increased maintenance and maintenance costs in relation to traditional structures. In relation to painted surfaces from medieval times, salt induced plaster or render peeling is very critical since original material is lost for the future, Figure 2.

Electromigration can be used to initiate an accelerated ion transport within the structure in both concrete and brick masonry structures.

In concrete, the main problem with salts is chloride ingress at depth (Castellote et al. 1999).

In brick masonry, however, desalination is carried out to protect the surfaces, meaning the aim is to transport the harmful ions out from the structures and away from the surface (traditionally only one side) which should be protected.

Documentation of the electrochemical method for desalination of an isolated brick (Rörig-Dalgaard

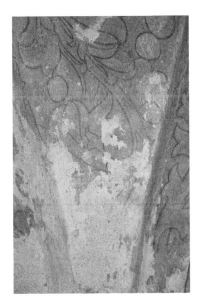

Figure 1.   Salt induced render peeling in a band above the ground.

Figure 2.   Salt induced plaster peeling of a painted vault in Our Lady's Abbey in Helsingør, Denmark.

2008) and in concrete systems (e.g. Tritthart 1999) and steel-reinforced concrete (Fajardo et al. 2006) has previously been investigated.

However, most Danish dwellings were erected as a combined system (brick masonry) consisting of both brick and mortar. Until around 1910, the use of lime mortar was dominating and later cement mortar or mixtures of cement and lime mortar dominated (Hansen2008). To predict ionic transport in brick masonry, a combined system of brick and mortar was investigated. To the authors' knowledge, no electrokinetic experiments have been carried out and this is the subject of the present paper. The aim was to clarify the influence of mortar and brick on the overall electromigration in this combined system.

2 ELECTROMIGRATION

When salts like sodium chloride (NaCl) are brought into contact with water they dissolve and in the case of NaCl into the ions sodium (Na^+) and chloride (Cl^-). When positive and negative electrodes are established, a current flow and the ions in solution will be attracted to the electrode of opposite sign and move according to their charge. This phenomenon is termed electromigration. By use of electromigration, a controlled ion transport is possible in porous wet materials like bricks. The idea is to transport the damaging ions out of the contaminated material and into an accumulating layer which can be removed after treatment has ended.

3 EXPERIMENTAL

3.1 Materials and sample preparation

The brick was of the type Falkenløwe red (handcrafted, burned in a circular kiln), with the saturation coefficient: 12.6 wt%, open porosity: 33.0 vol.-%, dry density: $1790\,kg \cdot m^{-3}$)

The mortar used was KC 60/40/850 dry mortar (60 kg lime to 40 kg basis cement to 850 kg sand). The water to solid ratio was 1:6. According to Aalborg Portland (2007) KC 60/40/850 is used for non load-bearing masonry in Denmark and KC 50/50/700 is used for load-bearing masonry. The mortar was mixed with a mixing machine for 2 minutes, rested for 5 minutes and finally mixed for 2 minutes. The bricks were pre wetted. After the bricklaying of the mortar and brick, the specimens were conditioned for 19 days at 20°C and 85% RH. According to Alborg, Portland (2007), basis cement obtains 95% of its 28 day strength within 14 days and 98% after 21 days.

The total length of all the specimens was 11 cm divided in 5 cm brick, 1 cm lime cement mortar and 5 cm brick (figure 2). Similar specimens only consisting of brick were 5 cm × 5 cm × 11 cm in size. The specimens were submerged into a sodium chloride solution (79.61 g NaCl/L distilled water) for 2

days. In species solely consisting of brick, 2 days of ingress resulted in a satisfying homogenous ion distribution; however in relation to mortar a longer ingress time would have been preferable and more representative but was unfortunately not possible. Subsequently the specimens were carefully wrapped in plastic film to minimize evaporation during the experiments

3.2 Setup and sampling positions

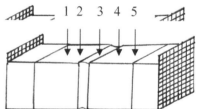

Figure 3. (Top) The setup consists from the left to the right of: the anode, accumulating poultice, brick, mortar, brick, accumulating poultice, cathode. (Bottom) Sampling positions.

3.3 Analytical

Simple extractions in distilled water were made to evaluate concentrations of Cl^- and Na^+. The drilling samples were dried at 105°C until equilibrium (one day). The extractions were made with the drilling powder from each position (around 5 g) and 12.5 ml distilled water. The suspensions were agitated for 24 hours. Afterwards 2 ml of the suspension was added to a mixture of 40 ml distilled water and 1 ml 1 M nitric acid, HNO_3. The chloride concentrations were measured by a titrator (Metrohm 716 DMS Titrino). After filtration, the sodium concentrations were measured with AAS (Atomic Absorption Spectroscopy).

3.4 Experimental overview

Table 1. Experimental overview.

Specimens nr.	Materials	Duration [Days]	Current [mA]	Voltage [V]
B4	Brick	4	20	3.6 → 4.7
B6	Brick	6	20	3.9 → 8.7
M4	Brick/Mortar	4	20	5.4 → 8.4
M6	Brick/Mortar	6	20	6.1 → 21.4
M14	Brick/Mortar	14	20	5.3 → 106.9
R14	Brick/Mortar	14	0	0

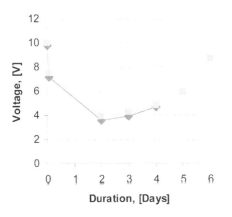

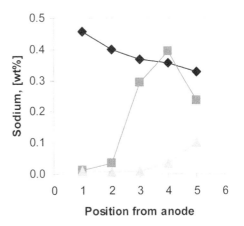

Figure 4. The change in voltage during the experiments as a constant current of 20 mA was used. ◆ M4 ▪ M6.

4 RESULTS AND DISCUSSION

4.1 *Electromigration in isolated systems*

During previous investigations (Rörig-Dalgaard, 2008) a coherence between the relative change in voltage and desalination degree in an isolated brick system was observed by use of constant current. The voltage during the present experiments in isolated systems is shown in Figure 4.

The relative high voltage in the beginning of the experiment decreased within the first 2 days and is supposed to be caused by a lower contact in the interface between brick and accumulating poultice as found by Ottosen & Rörig-Dalgaard (2008). Subsequently, a minor increase occurred between 2 and 4 days duration. After 6 days the voltage was the double compared to the lowest voltage and compared to previous experiments (Rörig-Dalgaard, 2008) almost total desalination was expected.

The transport of chloride and sodium ions in the electric DC field were evaluated on the remaining contents and compared to the measurement of a reference specimen without an applied electric DC field, Figure 5.

The profiles for the chloride content showed a decreasing tendency from position 1 to 5, whereas the sodium profile showed an increasing tendency from position 1 to 5. The content decreased throughout the experiment from the initial values (0.77 wt% in average for chloride and 0.38 wt% in average for sodium), to 0.22 wt% in average for chloride and 0.19 wt% on average for sodium after 4 days and almost total removal was obtained after 6 days for both chloride and sodium (0.003 wt% in average for chloride and 0.03 wt% in average for sodium).

4.2 *Electromigration in a combined system with brick and mortar*

The voltage during the experiments with combined systems was as in the case of isolated system recorded

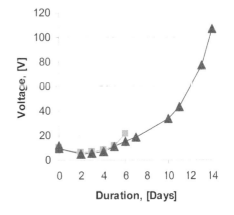

Figure 5. Electromigration in an isolated system solely consisting of a brick specimen (top) chloride profiles (bottom) Sodium profiles. ◆ Reference ▪ 4 days ▲ 6 days.

Figure 6. The change in voltage during the experiments as a constant current of 20 mA was used. ◆ M4 ▪ M6 ▲ 14 days.

to predict the desalination progress and is shown in Figure 6.

The recorded voltages during the three experiments were similar and likely reproducibility of the

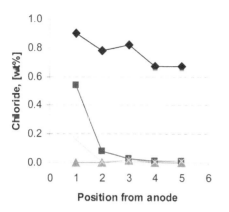

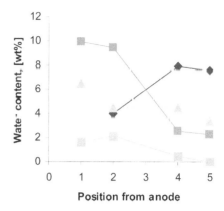

Figure 7. Electromigration in a combined system consisting of brick and mortar specimens. Chloride profiles ◆ Reference ■ 4 days ▲ 6 days ■ 14 days

Figure 9. The water content in the different positions after application of 20 mA for ◆ 4 days ■ 6 days ▲ 14 days ■ 14 days reference (I = 0 mA).

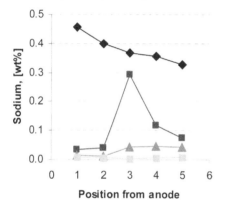

Figure 8. Electromigration in a combined system consisting of brick and mortar specimens. Sodium profiles. ◆ Reference ■ 4 days ▲ 6 days ■ 14 days.

experiments. As in the case with isolated brick system, a minor voltage increase was seen after 4 days and the voltage was doubled after 6 days experimental duration. However, between 6 and 14 days the voltage increased from 11 V to 107 V and such a high relative increase in voltage can be a result of both: almost total desalination and dewatering (Rörig-Dalgaard & Ottosen, 2008 B).

The results from the effect of the mortar on the overall transport of ions towards the anode and cathode are shown in Figure 7 and 8 for a combined system with both brick and mortar.

A decreasing profile for chloride content and an increasing tendency in the sodium profile from position 1 to 5 was observed. Also decreasing remaining chloride and sodium contents with increasing experimental duration were seen. In addition, continuously decreasing profiles are seen for both chloride and sodium with increased experimental duration. No significant higher chloride contents were seen in the mortar sample after 4, 6 and 14 days, compared with the brick samples next to the mortar. If there were different transport velocities in brick and mortar, a

change in the slope in figure 2 would occur as a function of the distance to the anode before and after the mortar. However, this was not seen.

One reason for the difference in chloride and sodium removal rate by the presence of the mortar might be caused by chloride binding to the brick/mortar and absence of sodium binding to the brick/mortar. Castellote et al. (1999) found that chloride ions remains unbound in cement until the concentration in the pore solution exceeds about 28 g/L. According to these findings chloride binding must be expected since the concentration of the solutions was 49 g/L Cl (79.2 g/L NaCl) in the present experiments. However, chemical chloride binding is only possible in OPC (with C_3A in it). Another reason could be ion adsorption to the solid matrix, which may happen to any stone/ concrete.

The measurements of the voltage during the 4–14 days experimental duration indicated both desalination and dewatering. The actual water contents after ended experiment is shown in Figure 9.

The results in Figure 9 show high deviation in the measured water content after ended experiment. However, there seems to be a tendency for higher water content in the experiments with relative short duration (4–6 days). Additional the water content after 14 days of experimental duration was significant lower than for the reference experiment without applied current. The final water content after 14 days duration is very low close to the cathode in position 4 (0.5 wt%) and 5 (0.0 wt%) but also low in position 1 (1.6 wt%) and 2 (2.2 wt%). The results after 6 and 14 days of duration might indicate a slower dewatering effect due to presence of the mortar but does not hinder the dewatering.

4.3 Comparison of isolated and combined systems

The effect of the applied electric DC field in isolated and combined systems is compared in Figure 6.

Figure 10 clarifies by direct comparison of the results for isolated and combined systems that the

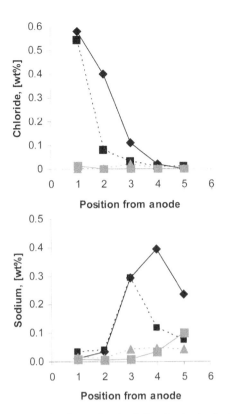

Figure 10. The electromigration induced reduction in the (top) chloride content (bottom) sodium content, in an isolated and a combined system. The black dots are the result after 4 days duration and the grey dots after 6 days duration. The continuous lines are for isolated brick systems and the dotted line for combined systems with both brick and mortar.

effect of the applied electric DC field seemed to be of the same magnitude.

After 4 days the remaining content of both chloride and sodium is lower in the combined system in the main part of the positions, whereas after 6 days the lowest content of both chloride and sodium was in the isolated brick system in some positions. There seemed to be a minor tendency for higher remaining sodium contents in the specimens with mortar than the specimens solely consisting of brick (e.g. after 4 and 6 days duration). However, the introduction of mortar in between the brick does not hinder reduction of the sodium and chloride even to a very low contents and electromigration in a combined system of brick and mortar is therefore considered as efficient.

5 CONCLUSION

A continuous reduction of the chloride and sodium content in the applied electric DC field was seen in isolated systems consisting solely of brick and also in combined systems consisting of both brick and lime cement mortar.

On the basis of the present results there is no indication that electromigration should be hindered by the mortar or decrease the electromigration effect compared to a system solely consisting of brick. These results are a very encouraging step towards a possible application of electromigration on existing structures.

6 FUTURE EXPERIMENTS

Electromigration in an isolated lime cement mortar sample should also be examined and specimens with pure lime mortar in combination with bricks are to be investigated to complete the typical situation in brick masonry structures both new and historical.

The presented results for chloride do only represent the water soluble part of the chlorides and does therefore not clarify the magnitude of bound chloride in the cement. Therefore determination of the total (acid soluble) chloride content Fajardo et al. (2006) in the mortar should also be carried out.

ACKNOWLEGDEMENTS

The Augustinus Foundation is acknowledged for their financial support. Thomas Friis is acknowledged for the measurements and Sara Laustsen is acknowledged for technical help with these very first experiences with cement.

REFERENCES

Aalborg Portland (2007). Cement og Beton – håndbogen om cement, beton og mørtel (in Danish). Cement and concrete – handbook dealing with cement, concrete and mortar. 18 th. Edition sept.: 131, 141.
Castellote, M., Andrade, C. & Alonso, C. (1999). Chloride-binding isotherms in concrete submitted to non-steady-state migration experiments, CEMENT and CONCRETE RESEARCH, 29: 1799–1806.
Fajardo, G. Escadeillas, G., Arliguie (2006). Electrochemical chloride extraction (ECE) from steel-reinforced concrete specimens contaminated by "artificial" sea-water, Corrosion Science, 48: 110–125.
Hansen, H. (2008). Mørtels udvikling I Danmark (in Danish). Mortars development in Denmark. Unpublished.
Ottosen, L.M., Rörig-Dalgaard, I. (2008). Desalination of bricks by application of electric DC field. Accepted for publication in Materials and Structures.
Rörig-Dalgaard, I. (2008). Preservation of masonry with electrokinetics – with focus on desalination of murals. Ph.D thesis.
Rörig-Dalgaard, I. & Ottosen, L.M. (2008 B). Electromigration versus electroosmosis in a clay brick under non steady laboratory conditions. Submitted.
Tritthart J. (1999). Ion transport in cement paste during electrochemical chloride removal. Advances in Cement research, 11(4): 149–160.

Concrete Solutions – Grantham, Majorana & Salomoni (Eds)
© 2009 Taylor & Francis Group, London, ISBN 978-0-415-55082-6

Long term behaviour of box girder sections with residual chloride after the application of ECE

U. Schneck

CITec Concrete Improvement Technologies GmbH, Dresden OT Cossebaude, Germany

ABSTRACT: In the Danube Bridge Pfaffenstein (Regensburg/ Germany) between 2003 and 2007 in total ca 200 m^2 box girder floor slab were rehabilitated non-destructively from chloride induced corrosion by electrochemical chloride extraction (ECE). These areas had become chloride contaminated by a leaking drainage system from the carriageway above (up to 4% of cement mass), and hence showed corrosion activity, but no considerable loss of reinforcement cross-section or concrete deterioration. After terminating the ECE, the related concrete areas, where more than 28 kg of chloride was removed, have been surveyed by repeated potential mapping for more than 3 years. It is evident, that chloride induced corrosion activity has been removed safely, even with some residual chloride in the concrete. The whole repair did not influence or limit the traffic on the highway or on the road that crosses the bridge underneath.

1 THE PRINCIPLE OF ECE

1.1 General remarks

The effect of chloride removal is caused by an electrical field between the reinforcement and an external, non-permanent electrode (see figure 1). This electrical field is controlled by the voltage between the electrodes, and all ions dissolved in the pore solution will be moved – negatively charged ions such as chloride or hydroxyl ions towards the outside anode; positively charged ions (mainly sodium in case of de-icing salt attack) towards the reinforcement, which acts as a cathode.

The process requires wet concrete, and related to different water adsorption to anions and cations, more water will be moved into the concrete than out of it during such a treatment. The higher the voltage can be set, the more intensive the chloride movement will be.

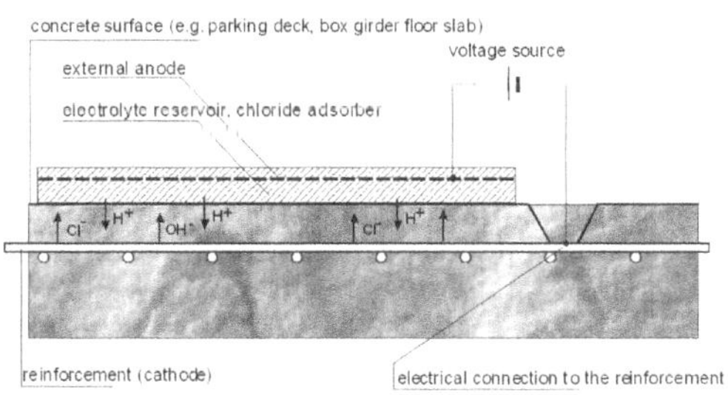

Principle of the electrochemical chloride extraction (ECE):

driving voltage between anode and cathode: ca. 30–40 V
max. current density, related to the rebar surface: ca. 1–5 A/m^2
duration: ca. 4–8 weeks

concrete surface (e.g. parking deck, box girder floor slab)

voltage source

external anode

electrolyte reservoir, chloride adsorber

reinforcement (cathode)

electrical connection to the reinforcement

Figure 1.   Principle of the electrochemical chloride extraction (ECE).

Usually, 40 V are chosen for a good chloride extraction progress under safe work conditions.

The migration of ions is a physical process which is forced along the field lines between the electrodes, indeed it can happen only within the capillary and shrinkage pores, which take other directions than the established field lines. Due to the different size, specific movability (Elsener 1990) and concentration of the dissolved ions they comprise varying percentages of the total ion movement. Practically, in the anion movement, the portion of chloride is largest at the beginning of the ECE and decreases over the duration of the treatment, whereas the portion of hydroxyl ion migration is increasing at the same time.

Electrochemical reactions take place on the reinforcement surface: the reduction of oxides, oxygen and water. All of them are related to the current and electrical charge which is impressed into the reinforcement/cathode by the ECE. Typical current densities (related to the reinforcement surface), range between 0.5 and 2 A/m^2, but can be much higher during the first hours/days of a treatment. Normally, the main process will be the reduction of oxygen and water, and according to equation (1a) it will generate hydroxyl ions, which raises the alkalinity of the concrete in the reinforcement vicinity and is the main target of a related method – the electrochemical re-alkalisation:

$$\frac{1}{2}O_2 + H_2O + 2e^- \rightarrow 2OH^- \quad (1a)$$

$$e.g.\, Fe_2O_3 + 3H_2O + 2e^- \rightarrow 2Fe(OH)_2 + 2OH^- \quad (1b)$$

$$2H_2O + 2e^- \rightarrow H_2 + 2OH^- \quad (1c)$$

Equation 1: Possible reduction reactions on the reinforcement surface forced by ECE.

Both processes – chloride migration and reduction of oxides, oxygen and water – run at the same time, but do not depend on each other in predictable terms. Whereas chloride migration depends on the applied voltage, the cover thickness and permeability of concrete and the water content, the current is controlled by the voltage, the temperature, the resistance of concrete (as a sum parameter of concrete cover, permeability, soluble ions and water content) and the charge transfer resistances on anode and reinforcement. This does not correspond with some other publications, e.g. (Polder & Walker 1993) but has got a practical backup when high amounts of chloride could be removed at rather low current densities and charges as well as only slow desalination progress being observed at high current densities and total charges.

So the ECE does not only reduce the chloride content of the concrete, but also raises its alkalinity as a result of the reduction reactions. This improves the corrosion protection additionally, since with a high OH^- content also an increased chloride content can be present in the concrete without triggering reinforcement corrosion activity.

Chemically on the cement matrix bound chloride (in the presence of C_3A) will be released partly during the ECE because of the dynamic relation between free and bound chloride. The total binding capacity does not exceed 0.5% (related to the cement mass) from our own experience, so no considerable amounts of chloride can be released.

On the external anode, which is usually of a dimensionally stable material, we find other electrochemical reactions that lead to very acidic conditions: the oxidation of water, hydroxyl ions and chloride as well as the formation of chlorine. According to Elsener, Molina & Böhni (1993) the reaction of water and chlorine can also cause an acidic environment.

1.2 Components for ECE

Generally, for an ECE the following components are needed:

- a dimensionally stable anode (DSA), usually activated titanium mesh
- an electrolyte reservoir that embeds and attaches the anode to the concrete surface
- a high power supply that establishes an electrical field between anode and reinforcement
- measuring and control units for recording and controlling voltage, current and - if present – the signals at installed reference electrodes

As electrolyte, simple tap water can be used. In order to prevent acidification of the electrolyte and of the concrete surface frequently an alkaline solution made of $Ca(OH)_2$ or of NaOH is used. Also from the SHRP comes the recommendation to take a lithium borate solution as electrolyte for a better alkaline buffering effect and to counteract expansive effects of ASR (Mietz 1998).

In Norway an application system has been developed and introduced (Noteby 1986), that has gained the

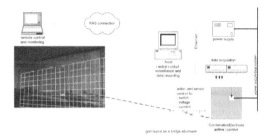

Figure 2. Schematic layout of the grid cell based ECE system by CITec.

biggest portion of ECE applications and uses sprayed, wet cellulose fibres as electrolyte reservoir, which provides an excellent connection between anode and concrete (see figures 2 and 3). It is also known as NORCURE and can be used on large surfaces. Other application systems use geotextile blanket anodes with circulating electrolyte or tanks that contain the anode and an aqueous electrolyte (Broomfield 2007). They can be mounted either horizontally or vertically.

A different technical solution has been developed in 2001 in Germany by CITec and is designed for the focussed treatment of smaller corrosion "hot spots". Since 2008, this patented technology has been used successfully on 14 structures, treating a total surface of ca. 1,700 m^2 and removing more than 70 kg of chloride – which equals an amount of 110 kg of NaCl. The basic layout is to be seen in figure 4 and has following key features:

- The electrodes used for the ECE have a size of 60×60 cm, are pre-manufactured, re-usable and have an ion exchanger for binding chloride
- According to the configuration (rebar spacing, concrete cover, chloride content, concrete permeability etc) the electrodes can be combined in groups (max 10 m^2).
- A chloride measuring unit (Schneck, Gruenzig, Vonau, Herrmann & Berthold 2006) signals the ion exchanger saturation and switches off the related electrodes.
- The control of the ECE is based on a uniformly provided voltage of 40 V and a pulse width modulation (PWR, Schneck, Mucke & Gruenzig 2001) that switches the electrodes or groups of electrodes in intervals – normally 12 hours. The on-time is reduced when defined parameters such as current or potential readings are exceeded.
- Chloride saturated electrodes are regenerated in a solution at pH = 14. So the bound chloride is replaced by OH-, can be analysed, and the electrodes get a fresh alkaline buffer capacity, and waste from the ECE application can be avoided.

1.3 Influencing factors and side effects

As already mentioned, the effect of the ECE is influenced by many factors, which are not easy to determine in practical cases, and these factors can vary strongly

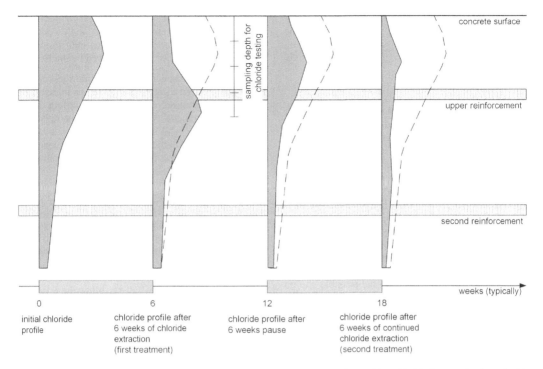

Figure 3. Dynamics of chloride migration during a multi-stage ECE application at high chloride ingress depths and with more than one reinforcement layer.

within a few square metres. The ECE itself influences the treated structural part not only by the removal of chloride and increase of alkalinity, but also in unwanted ways if the structure is not assessed and the ECE is not designed and applied appropriately. Some issues are:

1.3.1 Reinforcement layout

The reinforcement is the internal electrode for the ECE; concrete cover and rebar spacing have great influence on the desalination effect: The main part of the process takes place between the concrete surface and the upper reinforcement layer; if the concrete cover is low – less than 20 mm – the area directly above the bars will be desalinated almost totally, but only a negligible chloride removal can be achieved in concrete areas behind the upper reinforcement. Usually the chloride ingress has reached greater depths. Since the field lines get established from the anode mesh to the rebars, we will find a rather patchy chloride removal in case of a spacious reinforcement layout; the desalination is concentrated then on the area of the rebars. Of course short circuits between reinforcement and anode have to be avoided.

1.3.2 State of corrosion at the begin of treatment

If the reinforcement surface is entirely covered by corrosion products – which can be observed at high chloride concentrations in the reinforcement vicinity – the major initial reaction will be the reduction of oxides, which does not result necessarily in the formation of hydroxyl ions. Theoretically it can require

up to $500\,Ah/m^2$ to reduce tightly covering corrosion products from the steel surface until a larger portion of oxygen reduction (and related hydroxyl ion formation) can take place. This should to be considered at the definition of treatment targets.

1.3.3 Alkali-Silica reaction (ASR)

Alkali reactive siliceous aggregates show an expansive reaction in the presence of water and in a highly alkaline environment, forming cracks and alkali-silica gel. ECE can initiate or accelerate ASR because water is moved into the concrete – it can raise the water content up to about 8%, hydroxyl ions are generated by the oxygen and water reduction and sodium is moved into the concrete. Various literature about intensive research on that issue is available, such as (Page & Yu 1995), showing that no easy dependencies could be observed between impressed charge, time of application and effects on ASR, but it is advised to pay attention on the presence of alkali reactive aggregates in the concrete of a structure subject to ECE, and if conditions seem to be critical, to run trial applications beforehand.

1.3.4 Chlorine evolution and acidification of the concrete surface

If no precautions are taken, the anodic processes will generate large quantities of chlorine and very acidic conditions on the concrete surface. Chlorine is not only a health and environment hazard, but can establish a very corrosive atmosphere for neighboured metallic parts (installation equipment, cars etc.), especially because the ECE requires wet conditions that lead

37

to an increased air humidity adjacent to the application. Opportunities to limit these effects are alkaline and/ or buffering electrolyte solutions or an anodic ion exchanger that binds chloride and releases hydroxyl ions at the same time.

1.3.5 Bond strength

The possible reduction of bond strength as a result of ECE has been investigated under various test conditions. Negative effects – the reduction of bond strength up to 50% - have been found especially at smooth, corroded rebars after applying very high charges at high current densities, although no definite dependencies between the test parameters could be concluded. According to (Broomfield 2007) an increase of bond strength as a prestressing effect from corrosion products on the reinforcement has to be considered, which can be removed during ECE, but is no reduction of bond compared to the uncorroded state. Furthermore, if ribbed steels are used, the main bond is provided by the shape of the interface, so the possible slip on smooth surfaces cannot take place. Another influencing fact is the immense wetting of the concrete caused by the ECE, but this is a reversible effect which does not harm the concrete structure.

1.3.6 Use of ECE at pre-stressed concrete

A possible result of the cathodic reaction is the evolution of hydrogen from water reduction (see eq. 1c). It can be triggered at potentials (IR free) more negative than $-1,070\,mV$ vs. CSE (Copper/ Copper-Sulphate Electrode) at a pH of 13. If evolved in larger quantities, hydrogen may migrate into micro-joints of pre-stressing steel and lead to hydrogen embrittlement and failure of pre-stressing steel. In post-tensioned structures the pre-stressing steel is protected by tendon ducts that act as a Faraday's cage. (Gruenzig 2002) has shown that tendon ducts shield the duct inside even at larger defects (holes up to $3\,cm^2$) safely, so that in an aqueous solution an outside voltage of 40 V did not shift the potential inside the tendon duct more than $-50\,mV$.

1.4 Criteria for the termination of ECE

It is very difficult to define, predict or to calculate parameters on which an ECE application can be terminated. Several factors are influencing the treatment progress and its dynamics. For a practical approach the following factors are being used:

- Chloride content: the average content of the surveyed profile should not exceed 0.4%, related to the cement mass; the maximum content in the reinforcement vicinity should be max. 0.8%
- Impressed charge: depending on the extent of initial corrosion products on the reinforcement, the total impressed charge should range between 1,000 and $2,000\,Ah/m^2$, related to the reinforcement surface
- Rest potentials: during a repeated potential survey (done after recovery of the concrete and reinforcement from wetting and cathodic polarization) the general level of potentials shall be more

positive than during the initial measurement; the total spread of potentials over the whole concrete area shall not exceed 150 mV thus indicating that macro elements have been dissolved by the ECE

It has to be considered that the chloride content cannot be reduced to zero; often even an increase of chloride can be observed within the sampling depth during the treatment as chloride has penetrated far behind the upper reinforcement layer. Then, a multistage application is required. The ratio/ efficiency of chloride removal is less important than a generally low chloride content at the end.

Figure 3 shows how first the chloride within the concrete cover is mobilized, and as it is removed, more chloride arrives from deeper concrete zones. This may result in an increase of chloride content within the sampling profile and the (wrong) conclusion that the ECE might not work. If the amount of removed chloride can be analysed, this is very useful additional information to trace the ECE dynamics. Within a pause of about 6 weeks which is recommended to allow the concrete drying, more chloride is moved towards the concrete surface by capillary suction/evaporation. Furthermore, bound chloride can be released, but these are normally not large quantities. In a second treatment stage, the chloride which has now accumulated in the concrete cover zone can be reduced to an uncritical level.

The impressed charge is the second parameter for the termination of the ECE; the above suggested range may not be achieved in case the reinforcement is still in good condition and the chloride content of the respective treatment area is rather low. Then, ca. $400\,Ah/m^2$ shall be a sufficient value, also because little or no oxide has to be reduced on the rebar surface and hence almost all charge will be put into oxygen reduction.

The main task of ECE is to rehabilitate a concrete structure/part from corrosion activity. Thus, a repeated potential survey shall show considerably different results: a generally more positive level of measurement values and a reduced spread between the maximum and minimum potentials. As macro elements are being dissolved by ECE, formerly very negative potentials have to become much more positive, and formerly positive potentials shall become more negative. Overall, the potential spread shall stay within 150 mV. The repeated potential survey is not suitable to obtain the termination of treatment, because during ECE the reinforcement has been shifted deeply into the cathodic range. It has to recover from that polarisation back into a rest stage, and the high water content in the concrete has to be reduced to a normal level (from ca. 7% to 3%), until useful potential values can be measured. This requires at least 3 to 4 months.

2 THE ECE APPLICATION IN THE DANUBE BRIDGE PFAFFENSTEIN

2.1 The project

In preparation of a major repair damaged areas were investigated that had formed around leaking drainage

Figure 4. Side wall of the box girder with the damaged, temporarily fixed drainage inlet.

system inlets under the bridge deck. These inlets were to be found where the bridge deck projected widely above the hollow box girder. The damaged areas had undergone a preliminary repair, but high chloride concentrations did accumulate in the box girder floor slabs before that (up to 4% of the cement mass), but did not show considerable deterioration of concrete or loss of reinforcement cross-section. A conventional repair would have caused heavy impact on the traffic on the highway and on the road crossing the bridge underneath, because the whole cross-section of concrete would have had to be replaced in larger areas (see figure 4).

So a condition survey could identify the corrosion affected areas. With a trial desalination a suitable operation mode could be found for the ECE, and after all ECE was chosen to be the basic repair method.

2.2 The condition survey

With the condition survey, the approach of which has been introduced in (Schneck, Winkler & Mucke, 2001 and Schneck, 2005), chloride bearing and corrosion active areas ranging between 5 and 60 m² hadve been traced in 4 box girder sections. In total ca 200 m² were penetrated by chloride, and in 100 m² of that area, high corrosion activity was found, and chloride had migrated far behind the upper reinforcement layer.

2.3 Configuration of ECE

Figure 5 shows the areas on the box girder sections that were selected for the ECE treatment. Depending on concrete cover, reinforcement (= cathode) surface and chloride distribution, the pre-manufactured electrodes were combined in anode groups in a size between 2 and 10 m²; one electrode per group was defined as a master electrode controlling all measuring and switching of the group.

Due to the high chloride ingress depths a two-stage ECE application was chosen, which comprised a 6 week initial treatment, 6 weeks pause and a 6 week follow-up treatment. Figure 6 shows the electrode layout.

2.4 Results of ECE

A recording of the current densities is to be seen in figure 7 and shows how, especially in high chloride affected areas, high initial current densities could be observed (up to 10 A/m²), where on areas with relative little chloride typically ca 0.5 A/m² could be measured – at a constantly applied voltage of 40 V.

After some days, the current density generally was within 0.5 and 1.0 A/m², which corresponds with experiences from other projects. The initially high current densities certainly result from an intensive oxide reduction as the main cathodic reaction. Furthermore, the pulse width modulation is visible in figure 7.

In figure 8 a typical effect of a two-stage treatment of deeply chloride containing concrete can be seen: the initial chloride content is being reduced in the outer 8 cm, but increases later, when more chloride will be moved from deeper zones with capillary suction from drying processes (the concrete humidity has been increased during ECE to about 8% from 2.5%). The second application has been for the final treatment and could reduce the chloride content down to an uncritical level.

In the back row the original stage in 5 locations can be seen; each layer is 2 cm; the 0-2-cm-layer is on the right side of the blocks. The red/dark bars show the chloride content in reinforcement vicinity; after the second treatment the chloride values were ranging between 0.75 and 0.1%; within the concrete cover the chloride values were between 1.3 and 0.4%.

The applied electrical charges ranged between 674 and 2,165 Ah/m² (related to the reinforcement surface), which did support the chloride removal by alkalisation of the concrete in the reinforcement vicinity.

From the ion exchanger of the electrodes during the whole ECE treatment (that got a third stage of application in one of the areas in 2007) 28 kg of chloride were removed when the electrodes were regenerated in a NaOH solution at pH = 14. This corresponds to an amount of 43 kg NaCl that had entered the concrete. Across the single concrete areas the amount of removed chloride varied between 55 and 227 g/m².

After the regeneration, the electrodes are freshly loaded with OH^- and can continue to buffer acid anodic reactions and collect chloride from concrete.

2.5 Results of the repeated condition survey

In addition to the parameters of chloride and electrical charge, a repeated potential survey can give evidence about the success of an ECE treatment. Formerly macro corrosion elements should have been dissolved, and locations with initially very negative rest potentials should have much more positive readings then; formerly areas with positive rest potentials should be more negative, and the total spread of potentials should be limited to ca 150 mV. The latter value is a result of practical experience.

It may take some months until the decreased water content allows a useful measurement; the recovery

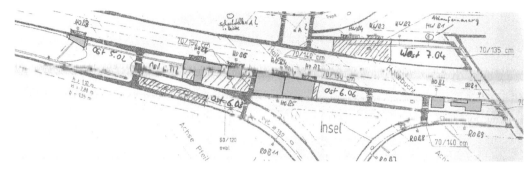

Figure 5. Overview of the areas treated with ECE across the hollow box girder floor slabs.

Figure 6. Electrode layout on the box girder surface.

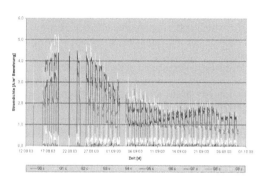

Figure 7. Recording of current densities during the ECE application.

of the reinforcement from the cathodic polarization should not take that long.

Figure 9 shows such repeated potential measurements for one of the ECE areas. Four months after the first treatment areas of possible corrosion activity could still be seen, but six months after finishing the second treatment, a repeated potential survey showed a very even potential distribution with low deviation from the average value ($-100\,$mV vs. CSE) and no signs of possible macro elements (and related corrosion activity) anymore. The highest potential shift of about $+400\,$mV was observed in the areas of initially most negative potentials.

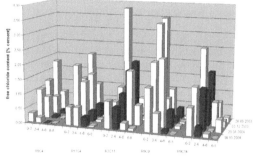

Figure 8. Chloride profiles at five surveillance locations before, during and after ECE.

In figure 10 is to be seen how the most negative potential readings and the average potentials develop within the single concrete areas of ECE application: all data get a shift towards more positive values. The figure shows also repeated measurement from 2007 – almost 3 years after finishing the ECE.

The more negative the potentials before starting ECE were, the greater was the potential shift that has been observed after the ECE.

3 CONCLUSIONS

The application of ECE has shown that this repair method is able to eliminate chloride induced corrosion activity safely and durably also in cases where the chloride ingress had reached deeper concrete zones or even the whole cross-section of a concrete slab. With a multi-stage ECE application chlorides can be removed widely.

Although some residual chloride remained (in this project up to 1% related to the cement mass in single cases), the corrosion protection could be re-established entirely, because the alkalisation of the concrete, caused by oxygen reduction on the reinforcement, enhanced the corrosion protection.

This technology is an interesting addition to the common repair techniques, especially because ECE is a one-off application, has only minor or no impact on the use of a structure and preserves the concrete.

Initial potential survey, May 2003 (before starting ECE)

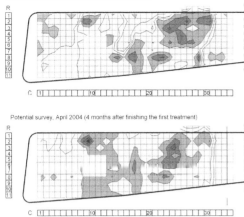

Potential survey, April 2004 (4 months after finishing the first treatment)

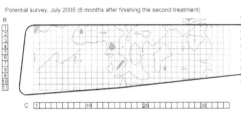

Potential survey, July 2005 (6 months after finishing the second treatment)

300 < x <= 1000	200 < x <= 300	[mV vs. CSE]
100 < x <= 200	0 < x <= 100	
-100 < x <= 0	-200 < x <= -100	
-300 < x <= -200	-400 < x <= -300	
-600 < x <= -400	-1500 < x <= -500	

Figure 9. Potential readings before, after the first and after the second ECE application stage.

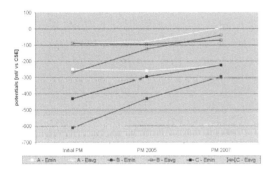

Figure 10. Development of most negative and average potentials within the single concrete areas from the initial monitoring before the ECE application, during a first series of repeated monitoring in 2005 and a second one in 2007.

4 NORMATIVE SITUATION

Following a CEN technical specification on electrochemical re-alkalisation (CEN TS 14038-1, 2004), a new work item on ECE is under progress (prCEN TS 14038-2). NACE has published the Standard Practice paper SP0107-2007 "Electrochemical Realkalization and Chloride Extraction for Reinforced Concrete".

REFERENCES

Elsener, B. 1990. Ionenmigration und elektrische Leitfähigkeit im Beton. *SIA Dokumentation D065: Korrosion und Korrosionsschutz, Part 5*

Polder, R, Walker, R. 1993. Chloride Removal from a Reinforced Concrete Quay Wall – Laboratory Tests. *TNO Report 93-BT-R1114, Delft*

Elsener, B., Molina, M., Böhni, H. 1993. Electrochemical Removal of Chlorides from reinforced Concrete Structures. *Werkstoffwissenschaften und Bausanierung Vol 420 Part 1 – Expert Verlag*

Mietz, J. 1998. Electrochemical Rehabilitation Methods for Reinforced Concrete Structures – A State of the Art Report. *European Federation of Corrosion Reports 24, Institute of Materials*

Noteby. 1986. European Patent Application No 86302888.2

Broomfield, J. 2007. Corrosion of Steel in Concrete – Understanding, Investigation and Repair. *Taylor and Francis, London*

Schneck, U., Gruenzig, H., Vonau, W., Herrmann, S., Berthold, F. 2006. Chloride Measuring Unit for the Improvement of Security and Performance of Electrochemical Chloride Extraction. *Concrete Solutions, Proc. of the 2nd Intl. Conference, BRE Press*

Schneck, U., Mucke, S., Gruenzig, H. 2001. Pulse Width Modulation (PWR) – Investigations for raising the efficiency of an electrochemical chloride extraction from reinforced concrete. *Proc. EUROCORR 2001, Riva del Garda*

Page, C.L., Yu, S. 1995. Potential Effects of Electrochemical Desalination of Concrete on Alkali Silica Reaction. *Magazine for Concrete Research 47*

Gruenzig, H. 2002. Orientierende Versuche zur abschirmenden Wirkung eines Spannstahlhüllrohres im elektrischen Feld. *CITec GmbH (unpublished)*

Schneck, U., Winkler, T. und Mucke, S. 2001. Integrated system for the inspection and the corrosion monitoring of reinforced concrete structures. *Proc. Quality Control and Quality Assurance of Construction Materials Conference, Dubai*

Schneck, U. 2005. Qualifizierte Korrosionsuntersuchungen an Stahlbetonbauwerken. *Bautechnik 82 (2005), 443–448*

41

Fire damage and repair

Concrete Solutions – Grantham, Majorana & Salomoni (Eds)
© 2009 Taylor & Francis Group, London, ISBN 978-0-415-55082-6

Approaches for the assessment and repair of concrete elements exposed to fire

E. Annerel & L. Taerwe
Laboratory Magnel for Concrete Research, Ghent University, Belgium

ABSTRACT: An adequate assessment of the residual strength of concrete structures exposed to fire is necessary for executing an appropriate repair. This strength can be determined indirectly by assessing the temperature history inside the concrete. Different techniques, such as colorimetry and water immersion are outlined in this paper. In the (L)ab colour space, an elliptical path is found as a function of the temperature. However, the strength is influenced by the test setup and storage conditions after fire. Storing the concrete in ambient air may introduce an additional strength loss up to 20 percent. Once the temperature distribution in the concrete is known, the residual capacity of a concrete element can be calculated. The use of the finite element program TNO DIANA in combination with simple calculations is illustrated for a concrete girder exposed to a real fire and afterwards repaired with shotcrete.

1 INTRODUCTION

Generally, concrete structures have a very good fire resistance. Although damage to the concrete gradually appears with increasing temperature, it is possible to repair the structure after an adequate assessment. To do this in a systematic way, knowledge is necessary concerning residual material properties and methods to assess this strength. Since the residual strength is temperature dependent, methods may be used to assess the strength indirectly by measuring the alteration in the material as a function of the temperature. A paper by Annerel (Annerel & Taerwe 2007) illustrates how cracks appear and the colour of the concrete changes from red (300–600°C) to whitish grey (600–900°C) and buff (900–1000°C). Once these temperatures are obtained, simple calculations based on EN1992 (EN1992-1-2-2004) may be used to determine the residual capacity of concrete elements, as elaborated in a paper by Taerwe (Taerwe, Poppe et al, 2006).

2 RESIDUAL STRENGTH

Table 1 summarizes the mix design of a self-compacting concrete (SCC), a traditional vibrated concrete with siliceous aggregates (TC) and calcareous aggregates (TCk), as well as a high strength concrete (HPC). One hundred and fifty millimetre cubes were cured for 4 weeks in an air-conditioned room at an RH >90% and a temperature of 20±1°C, after which they were stored at 60% RH and 20±1°C for drying until testing age (>17 weeks). Two cubes were heated for each of the examined temperature levels (till 800°C), occurring at a heating rate of 3.5°C/min. The target

Table 1.   Concrete mix design.

Component (kg/m³ unless stated)	SCC siliceous	TC siliceous	TCk calcareous	HPC siliceous
sand	782	640	663	650
gravel 2–8 mm	300	525	–	530
gravel 8–16 mm	340	700	–	720
limestone 2/6	–	–	450	–
limestone 6/20	–	–	759	–
CEM I 52.5	400	350	350	400
water	192	165	165	132
limestone powder	300	–	–	–
superplasticizer 1 [l/m³]	2.90	–	–	–
Superplasticizer 2 [l/m³]				16.5
W/C [-]	0.48	0.47	0.47	0.33
compressive strength 28d [N/mm²]	65.9	56.5	60.3	77.3

temperature was kept constant for 750 minutes. The cubes were allowed to cool slowly in the oven, after which they were immediately tested for compression. Figure 1 shows how the results are situated around the curves mentioned in EN 1992-1-2 [2].

During heating, the concrete expands, which is characterized by its thermal expansion coefficient. After cooling, a part of this expansion is residual (Fig. 2). For TC2k-5, which was tested for compressive strength at 28 days, the cubes were measured for both 0 and 28 days after cooling. During this storage period of 28 days under conditions 60% RH and 20 ± 1°C, the concrete had absorbed an amount of moisture and the volume had increased further. A linear relationship was found between both differences ($R^2 = 0.998$), whereas

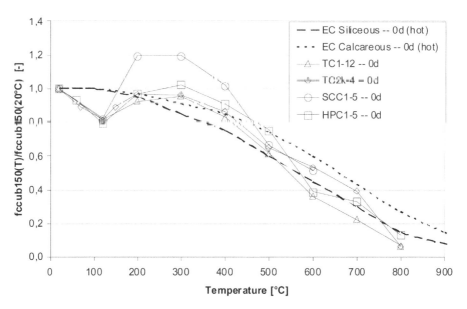

Figure 1. Residual compressive strength.

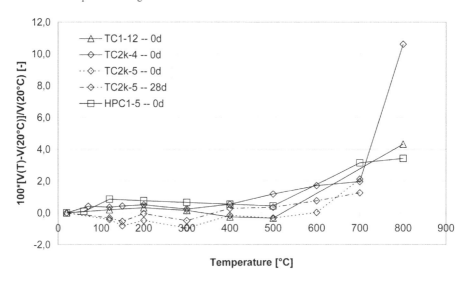

Figure 2. Increase in volume after cooling for 0–28 days after fire.

this difference was around 0.40% in the temperature range 200–400°C and 0.70% in the range 500–600°C.

Tests were executed to determine separately the influence of the heating rate, duration and cooling method on the residual strength at 0 days after heating to 350°C and 550°C (Annerel 2007). The heating rate was changed to 10°C/min for TC and 20°C/min for SCC. The duration at the target temperature was altered to 3600 minutes and the cooling regime was modified into a rapid cooling by quenching under water. Results showed that only the cooling method is an important parameter to consider, resulting in an extra drop of the residual strength of 30–35%.

Besides the test conditions, the storage after heating influenced the residual strength. One hundred

and fifty millimetre cubes TC2k were heated to different target temperatures according to the standard conditions as mentioned before. TC1 cylinders (Ø113mm × 320 mm) were heated at a rate of 1°C/min, kept for 750 minutes at the target temperature and slowly cooled in the oven. Figure 3 shows an additional strength decrease above 200°C compared to the strength loss due to heating (Fig. 1) of about 20% when testing the cubes and cylinders after a period (28 days; 12 weeks) of storage at RH 60% and 20±1°C. These experiments have been repeated for 350°C and 550°C, after which they were stored under water and ambient air (Annerel 2007) Again an additional strength decrease of 20–30% was perceptible around 7 days of storage, from where the strength slowly recovered.

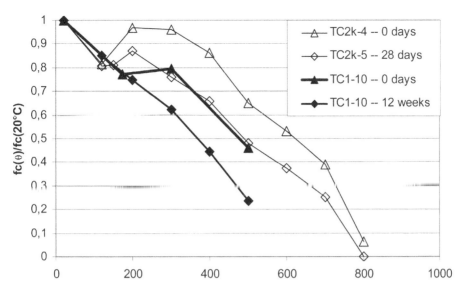

Figure 3. Further strength decrease and recovery of TC2k cubes and TC1 cylinders due to storage (60% RH and 20±1°C) after heating.

Strength recovery of up to 10% was found when storing for 56 days.

3 ASSESSMENT OF THE TEMPERATURE PROFILE

3.1 *Colorimetry*

The colour was measured with an X-rite SP60 spectrophotometer (aperture 8 mm) according to the CIE Lab-colour space. In this colour system 'L' is the lightness with values between 0 (black) and 100 (white), while 'a' is spread between magenta (positive values) and green (negative values) and 'b' is positioned between yellow (positive values) and blue (negative values). Figure 4 shows how the colour measured on cast surfaces describes an elliptical path in the a*b*-colour space. However, differences are noticeable between the 4 concrete types. The measurements were executed immediately after cooling down to ambient temperatures and before the compression test (Fig. 1). When storing these cubes under RH 60% and at temperatures of 20±1°C for several weeks, a shift towards the inner part of the ellipse was found (Annerel 2007) This shift can probably be attributed to moisture absorption, because a linear relationship between the colour change (L, a, b) and the weight increase exists with a R² of 0.7–0.8. Further studies were executed on TC and SCC samples cut from drilled cores, after polishing and masking of the colourful aggregates (Annerel 2007).

3.2 *Water immersion*

Heating of concrete introduces stresses, resulting in cracks. Meanwhile, chemical alterations, such as dehydration and decarbonation, lead to the disappearing of the hydration products, which increase the pore space

(Fig. 5). These two effects thus lead to an increase of the porosity when heating concrete. Immersion of concrete under water will fill the pores and cracks with water. The weight increase can then be used to assess the internal damage due to heating. The total water absorption can be defined as the difference in weight after storage under water and a reference weight, for instance the weight at uniform target temperature ($M_{0d,hot}$).

After 7 days, the weight increase flattens. $M_{0d,hot}$ can be determined from the heated sample itself by drying it till constant mass. Drying is necessary to eliminate the moisture due to climatic circumstances. Newly formed hydration products which may fill some small cracks and thus may hinder the water absorption are neglected in this method. In lab conditions, this weight is measured during the test when the concrete is at target temperature or after cooling down to 60°C. Figure 6 illustrates the water immersion of half cubes (TC1-8, TC2k-1) and small discs (TC1-10) with Ø80mm and 15mm height. These results are transformed in percentages by dividing with the reference weight $M_{0d,hot}$. For the discs, different cooling methods are used: 1) slowly cooling in the oven ('L, oven'); 2) cooling outside the oven at ambient air ('L, 20°C'); 3) cooling by quenching into water ('Water'). The results on half cubes and small discs are comparable, as is visible on the water absorption for specimens slowly cooled in the oven.

4 FINITE ELEMENT TEMPERATURE SIMULATION

In the 1970's an industrial hall in precast concrete elements was erected especially to submit it to a real scale fire test. This experiment was executed in order to gain knowledge about the behaviour of such a

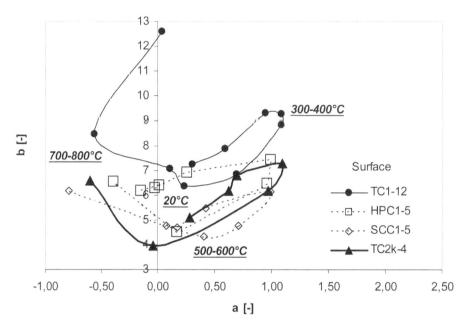

Figure 4. Colour development at concrete surface.

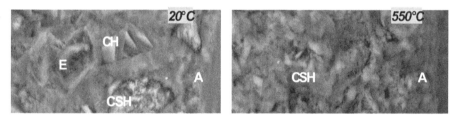

Figure 5. ESEM images of ordinary Portland concrete at 20°C (upper) and heated up to 550°C (bottom). E = ettringite, CH = portlandite and A = aggregate.

building. During the test, the temperature was recorded with several thermocouples. After the fire, a pretensioned concrete girder was lifted out of the building, was repaired by shotcreting and submitted to a static loading test up to failure. The girder collapsed at a maximum bending moment of 1632 kNm, corresponding to 4 point loads of 162 kN.

The roof girder was modeled in the finite element program TNO DIANA to obtain the temperature distribution inside the concrete. This 2D heat flow analysis was simulated by using the thermal material properties as outlined in EN 1992-1-2 for siliceous concrete. The fire load that occurred conformed to the ISO834 curve and also according to the surface temperature of the roof beam measured during the fire test. Figure 7 (left) illustrates the temperatures reached inside the concrete after 90 minutes of fire exposure. Rough interpolation yields that the real fire after 90 minutes had almost the same effect inside the cross section as the ISO curve after 85 minutes.

The maximum bending moment was calculated before, during and after the fire according to EN 1992-1-2 by using reduction coefficients $k_p(\theta)$. Before the fire, the calculated resistance capacity was 1855 kNm. During the fire, this capacity dropped to 57 %

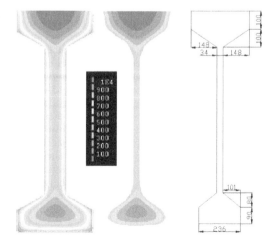

Figure 7. Temperature distribution in girder (left) and area with temperatures lower than 500°C (middle, right).

when using the 500°C method (Fig. 7 Middle and Right). After repairing and for the four point loads, the calculated capacity was 1604 kNm, which is comparable to the result obtained from the loading

48

test. This experiment also illustrates how after repair, a beam can regain 86% of its original strength.

5 CONCLUSIONS

- The residual strength is mainly influenced by the temperature during heating, the way of cooling and the time of storage after fire.
- Colour analysis and water immersion provide an adequate basis to assess the temperature history of concrete.
- Combination of finite element techniques and simple calculations show how the strength of a girder behaves before, during and after a fire.

ACKNOWLEDGMENT

The authors would like to thank FWO Flanders for the financial support through the research grant "Damage assessment and estimation of the residual strength of concrete members after exposure to fire".

REFERENCES

Annerel E., Taerwe L., Approaches for the Assessment of the Residual Strength of Concrete Exposed to Fire. *Proceedings of the International Workshop, Fire Design of Concrete Structures, From Materials Modelling to Structural Performance,* Coimbra, 8th and 9th November, 2007, 489–500.

EN 1992-1-2: 2004 - Eurocode 2: Design of Concrete Structures – Part 1-2: General Rules – Structural Fire Design, CEN, Brussels, 2004, 97pp.

Taerwe L., Poppe A.-M., Annerel E., Vandevelde P., Structural Assessment of a Pretensioned Concrete Girder after Fire Exposure. *Proceedings of the 2nd fib Congress,* Naples, 5–8 June, 2006.

Concrete Solutions – Grantham, Majorana & Salomoni (Eds)
© 2009 Taylor & Francis Group, London, ISBN 978-0-415-55082-6

Resistance of cracked concrete to fire and high temperatures

A. Badr

School of Built Environment & Engineering, University of Bolton, Bolton, UK

ABSTRACT: Fire may occur in concrete structures that have already been in use for some years. Therefore, concrete would have been loaded and stressed and, as a result, developed micro or macro cracks before being exposed to high temperatures. This paper reports the results of experimental research investigating the behavior of cracked and uncracked concrete exposed to high temperatures. Different types of concrete were investigated including normal strength concrete (NSC), high strength concrete (HSC) and fibre reinforced concrete (FRC). The effects of incorporating silica fume (SF) and polypropylene fibre (PPF) on cracked and uncracked concrete were also investigated. The results showed that the residual strength of cracked concrete was inferior to that of the counterpart uncracked concrete regardless of the presence of SF and/ or PPF. The effects of SF and PPF on the behavior of cracked concrete were similar to their effects on uncracked concrete. Particularly, the existence of micro and macro cracks did not alter the spalling or the explosive behavior of SF concrete. PPF slightly enhanced the residual strength of concrete manufactured using Portland cement (CEM1) but not that of SF concrete. PPF mitigated the spalling and the explosive behavior of SF concrete for cracked and uncracked concrete. The results of this investigation could have implications on the assessment of the residual strength of loaded concrete structures exposed to fire.

1 INTRODUCTION

Degradation of concrete due to fire and high temperatures has been widely reported. Fire and high temperatures cause severe damage to concrete structures such as spalling of concrete and reduction of concrete strength (Arioz, 2007, Chan et al. 2000, Dale 2000, Felicetti & Gambarova 1998, Jensen et al. 1997, Luo et al. 2000, Poon et al. 2004, Short et al. 2000). Compared to normal strength concrete (NSC), many investigations suggest that high strength concrete (HSC) is more susceptible to explosive spalling and loss of strength when subjected to rapid temperature rising (Castillo & Durrani 1990, Kalifa et al. 2000, Sanjayan & Stocks 1993). Explosive spalling is a dangerous type of failure and could impair the integrity of a concrete structure (Xiao & Falkner 2006). In some cases, spalling can be extensive and may cause an immediate failure of the structure (Anderberg 1997). Some investigations (Atkinson 2004, Dale 2000, Kalifa et al. 2001) reported that utilizing polypropylene fibre (PPF) in concrete may considerably control the spalling behavior and reduce the amount of spalling of HSC at high temperatures.

Even if a structural element shows only minor or no spalling, it may suffer severe reduction in concrete strength (Jensen et al. 1997) and, consequently, its load-bearing capacity could be significantly impaired. Therefore, the assessment of the residual mechanical properties of concrete exposed to fires or high temperatures is necessary for the evaluation of the serviceability of concrete structures.

Considerable research has been carried out to study the residual strength and behavior of NSC and HSC exposed to high temperatures (Badr et al. 2001, Chan et al. 1999, Felicetti & Gambarova 1998, Luo et al. 2000, Papayianni & Valiasis 1991, Phan & Carino 2002). However, the vast majority of this literature reports tests performed on new concrete specimens (not loaded or cracked). In practice, most fires occur in buildings that would have already been in use for some years and the concrete would have been loaded and stressed and, as a result, the concrete would have developed micro or macro cracks. Therefore, presently there is no available information about the behavior of cracked or stressed concrete.

The existence of micro or macro cracks within concrete before being exposed to fire may affect the behavior of concrete. Consequently, comparing the behavior of cracked and uncracked concrete exposed to high temperatures is of great interest to the construction industry.

2 OBJECTIVES

The main objectives of this research were to:

- Investigate the behavior of cracked concrete exposed to high temperatures and compare it to the behavior of uncracked concrete.
- Study the effectiveness of PPF in reducing the damage due to high temperatures in cracked and uncracked concrete.

• Investigate the effects of SF on the behavior of cracked and uncracked concrete exposed to high temperatures.

3 MATERIALS AND MIXES

3.1 *Materials*

Portland cement (CEM1), conforming to BS EN 197-1, was used as a main binder in this study. SF was used as a cement replacement material. The coarse aggregate used was Limestone natural gravel of 10-mm nominal maximum size. Siliceous sand which complies with zone M of BS EN 12620 was used as a fine aggregate. Coarse fibrillated PPF was used in this study. The fibre had a length of 50 mm and an average diameter of 500 μm, giving an aspect ratio of 100. Superplasticiser (SP) conforming to BS5075 Part 3, was used in all mixes.

3.2 *Mixes*

Four mixes were initially optimised from laboratory tests. The composition of these four mixes is given in Table 1. The name of every mix consists of two parts indicating the binder type and the fibre reinforcement. For example, CM00 is a control mix made with CEM1 as a sole binder and includes no fibre. Similarly, SFPP is a PPF reinforced concrete mix with part of the CEM1 replaced by SF. For mixes containing SF the replacement level was 10% on a weight-to-weight basis.

The control CEM1 mix (CM00) was designed as a normal strength concrete mix (NSC) with a nominal target compressive strength of 35 MPa. On the other hand, the SF mix (SF10) was aimed to be a high strength concrete (HSC) with a target compressive strength of 50 MPa.

The aggregate to binder ratio was kept constant at 4.2 for all mixes including NSC and HSC. The limestone was 40% of the total aggregate in all mixes. The water-to-cement ratio (w/c) for NSC and HSC was 0.38 and 0.35, respectively. A target consistence class S2 (between 40 and 110 mm, BS EN13250-Part 1) was achieved for all mixes by changing the SP dosage. As would be expected, mixes with PPF required higher SP dosages compared to counterpart mixes without fibres, as can be seen in Table 1.

3.3 *Mixing*

Coarse and fine aggregates were mixed with the binder in dry state for about one minute in a conventional rotary drum mixer of 50-litre capacity. Half of the mixing water was added and mixed for three minutes before adding the remaining mixing water and SP. The whole constituents were mixed for another three minutes. Where applicable, fibres were then added to the running mixer. A further two minutes of mixing were needed to achieve uniform distribution of the fibres.

3.4 *Specimens*

Concrete cubes (150 and 100 mm) were cast and a vibrating table was used to achieve full compaction of fresh concrete. In order to monitor the temperature profile inside the core of concrete specimens during the heating process, thermocouples had been set within the 100 mm cubes during the casting of the concrete giving intimate thermal contact with the concrete. The thermocouples were set at different depths of 25, 50 and 75 mm, as shown in Figure 1.

3.5 *Curing*

The specimens were covered with wet hessian and polyethylene sheets overnight. They were then demolded after 24 hours and submerged in a water-curing tank for 28 days. After which, the 150 mm cubes were cut into small cubes of 70 mm nominal dimension. The 100 mm cubes were kept intact. All specimens were then stored in room conditions for an additional 28 days to achieve moisture and temperature equilibrium with the room conditions before being exposed to high temperatures.

4 HEATING REGIME AND TESTING

4.1 *Preparation of cracked concrete*

Prior to starting the heating process, control specimens had been chosen randomly to be tested for compressive strength at room temperature. Once the compressive

Figure 1. Thermocouples set at depths of 25, 50 and 75 mm.

Table 1. Mix proportions (per m³) and consistence values.

Mixes	CEM1 kg	SF kg	Limestone kg	Sand kg	PPF kg	Water kg	SP liter	Consistence mm
CM00	440	–	740	1110	–	165	4.0	110
CMPP	440	–	740	1110	4.55	165	4.4	50
SF10	398	44	744	1116	–	154	4.4	70
SFPP	398	44	744	1116	4.55	154	5.2	80

strength of each mix was established half of the specimens were loaded with a load corresponding to 70% of the established control compressive strength of each mix. The load was applied using the compression testing machine which was manually stopped at the target load. The applied load was enough to develop microcracks within the concrete specimens and it was noticed that some specimens actually developed macrocracks.

4.2 *Heating regime*

The specimens were heated inside muffle kilns (oven) designed for firing pottery. Five digital controlled kilns were used to heat the specimens to target temperatures of 150, 300, 500, 700 and 900°C. The average rate of heating was 3.5°C per minute. Once the target temperature was reached it was kept constant for a specific time in order to maintain a uniform and stable temperature at various sections within the specimens. Mohamedbhai (1986) showed that this specific time depends on the target temperature. Therefore, in this study, the 150°C was maintained for 4 hours, the 300°C for 3 hours, the 500°C for 2.5 hours and all higher temperatures for 2 hours.

Once the heating regime was completed the specimens were allowed to cool gradually to room temperature inside the kilns. The gradual cooling was achieved by keeping the specimens inside the kilns with the doors of kilns shut during the cooling process.

The temperature profile inside the core of concrete specimens was monitored during the heating process at depths of 25, 50 and 75 mm using thermocouples

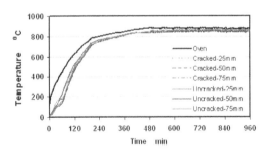

Figure 2. Temperature profile in cracked & uncracked concrete.

which had been set within the 100 mm cubes during casting. The temperature of the oven was monitored by an additional set of thermocouples.

4.3 *Testing of specimens*

The residual compressive strength of the heated concrete specimens was determined at room temperature. All compressive strength tests were carried out according to BS EN 12390: Part 3.

5 RESULTS AND DISCUSSION

5.1 *Temperature profile of cracked and uncracked concrete*

Figure 2 shows the temperature within concrete specimens heated in the oven. It can be seen that there was no difference between the behavior of cracked and uncracked specimens. In both cases, the concrete specimens had uniform temperature throughout the core, as monitored in sections at 25, 50 and 75 mm from the top surface of the concrete specimens.

It was interesting to notice that the temperature of the concrete specimens did not start to rise before the oven temperature approached 300°C. Up to this temperature neither cracked nor uncracked concrete absorbed enough heat to allow the core temperature of the concrete to rise significantly above the room temperature, as can be seen in Table 2.

5.2 *Behavior and residual strength of uncracked concrete*

Figure 3 shows the residual compressive strength of uncracked concrete heated in kilns up to 900°C. The behavior of concrete incorporating SF was significantly different from the behavior of concrete manufactured using CEM1 only, regardless of the presence or absence of PPF. For the latter mixes (CM00 and CMPP) the residual compressive strength-temperature relationship was characterized by an initial strength gain by about 5% between the temperatures of 100 to 200°C. Further heating resulted in gradual decrease in the residual strength. This behavior reflects a typical strength-temperatures relationship, for normal

Table 2. Temperatures (°C) of oven and concrete core.

Oven	Cracked 25 mm	Cracked 50 mm	Cracked 75 mm	Uncracked 25 mm	Uncracked 50 mm	Uncracked 75 mm
35	22	22	22	23	22	22
100	22	22	22	23	22	22
200	30	26	27	33	25	26
300	90	69	69	91	63	70
400	180	159	160	174	116	129
500	301	234	242	332	234	260
600	458	411	433	487	411	436
700	610	578	601	618	569	590
800	759	744	759	765	748	758

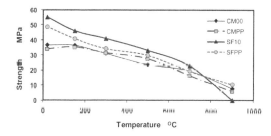

Figure 3. Residual strength of uncracked concrete.

Figure 4. Spalling and explosion of SF10 specimens.

Figure 5. SFPP specimens heated up to 900°C.

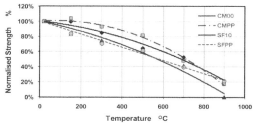

Figure 6. Normalized residual strength of uncracked concrete.

strength concrete exposed to high temperatures (Badr et al. 2001, Chan et al. 1999, Luo et al. 2000).

The residual strength at 700°C was about 50% of the original strength at room temperature. Between 700 and 900°C there was a sharp loss in strength and the average residual strength for the CEM1 mixes at 900°C was less than 20% of its original value.

For HSC containing SF, the loss of strength started once the temperature of the specimen started to rise. The loss of the compressive strength was proportional to the increase in the temperature up to 700°C. The rate of strength loss was higher than that of the CEM1 concrete. This is in agreement with Phan & Carino (2002). The average residual strength of the SF mixes at 700°C was only 40% of the original strength at room temperature. Specimens from the SF mix without fibre (SF10) started to spall when the temperature approached 500°C. The spalling increased with the increase of the temperature up to 700°C, after which the specimens suddenly exploded into small pieces (Figure 4). The tendency of HSC to spalling has been reported by many researchers (Castillo & Durrani 1990, Kalifa et al. 2000, Sanjayan & Stocks 1993).

PPF mitigated the spalling and explosive behavior of the SF concrete when exposed to high temperatures. This is in agreement with several investigations (Atkinson 2004, Dale 2000, Kalifa et al. 2001). At 500°C, SFPP specimens were sound with no sign of spalling or cracks. SFPP specimens heated to 700 and 900°C developed some minor cracks and spalling but they did not explode, as can be seen from Figure 5. The residual strength of the SFPP mix at 900°C was comparable to that of the concrete made using CEM1 only (CM00 and CMPP).

The curves in Figure 3 may suggest that the absolute values of residual strength of concrete containing SF (SF10 and SFPP) are higher than the corresponding values for mixes made with CEM1 only (CM00 and CMPP). This can be attributed to the initial high strength values of the SF concrete, which were 55.1 and 48.7 MPa for mixes SF10 and SFPP compared to 36.6 and 34.0 MPa for mixes CM00 and CMPP, respectively. The values of residual strength of the uncracked concrete are normalized with respect to the original strength at room temperature and are presented in Figure 6. Presenting the residual strength in the normalized format eliminates the effect of the higher initial strength of some mixes. It can be seen that the CEM1 mixes had higher normalized residual strength compared to the SF mixes.

Figure 6 also made the effect of PPF more noticeable. PPF slightly enhanced the performance of the CEM1 mix as can be seen from comparing mixes CM00 and CMPP. This is similar to observations made by Poon et al. (2004). However, the effect of PPF was not clear in the presence of SF. Nonetheless, PPF had a significant effect on reducing the spalling and the explosive behavior of concrete containing SF (Figs 4, 5).

5.3 *Behavior and residual strength of cracked concrete*

Figures 7, 8 present the residual strength and normalized residual strength of the counterpart cracked concrete. Comparing Figures 7, 8 to Figures 3, 6, it can be seen that the existence of micro and macro cracks within concrete before being exposed to high

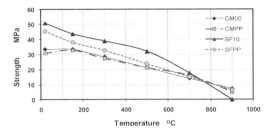

Figure 7.   Residual strength of cracked concrete.

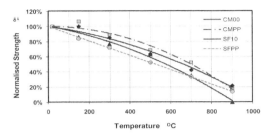

Figure 8.   Normalized residual strength of cracked concrete.

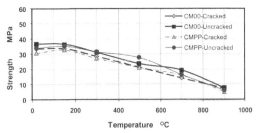

Figure 9.   Residual strength of CEM1 cracked & uncracked concrete.

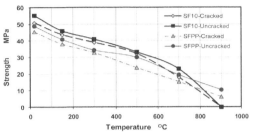

Figure 10.   Residual strength of SF cracked & uncracked concrete.

temperature did not change the characteristic features of the strength-temperature relationships. The behavior of cracked concrete incorporating SF was still significantly different from the behavior of cracked concrete manufactured using CEM1 only. The rate of strength loss in cracked SF concrete was higher than that of cracked CEM1 concrete. However, the spalling of cracked specimens from the SF mix without fibre (SF10) started at temperatures lower than 500°C and was more severe in nature compared to the spalling of the counterpart uncracked concrete. Just after 700°C the cracked SF10 specimens exploded into small pieces. As was the case with the uncracked concrete, PPF mitigated this spalling and explosive behavior of SF concrete when exposed to high temperatures.

In general, it can be stated that the effects of SF and PPF on the behavior of the cracked concrete were similar to their effect on the uncracked concrete. Of particular interest was the fact that the existence of micro and macro cracks did not mitigate the spalling or the explosive behavior of SF concrete specimens. Rather, the existence of the micro and macro cracks increased the risk of spalling of concrete specimens containing SF.

The residual strength of cracked and uncracked concrete is compared in Figures 9, 10. For clarity, mixes with and without SF were presented separately. Both figures indicate clearly that the residual strength of uncracked concrete was higher than the counterpart cracked concrete regardless of the presence of SF and/ or PPF. This outcome has been confirmed by plotting all the data relating the cracked concrete to the counterpart uncracked concrete in Figure 11. Almost all the points are plotted above the equality line indicating higher residual strength of the uncracked concrete compared to the cracked concrete.

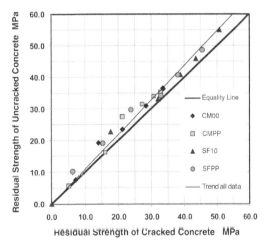

Figure 11.   Equality line for residual strength of cracked and uncracked concrete.

6   CONCLUSIONS

For the concrete and test conditions used in this investigation, the following conclusions are made:

- The residual strength of cracked concrete was inferior to that of the counterpart uncracked concrete regardless of the presence of SF and/ or PPF.
- The existence of micro and macro cracks did not mitigate the spalling or the explosive behavior of SF concrete specimens. Rather, micro and macro cracks could increase the risk of spalling of concrete specimens containing SF.

- The effects of SF and PPF on the behavior of the cracked concrete were similar to their effects on the uncracked concrete.
- CEM1 concrete exhibited an initial strength gain between 100 and 200°C followed by gradual decrease in residual strength. SF concrete suffered loss of strength once the temperature of the specimens started to rise.
- PPF slightly enhanced the residual strength of CEM1 concrete but not that of SF concrete. However, PPF mitigated the spalling and explosive behavior of SF concrete.

The results provide better understanding of the resistance of concrete structures to fire and could have implications on the assessment of the residual strength of loaded concrete structures exposed to fire.

REFERENCES

Anderberg, Y. 1997. Spalling phenomena of HPC and OC. *Fire performance of high-strength concrete; Proc. intern workshop, NIST, MD, USA, 13–14 February*: 69–73.

Arioz, O. 2007. Effects of elevated temperatures on the properties of concrete. *Fire Safety Journal* 42: 516–522.

Atkinson, T. 2004. Polypropylene fibres control explosive spalling in high performance concrete. *Concrete* 38(10): 69–70.

Badr, A., Brooks, J.J., Abdel Reheem, A.H. & El-Saeid, A. 2001. Residual impact resistance and flexural strength of concrete exposed to high temperatures. *Proc. 11th BCA Concrete Communication Conf., UMIST, Manchester, UK, 3–4 July*: 347–356.

Castillo, C. & Durrani, A.J. 1990. Effect of transient high temperatures on high-strength concrete. *ACI Materials Journal* 87(1): 47–53.

Chan, Y.N., Luo, X. & Sun, W. 2000. Effect of high temperatures and cooling regimes on the compressive strength and pore properties of high performance concrete. *Construction and Building Materials* 14 (5): 261–266.

Chan, Y.N, Peng, G.F. & Anson, M. 1999. Residual strength and pore structure of high-strength concrete and normal strength concrete after exposure to high temperatures. *Cement and Concrete Composites* 21(1): 23–27.

Dale, P.B. 2000. Fibres, percolation, and spalling of high-performance concrete. *ACI Materials Journal* 97(3): 351–359.

Felicetti, R. & Gambarova, P.G. 1998. Effect of high temperatures on residual compressive strength of high-strength siliceous concretes. *ACI Materials Journal* 95(4): 395–406.

Jensen, J., Hammer, T. & Hansen, P. 1997. Fire resistance and residual strength of HSC exposed to hydrocarbon fire. *Fire performance of high-strength concrete; Proc. intern workshop, NIST, MD, USA, 13 14 February*: 59 67.

Kalifa, P., Chene, G. & Galle, C. 2001. High-temperature behavior of HPC with polypropylene fibres - from spalling to microstructure. *Cement and Concrete Research* 31(10): 1487–1499.

Kalifa, P., Menneteau, F.D. & Quenard, D. 2000. Spalling and pore pressure in HPC at high temperatures. *Cement and Concrete Research* 30(12): 1915–1927.

Luo, X., Sun, W. & Chan, Y.N. 2000. Effect of heating and cooling regimes on residual strength and microstructure of normal strength and high-performance concrete. *Cement and Concrete Research* 30(3): 379–383.

Mohamedbhai, G.T. 1986. Effect of exposure time and rates of heating and cooling on residual strength of heated concrete. *Magazine of Concrete Research* 38(9): 151–158.

Papayianni, J. & Valiasis, T. 1991. Residual mechanical properties of heated concrete incorporating different pozzolanic materials. *Materials and Structures* 24(140): 115–121.

Phan, L.T. & Carino, N.J. 2002. Effects of test conditions and mixture proportions on behavior of high-strength concrete exposed to high temperature. *ACI Materials Journal* 99(1): 54–62.

Poon, C.S., Shui, Z.H. & Lam. L. 2004. Compressive behavior of fibre reinforced high-performance concrete subjected to elevated temperatures. *Cement and Concrete Research* 34(12): 2215–2222.

Sanjayan, G. & Stocks, L.J. 1993. Spalling of high strength silica fume concrete in fire. *ACI Materials Journal* 90(2): 170–174.

Short, N.R., Purkiss, J.A. & Guise, S.E. 2000. Assessment of fire damaged concrete. *Proc. 10th BCA Concrete Communication Conf., Birmingham, UK, 29–30 June*: 245–254.

Xiao, J. & Falkner, H. 2006. On residual strength of high-performance concrete with and without polypropylene fibres at elevated temperatures. *Fire Safety Journal* 41: 115–121.

Concrete Solutions – Grantham, Majorana & Salomoni (Eds)
© 2009 Taylor & Francis Group, London, ISBN 978-0-415-55082-6

Microscopic analysis of the connection between CFRP laminates and concrete at elevated temperatures

T.D. Donchev & S. Abouamer
Faculty of Engineering, Kingston University, London, UK

D. Wertheim
Faculty of Computing, Information Systems and Mathematics, Kingston University, London, UK

ABSTRACT: Carbon fibre reinforced polymers (CFRP) are a popular option for strengthening and retrofitting of reinforced concrete structures. The high strength, low weight and quick and easy installation of CFRP are factors that have led to their increasing use.

As relatively new materials, CFRP laminates have certain limitations based on insufficient knowledge of their properties and behaviour. The weakest point in a CFRP strengthened system usually is not the laminate itself, but the connection between the adhesive and concrete in the anchoring zone; this has been an area of intensive research. . The performance of this zone and the whole strengthened system when subjected to elevated temperatures and fire is still a relatively unknown area.

The aim of this paper is to examine microscopic changes in CFRP laminate, the adhesive and the boundaries between laminate, adhesive and concrete at elevated temperatures (up to 300°C). Three types of microscope were used; conventional optical, laser scanning confocal (LSCM) in transmitted light mode and a scanning electron microscope (SEM). The results are analysed and compared with mechanical testing of similar samples. Conclusions and considerations for future development of the research in this area are presented.

1   INTRODUCTION AND BACKGROUND

Fibre Reinforced Polymers (FRP) have been used in recent years for repair, strengthening and retrofitting of structural components. Their excellent properties of high strength, light weight and easy installation have led to beneficial applications in areas of building and bridge structures.

High temperature is considered one the biggest problems arising for FRP materials in the construction industry, especially in relation to fire resistance. FRP mechanical properties can be affected by high temperatures depending on which type of FRP and adhesive are used. Changes in microstructure of the materials, during and after heating, could be a means for estimating changes in their condition with increase of temperature. In our study microscopic analysis has been used to analyse the microstructure of layers of concrete, adhesive and CFRP laminate and corresponding changes at elevated temperatures.

1.1   CFRP at high temperatures

The research about high temperature resistant FRP materials and their behaviour is mainly directed to aerospace and naval applications (Papakonstantinou et al, 2001; Kobayashi et al, 2003) and for internal reinforcement for concrete (Abbasi et al, 2005). The fire behaviour of CFRP laminates used for strengthening of concrete is still a relatively unknown area.

Usually the mechanical properties such as strength, modulus of elasticity and volume stability of concrete are significantly reduced during exposure to fire. Exposing concrete to high temperature, considerably changes its chemical composition and physical condition. Dehydration from release of chemically bound water in calcium silicate hydrates becomes significant above about 110°C. Dehydration of the hydrated calcium silicate and the thermal expansion of the aggregates increase internal stresses and from 300°C significant amounts of micro cracks are induced in the material (Arioz, 2007).

The fire resistance of concrete reinforced with FRP laminates depends on the change in mechanical properties of FRP and concrete. The behaviour of these composites when subjected to fire is complex and not well established.

1.2   Microscopy for structural investigation

Three different types of microscope were used for the purposes of this work – optical microscope, scanning electronic microscope (SEM) and a confocal microscope used in transmitted light mode.

Optical microscopy with a relatively low level of magnification is widely used for different types of

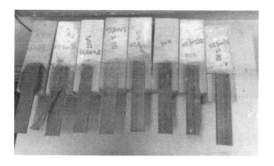

Figure 1. Samples heated to different temperatures.

Figure 2. Samples ready for optical microscopy.

Figure 3. Small cube samples.

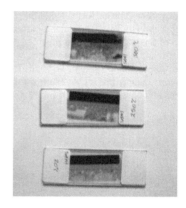

Figure 4. Thin section slide samples.

materials. Some FRP applications have been indicated in (Kobyashi et al, 2003; Koleva et al, 2007) The equipment used for this work was an Olympus BX51.

SEM is a significant supplement to the optical microscope when examining concrete; for example it can provide important information about the degree of hydration of cement, formation and distribution of hydration products and adhesion to aggregates (Koleva et al, 2007). SEM data and analysis of the effect of elevated temperatures for different types of FRP materials is given in (Mouritz et al, 2004), but the investigation did not include possible applications for strengthening of reinforced concrete structures.

The type of signals gathered in a SEM varies and can include secondary electrons, characteristic x-rays, and back scattered electrons. SEM Backscattered Electron Imaging is usually used to identify grey levels of different components; also it may provide high resolution imaging with elemental mapping. The SEM system used in this project was a Zeiss Evo 50.

The confocal laser microscope used in this study was a Leica TCS SP2. Confocal microscopy enables investigation of the structure of materials. Laser scanning confocal microscopy has been utilised to study the surfaces of cement-based materials. In this study we used the confocal microscope in transmitted light mode i.e. for generating a single plane image.

2 SAMPLE PREPARATION

The samples tested in this project were prisms of reinforced concrete with carbon fiber reinforced polymer (CFRP) attached using Epoxy Plus adhesive.

The samples were heated up to temperatures of 25, 100, 150, 200, 250 and 300°C as indicated in Fig. 1. An estimation of the bearing capacity of the system of CFRP laminate, adhesive and concrete at different levels of heating is conducted at (Donchev, T. et al, 2007). In this study the 25, 250 and 300°C samples were analysed in order to investigate the applicability of microscopic analysis for a series of phase changes.

In the initial stage of cutting the heated samples in transverse direction was done using a diamond blade cutting rig. The thickness of the processed samples is about 10 mm (Fig. 2) and after polishing they are used for optical microscopy.

Further cutting was done to produce the thin section samples for use with the Confocal Microscope and small cube samples for EM; two samples were produced for each of the three temperatures. The small cubes were prepared by using a smaller diamond ring blade with dimensions approximately $10 \times 10 \times 10$ mm (Fig. 3). At this stage section samples were prepared according to methods used for geological rock treatment and analysis and used for the confocal microscope (Fig. 4).

3 MICROSCOPE IMAGING RESULTS AND ANALYSIS

3.1 Optical Microscope

A temperature of 25°C is considered as 'room' temperature. The three layers of CFRP, adhesive and concrete can be seen in Fig. 6. Between the CFRP and adhesive there is a thin layer of glass FRP (GFRP) which

Figure 5. 2x, 250°C, CFRP/Adhesive/concrete.

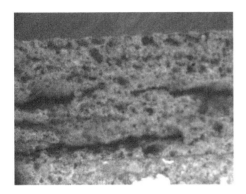

Figure 6. 2x, 300°C, CFRP/Adhesive/concrete.

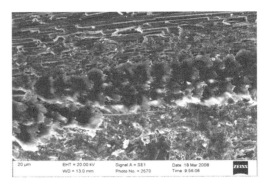

Figure 7. 20μm scale bar, 25°C, CFRP/ adhesive.

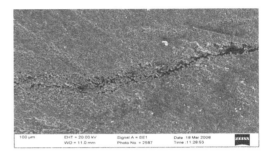

Figure 8. CFRP layer with major cracking through it at 300°C.

is used as a protective cover for CFRP laminates when they are manufactured.

All layers appear normal; different aggregates in the concrete and adhesive layer are visible.

At a temperature of 250°C some changes occurred to the specimen, the CFRP layer indicated some minor cracks, also the GFRP layer was being dissipated into the CFRP and adhesive layers (Fig. 5).

Minor changes occurred in the interface between the adhesive and concrete: the colour appeared to have become brownish in both layers. Some pinkish areas appeared in the concrete layer which were due to chemical transformations of the aggregate as a result of the high temperature. Small cavities into the adhesive appeared as well.

At 300°C a considerable change occurred in the specimen; Fig 8 shows that the adhesive layer had major cracks and many significant cavities. The GFRP layer had totally disappeared: it seems that the layer melted into the adhesive and CFRP layer. The color of the adhesive layer changed to dark brown, almost black and the concrete had a more intensive pinkish color (Fig. 6).

3.2 Scanning electron microscope (SEM)

Results from a sample under normal conditions at 25°C, can be seen in Fig. 7.

The CFRP matrix is not damaged, a GFRP layer below can be seen clearly as well. The texture of the adhesive layer allows identification of different substances used in its manufacture.

The interface between concrete and adhesive appears normal. Using the backscattered mode, the shape of different textures for concrete and the adhesive are visible otherwise the border between those two materials via SEM could not be easily identified.

At 250°C some changes occured to the specimen compared to the 25°C samples; in the CFRP layer some minor damage to the fiber matrix can be seen.

It is clearly visible that the GFRP layer appears to have melted into the adhesive. From the SEM images, the adhesive layer is relatively well preserved at this temperature apart from a few small cavities and some micro cracks.

Considerable changes occurred in the specimen at 300°C. There were massive cavities in the adhesive layer and visible cracking in the CFRP layer as shown in Fig. 8 and 9.

The GFRP layer had totally disappeared in the specimen. It is assumed that it melted into the adhesive, which caused discontinuity in the interface between the CFRP and adhesive in some parts.

Major cracking can be clearly seen in the adhesive layer which is corresponding to the reduced strength and other mechanical properties of the specimen.

The aggregates in the concrete layer seemed to have been affected by the high temperature as considerable micro cracking was visible.

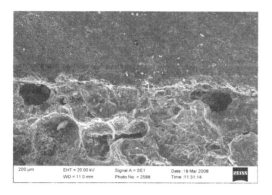

Figure 9. 200μm scale bar, 300°C, CFRP/adhesive interface.

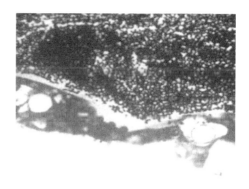

Figure 11. 20x, 300°C, CFRP/adhesive

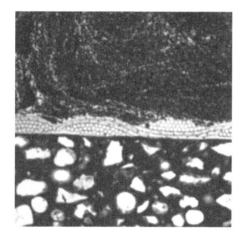

Figure 10. 10x objective, 25°C, CFRP/adhesive interface.

3.3 *Confocal microscope used in transmitted light mode*

The 25°C specimen apparently had not been damaged. The interface between CFRP and adhesive can be seen in fig 10: the top dark layer is CFRP which clearly represents the matrix of carbon fiber in normal conditions. The GFRP layer is visible as a thin layer below the CFRP.

The adhesive layer is slightly darker than the concrete layer and it has smaller aggregates.

The high temperature caused damage to different layers in the specimen. The carbon fibre matrix is slightly damaged compared to the 25°C sample. The GFRP layer is starting to dissipate and the glass fibre matrix appears damaged and some cavities are starting to appear on the border between the CFRP layer and GFRP.

The concrete layer aggregates start to change properties due to the high temperature which is visible as darker brownish colouring.

The changes that occurred at 300°C appear considerable in comparison with those at 250°C. The most dramatic change seen was at the CFRP/adhesive interface.

The CFRP layer developed damage in the fibres which can be seen as cavities and discontinuations between them (Fig. 11).

The GFRP layer cannot be seen at the interface between the CFRP and adhesive. The adhesive layer shows massive internal cracks. The concrete layer changes mainly in colour, which indicates that the chemical properties had changed compared to 250°C. The interface between concrete and adhesive did not get as damaged as the CFRP and adhesive interface.

4 CONCLUSIONS

From the above study the following main conclusions can be drawn:

- Optical microscopy is very good for examining the color of different layers.
- Scanning electron microscopy shows detailed texture and shape of different layers up to very high magnifications.
- Use of the confocal microscope in transmitted light mode appeared useful for investigating changes at the interface between different layers.
- The CFRP layer mechanical properties change with increasing temperature; internal cracking and discontinuity appear in the carbon fiber matrix at temperature above 250°C.
- The GFRP layer starts melting into the adhesive at 250°C and was not apparent at 300°C.
- The adhesive layer showed dramatic change in texture and color, significant internal cracking and appearance of cavities along the layer at a temperature of 300°C.
- Most of the cracking in the adhesive layer seemed to be in the upper section of the adhesive below the CFRP interface.

REFERENCES

Abbasi, A., Hogg, P.J. A model for predicting the properties of the constituents of a glass fibre rebar reinforced concrete beam at elevated temperatures simulating a fire test. Elsevier, Science Direct, *Composites part B 36*, 2005, 384–393.

Arioz, O., *Effects of elevated temperatures on properties of concrete*. Elsevier, Science Direct, *Composites part B 63*, 2007.

Donchev T., Wen, J and Papa, E. Effect of Elevated Temperatures on CFRP Strengthened Structural Elements. *10th International Conference Fire and Materials, 29–31 January 2007, San Francisco*, USA

Kobayashi, S., Terada, K., Takeda, N. Evaluation of long-term durability in high temperature resistant CFRP laminates under thermal fatigue loading. Elsevier, Science Direct, *Composites part B 34*, 2003, 753–759.

Koleva, D., Breugel, K., Correlation of microstructure, electrical properties and electrochemical phenomena in reinforced mortar. Elsevier, Science Direct, *Composites part B 63*, 2007.

Mouritz, A.P., Mathys, Z., Gardiner, C.P. Thermo-mechanical modeling the fire properties of fiber-polymer composites. Elsevier, Science Direct, *Composites part B 35*, 2004, 467–474.

Papakonstantinou, C.G, Balaguru, P., Lyon, R.E., 2001, *Comparative study of high temperature composites*. Elsevier, Science Direct, *Composites part B 32*, 2001, 637–649.

Concrete Solutions – Grantham, Majorana & Salomoni (Eds)
© 2009 Taylor & Francis Group, London, ISBN 978-0-415-55082-6

Mechanical and physicochemical properties of lightweight self-consolidating concrete subjected to elevated temperatures

H. Fares & A. Noumowe
Department of Civil Engineering, University of Cergy-Pontoise, France

H.A. Toutanji & K. Pierce
Department of Civil and Environmental Engineering, University of Alabama in Huntsville, USA

ABSTRACT: Exposing concrete to high temperature (due to accidental fire, etc) causes progressive breakdown of the cement gel structure and consequently severe deterioration and loss in the structure's load bearing capacity. This paper presents an experimental study on the mechanical and physicochemical properties of lightweight self-consolidating concrete (LSCC), subjected to high temperatures. Four LSCC mixes and one normal-weight (SCC) mix were tested. The specimens underwent two different tests: a fire test and a high temperature test. The first is the International Standard Organization (ISO-834) fire test, which consists of heating the prismatic specimens according to the standard fire curve up to 600°C. The second test is the characterization test, which consists of heating the specimens at a rate of 1°C/minute up to 400°C. Strength, loss of mass, density, water porosity, spalling characteristics and other physicochemical properties before and after the tests were recorded. LSCC performed as good as NSCC in terms of strength and spalling resistance.

1 INTRODUCTION

Self consolidating concrete (SCC) was introduced in Japan in 1986 (Ouchi 2003). Self-consolidating concrete (also called self-compacting concrete) differs from normal concrete in its high viscosity. SCC is defined as concrete that can flow and completely fill complex forms under its own weight, pass through and bond with reinforcement under its own weight, and has a high resistance to aggregate segregation. Because of its high viscosity, SCC is placed without vibration, which decreases placement time, allows for longer work hours (less noise), reduces labour costs, and also leads to a more uniform placement. It also allows for more complex shapes, and has a greater surface finish and quality. These advantages counteract the high cost of chemical admixtures and fine content in the concrete, making the end result a cheaper and more workable concrete. As advances in SCC continue and widespread use worldwide becomes common, the risk of exposure to fire increases. LSCC has many advantages over NSCC. It has a reduced mass, which in turn reduces the dead load of a building and stress on the foundation. The reduced weight also reduces energy demand during construction. LSCC also has high noise insulation and high thermal insulation, which will retard the spread of fire.

When exposed to high temperature, the chemical composition and physical structure of the concrete change considerably. As a result, severe microstructural changes are induced and concrete loses its strength and durability. The aim of this study is to analyze different concretes submitted to high temperature. Differential Thermal Analysis (DTA) and Thermo Gravimetric Analysis (TGA) were studied and SEM observation was also carried out to understand the behavior of self-consolidating concrete subjected to elevated temperature

2 EXPERIMENTAL PROCEDURE

2.1 Test program

In total, five different mixes were made and tested; four lightweight (LW) and one normal-weight (NW). The prism specimens were tested according to the ISO-834 (International Standard Organization) temperature curve (CEN). Another high temperature test, referred to as the characterization test, was performed on cylindrical and prismatic specimens. The cylindrical specimens were heated at 1°C/min until 400°C. At 400°C, constant temperature was maintained for 1 hour before cooling (RILEM TC-129 MHT 1996). The prismatic specimens were also heated at 1°C/min until 500°C, with a temperature held constant for 1 hour. The ISO-834 test is conducted to study the resistance of the concrete specimens to spalling and cracking when subjected to fire. The characterization test is conducted to study the residual strength of specimens subjected to high temperature. The mechanical properties were evaluated after the heating-cooling cycles on the specimens. The compressive strength, flexural strength, residual strength, porosity, density, mass

Table 1. Summary of different mixes.

	Coarse Light-wt. Aggregate I	Coarse Light-wt. Aggregate II	Coarse Normal wt. Aggregate
Fine Normal wt. Aggregate	LW-1	LW-2	NW-1
Fine Normal wt. and Fine Light wt. Aggregates	LW-3	—	LW-4

loss and physicochemical changes were evaluated after heating and compared to unheated specimens.

2.2 Mix properties

Four different LSCC mixes were prepared. Two mixes used lightweight coarse aggregate and fine normal weight aggregate (LW-1 and LW-2). The maximum size aggregate of mix LW-2 was 19 mm and that of LW-1 was 13 mm. The third mix was composed of lightweight coarse aggregate, normal weight fine aggregate and lightweight fine aggregate (LW-3). The last LSCC mix was made up of normal weight coarse aggregate and fine lightweight aggregate (LW-4). The normal weight SCC mix (NW-1) contained coarse and fine normal weight aggregates. For each test a minimum of 3 specimens was used. Table 1 summarizes the different mixes.

2.3 Specimen preparation

Each concrete mixture was cast in prismatic moulds with dimensions 75 mm × 75 mm × 350 mm and cylindrical moulds with dimensions of 75 mm in diameter and 150 mm in height. After curing for 90 days, the concrete specimens were subjected to fire testing. The prismatic specimens were subjected to the ISO fire test and cylindrical specimens underwent the high temperature characterization test. Figure 1 shows both the ISO fire test and high temperature characterization test (Noumowe et al. 2003 & 2006).

Before subjecting the prisms to the high temperature test, the specimens were weighed and their masses were recorded. The specimens were subjected to 600°C. Before cooling the temperature was held at 600°C for 2 hours in order to ensure uniform temperature throughout the specimens before they were cooled. The surface temperature of the specimens was recorded throughout the test, using thermocouples. After the test was completed, the mass was recorded again and the mass loss due to heating was calculated. In addition, the residual flexural strength was determined.

The cylindrical specimens were subjected to the characterization test, in which they were heated at 1°C/minute until 400°C was reached. At 400°C, the temperature was held constant for one hour before cooling. After the specimens returned to room

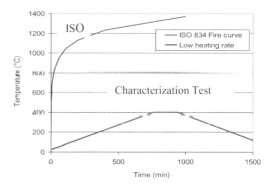

Figure 1. Heating curves (low heating rate up to 400°C and cooling and ISO fire up to 600°C).

temperature, their residual compressive strength was determined. The prismatic specimens were also subjected to a high temperature test until 500°C. After cooling, their residual flexural strength was measured. SEM was used to investigate the microstructure of the specimens and to study the morphology of the concrete.

3 EXPERIMENTAL RESULTS

3.1 The international standard organization (ISO-834) fire test

The ISO-834 test was conducted on concrete prisms.

3.1.1 Observation of specimens subjected to fire

In this fire test, most specimens showed cracking and some signs of spalling. Three kinds of observations were made:

- Spalling at the bottom: specimens groups LW-3 and LW-4., as shown in Figure 2.
- Cracking and porous on the surface: specimens groups LW-1 and LW-2, as shown in Figure 2.
- Severe cracking and some spalling: specimens of normal weight coarse and fine aggregate (NW-1), as shown in Figure 2.

It seemed that the LW-SCC performed as well as that of the NW-SCC in terms of spalling and cracking. Based on our observations, it is difficult to claim that LW-SCC performs better or worse than NW-SCC.

3.1.2 Mass loss

LW-SCC specimens showed a higher percentage of mass loss during the fire test than the NW-SCC specimens. The average mass loss of NW-SCC was about 8.0% and the mass losses of LW-SCC specimens ranged from 9.7% to 14.1%, as shown in Table 2. This is attributed to the fact that lightweight aggregate has a higher water absorption rate than normal weight aggregate, so more water is absorbed into the lightweight specimens before the fire test. In mix LW-4, mix which contained lightweight fine aggregates, the percentage loss was greater than the initial water content. This

Spalling

Specimen LW - 4

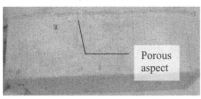

Specimen LW – 3

Porous aspect

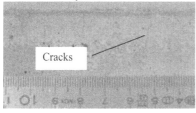

Specimen LW - 2

Cracks

Specimen LW - 1

SCC Specimens after fire test

Figure 2.   LSCC Specimens after fire test.

implies that substances other than water, such as CO_2 or other gases were released during heating.

3.1.3   *Density*

The density of the specimens subjected to high temperature was calculated. The density was measured

Table 2.   Mass losses after fire tests.

Mix ID	Mass before fire (kg)	Mass after fire (kg)	Mass loss (%)	Water content before heating (%)
LW-1	2.026	1.845	9.80	11.10
LW-2	1.819	1.637	10.01	11.7
LW-3	1.194	1.043	12.65	12.9
LW-4	1.229	1.08	12.12	11.7
NW-1	2.64	2.422	8.26	9.4

Table 3   Density values of all mixes

Temperature	LW-1	LW-2	LW-3	LW-4	NW-1
20°C	1.775	1.779	1.408	1.658	2.061
400°C	1.631	1.641	1.304	1.523	1.944
600°C	1.547	1.529	1.292	1.510	1.886

after specimens were cooled to room temperature ($\sim 20°C$). The apparent density, ρ_d was calculated by the following equation:

$$\rho_d = \frac{M_{dry}}{V} = \frac{M_{dry}}{M_{sat} - M_{sat}^{imm}} \qquad (1)$$

where M_{sat} and M_{sat}^{imm} are the mass of the saturated specimen, weighed in air and in water, respectively; and M_{dry} is the mass of oven dry specimen, weighed in air.

As shown in Table 3 the density of LW-SCC and NW-SCC decrease with temperature. This decrease pattern for both LW-SCC and NW-SCC is similar. The reduction in density is due to the evaporation of water resulting in weight loss with little reduction in volume.

3.1.4   *Water porosity*

The porosity to water, ε, of control specimens (not subjected to high temperature) and those subjected to elevated temperature was calculated. The porosity to water, ε, is calculated according to the following equation:

$$\varepsilon = \frac{M_{sat} - M_{dry}}{V} = \frac{M_{sat} - M_{dry}}{M_{sat} - M_{sat}^{imm}} \qquad (2)$$

As shown in Figure 3, the total porosity for both LWSCC and NWSCC decreases with temperature up to 400°C but the pattern is reversed beyond 400°C. All specimens followed similar behavior.

3.2   *SEM and BEI analysis*

Scanning Electron Microscopy (SEM) was used to study the microstructure of both heated and unheated specimens. Images of the control specimens (not heated), specimens heated up to 400°C, and specimens heated up to 600°C, are shown in Figure 4.

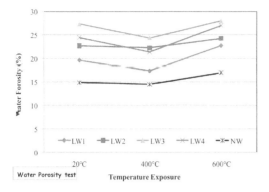

Figure 3. Water porosity vs. temperature, all mixes.

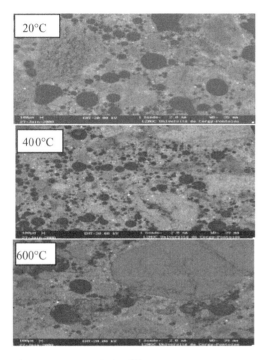

Figure 4. LW4 sample for different temperatures.

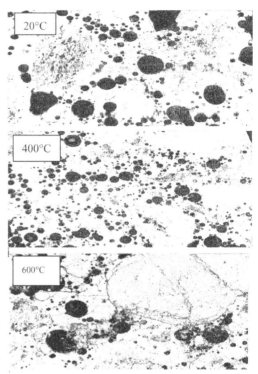

Figure 5. Backscattered Electron Images.

Table 4. Porosity values obtained by Backscattered Electron Images (BEI).

Mix ID	Original avg. before heating (%)	Avg. porosity after heating at 1°C/min up to 400°C (%)	Avg. porosity after fire test up to 600°C (%)
LW-1	16.48	10.64	21.63
LW-2	17.27	16.35	19.18
LW-3	26.25	16.89	35.14
LW-4	24.33	23.44	28.41
NW-1	12.25	11.54	17.00

Software such as "ImageJ" can be used to convert SEM images to Backscattered Electron Images (BEI) to determine porosity of the specimen (Scrivener 1988). It is another technique that can be used to determine porosity. With the BEI mode, chemical contrast in polished specimens can be observed (Bentz 1994). The different images obtained with this method are presented in Fig. 5.

From the BEI, cracks and pores in lightweight aggregates and in cement pastes can be observed. The bubbles represent the effect of AEA (Air entraining admixture). Based on the analysis for each image, the average porosity is shown in Table 4.

Two methods were used to determine porosity: experimentally: by using Equation 2 and by BEI. Based on the limited recorded temperature data (only at 400°C and 600°C), the porosity decreases up to 400°C and increases thereafter, beyond 400°C. This may be attributed to the fact that shrinkage takes place in the specimen, causing a decrease in volume and an increase in voids. However, the rate of change of voidage is much lower than of the volume of the specimen. At 600°C other ingredients beside water are evaporated causing a further increase in void content; on the other hand the specimen experiences insignificant volume change.

3.3 DTA and TGA results

The Differential Thermo Analysis (DTA) and Thermo Gravimetric Analysis (TGA) were carried out to

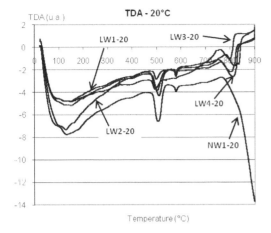

Figure 6. DTA results before heating.

measure the temperature at which the physicochemical changes take place and the consequence of mass loss that takes place during these changes.

Figure 6 presents typical DTA analysis curves of control specimens (unheated). The result shows that there is a difference between concrete containing lightweight and normal weight aggregates. The DTA peak at nearly 490°C for NW-1 is higher than that of LW. At a temperature of 580°C, a peak appears for LW-1, LW-2 and NW-1 concretes but not for LW-3 and LW-4 concretes. At a temperature of about 800°C, $CaCO_3$ is decomposed in LW-1, LW-2 and NW-1 but not in LW-3 or LW-4. We can observe the same evolution at 490°C and at 800°C. NW-1 presents a high peak. The first endothermic peak observed at 130°C indicates the dehydration of cement paste by driving out the free and the physically bound water. The second endothermic peak at about 490°C is the calcium hydroxide peak. The third peak at about 570°C and the fourth endothermic peak at 800°C can be an indicator for the α-β quartz transformation and the second step of C-S-H gel dehydration (Marsh 1988).

4 CONCLUSIONS

This study investigated the properties of lightweight self consolidating concrete exposed to high temperature. From the experimental results, the following conclusions can be drawn:

– LW-SCC specimens performed better than the NW-SCC specimens in the rapid heating test (ISO), but did worse in the slow heating test (characterization test).

– In general the LW-SCC specimens performed as well as that of the NW-SCC in term of spalling and cracking.
– LW-SCC mixes showed a higher percentage of mass loss during heating.
– The density of LW-SCC and NW-SCC decrease with temperature. The pattern of both LW-SCC and NW-SCC is similar.
– At elevated temperature up to 400°C, the mass loss is mainly due to water escape. Because of the water loss during heating, the solid phases shrink. The pore structure becomes more porous.
– The information generated by SEM and BEI gives a whole image of the microstructural evolution with increasing temperature.

ACKNOWLLDGEMENT

The authors would like to acknowledge the financial support of the visiting scholar program of the University of Cergy-Pontoise. Thanks to Annelise Cousture, Engineer at the University of Cergy-Pontoise, for the SEM tests.

REFERENCES

European Committee for Standardization (CEN). Design of concrete structures—part 102: general rules—structural fire design. *Eurocode 2.* Brussels: Belgium.

Noumowe, A. et al. 2006. High-strength self-compacting concrete exposed to fire test. *ASCE Materials Journal.* November/December: 754–758.

Noumowe, A. 2003. Temperature distribution and mechanical properties of high-strength silica fume concrete at temperatures up to 200°C. *ACI Materials Journal* 100(4): 326–330.

Ouchi, M. et al. 2003. Applications of self-compacting concrete in Japan, Europe and the United States. *5th International Symposium, ISHPC.* Tokyo-Odaiba.

RILEM TC-129 MHT 1996. Test methods for mechanical properties of concrete at high temperature. part 1: introduction; part 2: stress-strain relation; part 3: compressive strength for service and accident conditions. *Materials and Structures Journal* 28(181).

Scrivener, K.L. 1988. The use of backscattered electron microscopy and image analysis to study the porosity of cement paste; Proc. Materials Research Society Symposium, 137, Pore Structure and permeability of cementitious materials, p. 129–140.

Bentz, D.P et al. 1994. SEM analysis and computer modeling of hydration of Portland cement particles, Petrography of Cementitious Materials, ASTM STP 1215, American Society for Testing and Materials, Philadelphia, p. 60–73.

Marsh, B.K. et al. 1988. Pozzolanic and cementitious reactions of fly ash in blended cement pastes, Cement and concrete Research, Vol. 18, p. 301–310.

Concrete Solutions – Grantham, Majorana & Salomoni (Eds)
© 2009 Taylor & Francis Group, London, ISBN 978-0-415-55082-6

Loss of performance induced by carbonation of lime-based fire proofing material

K.S. Nguyen
MTRhéo – GCGM Laboratory, INSA & IUT de Rennes, Rennes, France
Extha SA, Route de Laval, Soulgé-sur-Ouette, France

C. Baux, Y. Mélinge & C. Lanos
MTRhéo – GCGM Laboratory, INSA & IUT de Rennes, Rennes, France

ABSTRACT: The carbonation of fire proofing materials changes the thermal resistance of cement, lime … based materials under fire attack. By using kinetic results identified with pure minerals, we have simulated the thermal response of a material, initially containing portlandite, at different rates of carbonation. The comparison of the different simulations enlightens the relationship between the microstructural evolution and the firebreak capacity of the material. These results allow us to consider the durability analysis of the performance of different fire protection solutions in real scale tests.

1 MODELING OF THE BEHAVIOR OF MINERAL MATERIALS UNDER HIGH TEMPERATURE SOLICITATION

Fireproofing materials delay the heat transfer from fire towards the protected structure by exploiting the combination of a low thermal conductivity and the endothermic effect related to the latent heat associated with mineral phase transition reactions (Féjean 2003, Firemat 2005). Such processes correspond, for example to dehydration and decarbonation of cement, gypsum, calcite-based products (Baux et al., 2008). The numerical modeling of the heat transfer problem through fire resistance materials has been previously studied (Mehaffey et al., 1994) (Axenenko & Thorpe 1996) (Bentz et al., 2006) (Baux et al., 2007. It appears that it is necessary to take into account the variation of the material characteristics according to the temperature evolution, for example density $\rho(T)$, conductivity $\lambda(T)$, specific heat $Cp(T)$ and phase transition to assure a good simulation.

We exploit a heat transfer model specially developed with the integration of phase transition kinetics (Baux et al., 2007). The balance equation of heat conduction is written in one-dimension (according to the Oz axis i.e thickness direction) as follows:

$$\frac{\partial}{\partial z}\left(\lambda(T)\cdot\frac{\partial T}{\partial z}\right)-\Phi_g = \rho\cdot Cp(T)\cdot\frac{\partial T}{\partial t} \qquad (1)$$

where T stands for the temperature and where $\Phi_g = \rho\cdot\Delta H\cdot\partial\alpha/\partial t$ is the source term conditioned by latent heat ΔH associated with chemical conversion of material.

Kinetic $\partial\alpha/\partial t$ of each reaction is generally expressed by multiplying an Arrhenius law of temperature k(T)

and a kinetic model $f(\alpha)$, that is the function describing the conversion rate α (varying from 0 to 1).

$$\frac{\partial\alpha}{\partial t} = f(\alpha)\cdot k(T) = f(\alpha)\cdot k\cdot\exp\left(-\frac{Ea}{RT}\right) \qquad (2)$$

Where the appropriated kinetic model $f(\alpha) = \alpha^m\cdot(1-\alpha)^n$ is an extended Prout-Tompkins model (Burnham 2000). k and Ea are two constant parameters of the Arrhenius law.

Both $f(\alpha)$ model and Arrhenius constants are identified by using thermal analysis data (DTA/TG) run on pure mineral samples. In the case of complex materials including various reactions of phase transition, we should define an average principle of thermo-physical properties (Baux et al., 2007), (Nguyen et al., 2007). In this paper, we apply numerical modeling to gradually simulate the effect of the carbonation phenomenon, by gradually replacing the hydrated lime (portlandite) initially present in a reference formulation by calcite (calcium carbonate). We conclude then about the significance of the carbonation phenomenon regarding the fire resistance capacity of the studied mineral material.

2 REFERENCE FORMULATION AND SIMULATION PARAMETERS

2.1 Composition

The reference product associates three compositions (cement, hydraulic lime and vermiculite) delivered by our industrial partner. The mixture of dry materials and water is directly made in a spraying machine. The pasty product thus obtained can be applied to shaped steel or concrete structures as a fire insulating layer.

Table 1. Thermo-physical properties of the composition de reference.

	Phase change	
	before	after
Density ρ (kg.m^{-3})	729	594
Thermal capacity		
$Cp(J.kg^{-1} \cdot K^{-1})$	1,295	1,295
Conductivity $\lambda(W \cdot m^{-1} \cdot K^{-1})$	7.86E-03exp(1.46E-03.T)	

Table 2. Variations of enthalpy associated with different phase change reactions.

	Enthalpy variation ΔH
Compound	$(J \cdot g^{-1}$ of compound$)$
H_2O	2,330
C-S-H	1,440
$Ca(OH)_2$	5,660
$CaCO_3$	3,890

The thermo-physical properties of such material evolve with the temperature. The values suggested in (Bentz et al., 2006), (Bentz et al., 2007) are summarized in table 1 and are used in our numerical simulation. One can notice that the thermal conductivity is influenced by the temperature but not by a possible phase transition during the exposure to fire. We consider that this exponential law remains valid in the presence of a carbonated mixture. The heat capacity (specific heat) is considered constant with temperature. Such an assumption is generally accepted for the studied mineral materials. Lastly, the change in density corresponds to the total mass loss of materials (cement, portlandite).

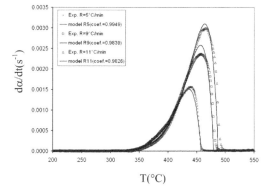

Figure 1. Fitting of kinetic model (tab. 3) with derived results $d\alpha/dt$ vs. T of portlandite conversion.

2.2 Modeling condition

The experimental conditions of fire tests are reproduced with our numerical simulation. A shield made from the reference formulation is exposed on the warm side to the interior of a furnace temperature that follows the ISO 834 curve. Meanwhile, the cold side stays at 20°C. The modeled shield has a 32 mm thickness. The heat transfer on boundary condition utilizes generally radiative and convective transfer on both fire side (hot side) and ambient side (cold side) (Mehaffey et al., 1994) (Axenenko & Thorpe 1996).

2.3 Mineral phase change during the elevation of temperature

2.3.1 Reaction of conversion
While the temperature in the element thickness increases, the decomposition of the hydrated binder (C-S-H and portlandite) and the decarbonation of calcite take place. These thermo-chemical processes become major in various ranges of temperatures. The theoretical enthalpy variations (latent heat) of these endothermic conversions are indicated in table 2 (Bentz et al., 2007).

$$
\begin{array}{ll}
H_2O_{free} \rightarrow H_2O_{gas} & (25 < T < 100°C) \\
C\text{-}S\text{-}H \rightarrow C\text{-}S + H_2O_{gas} & (100 < T < 300°C) \\
Ca(OH)_2 \rightarrow CaO + H_2O_{gas} & (300 < T < 600°C) \\
CaCO_3 \rightarrow CaO + CO_{2gas} & (600 < T < 1000°C)
\end{array}
$$

2.3.2 Kinetic analysis
DTA/TG analyses are carried out on powder samples of pure portlandite $Ca(OH)_2$ and pure calcite $CaCO_3$. During the tests, the heating rates remained constant at R = 3, 5, 7 and 11°C·min^{-1}. The kinetic results allow

the characterization of the thermal behavior of the material in arbitrary heat evolution (Galwey & Brown 1999), (Haine 2002). The fitting of the kinetic model (equation 2) on the experimental results helps us to identify the values of the triplet parameters ($f(\alpha)$, Ea, k). Figures 1–4 conclude that the mathematical model in table 3 is well adjusted with the experimental curves of portlandite and calcite decomposition.

3 SIMULATED EFFECT OF THE CARBONATION PROCESS

3.1 Carbonation reaction

The exposure of a portlandite containing material to an environment that naturally contains more or less CO_2 causes a normal process of carbonation. This phenomenon cannot be avoided for relatively porous and permeable materials such as fireproofing products. The carbonation reaction modifies the composition and the microstructure of the material, the portlandite being thus transformed into calcite.

$$
\left.
\begin{array}{l}
CO_{2\,gas} \rightarrow H_2CO_3 \rightarrow HCO_3^- \rightarrow CO_3^{2-} \\
Ca(OH)_{2\,solid} \rightarrow OH^- + Ca^{2+}
\end{array}
\right\} \Rightarrow Ca^{2+} + CO_3^{2-} \rightarrow CaCO_{3\,solid}
$$

To study the influence of the carbonation on the thermal performance of the reference sample, we have made a numerical study while varying in the model the values of characteristic parameters of the material. The

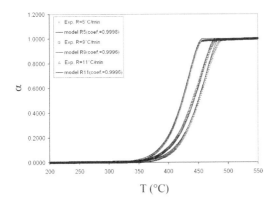

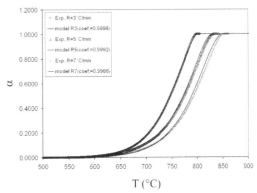

Figure 2. Fitting of kinetic model (tab. 3) with integrated results α vs. T of portlandite conversion.

Figure 4. Fitting of kinetic model (tab. 3) with integrated results α vs. T of calcite decarbonation.

Table 3. Kinetics of decomposition reactions of $Ca(OH)_2$ and $CaCO_3$.

	Portlandite $Ca(OH)_2$ ($\Delta m = 22.85\%$)	Calcite $CaCO_3$ ($\Delta m = 22.85\%$)
Conversion temperature (°C)	$300 \rightarrow 600$	$600 \rightarrow 1000$
Kinetic model $f(\alpha)$	$\alpha^{0.238}(1-\alpha)^{0.514}$	$\alpha^{0.075}(1-\alpha)^{0.476}$
Frequency factor $\ln(k)$, k in s^{-1}	12.858	12.064
Activation energy Ea(kJ·mol^{-1})	109.77	164.152

Table 4. Evolution of the thermo-physical parameters according to carbonation rates of portlandite.

	Carbonated portlandite (%)					
	0	20	40	60	80	100
ρ_{init} (kg·m^{-3})	729	751.1	772.8	794.5	816.1	837.8
$\rho_{at\,300°C}$ (kg·m^{-3})	669.4	691.1	712.8	734.5	756.1	777.8
$\rho_{at\,600°C}$ (kg·m^{-3})	594	631.1	667.8	704.4	741.1	777.8
ρ_{final} (kg·m^{-3})	594	594.4	594.4	594.3	594.3	594.3
ΔH_{CSH} (J·g^{-1})	118.5	115.0	111.8	108.8	105.9	103.1
$\Delta H_{Ca(OH)2}$ (J·g^{-1})	582.4	452.5	329.8	213.9	104.1	0
ΔH_{CaCO3} (J·g^{-1})	0	190.1	369.5	539.1	699.7	852.1

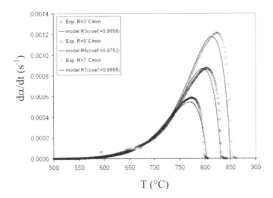

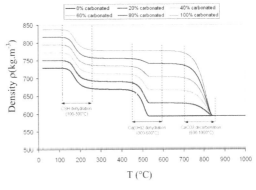

Figure 3. Fitting of kinetic model (tab. 3) with integrated results dα/dt vs. T of calcite decarbonation.

Figure 5. Kinetics of mass loss for various rates of carbonated portlandite.

rate of carbonation reaction is considered identical for all the thickness of the shield.

3.2 Evolution of the thermo-physical characteristics

For each rate of carbonated portlandite, density $\rho(T)$ and enthalpy variation ΔH of material are recomputed to take into account the changes of mineral composition (see tab. 4). Conductivity $\lambda(T)$ and specific heat

Cp are supposed to preserve the values given in table 1, for the initial mixture. The kinetics of mass loss associated with each rate of carbonated portlandite is presented in figure 5.

4 SIMULATION RESULTS AND DISCUSSION

For each rate of carbonation, the temperature versus time curves are identified in the thickness of the element at 16, 24, 32 mm depth. Figure 7 plots the

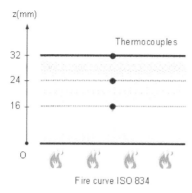

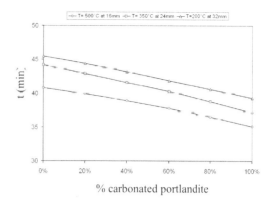

% carbonated portlandite

Figure 6. Thermocouples positions in the element thickness.

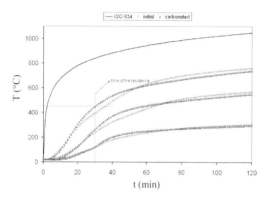

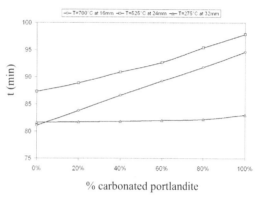

% carbonated portlandite

Figure 8. Carbonation effect on the duration of fire resistance of 32mm element: (up) low range of temperature and (down) high range of temperature.

Figure 7. Difference of temperature evolution between simulated elements of reference and of completely carbonated material.

5 CONCLUSIONS AND PERSPECTIVES

The simulation results obtained show that the carbonation of portlandite contained in certain firebreak products can considerably modify the characteristics of heat transfer. This report should push the professionals to certificate their fireproofing products with a validation protocol of mineral stability over their working life. This study carried out in the framework of portlandite carbonation can be also considered for the case of ettringite-based products.

However, it must be reminded that some simplifying hypotheses were used to produce the simulation results. Thermal conductivity will be modified regarding the evolution of mineral phases during thermal evolution. Mechanical effects (crack, geometrical deformation) are not taken into account in this paper. Further developments are expected to adapt our numerical model to take these types of effects into account in the evolution of the heat transfer parameters.

simulation result in the case of the reference composition (no carbonation) and the completely carbonated composition. It can be noticed that the temperature chart presents more or less visible isothermal plateaus at around 100 and 400°C, i.e. the temperatures of dehydration of C-S-H and portlandite respectively.

In figure 8, we compare the fireproofing ability of the considered element at different rates of portlandite carbonation by regarding the time needed to reach a given temperature. For the simulated case, one can note that if the allowed level of temperatures is not exceeding 500°C, the carbonation level induces a reduction of the "time of fire resistance" which can reach 10% on figure 8 (upper). If the allowed levels of temperature are greater than 500°C, the carbonation level induces an increase in the "time of fire resistance" which can reach 15% on figure 8 (lower).

These results show that the effect of the carbonation phenomenon can be complex to interpret. It seems thus difficult to guarantee a precise time of fire resistance of material during all the working life. A 20% variation of this fire proofing duration seems to be credible. Such influence is sufficient to significantly affect the results of an experimental study.

ACKNOWLEDGMENTS

The authors thank the Inorganic materials team (UMR 6226 – INSA de Rennes) for their contribution of thermal analyses TD-TG and thank Mr. Lebourgeois, our industrial partner, for the reference composition and fire test results of Firemat-R product (Extha SA).

REFERENCES

Axenenko, O. & Thorpe, G. 1996. The modelling of dehydration and stress analysis of gypsum plasterboards exposed to fire. Computational Materials Science 6: 281–294.

Baux, C., Lanos, C., Melinge, Y. & Nguyen K.S. 2007. Modélisation du comportement thermique haute température de matériaux minéraux. *Revue Européen du Génie Civil* (11-6): 787–800.

Baux, C., Melinge, Y., Lanos, C. & Jauberthie, R. 2008. Hydraulics binders based materials for fire protection. *ASCE's Journal of Materials in Civil Engineering MT/2005/023091* vol. 20:.

Bentz, D. P., Prasad, K. R. & Yang, J. C. 2006 .Towards A Methodology for the Characterization of Fire Resistive Materials with Respect to Thermal Performance Models. *Fire Mater.* (30-2006): 311–321.

Bentz, D. P. & Prasad, K. R. 2007. Thermal Performance of Fire Resistive Materials I : Characterization with Respect to Thermal Performance Models. NISTIR 7401.

Burnham, A. K. 2000. Application of the Šesták-Berggren Equation to Organic and Inorganic Materials of Practical Interest. *Journal of Thermal Analysis and Calorimetry*, 60(3): 895–908.

Féjean J. 2003. Développement des liants composites minéraux : application à la protection incendie. Thèse de doctorant, INSA de Rennes.

FIREMAT Project (EVG1-CT-2002-30006). 2005. Development of new materials with improved fire resistance for firebreak systems. Project final report.

Galwey, A.K. & Brown, M.E. 1999. Thermal decomposition of ionic solids. Elsevier.

Haines, P.J. 2002. Principle of thermal analysis and calorimetry. Royal Society of Chemistry.

Mehaffey, J.R., Cuerrier, P. & Carisse, G., 1994. A model for predicting heat transfer through gypsum-board/wood-stud walls exposed to fire. *Fire and Materials*, 18(5): 297–305.

Nguyen, K.S., Baux, C., Melinge, Y. & Lanos C. 2007. Simulation numérique du transfert thermique à travers des matériaux minéraux. *Annales BTP* N° 5: 37–43.

Concrete Solutions – Grantham, Majorana & Salomoni (Eds)
© 2009 Taylor & Francis Group, London, ISBN 978-0-415-55082-6

Improving fire resistance properties of structural concrete members by adding polypropylene fibres

L.A. Qureshi
University of Engineering & Technology Taxila, Pakistan

M. Ilyas Sh.
Bahauddin Zakariya University, Multan, Pakistan

ABSTRACT: In this paper, the performance of structural concrete members in actual fire situations is discussed. The fire resistance properties of a structural concrete member was improved by adding polypropylene fibres of special characteristics, manufactured in China. An experimental program was designed, in which 12 beams in 4 groups A, B, C and D were cast. In group A, no fibres were added, in B, PP fibres were added in 0.9 Kg/m^3 ratio, while in group C, PP fibres were added as 1.8 Kg/m^3. In the last group, D, local nylon fibres were tried with 0.9 kg/m^3 ratio. After curing and drying, 8 beams (2 from each group) were subjected to standard fire conditions. The temperature during the fire was measured with a digital infrared thermometer at different time intervals. Spalling was the most critical phenomenon that occurred in structural concrete members during the fire. PP fibres prevented spalling in the beam samples of group B and C. Significant spalling was observed in the beams having no fibres and those with local nylon fibre. A two point loading test was carried out on all unfired and fired beam samples. Deflections of these beams were measured at incremental loads. It was found that by adding PP fibres with the ratio of 0.9 Kg/m^3 the fire resistance of beams improved with a loss of 4 to 5% in cracking and failure loads after being exposed to fire. By adding PP fibres brittle behavior of the concrete also transited towards ductile.

1 INTRODUCTION

Fires in buildings have always been a threat to human safety. The threat increases as a larger number of people live and work in larger buildings throughout the world. Unwanted fire is a destructive force that causes many thousands of deaths and property loss each year. People expect that their homes and workplaces will be safe from the damage of an unwanted fire. The safety of the occupants depends on many factors in the design and construction of buildings, including the expectation that certain buildings and parts of buildings will not collapse in a fire or allow the fire to spread. Functionally structures provide living accommodation with safety and comfort; structures are also meant to provide protection against natural and man made disasters e.g. earthquake, fire, bomb blast, etc. Concrete is used as building material with its enhanced properties and with more effective form. Concrete is one of the best building materials to resist the fire than any other material, but still it needs improvements to withstand the high temperature during fires. Research has given us a greater understanding about the way that fire attacks buildings. Fire can be detected much earlier, but it still can destroy thousands of lives. So there is an ample need for consideration of fire hazards in the design of concrete structures (Pootha 2008).

As there is a great loss of human life and property associated with fire in buildings, the proper architectural and structural design parameters for concrete structures to resist massive fires should be adopted. This would minimize the risk of human and property loss. Moreover, the fire damaged concrete structures can also be utilized again by proper rehabilitation using modern techniques and materials. So it is vitally important that we design and construct buildings and structures that protect both people and property effectively. We can design buildings and structures with the help of research data regarding the behavior of concrete and fire. Moreover, different regulatory authorities such as ACI, ASTM, EURO Codes, and British Standards provide the proper guidelines for designing and testing the fire resistance of concrete structures and materials. As discussed above, generally concrete structural members perform well under fire situations; studies show, however, that performance of high strength concrete differs generally from that of normal strength concrete and may not exhibit good fire performance (Chana 2008).

At the high temperatures experienced in fires, hydrated cement in concrete generally dehydrates, reverting back to water (actually steam) and cement. This results in a reduction of strength and modulus of elasticity (stiffness) of concrete. In some fires, spalling of concrete occurs – fragments of concrete break loose from the rest of the concrete. Spalling results in the rapid loss of surface layers of concrete during a fire. It exposes the core concrete to fire temperatures thereby increasing the rate of transmission of heat to the core concrete and the reinforcement. Spalling is attributed to the buildup of pore pressure during heating. The pore pressure builds up due to release of combined water and density of concrete. In dense concrete this pressure reaches incredibly high levels and tensile stress develops in the cover concrete that finally breaks due to this pressure (Kumar 2003).

This research addresses the question that how can we improve the fire resistance property of structural concrete members? Fire resistance of concrete is measured in term of fire ratings. Ratings are given in hours. For example, the required fire resistance rating for columns of a high-rise hospital building is much more stringent than those for single story buildings used for storage of non combustible materials. The fire rating of structural materials (such as concrete) and assemblies is generally given by generic rating or it is determined through testing, such testing is conducted in accordance with ASTM E119 in the USA, BS EN 1992, Eurocode 2 and BS8500 in the UK and CAN/ULC-S101 in Canada. The fire resistance rating means "The period of time a building or building components maintain the ability to confine a fire or continues to perform a given structural function or both" (Khoury 2000). We know that concrete fares well in fire both as a material and as structural elements, but what can be done to improve its performance even further? We can enhance the performance of concrete as a fire resistant material by the following ways:

(a) By using Polypropylene fibres to reduce spalling.
(b) Using designed minimum dimensions and required cover.
(c) Addition of flame radiant additives
(d) Using carbonate aggregates instead of siliceous aggregates (Bailey, 2002).

2 EXPERIMENTAL PROGRAM

2.1 Materials used

2.1.1 Cement
The cement used in the experiment was ordinary portland cement arranged locally with the brand name "Lucky Cement". Different quality tests like compressive strength, setting time, consistency, soundness, etc. were carried out on the cement before mixing for the designed concrete.

2.1.2 Fine aggregates
Fine aggregate (sand) was arranged from the local river with water absorption of 0.8 % and specific gravity of 2.70. Grain size analysis was carried out conforming to the ASTM C33-93-170. Fineness modulus was calculated conforming to ASTM C127 and found to be 2.86.

Table 1. Specifications of polypropylene fibres used in concrete mix.

Density (g/cm³)	0.91	Elastic Modulus (MPa)	>3500
Length (mm)	20	Equivalent Diameter (mm)	100
Shape	Beam-like Net	Crack Elongation (%)	≥10
Acid & Alkali Resistance	Strong	Water Absorption	Nil
Tensile Strength (MPa)	346–560	Melting Point (°)	160–170

Table 2. Specifications of nylon fibres used in concrete mix.

Density (g/cm³)	1.2	Elastic Modulus (MPa)	>3500
Length (mm)	20–24	Equivalent Diameter (mm)	100
Shape	Fibres	Crack Elongation (%)	≥12
Acid & Alkali Resistance	Strong	Water Absorption	No
Tensile Strength (MPa)	300–400	Melting Point (°)	160–170

2.1.3 Coarse aggregates
The coarse aggregate used in the experiment was also arranged locally from Sakhi Sarwar. The main constituent was found to be limestone with a nominal maximum size of 19 mm. Water absorption was 0.6 % and specific gravity was 2.65. Grain size analysis was carried out as per ASTM C33-93-170.

2.1.4 Steel reinforcement
High yield strength deformed bars of 10 mm diameter and consisting of grade Fe 415 steel was used in the concrete as reinforcement. Mild steel bars of Fe 250 grade and 6 mm diameter were used as stirrups.

2.1.5 Polypropylene
Polypropylene fibres were imported from China. The products appeared as a net-like structure with many fibre monofilaments connected with each other.

2.1.6 Nylon fibres
Locally available nylon fibres were also arranged having matching characteristics with polypropylene fibres. These nylon fibres were different in color (blue) and these were cut into the required size and added in the concrete.

2.2 Casting program

12 RCC beams were cast with dimensions of 1200 mm × 160 mm × 230 mm conforming to the ISO 834/ BS 476. Each beam was provided with 3 bars of 10 mm diameter & grade Fe 415 as tension reinforcement and 2 bars of the same diameter were provided as hanger bars. Mild steel bars of grade Fe 250 &

Figure 1. Polypropylene fibres used in concrete mix.

Table 3. Concrete mix design.

Constituents	Contents			
	Plain	PP 0.9	PP 1.8	NF 0.9
	(A)	(B)	(C)	(D)
Cement (Kg/m^3)	309	309	309	309
Polypropylene (Kg/m^3)	0	0.9	1.8	0.9
Water (Kg/m^3)	206	206	206	206
Fine Aggregate (Kg/m^3)	595	595	595	595
Coarse Aggregate (Kg/m^3)	1380	1380	1380	1380
W/C Ratio	0.66	0.66	0.66	0.66
Slump (mm)	70	65	60	75

Figure 2. Nylon fibres used in concrete mix.

Figure 3. Fire is growing under the beams and from the sides of the beams.

6 mm dia @ 150 mm c/c were provided as shear reinforcement. Clear cover of 40 mm was provided as per code recommendations to all the beams. Beams were divided into four groups A, B, C and D. Each group consisted of 3 beams. In group A, all beams were cast without polypropylene. In group B, polypropylene was added at 0.9 kg/m^3, keeping other parameters constant. Similarly, in group C, polypropylene was increased to 1.8 kg/m^3 and in group D, local nylon fibres were added at 0.9 kg/m^3 again keeping other parameters constant. A slump test was carried out for all mixes to ensure proper workability of concrete. Room temperature as well as temperature of the fresh concrete was measured with the help of a digital thermometer.

2.3 Testing program

After 28 days, water from the curing tank was drained out and samples were dried for 44 days. After 44 days, eight beams (2 from each group) were taken out for the fire test, while four beams (1 from each group) were kept separate as control beams for strength test.

2.3.1 Fire test

A test pit was excavated and lined with bricks. Test beams were supported over the pit. Beams were exposed to fire from three sides as in the case of practical fire in a room.

Temperatures were recorded at different intervals of time. After 180 minutes, the fire was stopped and sample beams cooled down in natural conditions. Temperature was measured with the help of an infrared digital thermometer designated as Smart Sensor AR872S with measuring range (18°C to 1050°C). This thermometer had an accuracy of ±2°C.

2.3.2 Visual observations

Fired beams were observed visually during and after the fire test. Significant changes were observed in all four groups of concrete.

In group A beams (without fibres), spalling was observed after 90 minutes during fire. With the increase of temperature, color of concrete was changed from pink to red, then grey and finally to dark grey.

In group B beams (with polypropylene fibres of 0.9 kg/m^3 ratio), only change in color was observed and no spalling occurred until the fire was extinguished.

In group C beams (with polypropylene fibres of 1.8 kg/m^3 ratio), only change in color was observed

Figure 4. Spalling Observed after Fire Test in Group A Beams.

Figure 5. Explosive spalling observed after the fire test in group D beams.

like beams of group A and no spalling occured until the fire was extinguished.

In group D beams (with local nylon fibres @ 0.9 kg/m³ ratio), change in color was observed from pink to red, then grey and finally to dark grey. Explosive spalling was also observed during the fire after 75 minutes from the start of fire.

2.3.3 Two point loading test

Beams were tested for the loss of flexural strength during fire by arranging two point loading test. Eight Fired beams (2 from each group) and four unfired beams (1 from each group) were tested for flexural strength and deflections at different loads.

Figure 6. Two-point loading test and measurement of deflections.

3 RESULTS AND DISCUSSION

3.1 Time-temperature curve

Time-temperature data was plotted and compared with the standard time-temperature curve provided by ASTM E119. The experimental time-temperature curve was found to be closely in coincidence with the standard time-temperature curve validating the test data for our fire test. The temperature raised quickly in the start when the fire was spread out. Within 5 minutes of fire initiation, the temperature reached 506°C as compared to the standard E-119 temperature of 538°C. Experimental temperatures were approaching the standard temperatures in the middle of the time but the curve tended downward towards the end of the time.

This was because of wind blowing at the site even though the openings in the test pit were covered by the shuttering plates to minimize the effect of wind. In figure 7, the experimental time temperature curve was compared with both standard (ISO-834 and E-119) time temperature curves. By comparison it was found that the time-temperature curve obtained from the experimental data was fairly good to apply on the samples tested in the fire test.

3.2 Strength analysis of beams

Figures 8 & 9 show load versus deflection curves for unfired and fired beams of all groups respectively.

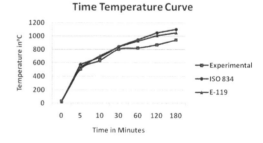

Figure 7. Comparison of Experimental, ISO 834, and E-119 time-temperature curves.

3.2.1 Strength of unfired beams

It was found that by adding the PP fibres and nylon fibres, the cracking strength as well as failure strength of group B, C and D beams (in which fibres were added) significantly increased as compared to group A beams (where no fibres were added).

The cracking strength of group B beams (cast with PP fibres with a ratio of 0.9 Kg/m³) increased by 10 % as compared to group A beams (cast without fibres). Similarly, the cracking strength of group C beams (cast with PP fibres with a ratio of 1.8 Kg/m³) was increased by 18.18 % as compared to the group A beams, and the cracking strength of group D beams (cast with nylon

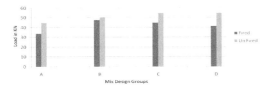

Figure 8. Cracking loads of fired vs unfired beams.

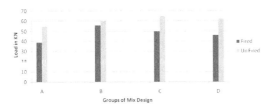

Figure 9. Failure loads of fired vs unfired beams.

fibres of ratio 0.9 Kg/m^3) increased by 14.3 % as compared to the group A beams. Failure loads were also analyzed and plotted as a bar chart in figure 9. Failure loads were also increased by 8.3 %, 15.38 % and 12 % of group B, C and D respectively as compared to those of group A beam.

3.2.2 Strength to fired beams

Strength loss occurred in all groups of beams after exposure to the fire. However, the strength loss behavior of different groups was found to be different.

In group A beams (without any fibres), there was a significant reduction, i.e., 25 %, in the cracking strength after being exposed to the fire. In group B beams, only 5 % loss in cracking strength was observed. While in the groups C and D, a significant loss of 18.18 % and 19.04 % respectively was observed in cracking strength. Similar trends were observed in beams for failure loads. A loss of 29 %, 4.16 %, 25 % and 24 % in failure strength was observed in the beams of group A, B, C and D respectively.

3.3 Deflections of beams

Figures 10 & 11 shows load vs deflection curves for unfired and fired beams of all groups respectively.

3.3.1 Deflections of unfired beams

Deflections of unfired beams are presented and compared in Figure 10. All results indicate that by adding fibres in concrete beams, their brittle behavior transit towards ductile. A significant increase in deflection was observed in group B beams as compared to group A beams. Further increase in deflection was observed in group C beams in which the PP fibre dose was kept double to group B. Beams with group D (casted with nylon fibres) showed greater ductility as compared to all other groups.

3.3.2 Deflections of fired Beams

Deflections of fired beams are plotted and compared in Figure 11. Group A beams showed increased deflection while deflections in groups B, C and D were

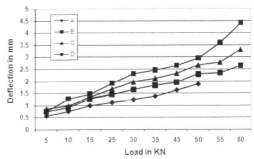

Figure 10. Loads vs deflections of unfired beams.

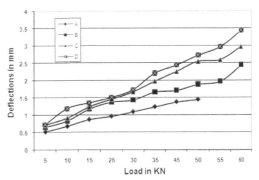

Figure 11. Loads vs deflections of fired beams.

reduced a little. One clear reason for the decrease in ductility of group B, C and D seems to be the melting of fibres close to the bottom during elevated temperature, while the increase in deflection of group A beams after fire may be due to physical changes occurring in the concrete due to chemical reactions at elevated temperatures.

4 CONCLUSIONS & RECOMMENDATIONS

4.1 Conclusions

1) The physical and load carrying behavior of structural concrete members changes with elevated temperatures.
2) The change in physical behavior of concrete at elevated temperatures is mainly spalling. By reducing spalling, we can increase the fire resistance of structural concrete members. We can increase the fire resistance of structural concrete members by adding polypropylene fibres of special characteristics in the concrete mix.
3) The cost of adding polypropylene in the structural concrete members is negligible as compared to its long lasting effects on the structure.
4) The fibres could also act as crack resistors by stopping the crack propagation by bridging the cracks.
5) When the polypropylene fibres were added into the concrete mix, they caused slump loss of the

fresh concrete. Fibres anchored mechanically to the cement paste and held the concrete together.

6) Both the beams of group B and C cast with PP fibres of 0.9 and 1.8 kg/m^3 ratio were prevented from spalling when subjected to fire, however a strength loss was measured in group C beams. A spalling phenomenon was seen both in beams of group A and D when subjected to fire.

7) Addition of nylon fibres in the concrete mix. gave very poor performance in the fire resistance, however the same fibres reflected good strength in unfired beams.

8) A significant increase in the strength of unfired beams of group B, C and D has been observed in load tests. The increase in cracking and failure loads is due the additions of fibres in the concrete mixture.

9) From the load verses deflection data it was concluded that by adding polypropylene fibres, the brittle behavior of concrete was reduced.

4.2 Recommendations

Polypropylene fibres may be added with the ratio of 0.9 to 1.0 Kg/m^3 for significant improvement in the fire resistance of structural concrete members.

Fire resistance of structural elements should be mandatory for design engineers. The buildings should be designed for fire resistance both ways by active control measures and passive control measures, more technically both in architectural and structural aspects.

REFERENCES

Bailey, Colin. August 2002. Holistic behavior of concrete buildings in fire: Proceedings of institution of civil Engineers, Structures and Buildings, 152.

Chana P. July 2008. Fire Safety with Concrete, Concrete for fire engineering: Proceedings of international Conference in university of Dundee, Scotland, UK.13–26.

Khoury G. 2000. Effect of fire on concrete and concrete structures: Proceedings of structural Engineering Material Journal, Vol.2. 21–34.

Kumar, A, Kumar, November 2003. Behavior of RCC beams after exposure to elevated temperatures: Institute of Indian journal, India, Vol 84, 44–56.

Potha Raju M. & Srinivasa Rao K. July 2008. Theoretical As sessment of Flexural Strength of Heated Reinforced Concrete Beams: Proceedings of international Conference in university of Dundee, Scotland, UK.27–34.

General repair

Concrete Solutions – Grantham, Majorana & Salomoni (Eds)
© 2009 Taylor & Francis Group, London, ISBN 978-0-415-55082-6

Repair and strengthening of reinforced concrete structures of industrial buildings from the beginning of the 20ᵗʰ century

P. Berkowski & G. Dmochowski
Wrocław University of Technology, Wrocław, Poland

M. Kosior-Kazberuk
Białystok University of Technology, Białystok, Poland

ABSTRACT: The methodology of evaluation of technical wear, the strategy of repair realization, and examples of strengthening methods of reinforced concrete structural elements of industrial buildings from the beginning of the 20th century are presented in the paper. Detailed estimation of the building's technical state and concrete examination should precede the design of constructional modernization. The paper also presents the results of an experimental investigation into the bearing capacity of old concrete slabs by using testing loads as a method of determination of the real range of strengthening required. The paper presents examples of technical solutions that permit adaptation of the selected building to new functional requirements. Some conclusions considering the preservation state of industrial RC buildings constructed in the beginning of the 20th century are given. Also presented are conclusions about the suitability of modern technical solutions for modernization and repair of old buildings made of concrete.

1 INTRODUCTION

The condition of old buildings is determined by the technical state of their elements, the type of structural construction, and the material used. Also the building maintenance, expressed by the frequency of interventions in conservation and repairing, has a significant influence on technical condition of its elements. However, the durability of the material of which elements are made plays the most important role (Konior et al. 1996). The type of structural system sets a building's functional flexibility, and at the same time, influences the technical possibilities for building adaptation to modernization and new functions.

This paper discusses some nearly one-hundred-year-old monolithic reinforced concrete structures which were under investigation. The durability of monolithic RC elements is qualified normatively for 150 to 200 years. The durability of buildings with a solid brick structure as an external bearing construction, and depending upon the material used for roof covering and the use of the building is qualified for 100 to 150 years. Differences in durability periods of building structural elements and of the building as a structure itself show relations in a co-operation of elements creating the building. In conditions of technically proper exploitation of the building, the technical condition of its elements (which are a significant part of the building static system) has the most important influence on the wear of reinforced concrete structure.

An evaluation of the condition of a building or its structural elements can be made by more or less detailed methods in respect to observation and description of the building. A macro-evaluation of the condition of a building can be worked out using the so-called time methods. Time methods describe the deterioration of a whole building or its elements, as a function of their durability and exploitation period. A mathematical form of such function is formulated appropriately according to the building maintenance, depending upon the care taken and frequency of conservation and repairs. Methods based on individual building observation are more penetrating than time methods. The range of observation depends on accessibility to the structure: the accessibility is limited when a building is in use. However, very often reliable conclusions can be drawn only after uncovering the structure. Usually, this happens after a decision to commence repair the building. A possible risk in evaluation of building deterioration considers is of course finance. Modern techniques used in the building industry permit a quick reduction of defects and can bring a building to a state that meets investor's demands.

The Ross and Unger formula, (Thierry et al. 1972) was applied to macro-evaluation of a selected RC structure and the degree of deterioration of a building. The degree of deterioration (s_z), assuming that there have been proper intervention and repairs realized, is as shown in Equation 1.

$$s_z = \frac{t(t+T)}{2T^2}100\%$$ (1)

where: T – standard durability, in years; t – period of exploitation, in years.

In the case of buildings constructed in the beginning of the 20th century, a sample calculation was made for an RC structure as an element of the building (T = 180 years, t = 100 years). It sets a degree of a structure deterioration $s_z = 43,2\%$. From the calculations made for the whole building (T = 130 years, t = 100 years) it appears that $s_z = 68,0\%$.

The proper determination of building deterioration and state is absolutely essential in case of necessity of structural strengthening. Nowadays, it occurs nearly in every situation during modernization of buildings, because they must be restructured to new functional goals and new loading regimes. In the first place, the following elements must be determined: structural system, loads, calculation methods used, material properties, and of the as-placed reinforcement. Trial holes are one of the best ways but, of course, modern equipment is absolutely necessary to obtain all information. It is also recommended to study historical documentation if it is accessible (Berkowski et al. 2001).

2 GENERAL DESCRIPTION OF EXAMPLE BUILDING STRUCTURE

A four-storey building with cellars and wooden roof, had a size in plan of 39.5 × 16.0 m and 20.0 m in height. It was erected in the years 1907-1910, so was in the beginning of the use of concrete for structural elements in that part of Europe. It must be stated that the first use of RC for industrial or shopping malls building structures took place in Wrocław in the first years of the 20th century. The building under investigation was dedicated for a cigar factory. It had a typical constructional system made of external brick walls and an internal concrete system of pillars, beams and slabs. Two monolithic staircases were the main stiffening structures. The external walls were made of ceramic bricks and reinforced concrete pillars 33 × 33/63 × 63 cm, dividing the building into two longitudinal courses, making a vertical bearing structure of the building (Fig. 1). Reinforced concrete beams of 7.35 m of span and an intersection of 29 × 65 cm (with slab thickness) bore on to the walls and pillars (Fig. 2). Monolithic reinforced concrete slabs of 15 cm of thick were laid on the beams. The slab span was 3.79 m, measured in axis. According to the archival documentation, the slabs were reinforced with ⌀12 bars every 20 cm. However, in the slabs there were no distributing bars and bars over the beams designed. All beams were reinforced individually, also with smooth bars. In the beams, there was no reinforcement for shear forces.

Internal moments in slabs were calculated as for partially fixed beams, while the beams were assumed to be simply supported.

The first inspection of the RC structure was conducted while it was covered with lime-and-cement plaster. The class and strength of the materials used in the structure were mainly ascertained with non-destructive methods. The reinforcement distribution

Figure 1. General view of RC bearing structure.

Figure 2. Detailed view of slab-beam-column connection.

was determined using an electronic metal detector. In characteristic places of the structure the reinforcement was broken out and checked. The concrete strength was examined with use of Schmidt's hammer type N and calibrated with normative probes drilled out from concrete elements. Chemical analysis of the concrete was also undertaken. The strength of bricks and mortar in the walls, supporting the RC elements, was determined with a comparative method. As a result of the research it was found that the structure was made of good quality materials. Concrete in columns and ribs was qualified as C16/20 class, and in ceiling slabs C12/15. The strength of the reinforcing steel was defined by examination of probes as 190 MPa. Bricks in an

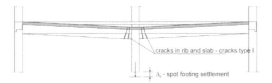

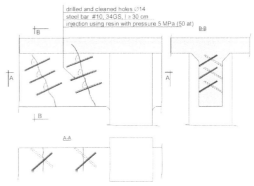

Figure 3. Formation of cracks type I – spot footing settlement.

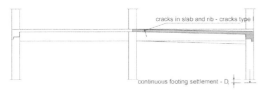

Figure 4. Formation of cracks type II – continuous footing settlement.

Figure 5. Formation of cracks type II – cracks along ribs.

Figure 6. Scheme of injection with an injecting resin and nailing with bars.

external wall were qualified as corresponding with a class 10/15 MPa, and a mortar resistance 2/3 MPa.

During structure examination numerous cracks and fractures were revealed in the beams and ceiling slabs. Two kinds of crack were observed. Cracks of type I, occurred in areas of beams bearing on pillars and propagating through the slab along a longitudinal axis of the pillars. A repeatable course of cracks is shown in Figures 2–3. No cracks were found along the beam span. For diagnostics, deflections of 11 beams were measured. The biggest measured deflection of a beam in the middle of the span was $f = 16$ mm $< f_{adm} = 25$ mm. Irregular building foundation settlements, especially difference in settlements of walls continuous foundations and pillars spot foundations, were considered as the main cause of this damage. The process was recognized as finalized.

Cracks of type II were in places where slabs were supported on beams – Figure 5. Crack widths were 0.2–1.2 mm. A detailed analysis of cracks pattern, supported by structural analysis, showed that cracks of type II were caused by the lack of plate reinforcement for a negative moment. During building design the reinforcement for a negative moment was not provided, with an assumption of a freely supported structure. Calculation checks on the limit state of bearing capacity for the slabs, beams and columns permitted a statement that it was preserved.

3 REPAIR AND STRENGTHENING OF CONCRETE ELEMENTS

No classical techniques were recommended to repair cracks in the structure, because they were considered

too work-consuming and uneconomical. And these techniques did not solve the problem of cracks of type II. An injection of cracks by modern injecting materials was chosen as an effective solution. These new materials are resistant to compression and tension about 4 to 5 times more than concrete. Methods of strengthening of all cracked places were identical and did not depend on crack causes. The wear of injecting materials is relatively small and the crack places after injection would be the strongest in the structure. An example scheme of injecting with an injecting resin and nailing with bars is shown in Figure 6.

The existing RC structure was designed to bear a load of 5 kN/m². The task of modernizing the structure to new demands required strengthening the slab above the 2nd floor and RC pillars to support a service load of 11 kN/m². A strengthening system was designed with a use of resin glues and strips made of carbon fibre. Also, the static scheme of the ceiling slab was changed. An existing reinforced concrete ceiling slab of 15.0 cm of thickness worked in a static scheme consisting of ten freely supported beams. Placing concrete on the slab to thickness equal to 25.5 cm was designed together with a change of a static system to a continuous beam with many spans. RC beams with an intersection of 25 × 80 cm worked in a static scheme of freely supported beams, however, initially were calculated as partially fixed. After strengthening, beams will work as T sections in a system of a continuous beam with two spans. Former intersections of pillars on the 2nd floor with the size of 33 × 33 cm were enlarged to 41 × 41 cm with an enclosure of 4 L80 × 80 × 6 joined by lacings made of sheet. A filling and a tacking layer was designed using modern composite materials. Beam strengthening was made by placing sheets 4 × 550 × 1500 mm in pillar zones. A proper integration of sheets and ribs was obtained by gluing and using press bolts. Beam strengthening in spans was made by gluing strips of carbon fibre to their undersides.

A strengthening scheme is shown in Figure 7. For securing work efficiency in the modernization project a detailed algorithm of realization and sequence of all the activities was prepared.

85

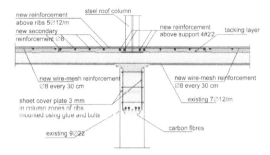

Figure 7. Strengthening scheme for slabs and beams.

Before placing CFRP strips checks of substrate strength and CFRP-concrete adhesion were made using the "pull-of" method. It was shown, that the old concrete had enough capacity and other strength parameters to permit the use this technique of strengthening.

4 CONCLUSIONS

Historical concrete structures have normatively 200 years of possible use, so the ones from the beginning of the 20th century are still in use and are possible to be repaired, modernized, and dedicated to new functions. It is possible to use new techniques and materials to strengthen structural elements made of concrete in structures erected in these years. However, it should be necessary to realize a proper technical description and deterioration determination of structures to be modernized. Also, it should be obligatory to make all examinations (chemical and strength) of concrete and reinforcement. Sometimes it is also advisable to make experimental loadings by using, for example, a hydraulic lift and in this way to determine the real bearing capacity of an old RC structure before taking decisions on its costly strengthening.

REFERENCES

Konior, J., Marcinkowska, E. 1996. Analysis of technical wear of old apartment houses in Wrocław. *International Conference on Developments in Building Technology.* Conference Proc. S.T.V. Bratislava, Slovak Republic: 23–25.

Thierry, J., Zalewski, J. 1972. *Repair and strengthening of buildings* (in Polish). Warszawa: Arkady.

Berkowski, P., Marcinkowska, E. 2001. Evaluation of technical wear and modernisation of R/C structure from beginnings of 20th century. *International Conference on Developments in Building Technology.* Conference Proc. S.T.V. Bratislava, Slovak Republic: 176–181.

Concrete Solutions – Grantham, Majorana & Salomoni (Eds)
© 2009 Taylor & Francis Group, London, ISBN 978-0-415-55082-6

How to clean soot deposits indoors?

M. Bouichou
Cercle des Partenaires du Patrimoine, Champs sur Marne, France

E. Marie-Victoire & D. Brissaud
Laboratoire de Recherche des Monuments Historiques, Champs sur Marne, France

ABSTRACT: The first step of any restoration operation is cleaning. If outdoor cleaning of ancient concrete is a topic now quite well defined, indoor cleaning is still a challenge. Effectively, the numerous abrasive projection, clay or stone-wool poultices or even water spraying techniques that can be efficient outdoors are difficult to apply indoors due to water and abrasive confinement difficulties. On another hand, indoors, dirt deposits are mostly encountered in churches, and they have more to do with candle soot than to thick black crusts. Therefore "softer" techniques could be considered. In that aim, a new water injection-extraction system and several latex poultices were tested. Their performances and potential side effects were evaluated by testing them inside a Parisian church dating back to the end of the 1930's, made of a pink bush-hammered concrete, covered in black soot deposits. If the water injection-extraction appeared very efficient, without deleterious impact, the cleaning efficiency obtained with the latex poultices varied from poor to excellent depending on the poultice. Some salt crystallization and some latex residues were also observed on the concrete with some of the poultices.

1 INTRODUCTION

Conservation and restoration of monuments made of concrete are an integral part of the preservation policy of cultural heritage in France. Numerous monuments are currently classified and require important restoration work, the first step of such operation being cleaning.

Several cleaning techniques such as wet or dry abrasive projection, poultices or even laser cleaning have shown their efficiency outdoors (Andrew 1994, Ashurt 1994, Marie-Victoire & Texier 1999, 2000, Vergès-Belmin & Bromblet 2000). But indoors, the problem is quite distinct. Firstly, dirt deposits are generally darker and less hardened than the black crusts frequently encountered outdoors. Secondly, indoor cleaning requires "clean" techniques, to avoid any decay of decorations, sculptures, wall painting or stained glass windows that can be encountered in churches, for example. So abrasive projection or water-based techniques which spread water and abrasives, imply protection and recycling measures that can be tricky to install indoors.

New techniques such as water injection-extraction and latex poultices (Stancliffe & De Witte 2005), that seem particularly suitable for indoor cleaning were recently developed. Both techniques seem all the more interesting since water is totally confined, and therefore the protection measures needed are minimal so that the monument can stay open during the cleaning operation. Therefore, several latex poultices and the water injection-extraction system were tested on the inner walls of a church made of concrete, dating back to the 1930's, and covered with a dark soot deposit. After a first evaluation of their ability, their efficiency and potential side effects were evaluated through colour measurements, microscopic observations and FTIR spectrometry.

2 CLEANING TECHNIQUES

2.1 Water injection

The water injection-extraction system (G) is based on a vacuum-washing technology. Low pressure water is sent on the surface through a sucking head under low vacuum, which also recovers the dirty water by a suction action (Fig.1). Thus water is confined. During the tests, two pressures were tested : 2 (G1a) and 20 bars (G2a).

2.2 Latex poultices

The technique consists of the application (brushing or airless spraying) of a natural latex dispersed paste on the surface to be cleaned. When a polymerized film is formed, it is peeled off (Fig. 2). The dirt deposit is then trapped in the film, which is removed by the mechanical peeling action. Depending on temperature and on relative humidity, the time before peeling can vary from 24h to 48h.

The major compound of the poultices is hevea sap, which was initially stabilised in a liquid phase by adding ammonia. As the ammonia fumes were

Figure 1. Water Injection–Extraction system.

Figure 2. Latex film peeling.

Table 1. Tested latex poultices.

Latex poultices	Ammonia	EDTA
F1	No	No
F2	No	+
F3	No	++
R1	No	3%
R2	No	No
E1	Yes	1%
E2	Yes	1%
E3	Yes	3%

problematical either for the applicator or for metals sensitive to ammonia, new stabilisers were recently developed. Some of the producers also introduced mineral additions or chemical agents (e.g. Diamine Tetra Acetic Ethylene). As a consequence, 3 families of products (E, F, R) were tested (Table 1), with or without ammonia or EDTA (which is generally used to dissolve calcium) (De Witte & Dupas 1992). As the testing areas were small, all the products were hand-brushed.

3 TESTING CONDITIONS

3.1 *Testing protocol*

Three points were examined : the ease of application, the efficiency and the potential side effects. Concerning the ease of application, the logistics and time necessary to obtain an optimum cleaning were considered. In order to optimise the application conditions, each manufacturer was asked to send a skilled operator. The efficiency was evaluated through naked eye and binocular observations (on cores sampled by dry drilling), but also through colour measurements performed before and after cleaning. Finally potential side effects were examined by optical and scanning electron microscope (SEM) observations, but also by Fourier Transformed Infra Red (FTIR) spectrometry.

3.2 *Testing site*

Tests were performed in *Sainte-Odile* church, built using concrete in 1935 and listed in 1979. The inner walls are made of pinkish concrete, obtained by mixing pink granite aggregates and red marble dust. The surface of raw concrete was worked with a bush-hammer, in such a way that the aggregates were exposed. Probably linked to decades of exposure to candle fumes, the very rough surface of these inner walls was covered with a thin and very dark dirt deposit.

As the *Sainte-Odile* Church is richly decorated notably with crystal paste and cement mortar windows, or a reredos (a screen or wall decoration at the back of an altar) made with enamelled copper plates; it seemed hard to introduce cleaning techniques using water or abrasives inside the church. Therefore it appeared to be an interesting site to test these latex poultices and the injection-extraction technique.

3.3 *Testing areas*

For each technique, a $0.16\,m^2$ surface had to be cleaned. The aim of the tests was to eliminate the dirt deposit, having a minimum impact on the bush-hammered surface. SEM observations and EDS analysis of a control sample revealed the presence of a calcite mat (probably linked to a carbonation process), combined with dust and gypsum crystals (Fig. 3); confirmed by FTIR spectrometry analysis (Fig. 4).

4 RESULTS

4.1 *Evaluation of ease of application*

A maximum duration of 5 minutes was necessary either to clean the $0.16m^2$ testing area with the G technique or to apply the latex poultices.

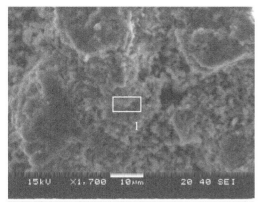

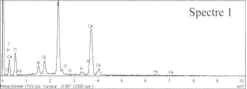

Spectre 1

Figure 3. SEM view (G=x1700) and EDS spectrum obtained on the testing areas before cleaning, revealing that a calcite mat combined with gypsum crystals was covering the concrete surface.

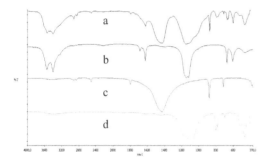

Figure 4. FTIR spectrum obtained on the control area (a), and reference spectra of : Gypsum (b), Calcite (c) and Quartz (d).

After a short time of installation (water and electricity supply) the G technique appeared easy to apply.

Concerning the latex poultices, it is to be noticed that the operators applied much higher quantities of products (around 3 kg/m^2 for E and F, 1.5 kg/m^2 for R) than the consumption recommended by the producers (1 kg/m^2 for E and F products, and 0.33 kg/m^2 for the R products). It also appeared that some of the poultices were too fluid (E2, F1) while others were too pasty (F3, E3), both being difficult to apply. Dealing with the coalescence duration (Table 2), all the latex poultices were ready to peel-off after 2 or 3 days. Only the E3 poultice was polymerized after 1 day. Finally, some of the poultices were too adhesive and therefore hard to peel-off (F3, R2), while the E3 almost did not stick to the surface.

Table 2. Latex poultices ease of application and removal.

Latex	Viscosity	Coalescence days	Adherence	Tears
F1	−	2	+	0
F2	−	2	++	0
R1	−	3	+	0
R2	−	3	++	+
E1	−	2	++	0
E2	−	2	+	0
E3	+	1	−	0

Table 3. Cleaning efficiency. Naked eye observations.

Cleaning techniques	Cleaning − insufficient +medium ++good + + +excellent
F1	−
F2	++
F3	+
R1	+
R2	++
E1	+
E2	+
E3	−
G1a	+ + +
G2a	+ + +

4.2 Cleaning efficiency

4.2.1 Naked eye observations

The cleaning efficiency varied from insufficient to excellent (Table 3). The best cleaning was obtained with the G technique (Fig. 5), with an homogeneous result. No difference in efficiency was noticed between the 2 tested pressures (2 and 20 bars).

The performances of the latex poultices were more uneven. The results varied from insufficient (F1 and E3) to medium (F3, R1, E1, E2) up to good (F2 and R2). The main differences between the G technique and the latex poultices concerned colour and cleaning homogeneity. The colour of the surface cleaned with the G technique was indeed pinker whereas with all the latex poultices the cleaned surfaces appeared whiter. It is to be noticed that some dark particles were observed on the areas cleaned with some latex poultices, in more or less high quantity depending on the poultice, leading to an heterogeneous result (Fig. 6).

4.2.2 Colour measurements

To quantify the cleaning efficiency, colour measurements (L, a, b system) were performed before and after cleaning, on each testing area and on a control area, using a Chroma Meter Minolta CR110©.

As the environmental conditions (temperature and relative humidity) during the two sets of measurements were comparable, the results obtained on the control area for the two sets of measurements were

89

Figure 5. Result obtained after G cleaning.

Figure 6. Result obtained after E3 cleaning.

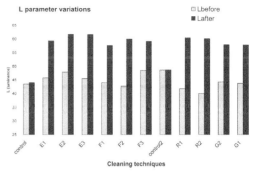

Figure 7. L parameter variations with cleaning. Each value corresponds to the average of 9 measurements.

Table 4. Particles (binder and small aggregates) observed in the latex films.

Cleaning technique	Particle observed in the latex film		
	+ a little	++ some	+++ a lot
F1	+		
F2	+++		
F3	+++		
R1	+		
R2	+++		
E1	+++		
E2	++		
E3	+		

4.2.3 Binocular observations

Cores were dry-sampled at the dirty/cleaned interface on each testing area. The least dirt residues were observed with the G technique, thus confirming the visual observations. On the contrary, the dirty/cleaned interface of the area treated with the F1 latex poultice was almost not distinguishable, so numerous were the dirt residues in the cleaned area. On the area treated with E3 poultice (which was the less adhesive), some dirt residues were also observed.

4.3 Cleaning impact on concrete

4.3.1 Observation of the latex films

A visual observation of the latex films (Table 4) revealed the presence of particles pulled out from the concrete skin (as shown on Fig. 8). It is interesting to note that the poultices F2, F3, R2 and E1 which were the most adhesive caused the most important aggregate loss.

This phenomenon could also explain the gap noticed between the visual observations and the colour measurements. Effectively, the mechanical action of the latex poultices, as it peels off not only dirt particles but also small aggregates, exposes a "new" concrete skin with a higher proportion of binder. As the colour of the binder is lighter than that of the aggregates for this concrete, the overall colour obtained after latex

very similar. Therefore, the two set of results can be directly compared.

The colour variations measured were mainly due to luminance variations (L parameter, Fig. 7), the "a" and "b" parameters being quite stable (except for the water injection-extraction technique that induced an increase of the b values).

But the L values obtained after cleaning were in the same range whatever the techniques (Lmin = 57.7 for F1, Lmax = 61.7 for E2, ΔLmax = 4, which is almost not perceptible to the human eye (Christment 2006)). So the correlation between the colour measurements and the naked eye observation is not very clear, as the highest luminance values after cleaning do not correspond to the best cleaning. This is probably due to the fact that the locally still very dark areas are averaged out with the overall cleaned areas so that the heterogeneity of the cleaning is not taken into account with those colour measurements.

Figure 8. Binocular view (G=x8) of R2 latex film, showing the presence of particles pulled-off from the concrete skin.

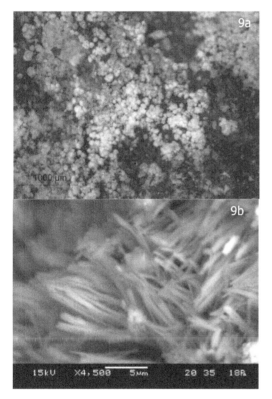

Figure 9. 9a (G=x25) : binocular view of E3 latex film revealing the presence of clusters of white efflorescence (80μm in diameter). 9b (G=x4500) : SEM view of E3 latex (G=x4500), showing the needle shape of those salts.

cleaning is lighter, even if some dirt particles are still present.

A large presence of white salts was noticed on Latex E3 (Fig. 9a). SEM observations (Fig. 9b) coupled with EDS analysis performed on this latex film revealed that these crystals were needle-shaped and Sulphur-Potassium-Calcium-based.

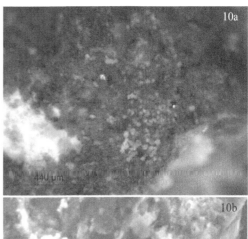

Figure 10. a: binocular view (Gx80) of the surface E3-cleaned, revealing the presence of white efflorescence. b: SEM view (Gx4500) of the surface F3-cleaned, showing the presence of latex residues.

4.3.2 Observation of the cores sampled

Binocular observation of the core sampled on the dirty/E3-cleaned interface revealed the presence of a few white crystals on the cleaned area (Fig. 10a) similar to those observed on the E3 film. SEM observation and EDS analysis confirmed that both their shape and composition are comparable.

Finally, some latex residues were observed on the areas treated with F3 and F2 (Fig. 10b) but in smaller quantity than with R2 and R3.

4.3.3 FTIR Spectrometry

FTIR spectrometry was performed both on the latex films, and on the dirty and cleaned areas of the cores sampled. No significant latex pollution was revealed on the cleaned area. This might be either due to the sampling process (soft scratching of the core surface) or to a very low latex pollution.

5 CONCLUSIONS

Two types of cleaning techniques, particularly adapted to indoor cleaning, were studied. Seven latex poultices and a water injection-extraction system with two

Figure 11. On Sainte-Odile church windows the lead network usually encountered in stained–glass windows was replaced by a cement mortar network, and the traditional cathedral glass was replaced by some crystal paste elements. Even the lines were realised with a cement grout.

Figure 12. Cleaning tests were performed with the water injection-extraction technique on the windows of the Sainte-Odile church. The cleaning efficiency obtained was interesting both on the crystal paste and on the cement mortar.

different pressures were tested on an ancient bush-hammered concrete, initially pink coloured, covered with black dirt deposits mainly due to candle soot.

Three evaluation criteria were considered: the ease of application and removal, the cleaning efficiency and the potential impact on the concrete. Regarding the ease of application and removal, if the water injection-extraction system appeared very easy to use, the latex poultices were more demanding, some products being too pasty or too fluid. Concerning the efficiency, the best results were obtained with the water

injection-extraction system. For the latex poultices, depending on the products, the results varied from insufficient (F1, E3) to good (F2, R2). But, regarding the impact on concrete, latex residues (F, R) and salt crystallisation (E3) were noticed with some of the poultices, when no impact was observed with the water injection-extraction system.

A previous study performed with the same techniques on another type of concrete lead to similar results (Bouichou 2008) thus indicating that for flat surfaces indoors, water injection-extraction seems to be the most suitable cleaning technique. Nevertheless, when complex reliefs are considered, water confinement might not be achieved with the injection-extraction technique. In such cases, some of the tested latex poultices appear to be an interesting alternative solution.

Finally, some conclusive tests were also performed on the windows of the Sainte-Odile church, which are very specific; as they mix crystal paste and cement mortar (Figs. 11 & 12). Therefore the development of a new, smaller, sucking head, is ongoing in order to deal with this last problem.

REFERENCES

Andrew C., 1994. *Stone Cleaning, a guide for practionners*. Edinburgh : Historic Scotland & Robert Gordon university,.

Ashurst N., 1994. *Cleaning historic buildings*. London : Donhead, Vol 1 & 2.

Marie-Victoire E., Texier A., 1999. Comparative study of cleaning techniques applied to ancient concrete, *In DBMC 8 : service life and asset management, Proceedings of the 8th international conference on durability of building materials and components, Vancouver, Canada, May 30-June 3 1999*, pp. 581–592.

Marie-Victoire E., Texier A., 2000. Historic buildings made of concrete: three comparative studies of cleaning techniques, *In Fifth CANMET/ACI , Proceedings of the international conference on durability of concrete, Barcelona, Spain, June 4-9 2000*, pp. 631–645.

Stancliffe, M., De Witte, I., De Witte, E.. St Paul's Cathedral, Poultice cleaning of the interior, *In Journal of Architectural conservation*, 2005, pp. 87–103.

Vergès-Belmin V., Bromblet P., Le nettoyage de la pierre, In *Monumental*, Paris, 2000, pp. 220–273.

De Witte E., Dupas M., 1992. Cleaning poultices based on EDTA, *In 7th International congress on deterioration and conservation of stone : proceedings, Lisbon, 1992*, Vol. 3, pp. 1023–1031.

Christment A., De la couleur à la colorimétrie,2006 *In Couleur et temps : La couleur en conservation et restauration, 12èmes journées d'études de la SFIIC, Paris*, pp. 22–27.

Bouichou M., Marie-Victoire E., Brissaud D., 2008. Comparative study of cleaning techniques to be used on concrete indoors, *In International conference on concrete repair, rehabilitation and retrofitting (ICCRRR08), Cape Town, November 24–26th 2008*, p.311.

Concrete Solutions – Grantham, Majorana & Salomoni (Eds)
© 2009 Taylor & Francis Group, London, ISBN 978-0-415-55082-6

Joint sealing on a 50 year old concrete tunnel, Santa Barbara Hydro Plant, Mexico

A. Garduno & J. Resendiz

Comision Federal de Electricidad, Mexico, D.F., Mexico

ABSTRACT: The Hydroelectric System located 60 km southwest from Toluca City, combines 6 hydropower plants, with regulator dams, tunnels and channels. The construction of the system started in 1938 and was finished in 1965. The intention is that all the structures work in optimum conditions, regardless their age. In 2007, during the inspection of a 2200 m concrete conduction tunnel a great amount of water coming through a lot of transversal and longitudinal construction joints was found. It was needed to seal them and reduce the loss of water during the operation of the tunnel, as well as to reduce the risk of geotechnical problems. This paper describes the materials (expanding polyurethane grout), equipment (and its specific adaptations) and procedures used for such works, including planning aspects and safety issues for the works. The tunnel was successfully sealed and it is now working.

1 INTRODUCTION

The construction of the Hydroelectric System Miguel Aleman, located in the State of Mexico, started by 1938 with the Ixtapantongo Hydro Plant and was finished in 1965. This system combines hydropower plants, regulator dams, syphons, pressured pipes, tunnels and channels. The intention is that all the structures work in optimum conditions, regardless their age, even though the maximum production of 1,400 GWh per year is not reached anymore.

In December 2007, in the conduction tunnel between the Ixtapantongo Regulator Dam and the Santa Barbara Hydropower Plant, a problem was found in the joint between the concrete conduction tunnel and the steel tunnel (syphon): the welding was broken. During the rehabilitation of this steel tunnel, it was possible to do a detailed inspection of the 2200 m concrete conduction tunnel which was joined to the steel tunnel. More than 200 leaking transverse and longitudinal construction joints were found. It was needed to seal them and reduce the loss of water during the operation of the tunnel, as well as to reduce the risk of geotechnical problems.

In the following, the materials (expanding polyurethane grout), equipment (and its specific adaptations) and procedures used for such works, including planning aspects and safety issues for the works are discussed. The tunnel was successfully sealed and it is working at the moment.

2 SITE INFORMATION

The Hydroelectric System Miguel Aleman is located 60 km southwest from Toluca City, and, within a

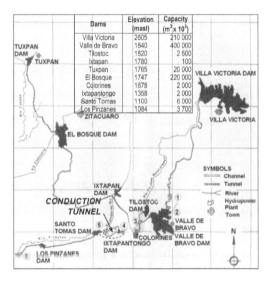

Figure 1. Location of the conduction tunnel from Ixtapantongo Dam to Santa Barbara Hydropower Plant.

3600 km² area, it has 6 hydropower plants, dams, tunnels, channels, etc. The advantages of the site are the difference in elevation from the first dam to the last hydro plant: about 2000 m, and the constant water supply. Figure 1 shows the location and Table 1 information about the hydropower plants.

The conduction system between the Ixtapantongo Regulator Dam and the Santa Barbara Hydro Plant has two concrete tunnels (diameter = 3.4 m), with 2200 and 1770 m length respectively, and between them a 600 m steel syphon. Broken welding on the steel tunnel was found in the joint between the conduction

Table 1. Information about the Hydropower Plants of the Hydroelectric System Miguel Aleman.

Hydropower Plants	Elevation (masl)	Capacity (KW)	Hydraulic head (m)
1 M. de Meza	2116	25,200	376
2 A. Millan	1840	18,900	276
3 El Durazno	1691	18,000	105
4 Ixtapantongo	1355	106,000	328
5 Santa Barbara	1094	67,575	262
6 Tingambato	685	135,000	380

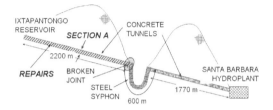

Figure 2a. Scheme of the conduction tunnels.

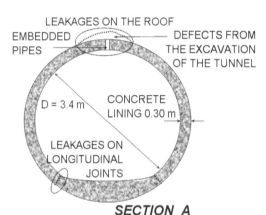

Figure 2b. Section of the concrete conduction tunnel.

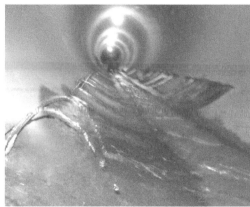

Figure 3. Leakage on longitudinal joints.

Figure 4. Leakage along a transversal joint.

tunnel from the Ixtapantongo Dam and the syphon. It is important to mention that in that length, the tunnel has access from the intake structure on Ixtapantongo, and on a 0.50 m diameter manhole on the steel tunnel, 2200 m after the first one. Figure 2a shows a scheme of the tunnels and siphon, and the location of the broken joint.

The thickness of the concrete lining is 0.30 m, and it is reinforced. The transversal joints are each 5 m and there are 2 longitudinal joints at the joint between tunnel walls and the floor. Figure 2b shows the section of the tunnel where the sealing was needed.

3 INSPECTION AND DAMAGE EVALUATION

In December 2007, in the conduction line between Ixtapantongo Dam and Santa Barbara Hydro Plant, the welding on the steel tunnel of the syphon was broken in the joint with the concrete conduction tunnel. This problem was noticed because of the amount of water that was being accumulated in the surface, just above the syphon joint.

While the welding on the syphon was being fixed, the concrete conduction tunnel from Ixtapantongo to the syphon (2200 m) was inspected and more than 200 instances of leakage were found on transversal and longitudinal joints (Fig. 3, 4). The transversal joints that presented leakage had to be sealed all over their perimeter. The leakage on longitudinal joints were in specific locations, but most of them presented infiltration of soil fines. This could represent a geotechnical problem, because those fines could be part of the contact material between the rock and the tunnel, so, during the operation of the conduction tunnel, the loss of material could produce cavities, endangering the safety of the tunnel.

The leakage also represent a loss of water that decreases the regular operation of the conduction tunnel, affecting the production of the Santa Barbara Hydropower Plant.

Table 2. Leakage found on the concrete tunnel.

Location	Number	Quantity of product
Holes on longitudinal joints	115	1 litre/hole
Transversal joints	117	2 litres/joint
Pipes on roof	4	3 litres/pipe
No contact material on roof	7	140 litres/hole

During the repairs of the steel joint, it was decided to seal the leakage of the joints on the concrete tunnel. For choosing the sealing procedures, a more detailed inspection was done. More joint leakage was found, and also some from the roof of the tunnel that came through 51 mm diameter pipes embedded in the concrete tunnel (used during construction), and in addition, some more from the roof due to the excavation of the tunnel where the concrete lining was not in contact with the rock. Table 2 shows an estimate of the leakage on the concrete tunnel.

4 REPAIR PROCEDURES

4.1 Longitudinal and transversal joint leakages

The sealing of joints was planned using a hydrophobic expanding polyurethane chemical grout (Sika type). The product expands 2 to 5 times in volume depending upon the amount of accelerator that is used. The chosen product has low viscosity, which allows injection into narrow cracks, and has excellent adhesion to wet and dry surfaces. The density of the product is about 1.1 kg/l, similar to water.

For developing the injection procedure, a few laboratory tests were done, in order to make adjustments to the equipment and to establish the injection system and the procedure. The injection nozzles usually used for this product, could not be used because they do not prevent the polyurethane grout leaking back before the reaction starts (because of the water pressure in the other side of the tunnel). A new nozzle system was developed using a 15 cm galvanized pipe of 0.635 cm ($\frac{1}{4}$") diameter with a spherical valve that could be closed, keeping the grout inside the joint, crack or hole, until the polyurethane grout had reacted. The tests were done on concrete blocks ($100 \times 100 \times 20$ cm) with a central crack; two injection nozzle valves were inserted on the side, and one in the upper face, as shown on Figure 5. From the upper nozzle, water was injected, simulating the condition of the tunnel, and the polyurethane grout was injected from the lower side nozzle until it started to leak from the next nozzle. The pressure for pumping the product should be higher than the pressure from the outside water (in some places, the water height was more than a 30 m).

The injection pressure was set at 5 kg/cm², which was easily supplied by the electric pump (karcher type) chosen for the job. A level of 3% of accelerator for the polyurethane grout was established, after a few

Figure 5. Laboratory test concrete block for developing the injection nozzles.

Figure 6 and 7. show the injection works, and on figures 9 and 10 the final phases of the sealing jobs can be appreciated.

trials, allowing about 10 minutes for injecting the product before the reaction was started. It was necessary to confine the grout while it started to react by temporarily sealing the joints (with oakum) as well as the nozzles perimeter. Figures 6 and 7 show the injection procedure on the concrete tunnel.

The procedure was: cleaning the surface near to the joint, drilling boreholes (0.953 cm, 3/8" diameter) crossing the joint as shown on Figure 8 (injection

95

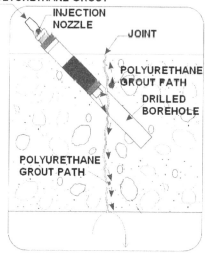

LOW PRESSURED
POLYURETHANE GROUT

INJECTION NOZZLE

JOINT

POLYURETHANE GROUT PATH

DRILLED BOREHOLE

POLYURETHANE GROUT PATH

Figure 8. Interception technique used for the injection of the polyurethane grout.

by interception technique, Emmons 1993), inserting injection nozzles each 50 cm or on the detected hole, then sealing the joint and borehole with oakum, so the polyurethane would not leak. Then, the grout was prepared combining the 3%, by volume, of the accelerator, mixing, and pumping on each nozzle until the product could not penetrate or it started to leak on the next nozzle (García 2002). For the transversal joints the injection was done starting from the bottom nozzles. The injection nozzles were taken out the next day, and finally the sealed joint was cleaned of excess polyurethane and covered with a thin coat of repair mortar (cementitious grout).

4.2 Roof leakage

The leakage that came through embedded pipes was plugged with a cylindrical wooden piece, which had an injection nozzle for the polyurethane grout in the middle section. It was established that an injection pressure of $5 \, \text{kg/cm}^2$, was needed to resist the water pressure.

To repair the cavities due to the defects between the concrete lining and the rock, left from the excavation of the tunnel, it was proposed to use a steel plate (about 80×80 cm), as a plug and support, with 25 mm diameter pipes for connecting valve and injector, and then injecting a commercial non-contraction mortar (Sika type). The plate was fixed to the roof of the tunnel with expansion anchors (Hilti type). The mortar was mixed on site using about 20% of water, because maximum fluidity was needed, and then it was injected with a manual pump (Hany type). After 3 days, the pipes and valves were removed, and the steel plate remained.

5 SITE WORKS

The sealing of the concrete tunnel was planned considering that the repairs on the steel tunnel had to be done (sandblasting, painting and welding ring steel reinforcement on the failure area) at the same time, in order to restart the power production of the plant as soon as possible, and this would increase the pollutants inside the tunnel. One solution was to work during the night and in the morning, when the wind direction helped to remove the gases from the tunnel.

The works had to be done wearing masks and avoiding, as possible, to work close to the sandblasting area. It was also needed to use the safety equipment in order to work with the polyurethane grout, such as gloves, and avoiding direct skin contact.

6 CONCLUSIONS

The sealing of joints, cracks, holes or any defects on the conduction concrete tunnel was necessary to reduce the amount of water that may reduce the efficiency of the conduction line, as well as to improve the safety of the tunnel. Otherwise, geotechnical problems could be generated.

The procedures and equipment used for sealing the concrete conduction tunnel had to be adapted from the usual ones. It was very important to consider the water pressure from the site, in order to establish the injection pressure. And producing a fast reaction with the polyurethane grout was a risk for the pumping equipment, so it had to be enough time for injecting it before the reaction started.

Adaptations for preparing plugs on the different leakages from the roof had to be improved, including wood plugs and the use of steel plates.

It is important to mention that the sealing of joints in the concrete tunnel was done in 2 months, and was planned for 6 weeks. The delays were associated to the low efficiency because of the difficulties of working at the same time with the sandblasting and steel sealing of the syphon on a closed environment.

The volume of water from the leakages decreased considerably after the sealing, about a 90%, and it is expected that the safety of the tunnel has been improved.

Further inspections are recommended, once a year, to check the behaviour of the concrete joints.

REFERENCES

Emmons, P.H., 1993. Concrete Repair and Maintenance Illustrated. RSMeans, USA, ch 5, 272–274
García Rodríguez F., 2002. Evaluación de estructuras, Técnicas y materiales para su reparación. IMCYC, Mexico, 2nd part, 141

Concrete Solutions – Grantham, Majorana & Salomoni (Eds)
© 2009 Taylor & Francis Group, London, ISBN 978-0-415-55082-6

Effects of admixture type on air-entrained self-consolidating concrete

N. Ghafoori

University of Nevada Las Vegas, USA

M. Barfield

Walter P. Moore & Associates, Inc., Las Vegas, Nevada, USA

ABSTRACT: Self-consolidating concrete (SCC) is a highly flowable construction material well-suited for repair applications that require thin overlays of concrete. In this investigation, mixtures were developed with a 635 mm slump flow utilizing admixtures from two sources in order to study the effects of admixture type on the air void characteristics of air-entrained self-consolidating concrete. The mixtures had a constant water-to-cementitious materials ratio of 0.40 and constant mixture proportions with the exception of admixture dosages. The main objective of this study was to compare different types of air-entrainment admixtures (AEA) and high range water reducers (HRWR) from two commonly available admixture manufacturers. The air void characteristics (specific surface and spacing factor) were measured on the fresh concrete using an Air Void Analyzer. The additional properties measured were: slump flow, J-Ring passing ability, rate of flowability, resistance to dynamic segregation and compressive strength.

1 INTRODUCTION

Self-consolidating concrete (SCC) is a highly flowable construction material that can be used for repair applications where a thin application of concrete is needed, or where site constraints prevent manual consolidation. In cold regions where concrete requires freeze-thaw durability, the air voids entrained throughout the concrete must be small and closely spaced. The highly fluid nature of SCC can make effective air entrainment more difficult to achieve and maintain than in conventional concrete. SCC can be used to quickly repair a structure by minimizing demolition work, for the concrete does not require manual consolidation and can flow easily around existing obstructions.

In addition to being highly flowable, self-consolidating concrete (SCC) must be cohesive enough to fill any size or shape without segregation or bleeding. SCC mixtures are characterized by both their unique flow properties and mixture proportioning. In terms of mixture proportioning, SCC mixtures typically have a higher amount of cementitious material and a higher fine-to-coarse aggregate ratio than conventional concrete, both of which increase the viscosity. The flow ability and viscosity of a SCC mixture are also controlled through the use of high range water reducers (HRWR) and viscosity modifying admixtures (VMA). The HRWR creates the necessary flow ability by adsorbing to cement particles and inducing an electrical charge, preventing cement flocs from forming (Rixom & Mailvaganam 1999). The VMA increases the viscosity of a mixture by affixing itself to water

molecules in the concrete and triggering a gel-like behavior (Khayat 1995).

The differences between self-consolidating and conventional concrete give rise to complications in producing a matrix that is durable in freezing-and-thawing situations. It is well documented that specific limits on the air void characteristics can greatly improve the frost durability of concrete when exposed to water, even in self-consolidating concrete (Khayat 2000). These limits are: specific surface, α, greater than $25\,mm^{-1}$, and spacing factor, $\bar{L}$, less than $200\,\mu m$.

2 EXPERIMENTAL PROGRAM

The main objective of this study was to optimize the admixture dosage requirements of two manufacturers and determine the effects of admixture type on the air void characteristics of the selected self-consolidating concretes. An optimized admixture dosage is the minimum admixture dosage required to obtain the target fresh properties (or performance characteristics) outlined below. Two mixtures were developed in this investigation utilizing two distinct admixture manufacturers and one slump flow of 635 mm. The mixture proportions were held constant with the exception of the admixture dosages. The volumetric air content of the fresh concrete was set at $6 \pm 0.5\%$. The water-to-cementitious-materials ratio remained uniform at 0.40, which is the maximum allowed by ACI 318-05 for concrete exposed to a severe frost environment.

2.1 Raw materials

The aggregates were obtained from a local quarry in Southern Nevada. The coarse aggregate had a nominal maximum size of 16 mm and was required to pass the #7 gradation limits defined by ASTM C 33. The fine aggregate also met ASTM C 33 gradation requirements and had a fineness modulus of 3.0. The coarse to fine aggregate ratio was set at 1.083, determined using the optimum volumetric density of the combined aggregate gradation. Fifty-two percent, or 864 kg m^{-3}, of the total aggregate was coarse, and the remaining 48% was fine aggregate, or 795 kg m^{-3}.

ASTM C 150 Type V cement and ASTM C 618 Class F fly ash were used. Class F fly ash was added at 20% by weight of cement in order to provide the trial self-consolidating concretes with sufficient cementitious materials. All mixtures contained 390 kg m^{-3} cement, 78 kg m^{-3} fly ash, and 196 kg m^{-3} water.

Admixtures were obtained from two different admixture manufacturers available in the United States, and labeled A and B to prevent endorsement of any company. The specific types of admixtures utilized can be seen in Table 1.

2.2 Mixing and testing methods

The concrete was produced in a horizontal pan mixer with 0.0283 m^3 capacity. For the mixing sequence, first the coarse aggregate, 1/3 of the mixing water and the AEA was added. Following two minutes of mixing, the fine aggregate and 1/3 of the water was incorporated and mixed for another two minutes. Finally, the cement, fly ash and remaining water was added. After mixing for three minutes, the HRWR and VMA was introduced and allowed to mix for an additional three minutes. At this point, the concrete was rested and mixed for two minutes each. The elapsed time of the total mixing sequence was 14 minutes, or 10 minutes following the first cement and water contact.

Four fresh properties of the concrete were conducted to assess the flow performance of a SCC mixture: 1) unconfined workability, 2) rate of flow ability, 3) passing ability, and 4) resistance to dynamic segregation. The unconfined workability was tested by measuring the slump flow (described in ASTM C 1611) and was required to be within 13 mm of the 625 mm target slump flow. The rate of flow ability was determined by measuring the T_{50} (also outlined in ASTM C 1611), which is the elapsed time from when the slump cone is lifted to when the concrete reaches a 50 cm mark on the testing plate. For this study, the T_{50} was required to be between 2 and 5 seconds. The passing ability was determined using the J-Ring, which is outlined in ASTM C 1621. The J-Ring measurement is done in conjunction with the slump flow, and had to be within 51 mm of the slump flow to indicate adequate passing ability of a SCC mixture. The stability or resistance to dynamic segregation of the concretes was evaluated based on the Visual Stability Index (VSI), as delineated in ASTM C 1611. The VSI essentially rates the stability of a concrete mixture on a scale of 0 to 3 (highly stable to highly unstable), based on the visual appearance of the slump flow.

The volumetric air content was determined on the fresh concrete using a roll-a-meter in accordance with ASTM C 173. The air void characteristics were determined using an Air Void Analyzer (AVA), which determines the air void specific surface and spacing factor on fresh samples of concrete. The results of the Air Void Analyzer are correlated to match those determined on hardened concrete using ASTM C 457 within a 95% confidence limit, and present an adequate assessment of the size and spacing of the air voids in concrete (Aarre 1998).

The compressive strength of the concrete was tested on 102 × 203 mm cylinders at 7, 28 and 90 days of curing, following ASTM C 39. The average of a minimum of three cylinders was reported as the compressive strength of the mixture. While there was no specified target compressive strength required of the mixtures, 34.5 MPa is recommended for severe freezing and thawing exposure under ACI 318-05.

3 DISCUSSION OF RESULTS

3.1 Optimized admixture dosages

The optimized admixture dosages are presented in Table 2. Each mixture is identified using the admixture source (A or B). Overall, source B was more volumetrically economical than source A.

The HRWR used from source A had a slightly different chemical composition than source B. Source A consisted of a polycarboxylate-ester (PCE) molecule, whereas source A was a polycarboxylate-acid (PCA) molecule. In general, the PCE molecule contains less anionic binding sites to adsorb to the cement particles, but more side chains that allow for better slump retention capability. The PCA molecule has more binding sites which allows for more dispersion of the cement particles, thus imparting greater flow ability to a mixture.

Table 1. Types of admixtures utilized.

Admixture	Source	
	A	B
HRWR	polycarboxylate-ester	polycarboxylate-acid
VMA	aqueous solution of polysaccharides	naphthalene sulfonate and welan gum
AEA	alkylbenzene sulfonic acid	tall oil and glycol ether

Table 2. Admixture dosages (ml kg^{-1} cementitious materials).

Source	HRWR	VMA	AEA
A	3.390	1.239	0.782
B	2.021	0.261	0.717

Trial-and-error procedures were used to achieve the optimum admixture dosages of the SCC mixtures. However, upon inspection of the HRWR and VMA dosage combinations, there was an ideal VMA-to-HRWR ratio for each admixture source to produce air-entrained SCC. Source A necessitated more VMA than source B to create a stable mixture, as evidenced by the VMA-to-HRWR ratio of 0.37. In contrast, source B had a VMA-to-HRWR ratio of 0.11.

The AEA dosage required to entrain $6 \pm 0.5\%$ air was similar for both admixture types, as seen in Table 2, even though source B was a wood-derived acid salt and source A was a synthetic detergent. The two classes of AEAs utilize different mechanisms to entrain air, and thus react differently with the other mixture constituents (i.e. cement, fly ash and admixtures). These mechanisms are discussed later in the air void characteristics section.

Dosages of VMA and HRWR can reduce the effectiveness of AEA to secure a proper air void system (Khayat & Assaad 2002). The admixtures interfere with the ability of AEA to stabilize air voids in the concrete through competition on a molecular level. The HRWR and AEA compete for adsorption locations on cement and fly ash particles. Additionally, dosages of VMA can prevent water molecules from forming air voids with the AEA.

3.2 Fresh properties

The actual slump flow, J-Ring passing ability, T_{50} rate of flow ability, VSI, and volumetric air content measured for each mixture design can be seen in Table 3. The measurements reported are the average of two or three trials for each test, depending on the consistency between trial batches.

The T_{50} flow times, which indicate the flow ability and viscosity (by inference) of a SCC mixture, suggested that both of the mixtures developed had a relatively low viscosity. This is due to the fact that the T_{50} times remained close to the lower limit of the suggested values of 2 to 5 seconds. The average T_{50} value for source A did not meet the standard of greater than 2.0 seconds. However, two of the three T_{50} trials conducted with source A were above 2.0 seconds. Due to the high operator error associated with measuring the T_{50} and the variance of data between batches, this mixture design was deemed acceptable.

Table 3. Fresh properties.

Property	Source	
	A	B
Slump Flow (mm)	638	648
J-Ring (mm)	600	610
SF – J-Ring (mm)	38	38
T_{50} (sec)	1.93	2.26
VSI	0	0
Air Content (%)	6.3	6.4

The VSI rating determined for each of the twelve mixtures indicated the mixture's dynamic stability, or resistance to bleeding and segregation. Both mixtures developed exhibited adequate stability and received a rating of 0 (highly stable).

The J-Ring test results of the SCC mixtures demonstrated that the passing ability was independent of slump flow. Although source A had a lower initial slump flow than source B, the passing ability was equal to 38 mm for both mixtures.

3.3 Air void characteristics

The results of the air void analyses can be seen in Table 4. Each value represents the average of at least four AVA trials. Overall, source A produced smaller and more closely spaced air voids than sources B.

The limits for specific surface (greater than 25 mm^{-1}) and spacing factor (less than 200 μm) were achievable with both admixture sources. All mixture designs were initially designed to meet the required air content of $6 \pm 0.5\%$ solely using the volumetric air meter. For all of the mixtures, the air void characteristics that resulted from the optimum AEA dosage met the air void standards.

The AEA from source A is a synthetic detergent, primarily constituted by alkybenzene sulfonic acid. These types of surfactants are influenced by increased fluidity due to their primary location at the air-water interface. Since the air voids are not necessarily anchored to cement particles, the bubbles produced by pure surfactant AEAs can move about freely in the matrix. Therefore, source A produced bubbles that were more likely to rupture on the surface and coalesce than bubbles produced by a salt-type AEA.

The air-entraining agent from source B was a form of wood-derived acid salt. The tall oil component of the source B AEA has been noted to generate the smallest air voids of all wood-derived AEAs (Kosmatka et al. 2002). The air voids generated by salt-type AEAs are primarily adhered or bridged to the cement particles due to the precipitates formed through a immediate reaction with calcium ions, resulting in similar air void characteristics regardless of slump flow. The mass of the cement particles acts like an anchor to disperse the air bubbles throughout the matrix and reduce the tendency of large air bubbles to float to the surface, regardless of the paste viscosity and fluidity (Du & Folliard 2005).

The high fluidity and low viscosity of the self-consolidating concrete makes it more difficult to entrain and stabilize air bubbles than in conventional concrete. The high deformability allows more coalescence of air bubbles, resulting in a decreased specific

Table 4. Air void characteristics.

Source	Specific Surface (mm^{-1})	Spacing Factor (μm)
A	44.9	120
B	37.5	145

Table 5. Compressive strength (MPa).

Source	7-day	28-day	90-day
A	32.6	41.0	55.5
B	29.3	37.8	48.0

surface and increased spacing factor. Additionally, as discussed previously, the increased dosage of HRWR and VMA at the higher slump flows can interfere with the mechanisms of air-entrainment.

3.4 Compressive strength

The compressive strength of each mixture was tested after 7, 28 and 90 days of curing, the results of which can be seen in Table 5. All mixtures met the recommended 28-day compressive strength of 34.5 MPa, required by ACI 318-05 for freeze-thaw durability under severe conditions. The significant increase in strength after 90 days can be attributed mainly to the 20% fly ash content, which is known to add delayed strength. Overall, source A produced concrete with higher compressive strength than source B, although it was limited to 7.5 MPa.

4 CONCLUSIONS

For the test results of this study, the following conclusions can be drawn about the optimization and performance of air-entrained self-consolidating concrete mixtures:

The admixture source primarily influenced the required admixture dosage of a SCC mixture. A polycarboxylate-ester (PCE) HRWR necessitated a larger dosage to impart the same flow ability as a polycarboxylate-acid (PCA) HRWR. Additionally, when using a PCE HRWR, a greater dosage of VMA was required to maintain stability, compared to mixtures developed using a PCA HRWR.

The air void characteristics of SCC were affected by both competition with other admixtures and the type of AEA utilized. Increased dosages of high range water reducer competed with air-entrainment for adsorption to cement and fly ash particles, and imparted fluidity that increased bubble coalescence. Increased dosages of viscosity modifying admixture competed with air-entrainment by preventing water molecules from forming bubbles. Surfactant-type air-entrainment (i.e. synthetic detergents) secured better air void characteristics in SCC than salt-type air-entraining admixtures.

ACKNOWLEDGEMENTS

The contributions of a number of materials manufacturers are greatly appreciated, and without their cooperation this research would not have been possible.

REFERENCES

Aarre, T. 1998. Air Void Analyzer. Portland Cement Association, *Concrete Technology Today* 19(1): 1–3.

ACI 318-05. 2005. *Building Code Requirements for Structural Concrete and Commentary*, Farmington Hills, Michigan: American Concrete Institute.

ASTM standards. American Society for Testing and Materials. http://www.astm.org.

Du, L. & Folliard, K.J. 2005. Mechanisms of air entrainment in concrete. *Cement and Concrete Research* 35: 1463–1471.

Khayat, K.H. 1995. Effects of Anti-Washout Admixtures on Fresh Concrete Properties. *ACI Materials Journal* 92(2): 164–171.

Khayat, K.H. 2000. Optimization and Performance of Air-Entrained, Self-Consolidating Concrete. *ACI Materials Journal* 97(5): 526–535.

Khayat, K.H. & Assaad, J. 2002. Air Void Stability in Self-Consolidating Concrete. ACI Materials Journal 99(4): 408–416.

Kosmatka, S.H., Kerkhoff, B. & Panarese, W.C. 2002. *Design and Control of Concrete Mixtures*. Skokie, Illinois: Portland Cement Association.

Rixom, R. & Mailvaganam, N. 1999. *Chemical Admixtures for Concrete*, New York: E & F.N. Spon Ltd.

Concrete Solutions – Grantham, Majorana & Salomoni (Eds)
© 2009 Taylor & Francis Group, London, ISBN 978-0-415-55082-6

Remediation of slump flow loss of fresh self-consolidating concrete induced by hauling time

N. Ghafoori & H. Diawara

Department of Civil and Environmental Engineering, University of Nevada, Las Vegas, USA

ABSTRACT: This investigation was devoted to study the effectiveness of an overdosing remediation technique which was used to provide initial optimum admixture dosages resulting in suitable self-consolidating concretes (SCC) for different hauling times. The selected SCCs were made with a constant water-to-cementitious materials ratio, uniform cementitious materials content, and constant coarse-to-fine aggregate ratio that provided the optimum aggregate gradation. Polycarboxylate-based high range water-reducing admixture and viscosity modifying admixture were used to produce matrices with slump flow of 635 mm, and visual stability index (VSI) of 0 (highly stable concrete). Five different hauling times, namely: 10, 30, 50, 70, and 90 minutes, were used. The test results indicated that the selected remediation method was successful in producing SCCs with a similar unconfined workability, flow ability rate, dynamic stability, and passing ability to those obtained at the control hauling time of 10 minutes.

1 INTRODUCTION

Self-consolidating concrete (SCC) can be defined as a high performance concrete in fresh state. It is a highly flowable, non-segregating concrete that can spread into place, and encapsulate the reinforcement without any mechanical consolidation; making it a recommendable construction material for repair applications of various structural elements. SCC requires special attention in the mixing and delivery method due to its low water content relative to the high cementitious materials content. Like any other concrete, the mixing of SCC for a long period of time can result in reduction of slump flow, leading to an unusual rate of stiffening of fresh concrete; loss of entrained air, strength and durability; difficulty in pumping and placing; and excessive effort in placement and finishing operation.

To overcome the slump loss, two main remediation methods are generally practiced. There are: overdosing (using a higher initial slump) or retempering (adding extra water or admixture before placing concrete). In recent years, the increased use of chemical admixtures in the concrete industry has facilitated the control of slump loss. Superplasticizers or high range water-reducing admixtures (HRWRA) were developed in order to improve the dispersibility and the slump retention of melamine and naphthalene type admixtures. Their extended life can impart up to 2 hours longer working life to concrete. Overdosing the admixture amount in attaining the target slump at the job site or retempering with admixture instead of water are the preferred methods in remediation of the slump loss, since the use of extra water in retempering or in making a higher initial slump can induce side effects on the properties and serviceability of the hardened concrete

(i.e. decrease in strength and durability, increase in permeability and drying shrinkage, etc.) (ACI 237-07, Kosmatka et al. 2002)

2 RESEARCH OBJECTIVES

The study presented herein was intended to remediate the adverse effect of hauling time on freshly-mixed self-consolidating concretes made by an overdosing method. This method consisted of using sufficient initial admixture dosages to obtain the target fresh properties of the trials matrices at various hauling times. Five different hauling times, namely: 10, 30, 50, 70, and 90 minutes, were used. The unconfined workability (slump flow), flow rate (T_{50}), and dynamic segregation resistance (VSI) of the remediated SCCs were evaluated at the end of each hauling time, and compared to the equivalent fresh properties obtained at the control hauling time of 10 minutes.

3 EXPERIMENTAL PROGRAMS

3.1 Raw materials

The cementitious materials used in all mixtures consisted of ASTM C150 Type V Portland cement and ASTM C 618 class F fly ash. The Type V Portland cement had a Blaine fineness of $423\,m^2/kg$ and the following percentages of chemical constituents: $SiO_2 = 20.1\%$, $Al_2O_3 = 4.0\%$, $Fe_2O_3 = 3.6\%$, $CaO = 63.5\%$, $MgO = 2.8\%$, $SO_3 = 2.9\%$, $C_3A = 4\%$, $C_3S = 58\%$, $C_2S = 14\%$, Na_2O equivalent $= 0.57\%$, loss on ignition $= 2.3\%$, and

Table 1. Fresh properties of hauled SCC.

Hauling time (min)	Slump flow (mm)	T_{50} (sec.)	VSI	Slump flow loss (mm)
10	651	2.79	0	0
30	540	3.10	0	−111
50	493	*	0	−156
70	432	*	0	−219
90	406	*	0	−245

*T_{50} can not be measured since the slump flow is less than 508 mm.

insoluble residue = 0.44%. The fly ash had the followings chemical composition: $SiO_2 = 58.2\%$, $Al_2O_3 = 17.4\%$, $Fe_2O_3 = 4.8\%$, $CaO = 7.9\%$, $SO_3 = 0.6\%$, moisture content = 0.0%, and loss on ignition = 4.2%. The fine aggregate used had bulk and saturated surface dry specific gravity, absorption, and fineness modulus of 2.75 and 2.78, 0.8%, and 3.0, respectively. The coarse aggregate had a nominal maximum size equal to 12.50 mm and complied with ASTM C 33 size number 7. Its bulk and saturated surface dry specific gravity, absorption, and dry rodded unit weight were 2.77 and 2.79, 0.6%, 1634 kg/m³, respectively. Other concrete constituents were tap water, polycarboxylate-based high-range water reducing admixture (HRWRA) and viscosity modifying admixture (VMA) complying with the ASTM C 494 Type F requirements.

3.2 Mixture proportion

All matrices were prepared with a constant water-to-cementitious materials ratio of 0.4, a uniform cement factor of 391 kg/m³, and a constant amount of fly ash representing 20% of the cement weight. In proportioning the aggregate contents, particular attention was given to the coarse-to-fine aggregate ratio due to its critical role in generating a sufficient amount of mortar for the selected self-consolidating concretes. The ASTM C 29 was used to determine the compacted bulk unit weight and the calculated void content using different ratios of the combined coarse and fine aggregates. The optimum volumetric coarse-to-fine aggregate ratio, utilized in the proportioning of the concrete constituents, was found at 0.52/0.48. The quantities of coarse and fine aggregates used in the matrices were 923 and 850 kg/m³, respectively. The optimum (minimum) dosages of the high range water-reducing admixture (HRWRA) and viscosity modifying admixture (VMA) used at the control hauling time of 10 minutes were 209 and 26 ml/100 kg, respectively. These dosages were obtained by evaluating the consistency and stability of concrete using different trial batches until a satisfactory slump flow of 635 ± 25 mm; and a visual stability index of 0 were attained. Table 1 displays the measured fresh properties at the selected hauling times.

3.3 Mixing, sampling and testing

Laboratory trial mixtures were used to produce self-consolidating concretes. An electric counter-current pan mixer with a capacity of 0.028 m³ was used to blend concrete components. In simulating the influence of hauling time on the fresh SCCs, a realistic concrete mixing tool with changeable velocity was needed. Therefore, a speed control box, designed and mounted to the mixer in the laboratory, was used to adjust the mixer's rotational velocity during hauling. The mixing sequence consisted of blending the coarse aggregate with 1/3 of the mixing water for two minutes, followed by the fine aggregate with 1/3 of the mixing water for another two minutes, and the cementitious materials with the remaining 1/3 of the mixing water for three minutes. Finally, the HRWRA and VMA were added and blending of the matrix continued for an additional three minutes, followed by a two-minute rest and resumption of mixing for two additional minutes. From that point on the mixing speed (14.5 rpm) was changed to an agitating speed (7.25 rpm) until the desired hauling time was achieved. The hauling time was defined as the elapsed time between the first contact of water and cementitious materials to the beginning of concrete discharge. The concrete mixtures at the end of hauling time were used to determine the unconfined workability, T_{50} flow rate, dynamic stability, and J-ring passing ability in accordance with ASTM C 1611 and C 1621.

4 DISCUSSION OF RESULTS

The hauling time affected the fresh performance of self-consolidating concretes in the form of decrease in unconfined workability, and gain in flow rate or viscosity per inference. The dynamic stabilities of the fresh concretes remained unchanged. Table 1 presents the changes in the fresh performance of self-consolidating concrete as affected by hauling time.

An overview of slump flow loss and the involved mechanism of action is necessary before proceeding with the discussion on remediation. The fundamental mechanism of slump flow loss of concrete during its hauling has been established and reported by several, studies (Kosmatka et al. 2002, Jolicoeur and Simard. 1998, Flatt et al. 1997). It involves mainly the additional fines brought to the concrete mortar by the grinding of aggregates and cement particles, the growth of the cement hydration products, and the competitive adsorption between the superplasticizer and the sulfate ions (SO_4^{2-}) on the cement hydrated products throughout the hauling time[3]. The flow chart of Figure 1 presents the phases and actions involved in aggregate-cement-admixture mechanical interaction during mixing and hauling which resulted in a decrease in workability. Since the fluidity of concrete is mostly controlled by the fluidity of the mortar portion[1], the slump flow losses recorded during the present investigation can be explained through the

Table 2. Required optimum dosages of admixtures for remediation.

Hauling Time (min)	HRWRA (ml/100 kg)	VMA (ml/100 kg)
10	209	26
30	255	26
50	275	26
70	301	26
90	327	26

Table 3. Fresh properties of remediated SCC at various hauling times.

Hauling Time (min)	Slump Flow (mm)	T₅₀ (sec.)	VSI	J-ring value (mm)
10	651	2.70	0	22
30	641	2.22	0	32
50	645	2.24	0	19
70	645	2.30	0	35
90	641	2.30	0	29

1 ml/100 kg = 0.0153 oz/cwt.
1 mm = 0.03937 inch.

increase in specific surface area of concrete mortar (ΔSSAm) and the change in the adsorption amount of chemical admixtures (ΔAds). The ratio Ads/SSAm was used to characterize the SCC fresh performance.

In order to overcome the abovementioned adverse effects, an overdosing remediation method was used. This technique consisted of using sufficient initial admixtures amount to attain the target fresh properties at the end of the selected hauling times. Table 2 displays the required optimum dosages of admixtures and Table 3 documents the measured slump flow, T₅₀ times, VSI ratings, and J-ring values of the remediated matrices at different hauling times.

4.1 HRWRA requirement for remediation

The optimum dosage of HRWRA in attaining the required workability increased as the hauling time increased. In comparing to the optimum dosage at the reference 10 minutes hauling time, the selected SCCs required 22, 31, 44, and 56 %, more HRWRA at 30, 50, 70, and 90 minutes hauling times, respectively. The higher demand for superplasticizer in contesting slump flow losses can be explained through Figure 1. The idea behind the adopted remediation technique was to find by trial and error an initial admixture dosage in which $(Ads/SSAm)_{ht}$ at the end of the hauling time became identical or nearly identical to the $(Ads/SSAm)_{10}$ of the control hauling time at 10 minutes. The term h_t refers to the hauling at time t = 30 to 90 minutes. This was achieved by overdosing admixtures, and is explained through the following equations (1) and (2).

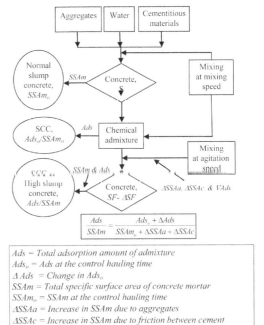

$$\frac{Ads}{SSAm} = \frac{Ads_o + \Delta Ads}{SSAm_o + \Delta SSAa + \Delta SSAc}$$

Ads = Total adsorption amount of admixture
Ads₀ = Ads at the control hauling time
ΔAds = Change in Ads₀
SSAm = Total specific surface area of concrete mortar
SSAm₀ = SSAm at the control hauling time
ΔSSAa = Increase in SSAm due to aggregates
ΔSSAc = Increase in SSAm due to friction between cement particles and growth of cement hydrated products
S = Concrete slump, SF = SCC Slump flow, ΔSF = Change in SF

Figure 1. SCC production and slump flow loss mechanism during hauling.

- *At 10 minutes hauling time:*

$$\left(\frac{Ads}{SSAm}\right)_{10} = \frac{Ads_o}{SSAm_o} \tag{1}$$

- *During the remediation, at hauling time h_t:*

$$\left(\frac{Ads}{SSAm}\right)_{ht} = \underbrace{\frac{Ads_o + \Delta Ads_{ht1}}{SSAm_o + \Delta SSAa + \Delta SSAc}}_{\text{Ⓐ}} +$$

$$\underbrace{\frac{\Delta Ads_{ht2}}{SSAm_o + \Delta SSAa + \Delta SSAc}}_{\text{Ⓑ}} \approx \left(\frac{Ads}{SSAm}\right)_{10} \tag{2}$$

Ⓐ Characterizes the slump flow loss

Ⓑ Characterizes the slump flow restoration

An ultraviolet-visible spectroscopy test revealed that the concentration of free admixture in the cement-water solution was augmented as hauling time increased up to 80 minutes. Beyond that time, gradual decreases were recorded as the hauling time was increases to 90 minutes and beyond. The increase or decrease in admixture concentration led to an increase or decrease in admixture adsorption on cement particles.

ΔAds_{ht1} and ΔAds_{ht2} in Equation (2) correspond to the increases in adsorption amount of admixture during the hauling time and brought by the additional superplasticizer used for the remediation purpose, respectively. The term $(\Delta SSAa + \Delta SSAc)$ reflects the increase in specific surface area of concrete mortar due to the additional fines brought by the grinding of aggregates and cement particles, and the growth of the cement hydrated products. While Ads and SSAm increased with hauling time (up to 80 min for the Ads), the contribution of SSAm on the slump flow loss was greater than that of Ads.

In remediating the slump flow loss, the designed optimum dosages of HRWRA at h_t were sufficient to maintain the solution concentration of free admixture and sulfate ion at the level that produced adequate amount of adsorption to meet the target fluidity at the end of the hauling time h_t. ΔAds_{ht2} generated additional repulsive electrostatic and steric hindrance forces between the cement particles to further disperse the cement agglomerations provoked by the grinding and hydration of cement particle in the course of hauling time.

4.2 VMA requirement for the overdosing remediation

As shown in Table 2, the selected self-consolidating concretes did not require any adjustments in their initial VMA dosage in reaching the target fresh properties. The viscosity modifying admixture was mainly used in SCC to increase its viscosity and stability. Despite a higher demand for HRWRA, the increase in $SSAm$ during hauling was sufficiently effective in enriching and thickening the paste, resulting in a greater viscosity (T_{50} time between 2 and 5 seconds) and a higher VSI (0) of the trial self-consolidating concretes.

4.3 Fresh properties of remediated self-consolidating concrete

As reported in Table 2, the test results showed that all remediated self-consolidating concretes were within the target slump flows ± 25 mm, VSI of 0 (highly stable), T_{50} time between 2 and 5 seconds, and J-ring values between 0 and 50 mm. The test results indicate that the overdosing method was effective in obtaining unconfined workability, flow rate/viscosity, resistance to dynamic segregation, and passing ability

which were similar to those of the control hauling time (10 minutes).

5 CONCLUSIONS

The fresh performance of self-consolidating concrete was affected by hauling times. The slump flow loss was explained through the decrease in adsorption amount of chemical admixtures (ΔAds) per specific surface area of concrete mortar ($\Delta SSAm$). The changes were manifested in the form of loss in flow ability, and gain in flow rate/plastic viscosity and dynamic stability. The alterations in fresh properties were reverted by way of admixtures overdosing which produced self-consolidating concretes with similar fresh properties to those obtained at the control hauling time. The additional amount of admixtures generated supplementary repulsive electrostatic and steric hindrance forces between the cement particles to assist in dispersing the cement agglomerations generated by the grinding and hydration of cement particles during hauling times.

ACKNOWLEDGEMENTS

The authors would like to acknowledge the financial support of the Nevada Department of Transportation, Grant number P 077-06-803. Thanks are also extended to a number of admixture manufacturers and concrete suppliers who contributed materials used in this investigation. Their names are withheld to avoid any concern of commercialization or private concern.

REFERENCES

American Concrete Institute. 2007. Self-Consolidating Concrete. *Reported by the Committee 237*: 30 pp.

Flatt, R.,J., Houst, Y., F., Bowen, P., Hofmann, H., Widmer, J., Sulser, U., Maeder, U., and Burge, T.,A. 1997. Interaction of Superplasticizers with Model Powders in a Highly Alkaline Medium. *Proceedings of the 5th Canmet/ACI International Conference on Superplasticizers and Other Chemical Admixtures in concrete SP-173*: 743–762.

Jolicoeur, C., and Simard, M., A. 1998. Chemical Admixture-Cement Interactions: Phenomenology and Physicochemical Concepts. *Cement and Concrete Composites. Vol. 20: 87–101.*

Kosmatka, S. H., Kerkhoff, B., and Panarese, W. C. 2002. Design and Control of Concrete Mixtures. 14th *Edition, Portland Cement Association, Skokie, Illinois: 358 pp.*

Concrete Solutions – Grantham, Majorana & Salomoni (Eds)
© 2009 Taylor & Francis Group, London, ISBN 978-0-415-55082-6

Study of new methods for restoration and repair of concrete in dams

M. Hokmabadi Ghoshouni, Seyyed Mohammad Ali Seyyed Hejazi &
Bahareh Salmani Ghabel
Engineering Faculty, Islamic Azad University, Iran

ABSTRACT: Repairs & Removing the Defects in Dam-Construction Projects are divided into four main categories:

- Repairs of Concrete Parts (dam body, foundation, spillways and related structures)
- Repairs in Steel Parts (gate, hydraulic installation, steel structures)
- Repairs in Equipments & Instruments
- Repairs in Soil & Stone (stability of embankments, rocks, supports and foundation improvement)

New methods for repairing the concrete parts in dams are discussed in this paper, as the operating personnel have no expertise in repairing the steel parts, equipments and instruments and also the general services, soil & stone repairs and foundation improvement and dam area are related to construction stage of dams and they are not usually faced in operation stages.

1 INTRODUCTION – CRACKS IN CONCRETE

In case of any delay in repair, a defect may expand and the system may fail or there may be a danger and the maintenance and inspection personnel will not be able to remove the defect. In these cases and in the case of basic defects in the foundation, body, steel installations or other parts, the contractor and consulting company will be responsible for removing the defect. Concrete shows a good behavior under compressive stresses but it is very weak under tension and considering that reinforcement is not used in the body of concrete dams in order to bear the tension, most of the cracks in the body of dams are tensile cracks. Other factors like chemicals, heat, and stress concentration, non-continuity of surfaces, section variation and permeability are among other causes of crack in dams. Cracking can take various forms.

1.1 *Cracks before hardening of concrete*

Plastic settlement cracks, plastic shrinkage cracks, cracks due to formwork movement.

1.2 *Cracks after hardening of concrete*

Physical Cracks:

- Cracks resulting from shrinkage of concrete as a result of desiccation
- Cracks resulting from carbonation of concrete

Chemical Cracks/damage:

- Erosion resulting from acid attack
- Erosion resulting from base attack
- Cracks resulting from reaction of aggregates and bases (ASR)
- Cracks resulting from sulfate attack

  Thermal Cracks – due to shrinkage following expansion due to heat of hydration
  Cracks due to freeze-thaw damage
  Corrosion induced cracks
  Cracks resulting from the effect of chlorides or carbonation on buried metal parts.
  Structural Cracks

2 REPAIRS IN CONCRETE PARTS

Concrete is one of the most stable and durable construction materials and has a long life. We should note that the concrete will have weak and susceptible points, which should be taken into account in design and execution; otherwise the concrete will be destroyed after some time and it is necessary to repair it (Construction & Housing Researches Center, 2003, Varshney, 1998). Although there are different methods to repair the concrete, the principles are the same in all methods and they are used to reach the following specifications:

- The new concrete should be completely and permanently adhered to the old concrete.
- The new concrete should be sufficiently permeable.

- The new concrete should not be excessively cracked by shrinkage.
- It should have enough stability against chemical factors.

Different execution conditions dictate different methods to reach the above-mentioned specifications.

2.1 *Preparation of concrete surface*

The first step in the preparation of a concrete surface is to remove the low-quality damaged layers. More care should be taken while removing the concrete. The most usual method to remove the concrete is to use a pneumatic hammer and saws are used in cases that more irregular lines and surfaces are needed. After this stage, a primary washing is carried out in order to remove loose material. A second washing is carried out with the related sandblast and then washing is completed with an air and water jet and finally an air jet is used to clean and dry the surface. It should be noted that no water film layer should be present between two layers of concrete because it will decrease the adhesion between the two layers and drying the surface of the old concrete with an air jet is done for this purpose. By drying the surface of the concrete we want to create the concrete saturation conditions with the surface of dry concrete and the aggregates absorbing no moisture from the new concrete. When the surface of the old concrete is completely dry, some of the water from the new concrete will be absorbed by the old concrete, which can cause a lack of cement hydration and destroy bond.

Materials used in repair include:

Cement
The cement used in old concrete should be used for repairing the concrete so that there will be no difference between the electrochemical properties of the old and new concrete.

Aggregate
The performance of aggregates in the repaired concrete should be the same as those in the base concrete. In case of small surfaces or volumes, the gradation of concrete should be in accordance with the ordinary concrete and coarse aggregates should be avoided.

Epoxy Resins
Epoxy resins with special curing materials create an adhesive and adhere the aggregates to each other and if epoxy adhesives are used, the new epoxy mortar or concrete can adhere to the old concrete after including the two materials and applying them on the dry surface of the old concrete.

Latex Adhesives & Latex Mortar
These materials usually exist as water emulsions and are added to the mixtures of Portland cement. The water reacts with cement and hydration takes place and the latex particles improve the stability and properties of the mortar. Latexes decrease the compressive strength and increases the tensile and bending strengths. Latex mortars increase the primary adhesive properties on most surfaces.

Polyester Resins
Performance of these materials is similar to that of epoxies but their adhesion on smooth surfaces is weak.

Bituminous Materials
These materials are used as protective layers on the concrete.

Materials Penetrating into the Surface
These materials are used for protection of newly repaired surfaces against moisture and harmful chemicals.

2.2 *Repair methods*

Methods used for repairing are standard methods and suitable methods are used depending on the shape of the place to be repaired. Common methods include:

- Dry Pack Method
- Concrete Replacement
- Pre-Packed Aggregate With Intruded Grout
- Shotcrete Repair

(Construction & Housing Researches Center, 2003)

2.2.1 *Dry pack method*
This method is usually used for repairing holes (deep with a small area) and it is not used when the repair area is not deep, where the mortar cannot be well in contact with the old concrete. The dry pack method is used for filling the back parts of long reinforcements or holes existing in deep sections or walls. After preparation of the surface of the old concrete, a thin layer of rich cement mortar, which is often polymer modified, should be applied on the surface of the old concrete using a brush. Before the mortar dries, the adhesives should be placed at the place to be repaired.

2.2.2 *Concrete replacement method*
This method is recommended by U.S.B.R and is used for repairing deep or large holes or holes that continue along the depth of the cross section. When the hole includes all the depth of the cross section, form working is necessary on both sides; otherwise the formwork of one side is omitted but preparation of the vertical surface of the old concrete will be necessary.

The quality of this repair mostly depends on form working and quality of form working. Before pouring the concrete in each stage, the surface of the old concrete, which will be in contact with the new concrete, should be covered by a mortar with a thickness of 3 mm. The amount of sand and cement and water-cement ratio in this mortar should be equal to the amounts of the new concrete. The surface of the old concrete should be moist (not wet) and in order to decrease the shrinkage, the lowest water/cement ratio and the largest size of aggregate should be used in the new concrete. Furthermore, the new concrete should have the lowest temperature, so in high temperatures the constructional materials should be kept in the shade. When the intended area is completely filled, the concrete should be pressurized and vibration should be started. When the exit section is filled and the compressive cap is put in place, vibration and pressure should be simultaneously applied on the concrete.

In case of using vibration without pressure, a layer of water is put on repaired part and will decrease the adhesion in the other parts. On the whole we can say that, the concrete should be under pressure and vibration on a limited area of the mould in order to be involved with the old concrete in this method.

2.2.3 *Pre-packed aggregated with intruded grout*

This method is used for massive repairs, especially for repairing the bases and marine structures. First the damaged areas are removed and formwork is erected and then the mould is filled with coarse aggregates (sand) and struck. Finally the cavities between the aggregates are filled by intruding the mortar. The largest size of the concrete depends on the size of the part to be repaired. Striking and vibrating the aggregate should be possible in the mould and the mould should tolerate the pressure of the mortar. Intrusion of mortar is started from the lowest point of the form working in order to be transferred to other parts of the mould. Mortar intrusion continues until the appearance of concrete in the exit section of the mortar at the highest point of the cavity. Pressure will not be stopped after this stage and will continue for a short period until improvement of the connection between the old and new concrete. The shrinkage of this kind of concrete will be less and intrusion together with the pressure of the mortar will create a good adhesion between the old and new concrete. As the mould is closed and resistant against pressure in this method, concreting operations can be easily done against water flow.

2.2.4 *Shotcrete method*

This is the most common and prevailing method for repairing vertical and overhead surfaces. In this method, the mortar and concrete is sprayed on the intended surface. Spraying is done by either dry spraying or wet spraying.

2.2.5 *Dry spraying*

In this method, the concrete is mixed with dry aggregates and then it is put inside the machine. This mixture enters inside a flexible hose with the air pressure and the water is added to the dry mixture in the nozzle. The operator of the nozzle adds water to the mixture using a tap and considering the importance of the amount of in the concrete, the skill of the operator is very important in this method.

2.2.6 *Wet spraying*

In this method, all the components and the water are mixed together and then enter the flexible hose under pressure and the air pressure in the nozzle increases the speed of the concrete and pushes it toward the intended surfaces. Basically, shotcrete is a kind of concrete and the rules of concrete are valid for shotcrete. Because of the special execution method, the way of execution and training of the crew are of more importance. On the whole, shotcrete is a strong and durable material and its adhesion to concrete, masonry, steel and stone is excellent.

2.2.7 *Cathodic protection*

Cathodic Protection of concrete structures was used for controlling the corrosion of the reinforcements in embedded concrete pipes. The 1970's was the starting period of cathodic protection and in 1980's, the cathodic protection systems appeared, which covered all the surface of the concrete. By progress of protection systems and examining the executed cases, it was observed that cathodic protection systems are very effective in corrosion control, as the American Concrete Institute (ACI) expressly announces in SPIO2 journal (1989) that: cathodic protection is the only method that stops the corrosion of the concrete structures (Construction & Housing Researches Center, 2003). In the cathodic protection method, the corrosion is completely stopped by lowering the potential of the metal from an active to an inactive corrosion potential.

2.2.8 *Protective layers*

In case that cathodic protection is not considered for any reason, we can optimistically use impermeable layers, intrusion of epoxy and saturation of concrete surfaces. In case of covering the area with impermeable layers or intrusion of epoxy in it, the cracks will be covered and the moisture and air will not penetrate. But this area is contaminated and some oxygen and moisture have remained there and the corrosion will continue until all the moisture and oxygen are consumed. This method is used for blocking the cracks of the concrete and filling the cavities spread all over the low-quality porous concrete. By saturating the surface of the concrete using special kinds of polymers we can decrease the corrosion and increase the life of the structure up to 15 years.

2.2.9 *Repairing the cracks*

Where rust is formed and causes clear and visible cracks in the concrete cover of the reinforcements or when the concrete cover is completed destroyed and collapsed, the following actions should be taken:

- The contaminated concrete should be removed up to the required depth
- The rust on the reinforcement should be removed
- In case that the rust of the reinforcement has decreased the cross section of reinforcement up to 20%, the reinforcement should be cut and new reinforcement should be placed (welded).
- Preparation of the surface of the old concrete should be carried out

Concreting for repairing should be carried out following the operations.

3 CONCLUSIONS, RECOMMENDATIONS & SUGGESTIONS

Considering the points mentioned in the repair section, Portland cement type I, is recommended adding silica fume for repairing the concrete. Using silica fume in concrete repair improves most of the properties of concrete and considerably decreases the permeability against water. It also creates good protection against the recurrence of corrosion. Studies show that the when the ratio of water-cement is low in the concrete, with a high cement content and additives which decrease the amount of water, the concrete will have an electrical resistively of more than 8500 ohm/cm which hinders the corrosion.

REFERENCES

Concrete Damages, Its Causes & Reasons, Construction & Housing Researches Center, 2003

Concrete Dams: R.S.Varshney. Oxford & IBH, New Delhi, 1998

Controlling The Stability of Concrete Dams, Dr. Nemat Hassani, Dr. Reza Rastı Ardakani, Eng. Farshad Farzinram, 2007

DD ENV 1504-9:1997 Products and Systems for the Protection and Repair of Concrete Structures - Definitions, Requirements, Quality Control and Evaluation – Part 9: General Principles for the Use of Products and Systems

Concrete Solutions – Grantham, Majorana & Salomoni (Eds)
© 2009 Taylor & Francis Group, London, ISBN 978-0-415-55082-6

Active repairs of concrete structures with materials based on expansive cement

A. Halicka
Lublin University of Technology, Poland

ABSTRACT: In the paper the repair of concrete structures using expansive concrete are presented. These repairs are active due to the initial stress-strain state generated by expansion in the structure. The features of expansive concrete and the author's tests of bond strength on the expansive concrete are described. Examples of the utilization of expansive concrete to repair concrete structures are presented, especially strengthening of columns and slabs and improving the water resistance of concrete tanks.

1 INTRODUCTION

1.1 Active repairs

Repairing and strengthening of concrete structures, depending on the applied technique or material, can be passive or active in relation to the structure.

Passive repairs consist of replenishing concrete lost from spalls, adding new section parts, or structural repairs (injection) by using a material which only fills the required space and cooperates with the original structure in bearing loads, but itself enters the cooperation only when external forces are applied.

The essence of active methods, on the other hand, is the application of a material or structural solution which will cause a strain distribution different from that obtained in the structure before the repair. This change should have a positive effect on the load bearing capability of the structure. Examples of active repair are: prestressing, change of the static scheme or application of active materials.

1.2 Repair material requirements

Repair of building structures demands repair materials complying with specific requirements: compatibility with repaired material, high adhesion to the substrate and durability.

In the case of concrete structures, the requirement of compatibility of repair material is met by materials based on mineral binders. While searching for materials satisfying the conditions of adhesion and activity, one can consider a special kind of mineral binders, i.e. expansive ones.

2 EXPANSIVE CONCRETE

2.1 Basics of expansive concrete

Concrete made with expansive cement binders belongs to the group of special concretes. Its special feature is an increase in volume during setting and curing. A distinction is made between shrinkage-compensating concretes, in which a small strain of free expansion (up to 0.1%) only compensate the shrinkage taking place, and expansive concretes, i.e. those with greater strains.

2.2 Expansive binders

The most frequent among expansive cement binders are those, in which reactions of aluminosulfate compounds are used, leading to the formation of ettringite. Since the increase in volume is a result of the growth of ettringite crystals, the character of the expansion can be described as structure-forming and hard phase building. It occurs first of all 'inwards', filling the available pore space, and only then outwards.

Apart from expansive cements, more and more often additions to Portland cement are used, increasing the volume of the grout, mixed with cement usually at the stage of mortar or concrete mix (Fu et al. 1995, Nagataki & Gomi 1998). It seems that for practical reasons these additions will find the broadest application.

The basic feature characterising binder expansion is the free linear strain $\varepsilon_{w,CE}$, subject to which is a cuboid specimen not confined during its curing by any external constrains. If the expansion takes place in confined freedom conditions, then smaller, so-called non-free, strains $\varepsilon_{n,CE}$ will occur, while in the specimen there is self-stress $\sigma_{c,CE}$ generated. The two parameters mentioned above are the measure of the power of chemical expansive reactions. They are examined in a standardised manner, described in e.g. Król & Tur 1997, on specimens made of expansive mortar, kept in water for 28 days. It is to be emphasised that the characteristics of expansive binders, including the rate of strain growth during the curing of the cement material, are the result not only of the type of cement

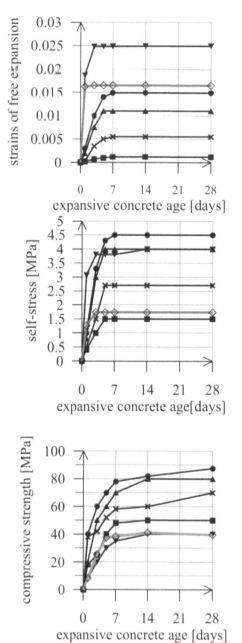

Figure 1. The characteristics of expansive cements produced in Poland (author's tests).

or expansive additive, but also heavily depend on the chemical composition of the components used in their production.

Figure 1 shows the characteristics of Polish expansive cements made as a ready product in Poland.

2.3 Features of expansive concrete

As with expansive binders, the parameters characterising concrete containing them are: the value of free strains and self-stress. The value of the achieved free expansion of concrete according to Neville (Neville 1997) depends not only on the type, power of expansion and amount of the applied cement, but also on the amount of aggregate used, its modulus of elasticity and grain size. The curing conditions are also of fundamental importance (Tazawa et al. 2000).

In turn, the value of the self-stress, occurring due to the confinement of the freedom of expansion by both internal and external constraints, is the result, on the one hand, of cement free expansion, and on the other of the rigidity of the structure and of the constraints limiting the expansion. These parameters (free expansion and self-stress of the concrete) are examined in a standardised fashion.

The expansion begins at the moment of supplying the concrete with a suitable amount of water, sufficient to commence chemical reactions. According to Król & Tur 2000 intensive moisturising, tantamount to initiating the expansion, should be started when the strength of concrete reaches $7,5 \div 9$ MPa. If this strength is not achieved, the structure of the concrete will not be rigid enough and the expansion 'inwards' will not take place; neither will any self-stress occur. If, on the other hand, the hydration products, causing the expandability of the structure, come into being after reaching too much instantaneous strength, then, under their influence, considerable self-stresses will occur, in an extreme case able to cause the destruction of the concrete's structure. That is why the ultimate value of both free and non-free expansion depends on the strength of the concrete at the moment of starting the expansion.

In structural elements made of expansive concrete, due to the confinement of the freedom of expansion by internal constraints (structure, reinforcement) as well as external ones (abutments, supports, adjacent structural elements), there occur internal stress (so-called self-stress). That is why, apart from the idea of 'expansive concrete,' there are such ideas as 'self-stressed concrete' and 'chemical prestressed concrete'.

Expansion 'inwards' causes an increase in the concrete's density and tightness, and following that an increase in the resistance to damp, water, gasses, corrosion and frost. External confinement, as well as reinforcement, may generate in the concrete itself and the adjacent structural elements an advantageous state of initial stress. Rational application of expansive concrete consists of consciously taking advantage of those properties.

3 ADHESION OF EXPANSIVE CONCRETE

3.1 *Singularity of the interface between shrinkable and expansive concrete*

It was mentioned above that the increase in volume of materials based on mineral expansive binders is a result of the growth of ettringite crystals. In the case of repair, the ettringite crystals grow also in the interfacial zone, filling roughness of 'old' concrete surfaces and penetrating open pores. This has been confirmed by investigation of the microstructure of the interfacial zone between a concrete based on Portland cement and an expansive one, carried out by means of a scanning electron microscope (Halicka 2007, Li et al. 2001). Owing to the effect described above, the effectiveness of both kinds of bond mechanisms: mechanical interlocking and chemical adhesion, grow. Additionally, the expansion suppresses the possibility of shrinkage cracking in the interface.

In the case of replacing spalled concrete or adding new section parts of expansive materials, another additional agent appears. Self-stress generated due to confinement of expansive freedom by the adjacent part of the structure exerts pressure on the interface, considerably increasing the bond strength of the joint.

3.2 *Tests of bond strength of joint between two different concretes*

In order to evaluate the influence of mineral expansive binders on the bond of repair materials, a number of tests precisely described in Halicka 2007a, Halicka 2007b and Halicka 2008 were carried out by the author. There were tests of bond strength of the joint between Portland cement based concrete and the expansive one working under different stress-strain states (Figure 2):

- axial tension,
- tension realised by splitting,
- shear in the joint shaped as a cylindrical surface,
- combined stress state of shear and compression in a 'slant-shear test',
- combined stress state of shear and tension in the joint shaped as a truncated cone,
- combined stress state of shear and tension in a modified 'bi-surface shear' specimen.

All specimens were made in two stages. First the 'old' concrete was placed into a mould and the remaining space was filled by foamed polystyrene. After 28 days of curing, the surface of the concrete was moistened and roughened by wire-brushing and the specimen was complemented by 'new' concrete. Tests were executed after the next 28 days in a hydraulic testing machine (some of them are shown in figures 3–4).

In Table 1 the characteristics of 'old' and 'new' concrete used in the tests are compiled. There are the values of compressive f_{cm} and tensile strength f_{ctm} determined on the day of the main tests. The characteristics of the expansive concrete are complemented by the information on the value of free expansion and whether the freedom of strain was confined during its curing.

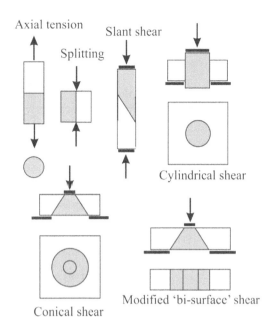

Figure 2. Specimens used to quantify the bond strength of joints of two concretes.

Figure 3. Modified 'bi-surface shear' specimen during test.

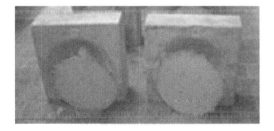

Figure 4. 'Conical shear' specimen during test.

3.3 *Discussion of results*

Table 2 presents the results of the tests, as mean values of three specimens. The values of the failure forces and the calculated values of the failure stress are given. As the main parameter enabling the appreciation of bond strength and comparison of results the coefficient

111

Table 1. Characteristics of concrete used in bond tests.

Test	Type of 'new concrete'	Free expansion strains	Strains during curing of 'new' concrete	'Old' concrete strength compressive f'_{cm} MPa	'Old' concrete strength tensile f'_{ctm} MPa	'New' concrete strength compressive f''_{cm} MPa	'New' concrete strength tensile f''_{ctm} MPa
Axial tension	Expansive	0.0019 (shrinkage compensating)	Confined	18.6	1.96	39.6	7.48
Splitting	Shrinkable		Free	28.81	2.22	29.03	2.20
	Expansive		Free	28.81	2.22	29.80	2.32
	Expansive	0.005 (low level of expansion)	Confined	28.81	2.22	44.51	3.50
Cylindrical shear	Shrinkable		Free	41.83	3.10	38.49	2.91
	Expansive	0.016 (high level of expansion)	Confined	41.83	3.10	50.33	3.62
'Slant shear' test	Shrinkable		Free	27.07	2.11	27.64	2.30
	Shrinkable		Confined	26.83	2.56	27.40	2.15
	Expansive	0.0024 (shrinkage	Free	27.07	2.11	40.20	3.01
	Expansive	0.0024 compensating concrete)	Confined	26.83	2.56	41.36	3.59
'Conical shear' test	Shrinkable		Free	38.40	2.92	34.50	2.66
	Expansive	0.0014 (shrinkage compensating)	Confined	38.40	2.92	48.75	3.50
Modified 'bi-surface shear' test	Shrinkable		Free	39.03	2.76	36.84	2.47
	Expansive	0.005 (low level of expansion	Confined	25.97	1.27	36.33	2.45

Table 2. Results of bond tests.

Test	Type of 'new concrete'	Free expansion strains of 'new' concrete	Strains during curing	Failure force F KN	stress σ MPa	Coefficient of joint effectiveness
Axial tension	Expansive	0.0019 (shrinkage compensating)	Confined	10.13	$\sigma_{jt} = 1.8$	$\sigma_{jt}/f_{ctm,min} = 0.95$
Splitting	Shrinkable		Free	38.5	$\sigma_{jt} = 1.09$	$\sigma_{jt}/f_{ctm,min} = 0.49$
	Expansive		Free	59.0	$\sigma_{jt} = 1.65$	$\sigma_{jt}/f_{ctm,min} = 0.74$
	Expansive	0.005 (low level of expansion)	Confined	68.67	$\sigma_{jt} = 1.92$	$\sigma_{jt}/f_{ctm,min} = 0.86$
Cylindrical shear	Shrinkable		Free	100.67	$\sigma_{jt} = 1.62$	$\sigma_{jt}/f_{ctm,min} = 0.30$
	Expansive	0.016 (high level of expansion)	Confined	430.48	$\sigma_{jt} = 7.08$	$\sigma_{jt}/f_{ctm,min} = 1.24$
'Slant shear' test	Shrinkable		Free	188.3	$\sigma = 18.8$	$\sigma/f_{cm,min} = 0.70$
	Shrinkable		Confined	220.0	$\sigma = 22.0$	$\sigma/f_{cm,min} = 0.82$
	Expansive	0.0024 (shrinkage	Free	222.5	$\sigma = 22.2$	$\sigma/f_{cm,min} = 0.82$
	Expansive	0.0024 compensating)	Confined	235.0	$\sigma = 23.5$	$\sigma/f_{cm,min} = 0.87$
'Conical shear' test	Shrinkable		Free	51.0	$\sigma_{jt} = 0.52$ $\tau_j = 0.78$	$\sigma_{jt}/f_{ctm,min} = 0.20$
	Expansive	0.0014 (shrinkage compensating)	Confined	65.67	$\sigma_{jt} = 0.67$ $\tau_j = 1.0$	$\sigma_{jt}/f_{ctm,min} = 0.23$
Modified 'bi-surface shear' test	Shrinkable		Free	12.5	$\sigma_{jt} = 0.27$ $\tau_j = 0.47$	$\sigma_{jt}/f_{ctm,min} = 0.11$
	Expansive	0.005 (low level of expansion)	Confined	13.0	$\sigma_{jt} = 0.28$ $\tau_j = 0.48$	$\sigma_{jt}/f_{ctm,min} = 0.22$

of joint effectiveness was adopted. It was defined as follows:

$$\alpha_j = \frac{f_j}{f_{min}} \tag{1}$$

where f_j = value of failure stress of the composite specimen (bond strength of the joint), f_{min} = strength of the weaker of the joined concretes, tested in the same stress-strain state as the composite specimen.

Thus for each type of specimen the coefficient fixes the ratio of different stresses (normal or shear).

The effectiveness of split connection between a shrinkable concrete and an expansive one, defined as the ratio of the failure stress of the composite specimen and the tensile strength of the weaker concrete, was over 50% higher in the case of concretes unconfined during curing than the joint between the two shrinkable concretes. When the freedom of expansion was confined, this effectiveness was higher: about 75%.

In a 'slant-shear' test, besides the increase of the ratio of failure stress to the compressive strength of the weaker concrete, a change of the mode of failure was observed. The specimens made of two shrinkable concretes failed at the interface, whereas most of the specimens with one half made of expansive concrete failed as monolithic specimens.

In the specimens with the interface shaped as a cylinder the effect of shrinkage or expansion was revealed. It was apparent that the strength of such a joint, if the cylinder tested was made of expansive concrete, was four times higher than in the case of a shrinkable concrete cylinder. It should be noticed that in these investigations an cement of high expansion was used (free expansion 1.65%, self-stress 4.0 MPa).

In the specimens with the interface working in a combined stress-strain state of shear and tension, the tensile stress under test depends on the slope of joint surface. The increase of tensile effectiveness of the joint between expansive and shrinkable concrete in relation to the joint between the two shrinkable concretes was also observed in such specimens. An increase of 15% was observed in the conical shear specimens with the joint inclined to the specimen axis at the angle 45°. The strength of the joint with expansive concrete, was almost two times higher than in the case of a shrinkable concrete in the modified 'bi-surface shear' specimens, where the slope was 60^0.

The obtained effectiveness values were compared by calculating the ratio of the effectiveness coefficient of the joint between expansive and shrinkable concretes to the corresponding coefficient of the joint between two shrinkable concretes:

$$\beta = \frac{\alpha_{j,exp}}{\alpha_{j,shr}} \qquad (2)$$

where $\alpha_{j,exp}$ = the coefficient of effectiveness of joint between expansive and shrinkable concretes, $\alpha_{j,shr}$ = coefficient of effectiveness of joint between two shrinkable concretes.

The ratios calculated in this way are presented in Table 3. It should be emphasised that joint effectiveness depends on the expansion power of the cement. Expansive cement, characterised by a bigger free expansion, caused a higher bond strength than cement of lower free expansion.

3.4 Conclusion of experiments

On the basis of the above results the following rule can be recommended: the bond strength of a non-reinforced joint between expansive and shrinkable

Table 3. Compilation of effectiveness of joints.

Test	Expansion	Inclination of joint to specimen axis	Ratio β^*
Splitting	low		1.76
Cylindrical Shear	high		4.13
Slant-shear	shrinkage compensating	60^0	1.24
Conical shear	shrinkage compensating	45^0	1.15
Modified 'bi-surface shear'	low	60^0	2.0

* the ratio of the effectiveness coefficient of the joint between expansive and shrinkable concrete to the corresponding coefficient of the joint between two shrinkable concretes, according to equation (2).

concrete can be safely calculated as the value for shrinkable concretes according to e.g. code EC2 (product of coefficient characterising the roughness of surface and tensile strength of weaker concrete) multiplied by 1.1 in the case of shrinkage-compensating concrete and by 1.5 for expansive concrete.

4 UTILIZATION OF EXPANSIVE CONCRETE TO REPAIR CONCRETE STRUCTURES

4.1 Strengthening of concrete column by jacketing

In strengthening and repair of reinforced concrete columns, which have suffered from destruction or degradation or which demand adaptation for increased service loads, the classical method of jacketing is used.

As a result of jacketing a structural member of composite cross-section is obtained. It is made up of two types of concrete of different mechanical and rheological properties and of steel rods (longitudinal and spiral reinforcement) as shown in Figure 5. The essence of such a solution is an increase in the load capacity of the column through: an increase in the cross-section area, a decrease in its slenderness and an increase in the concrete strength of the primary column in which, due to spiral reinforcement, a state of triaxial compression can be found. If the jacket is made of concrete based on Portland cement (shrinkable concrete) a passive repair is obtained.

In the paper by this author (Halicka 2005) the detailed calculations of distribution of internal forces in the strengthened columns were presented. Two types of concrete used to make the jacket were considered: shrinkable concrete and expansive concrete.

It was proved that in the case of a shrinkable concrete jacket, the original column (core) carries the greater load than force calculated assuming the distribution of forces according to stiffness of jack and especially by shrinkage of the jacket. In the case of

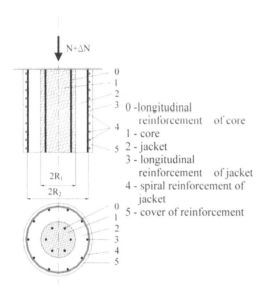

Figure 5. Model of column strengthened with a concrete jacket.

0 - longitudinal reinforcement of core
1 - core
2 - jacket
3 - longitudinal reinforcement of jacket
4 - spiral reinforcement of jacket
5 - cover of reinforcement

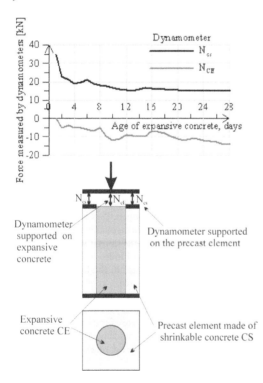

Figure 6. Redistribution of forces in composite column with expansive cement (tested at Lublin University of Technology).

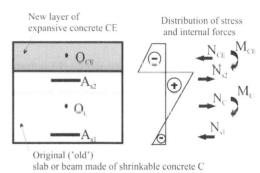

Figure 7. Initial stress-strain state in the cross-section of slab or beam strengthened by a layer of expansive concrete.

the case of a shrinkable concrete jacket, the original column (core) carries the greater load than force calculated assuming the distribution of forces according to stiffness of jacket and core. This is caused by rheological phenomena and especially by shrinkage of the jacket. In the case of a jacket made of expansive concrete the original column is 'relieved' in comparison to the distribution of forces according to stiffness. The disadvantage of this solution is the necessity of utilization of strong spiral reinforcement ensuring the adhesion of the expansive jacket to the original column.

So it is rational to create a jacket of shrinkable concrete (this will ensure triaxial compression of the original concrete) while leaving under the floor a gap subsequently filled with expansive concrete, allowing an effective redistribution of the longitudinal force.

The redistribution of internal forces in a composite column with expansive concrete component was proved by laboratory investigations (Król & Tur 1998). The cross section of the precast column made of ordinary concrete was rectangular 150×150 mm, with circular hole in the middle. This hole was filled with expansive concrete. The forces exerted by two parts of the column onto the support were measured by dynamometers during the time of curing of the expansive concrete. The expansive concrete progressively took the load.

4.2 Active strengthening of concrete slabs

Strengthening of reinforced concrete slabs or beams are realized by putting a layer of concrete on the existing element. This results in enlarging the cross-section height, but the shrinkage of the new concrete causes the tensile stress of the lower part of beam. If the expansive 'new' concrete is used an advantageous distribution of stress along the height of the cross-section is obtained. In the lower part of the beam compression is generated, so one can say that the element is prestressed (Figure 7).

4.3 Improvement of the water resistance of concrete tanks

The water resistance of precast water tanks may be improved by filling the gaps between precast elements

a jacket made of expansive concrete the original column is 'relieved' in comparison to the distribution of forces according to stiffness. The disadvantage of this solution is the necessity of utilization of strong spiral reinforcement ensuring the It was proved that in

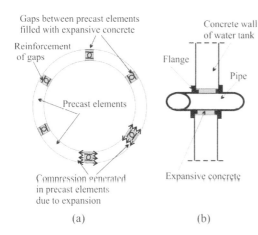

Figure 8. Improvement of water resistance of the concrete water tanks: (a) filling the gaps between precast element by expansive concrete, (b) Filling the hole with expansive concrete enabling the pipe to cross the wall.

with expansive concrete. The high adhesion of the expansive concrete ensures the absence of shrinkage cracks in joints and the stress generated due to expansion causes prestressing of precast elements (Figure 8a).

The water resistance of the junction of the tank wall and the pipe is improved by filling the hole with expansive concrete enabling the pipe to cross the wall with no leakage (Figure 8b). This improvement is possible due to the high adhesion and lack of shrinkage in the expansive concrete.

5 FINAL CONCLUSIONS

On the basis of above considerations the following general conclusions can be formulated.

1. The expansive concrete may be successfully used for repair and strengthening of concrete structures due to its activity and high adhesion to 'old' concrete.
2. The author's tests proved that the adhesion of expansive concrete is better than that of shrinkable concrete.
3. In analysis of bearing capacity of repaired or strengthened concrete structures the bond strength of a non-reinforced joint between expansive and shrinkable concrete can be safely calculated as the value for shrinkable concretes multiplied by 1.1 in the case of shrinkage-compensating concrete and by 1.5 for expansive concrete.
4. Expansive concrete used as a repair material causes an advantageous initial stress state in the structure. So it is reasonable to use it in strengthening of concrete column and slabs or beams.
5. The positive redistribution of internal forces in columns strengthened by expansive concrete was proved by tests.

REFERENCES

EC2. Design of concrete structures. Part 1-1: General Rules and Rules for Buildings.

Fu, Y. Ding, J. & Beaudoiu, J.J. 1995. Expansion characteristics of a compounded-expansive additive and pre-hydrated high alumina cement based expansive additive, *Cement and Concrete* Research Vol.25 No.6: 1295–1304.

Halicka, A. 2005.Redistribution of internal forces in reinforced concrete columns strengthened by jacketing. *Archives of Civil Engineering*, LI.1.:.43–63

Halicka, A. 2007. Adhesion of mineral repair materials based on expansive binders. In L. Czarnecki & A. Garbacz (ed.), Adhesion in interfaces of Building Materials – a Multiscale approach. Aedificatio Verlag: 175–183.

Halicka, A. 2008 Properties of Interface in Concrete Composite Structures. In R. K. Dhir, M. D. Newlands, M.R. Jones & J. E. Halliday (ed.), *Precast concrete: towards leaning structures; Proc. Intern. Congress: Concrete: Construction's Sustainable Option, Dundee 7–10 July 2007:* 147–158

Król, M. Tur, W., 1997, Expansive concrete (in Polish), Arkady, Warsaw.

Król, M., Tur, W. 1998. Redistribution of forces in columns strengthened with expansive concrete [in Polish]. *Concrete and Prefabrication; Proc. of XVI Scientific and Technical Conference, Serock-Jadwisin 20–23 April 1998*

Li, G. Xie, H. Xiong, G. Transition zone studies of new-to old concrete with different binders. 2001 Cement and Concrete Composites 23: 381–387.

Nagataki, S. & Gomi, H., 1998, Expansive admixtures (mainly ettringite), Cement and Concrete Composites, 20, 163 170.

Neville, A.M. Properties of Concrete, 1997, Longman.

Tazawa, E. Kawai K. & Miyaguchi, K. 2000, Expansive concrete cured in pressured water at high temperature, Cement and Concrete Composites, 22, 121–126.

Concrete Solutions – Grantham, Majorana & Salomoni (Eds)
© 2009 Taylor & Francis Group, London, ISBN 978-0-415-55082-6

Protection of a WW2 concrete defence wall at St. Ouens Bay, Jersey, Channel Islands

Steve Hold
Arup, Cardiff, UK

ABSTRACT: St. Ouen's Bay on the Island of Jersey in the English Channel is situated at the west of the Island. There is no land mass between St. Ouens Bay and the United States of America and this part of the island's coast line is therefore susceptible to the full force of the Atlantic waves. This paper describes the approach taken to the repair and renovation of this important sea wall, considering the economic and engineering considerations of the task.

1 INTRODUCTION

1.1 *Location and description*

St. Ouen's Bay on the Island of Jersey in the English Channel is situated at the west of the Island. There is no land mass between St. Ouens Bay and the United States of America and this part of the island's coast line, is therefore susceptible to the full force of the Atlantic waves. A sea defence wall had been in place at St. Ouen's Bay since the 19th Century but during the Second World War, the German Occupation Forces constructed two additional lengths of defence wall. These were primarily as anti tank defences, rather than as protection of the coastline from erosion by the sea.

1.2 *Historical context*

The WWII concrete wall was built as a defensive structure and formed from mass concrete with a layer of reinforcement placed towards the front face of the concrete section to provide blast resistance and not as structural integrity reinforcement. The wall varies in height but generally is 7 m. It is approximately 2 m wide at the base tapering to 0.5 m width at the top below a recurve capping beam. The older sections of the wall are faced in granite and were constructed in the 1880s. The German Garrison's defenders required two further sections to be constructed; one from the northern bunker to the Cutty Sark and in the southern section of the Bay from El Tico to the Le Bray slip way. The section of concrete wall extend to a total of approximately 4 km.

1.3 *Heritage of the wall*

St. Ouens Bay is a well known beach and therefore an important tourism asset for the Island of Jersey. Furthermore, the land that is protected behind the sea defence wall provides amenity and other vital land based assets for the island. In addition to this

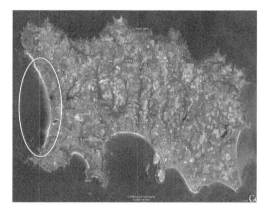

Figure 1. St Ouens Bay.

the heritage aspects of the World War II concrete defences i.e. the bunkers, etc. require conservation and protection in terms of their historical value.

1.4 *Description of the problem*

Both concrete sections of the St. Ouens Bay sea wall suffer from similar problems related to the extreme environment in which they sit:-

- There are large wave energy impacts.
- Abrasion problems from sand, gravel and rocks suspended in, and thrown by the waves.
- The threat of overtopping during high tides and high seas.
- The threat of scour from beneath the wall as the wall was constructed directly from the beach sand level with no additional foundation.
- It is also recognised that the very nature of the beach materials available and construction methods at the time produced a variable mix in terms of resistance to surface abrasion, i.e. there are some areas of "softer" concrete than others.

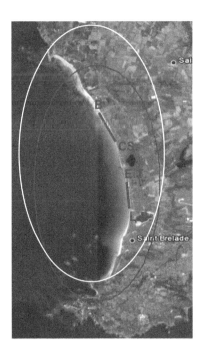

Figure 2.   Two long sections of concrete wall.

Figure 3.   Original granite faced sea wall.

2   FEASIBILITY STUDY OF REPAIR AND PROTECTIVE OPTIONS

The States of Jersey Engineers Department commissioned a feasibility study. During this feasibility study which was completed early in 2003, it was necessary to study maintenance records, archive materials and other historical data about the construction and maintenance of the wall.

The large potential scope of repair and protection works also needed careful consideration as several kilometres of repair or remedial protection work were required to maintain both the concrete wall itself and its sea defence capabilities.

The Jersey Engineers Department had previously carried out an Island wide review of all sea defences where the coast was vulnerable, they determined that St. Ouens Bay and its defences were the top priority in terms of obtaining funding for protection and maintenance work at this location.

Figure 4.   Severe abrasion damage.

The feasibility study was essentially to determine how to extend the life of the wall for 25 years with reduced maintenance, improved stability if possible, improved protection to the rear of the wall, reduce negative effects on the beach, maintain or improve the amenity value of the wall and to improve the appearance of the wall. The study also considered flexibility of design for the repairs and protection.

Five generic options were considered in more detail:-

- Cladding the wall in granite to blend in with the Victorian section of the wall.
- The application of sprayed concrete overlay to the existing wall surface.
- Placing pre-cast or cast in situ concrete protection panels over the existing wall.
- Constructing a stepped terrace protection in front of the wall.
- Construct new geometry and wall as part of a partial retreat.

The last option had been put forward as part of the Island wide coastal study and was thought viable by the consultants engaged to carry out the study, but not thought politically acceptable or viable by the States of Jersey Engineers Department.

The five generic options were costed for the two lengths of concrete wall and the only two options viable in terms of the budget available were either in situ

Figure 6. Delaminating concrete cover.

Figure 5. Wave impact and reflected waves.

overlay concrete panels or sprayed concrete. In the mid 1990s the Jersey Engineer's Department had trialled several options of sprayed concrete, but the samples had not been able to prove the integrity of this method over the longer term. Therefore the policy decision was taken by the Jersey Engineer's Department to place a concrete "armour" protection panel on the vulnerable section of the lower part of the wall.

3 ENGINEERING ISSUES TO RESOLVE

The structural stability of a large section of the wall was vulnerable due to both undermining and overtopping as a result of the high tidal range and wave impact. The high tidal range and reach of waves in St. Ouen's Bay are likely to increase with global warming sea level rises and present a significant long term maintenance problem.

The single most damaging aggressive action in the Bay acting upon the wall is that wave action also carries shingle, boulders and suspended sand. Therefore the abrasion resistance of any overlay concrete protection must be a high priority together with the fixity of the protection panel to the substrate.

The reinforcement placed within the front section of the concrete wall was not intended in the design to act as structural reinforcement. Its function was to improve blast resistance from ordnance impact, but over a 60 year period the reinforcement had corroded significantly with the resultant rust expansion spalling large sections of concrete cover. The porosity

of the original concrete was such that salt laden water became in contact with the reinforcement accelerating the corrosion and delamination processes.

Environmental issues dictated that it would not be practical or sound policy to remove the gravel beds, boulders and other loose elements on the beach in order to reduce the abrasion. Studies of the beach sand in the Bay had revealed consistently large migrations during storm events. A peat layer just beneath the surface of the sand is soft and vulnerable to erosion if revealed by a major beach sand migration which could occur if the larger loose covering materials were removed from the beach.

4 INTRUSIVE INVESTIGATIONS

As part of the feasibility study in 2003, trial pits were excavated at the rear of the wall to determine the cross section geometry of the walls. The pits confirmed that the sections of concrete wall were constructed directly on beach sand and that there was also a retention of groundwater behind the wall. This was evidenced by several drainage outlets and weepholes throughout the length of both sections of the concrete wall. (One unusual aspect of the trial pitting was that care had to be taken whilst excavating such a large and deep trial pit due to the possibility of unexploded ordnance behind the wall!)

Three cores were taken to determine the concrete cross-section of the wall; at the top, middle and bottom, so that the concrete materials and the integrity of the concrete could be assessed through the three cross-sections of the wall.

4.1 The BRE analysis of the concrete

A petrographic analysis was carried out on the concrete cores by the Building Research Establishment and the conclusions of their examination was that the cross-section of the wall did suffer from localised poor compaction (only to be expected). It was formed from dense sand material containing large rounded boulders. The Portland cement matrix had largely hydrated. Some of the aggregate problems experienced elsewhere in Jersey such as ASR were not present to a significant extent in the concrete examined.

119

Figure 7. Trial pit behind the sea wall.

Figure 8. Sprayed concrete trial panels – note unsatisfactory crazing.

Both concrete sections of the sea wall were seen to be of a similar cross-section geometry and experiencing similar problems. The beach levels in St. Ouens Bay drop in the middle of the Bay where the Victorian section of sea defences dominate and therefore beach levels increase, moving outwards from the centre of the Bay, reducing the levels of the wall and exposure, i.e. the environment is most aggressive at the centre of the Bay.

The intrusive investigations produced a series of defined requirements for a concrete protection overlay to the zones of the concrete wall subject to the worst abrasion. With the unsatisfactory long term performance of the simple spray concrete solutions a more sophisticated designed concrete overlay panel was required.

5 DESIGN OF CONCRETE OVERLAY PANELS

A three stage design process was based on the known requirements:-

- An initial scheme design cross-section was produced.

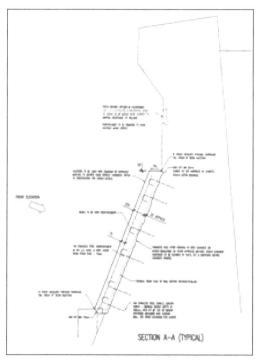

Figure 9. Designed cross section.

- A trial panel was designed and cast on the wall.
- The trial panel ideas were then adjusted and refined and further panels constructed.

The States of Jersey Engineers Department has a direct labour unit experienced in carrying out sea defence works and this team was also to carry out the construction of the concrete panel sea wall overlay protection. They were already involved in installing sheet pile protection to the base of the wall to reduce the risk of scour. The direct labour concept overcame potential contractual issues related to the difficult access, time penalties in contracts and other related issues, when working on the beach at St Ouen's Bay.

However, a specialist contractor was engaged to carry out the smaller, higher level, concrete patch repairs to the recurve that were also necessary as part of the overall repairs to the concrete sections of the wall. This mini contract was undertaken as a separate exercise to the panel overlay work. The concrete repair work involved both localised cutting out and patch repairs, but wherever possible, the contractor utilised the scaffolding and access already in place as part of the main overlay repair works by the direct labour unit.

5.1 *Design of panel*

- Maximum impact and/or abrasion resistance of the new concrete: this was to be achieved using a low water cement ratio and the use of Microsilica to give good matrix properties and enhanced cement/ aggregate band. Together with a crushed rock hard aggregate to maximise the cement aggregate band and give good inherent abrasion resistance.

- The substrate of the original wall needed to be sound therefore it was necessary to remove corroding reinforcement and delaminating, or soft concrete surface areas. (Possibly to a depth of 200 mm in some locations).
- The surface of the overlay panel was to have a minimum of blow holes and minimise porosity at the concrete surface so as to eliminate 'weak' points for the initiation of abrasion loss. This was achieved by placing a controlled permeability forum fabric on the surface of the shutters to induce water extraction from the concrete surface and avoid 'blowholes'. The avoidance of blowholes being particularly important with the reverse slope of the wall because very large instantaneous wave impact forces can also be generated in such cavities
- To ensure resistance to the pull off and impact forces (in the region of 500 kN/m^2) it would be required to fix a new mesh and to tie this mesh back by use of dowels into the original wall. Cuts were made at both the head and base of the new overlay panels to form a square edge with depth to ensure a firm key at these edges. Large concentrated forces can be generated at these locations.

Figure 10. Cracking of trial panel and core taken through crack.

6 CONCRETE TRIAL PANELS

Arup produced a performance specification in October 2003 and the local concrete manufacturer Granite Products Ltd was selected to develop the mix and carry out trial mixes at their quarry. The distance from the quarry to the St Ouen's Bay site would mean a maximum travel time to the site of approximately 1 hr with a marginal allowance for traffic and access problems during the busy summer months.

6.1 Construction of the trial panel on the beach

The thickness of the overlay panel was to be approximately 250 mm and the size of the panel would be determined by being able to deliver 1 truck mix of 7 m^3. Although several combinations of mix and aggregates were evolved the general components of the F1S mix were;

- Coarse aggregate; 10 and 20 mm aggregate,
- Fine aggregate; Simon Dune Sand and granite sand.
- Cementitious content; CEM1, PFA and Microsilica
- Water, Super plasticiser; SP8 to achieve high workability.

The first panel cast on the St Ouens wall was a full scale trial. On striking the formwork, it was found to exhibit an array of fine recti-linear cracks that 'split' the panel into approximately 1 m × 1 m squares. It was initially presumed that the cracking was caused by restrained early age thermal contraction strains – being accentuated by the high cement content, the early age of form striking for quick turnaround and the high restraint

from the good interlock with the existing wall. Information from this trial was used in further development work and trials off-site.

The further development work included thermocouple measurements of in situ temperature and cross section temperature profile, prolonged retention of forms to mitigate thermal strains and give enhanced moisture curing, application of a post striking curing membrane, modification of reinforcement quantity and modification of the mix design to reduce cement content and hence heat of hydration (the target C60 strength requirement being easily met).

Although some benefit was seen, the fine cracking persisted. Because of the time of the cracking, conventional drying shrinkage was discounted as a cause. The reduction in temperature build-up was not entirely convincing as the primary cause although the high restraint appeared to be a strong contributing factor.

However, further consideration of the potential causes of contraction strains in a high quality, very low water cement ratio matrix, concrete, highlighted the possible contribution of autogenous shrinkage. This autogenous shrinkage, unlike conventional drying shrinkage, can cause volume loss at early age and therefore fitted the observed phenomenon. Reduction in conventional drying shrinkage of concrete can be achieved through mix water reduction using water-reducing admixtures (plasticisers).

More recently a new family of shrinkage-reducing admixtures (SRA) have become available. These provide beneficial reduction of shrinkage by acting on the pore pressures in the cement gel as water is lost.

Figure 11. Increased reinforcement and bond break between wall and new panel.

It was thought that such a mechanism might also be helpful in reducing autogenous shrinkage. The adoption of this SRA approach substantially avoided the previously encountered early age fine cracking.

7 CONSTRUCTION PHASE

As the States of Jersey direct labour unit was involved, the sensitive nature of both access and the site works could be handled by the local experienced workforce. In addition, public notices and briefings of what was to happen assisted the Public's understanding for the need for the repair and protection work on this sensitive site.

An important example was that there was a need to cut out large areas of soft concrete from the old wall as well as delaminating concrete, along with the removal of the corroding reinforcement. This was judged to be best carried out by using a hydraulic lance, which by its very nature required cordoning off large sections of the tourist beach area above and below the section being worked on.

The overlay panels were given individual numbers so that they could be monitored for both heat of hydration deterioration and any cracking, should it occur. Minor cracking continued in several of the panels but was not viewed as a performance failure as the

Figure 12. The protected wall in 2008 with shingle and boulders beneath.

cracks were very fine and the surface erosion and the structural integrity of the panel was not threatened. The decision was also taken to leave the trial panel (which was subject to initial cracking but performance wise was not seen to deteriorate further) permanently in place.

The site works were completed in 2004 and since that period there have been a significant number of storms and tests for the protection overlay panels. A survey carried out in August 2008 showed that all panels were performing satisfactorily and no deterioration of the panels had taken place.

8 SUMMARY

As with any sea defence, the St. Ouens Wall is vital to the island of Jersey. This particular wall was recognised as the most vulnerable and given the highest priority for repair and protection. Furthermore, its historical, environmental and financial importance needed to provide protection or "armour" to the WW2 wall beneath and was given priority by the States of Jersey Engineer's Department. The hardness of the armoured protection provides defence against future abrasion and impact resistance together with a minor enhancement to the structural integrity of the wall. Overtopping will continue and increase but the toe sheet piling carried out by The States Engineers Department in combination with the overlay panels is likely to achieve the 25 years further life in use for the wall.

Concrete Solutions – Grantham, Majorana & Salomoni (Eds)
© 2009 Taylor & Francis Group, London, ISBN 978-0-415-55082-6

Restoration of St. Catherine's breakwater roundhead, Jersey, Channel Islands

Steve Hold

Arup, Cardiff, UK

ABSTRACT: This paper describes problems encountered with void formation and washout beneath and behind protective stone blocks and a concrete slab at St. Catherine's breakwater, Jersey, in the Channel Islands. It discusses the unique problems encountered when designing and funding the repair of a heritage structure. Details of the testing, the considerations of the design and the installation of the remedial works are given.

1 INTRODUCTION AND BACKGROUND

1.1 Location and description

St. Catherine's Breakwater is situated at the eastern end of the Island of Jersey in the English Channel. The breakwater was conceived in the 1850's as part of a harbour to form a series of Napoleonic War harbours to blockade the French Fleet. The southern arm of the harbour from Archirondell was never completed and therefore the completion of the northern arm of the harbour at St. Catherine's some 600 m long has been viewed as an expensive engineering "folly" of that era.

1.2 History of the repair project

The breakwater, and in particular the roundhead, has been the responsibility of Jersey Harbours for many years but day to day maintenance and isolated repairs were becoming more frequent, punctuated by emergency repair works to the face of the roundhead at the extreme end of the breakwater. Lost facing stones were filled with cement bags and voids that appeared were also historically filled with concrete.

In December 2005 after a severe storm, a major loss of facing stones was discovered towards the lower section of the roundhead and required emergency cement bagging of the breach and filling of the void behind it.

1.3 Description of the roundhead

The roundhead, as its name implies, offers a rounded face to the end of the breakwater so that wave impact is only imparted upon a small area at any one time and not striking with maximum force at right angles over a larger area which would be the case on a straight wall. There is a mass gravity masonry wall with secondary core infill typical of the era which relies for its structural integrity upon the combined mass of both facing stones and core fill behind.

Figure 1.   St Catherine's Breakwater.

1.4 The structural problems at the roundhead

For many years prior to the recent repair and restoration work there had been differential slab settlement observed at deck level at the roundhead (the remainder of the breakwater has no concrete deck). Large sections of slab had settled in the region of 15–20 mm and required anti-trip painting marks at the edges for the safety of tourist and walkers. In addition to the evidence of movement at deck level, distress was also seen to be accelerating on the roundhead vertical walls with facing stones being pulled out more frequently. The remainder of the 600 m length of the breakwater remained undamaged.

1.5 Site constraints and conditions

The roundhead is situated some 600 m out to sea at the end of the breakwater in an approximate 15 m tidal

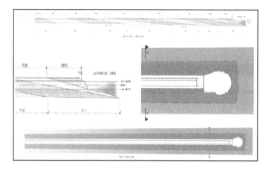

Figure 2. The Roundhead construction from archive drawings.

Figure 3. More frequent damage occurring.

range and an area of strong current. The concrete decks were provided at the roundhead to combat the problem of wave overtopping during storms at this exposed location. The combination of wind, tide and current during storms produces significant overtopping and wave impact. This is recognised to be in the region of $500 \, kN/m^2$ (or more in isolated cases) and therefore the rounded shape at the end of the breakwater is particularly important for the long term life and protection of the structure as a whole.

1.6 *A heritage monument*

Although never able to function as a harbour, the breakwater is a valuable tourist asset to Jersey, providing a magnificent walk and also affords shelter and landing areas for pleasure craft. It is also a favourite location for fishing. Directly to the west of the breakwater are very sensitive scallop farming beds and this area too is afforded some protection by the presence of the breakwater.

1.7 *A custodial asset only*

As with all too many heritage structures, they provide no direct operational use and to Jersey Harbours, who maintain the structure, the breakwater is a financial liability. In the case of St. Catherine's Breakwater, Jersey Harbours' budget is limited to annual minimal or small scale maintenance on a long term basis. No revenue is produced by the structure and therefore in terms of the overall assets and liabilities of Jersey Harbours, the funding available to carry out significant maintenance and/or emergency repairs on this large structure is not readily available.

1.8 *The storm of December 2005*

A combination of a particularly strong storm, large waves and tide in December 2005 resulted in a significant loss of facing stones at the roundhead. A concrete pump and ready-mix wagons were sent to the end of the roundhead to fill the void that was exposed and to fill the breach in the missing roundhead stones with cement bags. This work was carried out expeditiously, in very difficult weather conditions in order to reduce the amount of further damage to the roundhead.

2 ASSESSMENT OF LIKELY CAUSES OF DETERIORATION

2.1 *Data gathering*

As part of a "health check survey" in November 2005 (just prior to the severe storm) searches were made to obtain information and data relative to the original construction and maintenance history of the breakwater. The information provided a desk study which helped in understanding the construction principles and method used in the construction. A rock bund was formed by transporting rocks by suspended trestles. Outer dressed stone walls were then constructed on the bund with an infill core of secondary material behind. The works took approximately 5 years to complete and the diary of the engineer responsible at the time, Hammond Spencer, revealed that the secondary core material inside the roundhead was being "pulverised" by the waves on the outside face of a dressed stone wall.

This comment was later to become more significant following the intrusive investigations. It is also possible to observe a physical example of the structural cross-section of the St Catherine's breakwater when viewing the abandoned southern Archirondel arm of the proposed harbour. This part of the harbour wall was halted on cost grounds and due to storm damage problems.

An indication of the size of the outer dressed stones and the core material at Archirondel has given an insight to the significant feat of marine engineering that produced the St. Catherine's breakwater.

2.2 *Sustained storm damage*

In recent years, the lost facing stones at the roundhead were repaired using cement bags and then infilling the void behind with concrete. However, the scale of the damage in December 2005 resulted in not only a large loss of facing stones with cement bag replacement, but a noticeable 'settlement' of the stones forming an inverted wedge at the extreme end of the breakwater roundhead. The settlement indicated a loss

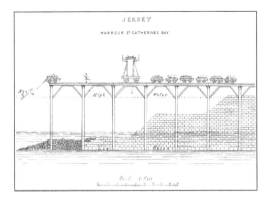

Figure 4. Method of construction from archives.

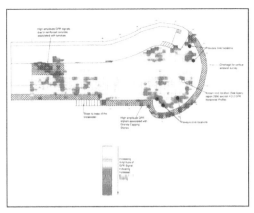

Figure 6. Radar Scan.

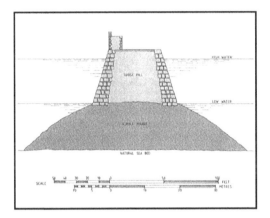

Figure 5. Cross section from archives.

of vertical support beneath this location of the original rock bund forming the foundations to the roundhead facing wall above. It was feared that a 'triggered' wedge failure would rapidly open the end of the roundhead during a storm and major wash-out would then occur and the damage escalate.

2.3 Investigations

From the desk study, a series of investigations were planned so that the cause or causes of the accelerating level of distress at the roundhead could be identified. Initially, non-destructive testing, using an impulse radar survey, was carried out on the whole length of the breakwater. Following this, a further targeted radar survey upon the breakwater roundhead deck and walls was undertaken. The interpretation of the information indicated that there were several areas of potential voiding beneath the concrete deck slabs and the walls at the roundhead, whereas the length of the main breakwater which was unpaved had historically been "topped up" when depressions and wash out had occurred. The impulse radar survey information at the roundhead enabled areas to be targeted where boreholes could be sunk to assess the extent of the possible voiding as well as the condition of the breakwater core material.

During this time, a newspaper article focused humorously upon a dog being taken for a walk on the breakwater and refused to walk upon the end roundhead concrete slabs. This sensitive canine, like the original engineer Hammond Spencer 100 years before, was also proved to be correct in its interpretation of the lack of structural integrity beneath the end of the roundhead!

2.4 Sea state conditions

Another significant part of the data gathering was to assess the information produced by HR Wallingford for the States of Jersey Public Services Department, who had mapped the island with predicted wave, current and tide movements. This information was used to understand the conditions to be resisted at the roundhead and to predict the maximum wave heights and impacts that could be expected on the face of the roundhead.

2.5 Discovery of the large void

A borehole at the end of the roundhead found a void approximately 1 m deep beneath the roundhead slab. A trial hole through the slab was instructed, and revealed a very large cavern beneath the unreinforced concrete slab. Neither the NDT nor the borehole information had identified that such a significant amount of wash out had taken place or that the integrity of the roundhead structure was severely weakened. Upon exploration of the cavern, it was also discovered that the void was even larger than at first appeared. The 400 mm thick concrete deck slab had separated in half horizontally into two laminations with the lower section wedged in the tapering void, therefore disguising the further voiding beneath it. The discovery of this very significant loss of integrity behind the roundhead masonry wall, enabled theories to be put forward as to why there was the acceleration of damage in recent years at the roundhead and accelerating loss of facing stones. As a result of 'tightness of fit' it is usually very difficult to extract masonry that is cut square

Figure 7.   Inverted wedge failure (shaded).

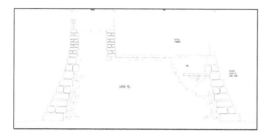

Figure 8.   Discovery of a large void beneath concrete slabs.

Figure 9.   Graphical representation of failure mode.

from a square hole. However, at the roundhead it was thought that when water in the void of this free draining structure was agitated by external wave impact on the outside face of the stone wall, a reflected wave was produced within the body of the water inside the void. This reflected wave from inside the cavern produced an outward force on the facing stone corresponding with wave suction on the outside face sufficient to extract the 3 tonne dressed facing stones. The release of support in the area of the wedge failure (i.e. lack of foundation support) also contributed to this weakness.

2.6   Obtaining emergency funding for repairs

Upon discovery of the large void beneath the concrete slabs urgent efforts were made to convince the States of Jersey Treasury to provide emergency funding to save and restore this heritage monument. A feasibility study was produced with one of the focal points being a graphical interpretation of the predicted failure method anticipated at the roundhead and then what was likely to take place after the inverted wedge breach.

The initial 'wedge failure' of the facing stones would result in a large localised breech for the full height of the roundhead wall and during a storm rapidly increasing wash out would occur making the facing stones more vulnerable. The unbuttressed facing stones would also have no resistance behind and were likely to be washed out reducing the length of the breakwater dramatically in a very short time period.

3   THE FEASIBILITY OF REPAIR AND FUTURE PROTECTION

3.1   Feasibility study

The feasibility study was necessary to define and evolve a scope of repairs to not only repair but to preserve the roundhead.

The document considered options for providing protection against future wave energy impacts by considering different levels and combinations of rock armour protection after repairs to the roundhead. A bathometric survey identified a significant loss of original foundation bund material, in particular, at the end of the breakwater when compared to the archive drawings. A graphical interpretation of the bathometric survey showed the loss of material and the consequent vulnerability of the breakwater facing stones immediately above. The loss of the bund material was at its worst at the location beneath the inverted wedge failure.

The same model was used to consider options for the protection of the roundhead in future using rock armour. A variety of slopes were studied, i.e. 1 in 2 and 1 in 1½ and springing levels down from deck level from the roundhead, e.g. 4 m from deck level and 7 m down from deck level. The latter was approximately the original constructed bund level and the models showed what the restored roundhead would look like with rock armour reinstated to the same original bund level or higher up the roundhead wall.

3.2 Initial design principles

The feasibility study identified objectives which then allowed targeted funding to be sought. The scope of the works was defined as:

- Replacement of the voided and soft core material at the end of the breakwater to reinstate the buttressing restraint for the facing stones.
- Replacement of the facing stones and restoration of the integrity of the outer "shell" of the dressed masonry.
- Consideration of how to reinstate the loss of foundation material at the supporting bund beneath the potential wedge failure, i.e. install piles, underpin and basic engineering methods were considered.
- Reinstatement of the surface of the roundhead so that a water resistant surface was constructed at deck level to ensure water shedding when overtopping occurs.

A solution to the foundation problem was also evolved, whereby the reinstatement of the loss of bund material coupled with rock armoured protection following repairs would; not only support the replaced facing stones, but if designed to modern standards, would reduce the effects of wave energy impact on the repaired structure for the future. If this could be achieved then the long term of the structure would be assured for future generations.

3.3 Initial cost estimates evolved

Only best guess estimates could be made of the volumes for replacing soft and missing core materials, despite further trial pits and opening up in the area of the roundhead core. In addition to the repair and restoration work, much consideration and thought was given to the geometry necessary to provide long term wave energy protection to the structure. The two critical factors were the springing of the rock armour protection level at either 4 m or 7 m below deck level and the slope of the rock armour. A slope of 1 in 1½ required less rock material but was not able to reduce the wave energy impact significantly but a shallower slope of 1 in 2 was far more expensive because a much larger volume of rock was required.

4 DETAILED DESIGN AND CONTRACTOR PROCUREMENT

4.1 Further intrusive tests

It was a matter of utmost urgency to obtain as much data as possible in order to quantify volumes of materials that may be needed. Additional trial holes were dug through the concrete slabs and some other areas of voiding found. It was also necessary to determine the integrity of the core material and the bund material beneath to confirm that sea salt and other wave or sea water degeneration mechanisms were not reducing or deteriorating the bund or core materials that were to remain. Some rocks are susceptible to breakdown in sea water conditions, therefore a further borehole was sunk through the end of the structure to bedrock to determine the whole support structure at the end of the roundhead.

4.2 Wave energy studies

This information was gleaned from local data, the HR Wallingford study and other local and available marine texts. A large number of combinations of sea state conditions were analysed in order to produce a maximum wave height that could be expected at the roundhead. Determination of the maximum wave height allowed predictions of resultant wave impact forces to be made and in order to design the rock armour at different slopes and of different sizes. Some 500 combination of load cases were considered with a final result requiring a 1 in 2 slope of 50 tonne rock armour commencing 7 m down from the deck.

4.3 Value engineering

This abnormal size of rock armour, i.e. 50 tonnes, was thought to be at the upper range of buildability and following consultations with the client, value engineering dictated the size of rock armour down to 30 tonnes, with the client's recognition and acceptance that in occasional severe storms there was a likelihood of some movement of the 30 tonne rocks which would then have to be repositioned.

4.4 Contract procurement

With both the geometry and size of the materials now identified, a fast track procurement programme was initiated to execute the works. The States of Jersey required an NEC3 form of contract to be used and a contractor, Trant Ltd, was appointed with approximately £1.4 M required for the repair and restoration work at the roundhead and £3 M required for the additional rock armour protection.

5 THE CONSTRUCTION PROCESS

5.1 Commencement

The contract began in September 2006 with a programme of 3 months. This was acknowledged to be a risky weather window, but was just before the onset of the predicted severest of winter storms in 2007. It was felt that with the momentum gained in terms of knowledge and finance to restore the structure, to delay the project for 6 months until the following summer may have resulted in significant further loss and damage to the roundhead structure.

5.2 Initial break out

The concrete deck over the void was broken out and the material broken up and temporarily backfilled to provide buttressing protection to the exposed surface masonry wall. To temporarily counteract the effects of overtopping, a tarpaulin was installed over the backfilled material to watershed overtopping waves.

5.3 Excavation and backfill

The materials were then re-excavated in a controlled manner from behind the facing wall to an approximate depth of 7 m and terraced back as the distance from the end of the roundhead increased.

5.4 Backfill and protection

A crushed, graded rock from a nearby Jersey quarry was tipped into the excavations which had the sides and base of the excavation wrapped in geotextile membrane to protect against future fines wash out.

The facing stone wall was seen to have a peripheral ring beam of concrete at just below deck level and this was used to back prop against high tides. As it was possible to break up and re-use the old concrete deck materials and recycle them into the rock fill, this saved money on both imported rock material and unnecessary journeys away from the site to the tip. (These financial cushions would be required later!).

5.5 Replacement facing stones

This operation was carried out over the side from deck level with suspended cradles and baskets. The concrete bags were broken out and 2–3 tonne granite replacement facing stones were then lifted into position and dowelled into place with stainless steel fixing pins into the concrete and now stable material behind.

5.6 Replacement surfacing

The restoration of the breakwater required that a water resistant deck be re-provided and not the rigid deck that the concrete paving provided which had disguised the creation of the large voids beneath. Therefore, the top layers of rock fill were graded with sub-base material, base course and a wearing course of asphalt. However, to achieve the required aesthetics, a resin bonded "granite look topping" was placed over the asphalt in order to blend in aesthetically with the pink granite of the original structure.

5.7 Protection with rock armour

The sourcing of the 30 tonne Rock Armour blocks was from a quarry in Norway as no Island resources could provide such sizes and quantities. There were also issues with respect to Jersey's sustainability goals, road trafficking and other considerations with such narrow roads on a small island. A quarry in Norway specialised in large rock sizes and re-used their extraction as filling sites.

5.8 Transporting the rock armour

This proved to be the most difficult part of the project as two barges 40 m long and 10 m wide were commissioned to transport the rock in three deliveries from Norway to Jersey, towed by tugs. The programme was tight and the window of opportunity small where the returning first barge would be filled whilst the second barge was on site. However, as in many civil

Figure 10. Rock armour from Norway.

engineering projects, the weather conditions turned adverse and the second barge broke free in a severe storm in the North Sea. This second barge was towed into shelter in a German harbour until the weather window cleared and onward transportation could be made. This was a significant compensation event in terms of the NEC3 Contract Form and ironically the compensation events approximately balanced the cost savings gained from recycling concrete as replacement rock core material.

5.9 Placing of rock armour

The profile of the designed rock armour required head and toe platforms with the latter keying into the sea bed with a geotextile placed on the sea bed to ensure against rock settlement into the sand. When they arrived in Jersey from Norway, the large rocks were transported from the mother barge by a smaller barge to the site. The small barge had a gib mounted grab with GPS attached to enable precise placing of the rock 30T armour stones.

5.10 Material weathering

In a very short space of time in view of the extremely harsh environment, both the rock armour and new facing stones had weathered in terms of algae and other staining to blend in with the existing structure. All of the other minor elements of the repair works were completed in a sympathetic way and the resin bonded surfacing materials proved to be a satisfactory finish aesthetically.

6 CONCLUSIONS AND LESSONS LEARNED FROM THE CONSTRUCTION PROCESS

6.1 Funding considerations

Heritage structures are difficult to obtain funds for in terms of repair and/or restoration. It is even harder in the current climate to obtain additional funding to protect heritage assets for future generations. Therefore, this particular project owes a debt of gratitude to the States of Jersey Treasury for supporting the rock armour protection principle.

6.2   Hidden voids

The phenomena of concreting over free draining gravity structures has resulted in a potential dramatic failure here. It has also been found to be responsible for several other marine failures throughout the UK. It would therefore be prudent for custodians and/or owners of marine structures of this type of structure with concrete decks over old free draining marine quays or piers to survey them and check that voids have not been forming beneath the concrete.

6.3   Lobby for future protection

One of the unsung aspects of this project was the modern day use of graphics and 'what if' scenarios, that enabled further significant funding to be obtained in order to protect this structure for future generations.

6.4   A goal for any heritage restoration

With the exception of exposure of the large rock armour at low tides, the restoration of the roundhead and end of St. Catherine's Breakwater shows little or no signs of the significant civil engineering and heritage restoration work that was carried out on this structure. This is a testament not only to the dedication of the design team but to the skills of the contractor as well as the forward thinking of the client and custodians of this monument.

Concrete Solutions – Grantham, Majorana & Salomoni (Eds)
© 2009 Taylor & Francis Group, London, ISBN 978-0-415-55082-6

Repairing a sinking bridge over Guadalquivir river

C. Jurado

Polytechnic University of Madrid, Ingecal Engineers, S. L. Spain

ABSTRACT: The bridge over Guadalquivir river, near Alcalá del Rio in Seville (Spain), has nine spans of 30 m, with a total length of 270 m. In 1993 it suffered a sinking of 14 cm and 6 cm on two of the central piers located in the centre of the river. The superstructure has six prefabricated concrete T-beams 150 cm. in height. The foundations comprise two circular pier-piles of 1.50 m diameter in each support.

The bridge was constructed between July 1990 and October 1992. In the first week of April 1993 it was observed that the handrails formed a vertical angle in the centre of the bridge, which indicated a sinking of piers P-4 and P-5 just in the middle of the river. The topographic observation confirmed this settlement.

After studying different solutions the best technically and economically solution concluded, was to use Jet Grouting injections.

1 INTRODUCTION

The JUNTA DE ANDALUCIA Administration has carried out between July of 1990 and October 1992 the construction of the Variant of Alcalá del Río, which includes a nine-span, isostatic bridge over the Guadalquivir river, whose geometric definition is described later.

The structure consists of nine equal spans of 30 m. each , which amounts to a total length of 270 m. The deck consists of 10 m. of pavement plus two sidewalks of 1 m., totalling a width of 12 m (Figure 1).

The bridge was designed with deep foundations for all its supports, due to the existing soft ground in the zone.

The foundations of the bridge comprise two circular pier-piles of 1.5 m. of diameter on each support and 25 m. in length, in the case of piers. Of the 25 m. mentioned in case of the pillars, 2 m. were embedded in gravels and 12 m. (8 diameters) in blue marls; the 11 m. remaining corresponds to the exposed part of the pier, of which the lowest 2 m. corresponds to the average depth of the Guadalquivir river in the above mentioned zone (Figure 2).

Figure 1. Bridge in phase of construction.

2 DETECTED PATHOLOGY

In the first week of April, 1993, a vertical angle in the metal vehicles barriers and in the handrails of the bridge was observed, that indicated the possibility of a descent of the deck, at the level of piers P-4 and P-5. (Figure 3).

It was decided to verify the level of these piers, and average settlements of 13.9 and 5.4 cm were observed. After this it was decided to carry out a follow-up of the settlements by means of a topographical survey,

Figure 2. Finished bridge.

131

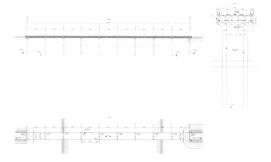

Figure 3. Plan, longitudinal and transversal sections of the bridge.

installing fixed levelling bases on both sides of the bridge.

From the 4th of April follow up measurements were carried out, in which it was observed that until 17th of May both abutments and the rest of the piers did not have any settlement, but the piers P-4 and P-5 continued descending reaching an average settlement of 14.6 and 6.0 cm, respectively. The settlements of both pillars were similar.

3 LOADS ON THE PIER-PILE

A review of the bridge project was carried out verifying the validity of the design. The actions over the pier-piles were calculated considering the following loads:

• Permanent loads.
• Live loads due to a uniform load of 0.40 Ton/m² and to lorry loads of 60 Ton, according to the forces in the Spanish Instruction of Bridges, considering diverse positions of the loads

The maximum axial load characteristics that were transmitted from the piers to the ground (assuming that both pillars were working equally), without considering the action of wind and the earthquake, proved to be of 503.02 Ton for the pile (Figure 4).

Allowing for the load cases of wind and the earthquake the maximum and minimum loads were:

$Q_{max} = 554.22$ Ton. $Q_{min} = 451.82$ Ton.

The scheme of maximum horizontal and vertical actions in the piers was:

During the bridge project a site investigation had been carried out consisting of soil borings, that indicated the existence of one alluvial layer of several meters thickness and below it, the Mioceno substrate of marls of unlimited thickness. None of the borings were under piers P-4 and P-5. The alluvial was proved to show an average compaction and the marl with a compression resistance of $q_u = 6.3$ kp/cm². S.P.T. tests were also performed, giving an average value of N = 62.

With these values the allowable load capacity for pile was calculated as:

$$Q_h = Q_p + Q_f = A_p \times r_p + A_f \times r_f$$

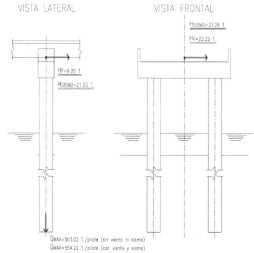

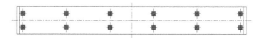

Figure 4. Schematic loads.

P. L.(T)	66.82	79.02	79.02	79.02	79.02	66.82
+	+	+	+	+	+	+
L. L.(T)	20.16	25.92	79.27	79.27	25. 92	20.16
TOTAL	86.98	104.94	158.29	158.20	104.94	96.98

P. L. Permanente Load   L. L. Live Load

*Total P.L + L. L. = 700.42 Ton		TOTAL =1006.04 T. per 2 piles
*Weigth lintel beam = 72.00 Ton		So 503.02 Ton per pile
* Weigth pile – piles = 233.62 Ton		

Figure 5. Table of loads over the bridge.

A_p, A_f = Area of the tip and perimeter of pile
r_p, r_f = unitary ground resistance of tip and on shaft friction
$q_u = 6.3 \, k_p/cm^2 = 63 \, Ton/m^2$
The following was adopted in the project:

$r_p = 5 \times q_u = 315$ Ton/m²

$r_f = 0.25 \times q_u = 15.75$ Ton/m²

The maximum load capacity of pile before collapse, ignoring the contribution in the resistance of shaft friction of the gravels showed:

$Q_h = 1.766 \times 315 + 56.52 \times 15.75 = 1446 \; Ton$

132

The safety coefficient to the maximum load of service without factoring this was:

$$F = \frac{Q_h}{N_{max}} = \frac{1446\,Ton}{503\,Ton} = 2.88$$

That was considered in the project to be acceptable.

Allowing for the possibility of earthquake and transverse winds, this coefficient of safety, was diminished to F = 2.51 in the most loaded pier, a value that also was considered sufficient.

1 INVESTIGATIONS AND RESULTS

In view of the detected pathology it was decided to carry out an additional geotechnical investigation consisting of two mechanical borings to rotation, on the basis of the following hypotheses:

4.1 Hypothesis

Due to the similar settlement in the couples of pier-piles of pillars P-4 and P-5, an hypothesis was made that the marl could not be homogeneous, containing some soft nodules under the tip of the pile or some cavity.

The possibility of imperfection or bad execution of the pile and specially the tip was discarded, because it seemed to be strange that 4 contiguous piles were badly executed. By the control taken during the construction heterogeneities were discarded also along the pillar.

4.2 Investigation

The investigation consisted of the execution of 2 mechanical rotation soil borings, one between the piles of the pier P-4 and other between those of pier P-5.

The soil borings were executed on a boat since the piers are placed in the centre of the river.

In situ standard penetration tests (S.P.T.) were made every 2.00 m. of vertical distance in each boring, and soil-samples were recovered for laboratory tests.

4.3 Results obtained

No hollow or soft zones were detected below the tip of the piles; nevertheless, the investigation suggested a consistency of the marl smaller than expected.

The profiles obtained of soil were the following:

BORING	PIER	MOUTH ELEV.	DEPTH (m)	TYPE OF SOIL
S-1	P-4	−1.98	From 0 to 0,25	Lime
			From 0.25 to 2.25	Gravel
			From 2.25 to 28.50	Blue Marl
S-2	P-5	−1.93	From 0 to 1.30	Fine Gravel
			From 1.30 to 21.30	Blue Marl

In effect, the geotechnical investigation realised in the project indicated that the marl had an N(SPT) = 62 and resistance to compression of at least $q_u = 6.3\,kg/cm^2$.

The S.P.T. test performed in the last investigation were registering blows less than half of the expected values, along the pile and the under the tip.

The average blow N turned out to be of 36 (in S-1) and of 35 (in S-2). All these values lower than N = 62 suggested by the original design of the project from investigations carried out in the zones of the abutments. Up to 20 to 23 meters depth blows in the S.P.T. test of 60 or better were not obtained.

Applying the relation that suggested in the literature between the S.P.T. test and the resistance of compression of $q_u = N/7.5$, values as low as $q_u = 21/7.5 = 2.8\,kg/cm^2$, were obtained, very far away from the value of 6.3 kg/cm² given in the initial geotechnical report of the base project.

The simple compression tests performed on samples gave low values, between 3.40 and 4.91 kg/cm², in the first 12 meters, except one value of 10.08 kg/cm². Major values of 6 kg/cm² were found below 16 m. of depth, corresponding to values of N better than 50.

In conclusion, the following points were indicated:

a) The existence of large soft zones of soil or caves below the tip of the piles was not confirmed.

b) Nevertheless, it was apparent that in the first 16 meters investigated, the marl showed a much lower consistency than expected.

c) Due to the lower geomechanical resistance of the soil, it was decided to verify the safety coefficients of piers P-4 and P-5 with new geomechanical parameters.

d) For the calculations that are developed later it was supposed that the resistance to simple compression along the pile and in the tip was 5 kg/cm², that is equivalent to one blow of N = 38 ($q_u = 38/7.5 = 5\,kg/cm^2$).

5 ANALYSIS OF THE FOUNDATION AND POSSIBLE REASONS OF THE PATHOLOGY

5.1 Piers 4 and 5

The foundation design of the project was checked, considering the caps of gravels and existing blue marls for piers 4 and 5 with the thickness obtained in the soil borings.

The admissible load of the pile was calculated from the ultimate pile capacity of collapse. The maximum load capacity of pile before collapse was calculated as the sum of the resistance by the tip more the shaft friction resistance.

$$Q_h = Q_p + Q_f = A_p \times R_p + \Sigma\,Perimeter \times L_i \times R_{fi}$$

The admissible load Q_{adm} is equal to the maximum load of collapse Q_h divided by the safety coefficient F. The value of the safety coefficient in the normal cases is F=3. In the cases in which there is sufficient knowledge of the site area and its resistant properties, the value of the coefficient might be reduced according to some authors up to 2. The admissible load must be compared with the maximum characteristics axial load corresponding to the maximum axial load of service, without magnification.

In this case, a tip resistance was adopted:

$$R_p = 9 \times C_u = 4.5 \times q_u \text{ (marls)}$$

As resistance for shaft friction in the zone of marls, the value was:

$$R_f = \alpha \times C_u \text{ (marls)}$$

According to Woodwaver mentioned by J.A. Jiménez Salas (1972), for a value:

$C_u = q_u/2 = 5/2 = 2.5 \text{ kg/cm}^2$, $\alpha = 0.3$ from which is obtained:

$$R_f = 0.3 \times C_u$$
$$R_p = 4.5 \times 5.0 = 22.5 \text{ kg/cm}^2 = 225 \text{ Ton/m}^2$$
$$R_f = 0.3 \times 2.5 \text{ kg/cm}^2 = 0.75 \text{ kg/cm}^2 = 7.5 \text{ Ton/m}^2$$

Resistance under the tip: pile $\emptyset = 1.50$ meters:

$$Q_p = A_p \times R_p = 1.766 \times 225 = 397 \text{ Ton}$$
$$A_p = 1.5^2 \times \pi/4 = 1.766 \, m^2$$

Resistance for shaft friction: pile of 12 meters. Resistance for shaft friction in marls:

$$R_f = 7.5 \text{ Ton/m}^2$$

Resistance for shaft friction in gravels: $R_f = 5 \text{ Ton/m}^2$

$$Q_f = \pi \times 1.5(2 \times 5 + 9.7 \times 7.5) = 4.712 \times 83.1 = 392$$
$$Q_f = 392 \, Ton \, (1)$$
$$Q_f = \pi \times 1.5 \times 9.75 \times 7.5 = 344 \, Ton \, (2)$$

1. Considering the contribution of the gravels.
2. Not considering the contribution of the gravels.
   Ultimate loads of collapse.

$$Q_h = Q_p + Q_f = 397 + 392 = 789 \, Ton \, (1)$$
$$Q_h = Q_p + Q_f = 397 + 344 = 741 \, Ton \, (2)$$

Values lower than that obtained in the base project and for which the safety coefficient was turning out to be:
Real coefficient of safety piers P-4 and P-5.

$$F = Q_h / N = 789/503 = 1.57 \, (1)$$
$$F = Q_h / N = 741/503 = 1.47 \, (2)$$

These coefficients of safety were considered to be totally insufficient. The working load was near the ultimate load and the soil was in an elastoplastic phase, justifying the observed settlements, or even worse if compared with the maximum load of design (with wind and earthquake).

With the previous parameters r_p and r_f, supposing a module of deformation of the soil of $E = 100 \, C_u$ it was estimated, by means of an elastoplastic program

a settlement of 16.3 cm. was obtained, very similar to the observed ones.

From this result it was proved, that the resistance by shaft friction was almost totally exhausted. This result also can be estimated seeing the values of maximum resistance for shaft friction $Q_s = 344$ to 392 Ton that is the first one that is exhausted.

6 FIRST HYPOTHESIS OF SOLUTION

6.1 *Piers 4 and 5*

In the previous paragraph it's appeared that piers P-4 and P-5, according to the calculations, had a very low coefficient of safety.

Therefore it was considered necessary to increase this coefficient of safety, and three possible procedures were considered.

a) The first consisted of transmitting the load of the structure by means of a footing. This lead up to a footing of 5m × 10m with big problems of connection to the existing piers and execution under 2 meters of water in the middle of the river.
b) The second solution would consist of performing a series of piles or micropiles that would join by means of a concrete pile cap, the piles and the existing pier-pile. Besides the possible limitations of vertical space for the execution of these, there would exist the same problems of execution mentioned previously for the concrete pile cap. It this case it would be necessary to have 5 piles $\emptyset = 450$ mm. for each pier-pile.
c) The third solution would be to improve the soil. The only suitable technology is jet grouting.

In the solution that was proposed, the soil was improved along the shaft of the pier-pile and the superior active zone, the inferior and the security zone were injected, to obtain a "rigid element" with a diameter of approximately 3.50 m. Without considering the shaft friction, simply as tip resistance this element had a coefficient of safety larger than 3, with the service load of 503 Ton.

$$Q_p = 8 \, c_u \times A_p = 8 \times 25 \times \pi \times 35^2/4 = 1920 \text{ Ton}$$

$$F = 1920/503 = 3.8$$

A coefficient of 8 instead of 9 was considered since in piles of so great a diameter it is recommended to reduce N_c.

This solution there had the advantage that it was possible to undertake from a floating pontoon and it was not necessary to construct any element of union to the existing pier.

7 ADOPTED SOLUTION

Following an economical analysis of all three solutions it was found that the solution of Jet Grouting had the least cost and the least problems of accomplishment. Besides, this solution could obtain a reduction in the settlements of piers P-4 and P-5.

This solution consisted of the execution of 18 columns of injection, nine vertical from a distance to the existing pile of 0.50–1.00 meter and the other nine inclined approximately 6.5° degrees, according to the disposition that is indicated in Figure 3 at the end of this article.

The injection would produce a column of soil-cement of approximately 40–50 cm. of diameter, consolidating and refilling the zones of the marls that surround the pile.

The working pressure recommended for the injection pump was 350–400 kg/cm^2 and the cement injection was fixed to 250–300 kg of cement per linear meter of treated zone.

The dosing of the cement grout in weight was water/cement = 1/1, with which is achieved a density of the mixture of 1.5 Ton/m^3.

The columns were executed alternatively, so that between 2 adjacent columns executed in the same day, there was at least 1.50 m, or 24 hours of consolidated cement of the adjacent columns, to that which was being executed at the time. The object of this gap in space and time was to avoid the possible fluidity of an important area during the injection of the mortar, which could give place to an unforeseeable increase of the settlements. Also, during the execution of the treatment the traffic was not interrupted, though it was stopped in the lane that was over the pier-pile in treatment in every case and diminishing the speed to 30 km/h.

The worst moment of the procedure is when one injection is executed. Therefore, it was recommended to make frequent levelling of the affected piers controlling the raising of the pier when the cement was injected and also reducing the speed of the vehicles on the bridge to avoid important dynamic loads. With the effected treatment, there was corrected the settlement of the piers P-4 and P-5 obtaining a raising of the twin pillars of approximately 10 cm as maximum (Figure 6).

INITIAL SEAT	FINAL SEAT
P-4: 14.6 cm.	6.0 cm.
P-5: 6 cm.	3.5 cm.

The work was carried out during June and July of 1993. In Figure 6 at the end a scheme is attached, representing the initial and final settlements.

8 RESULT OF THE PERFORMANCE

The principal result was the total correction of the pathology which had happened in piers P-4 and P-5, restoring as minimum the same conditions of safety against the collapse of the foundation predicted in the project, turning these coefficients to be acceptable. On the other hand, it was possible to correct a great part

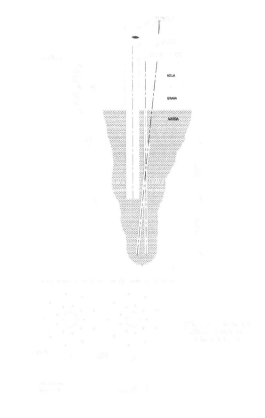

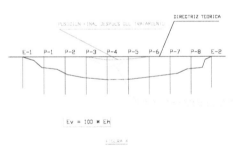

Figure 6. Scheme of jet grouting and recovering settlements.

of the settlements produced in the piers, which largely reduced the angular distortion of the deck.

Though in an isostatic bridge of these characteristics it is possible at first to perfectly maintain the functionality of the structure, in spite of the fact that differential settlements had taken place between piers, with the adopted solution, these and the angular distortions diminished to minimal values, recovering the bridge again to an acceptable aspect from the aesthetic point of view.

REFERENCES

Ministerio de Fomento. *"Spanish rules relative to the actions to consider in the project of roadway bridges". IAP-98.*

Ministerio de Fomento. *"Norm of seismic construction: General part and Building Construction"*. NCSE–O2.

Ministerio de Fomento. *"Bases of project and design of parapets in roadway bridges"*

Eurocódigo 1.1 *"Bases of project and actions in structures"*

Eurocódigo 2. *"Project of structures of reinforced concrete"*

Eurocódigo 3. *"Project of structures of steel"*.

Geotecnia y Cimientos. Jiménez Salas y otros. Ed. Rueda.

Jiménez Montoya, P., García Meseguer, A. & Morán Cabré, F. Hormigón armado. 14a edición. Ed. Gustavo Gili.

Bowles, J. Foundation Analysis and Design, Mc Grow Hill. 5th edition.

Concrete Solutions – Grantham, Majorana & Salomoni (Eds)
© 2009 Taylor & Francis Group, London, ISBN 978-0-415-55082-6

The underpass in O'Donnell street in Madrid

C. Jurado
Polytechnic University of Madrid, Ingecal Ingenieros, S. L. Spain

ABSTRACT: At the end of 1997, due to the increasing traffic in O'Donnell street towards Madrid airport, the Town Hall of Madrid decided to construct an underpass in O'Donnell street, below the existing underpass in Doctor Esquerdo street and above the underground tunnel of line 6. The new underpass needed to thread between the other two with a limited height clearance, so that the floor of Doctor Esquerdo's tunnel was the roof of O'Donnell's tunnel.

Owing to these circumstances, besides the technical difficulty, the tunnel was, at the time of construction, the deepest (14 m under the surface of the street) and the longest (547 m) of the existing tunnels in Madrid.

Additional considerations in the project and the execution were the numerous services affected, since the site corresponds to an urban central area: telephony, electricity, supply of water, sewage, etc.

1 INTRODUCTION

At the end of 1997, the Town hall of Madrid, motivated by the black spot caused by the density of existing traffic in O'Donnell' street, one of the principal arteries of Madrid where it exits to the surrounding M-30/M-40 motorways, especially in the crossing with Doctor Esquerdo street, decided to call for a bidding of project and execution of work, for the construction of a dual carriageway underpass under O'Donnell street.

The bidding was won by the UTE CORSAN-RODIO with the project made by D. Carlos Jurado, author of this communication, who was the Technical Director of CORSAN's contractor and engineering firm at the time of bidding. The Director of the work was D. Jorge Presa Mantilla. The work had as the most important technical difficulty the existence of one tunnel under Doctor Esquerdo street and below this Line 6 of the underground and it was mandatory to project the new tunnel threading it between other two and with a height such that the floor of the Doctor Esquerdo's tunnel was to be the roof of O'Donnell' tunnel (see figure 1).

Owing to these circumstances, besides the technical difficulty previously mentioned, the tunnel was the deepest at the time of construction (14 m under the road surface) and the longest (547 m) of the existing ones up to that date in Madrid (see figure 2).

The compulsory condition of design imposed was that the maximum longitudinal slope of the tunnel was 7% and that the beginning of the tunnel in the exit direction of Madrid, was designed allowing the access of the traffic from Maiquez street, which imposed very restrictive conditions for the turning of high tonnage vehicles.

An additional point, determinant to the project and to the execution of the work, was the numerous existing

Figure 1. Aerial view of O'Donnell's tunnel.

services affected, since the site corresponds to an urban central area: telephony, electricity, supply of water, waste water, etc.

Among them, it is necessary to mention the existence of two service galleries that circulate on each side of Doctor Esquerdo street, made in reinforced concrete with transversal section in vault, with an exterior width of 3.10 m and a maximum interior height in the middle of the arch of 2.65 m. The walls were 0.40 m. of thickness, (see figures 3 and 8).

Another particularly affected service was a large sewer collector which was circulating along

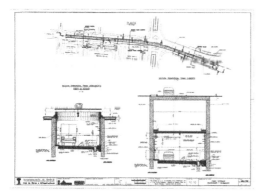

Figure 2. Plan and sections of the underpass.

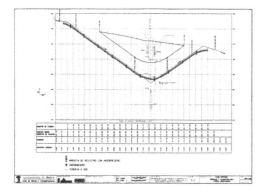

Figure 3. Longitudinal profile of the O'Donnell tunnel.

Figure 4. Aerial view of the link of O'Donnell and Alcalde Sainz de Baranda.

O'Donnell street, constructed by hand and affected by the "in situ" concrete sheet pile walls of the new underpass. It was necessary to divide this into two collectors on both sides of the works, and to rebuild it by the same constructive procedure in a length of approximately 1500 meters.

The work included the remodeling of the intersection with Alcalde Sainz de Baranda street, as well as the entry from the neighbourhood of the Elipa district, close to O'Donnell street, which implied the construction of a of prestressed concrete bridge, of

two spans with 44 m. total length, over the link between O'Donnell and Sainz de Baranda streets (see figure 4).

2 ALTERNATIVES OF EXECUTION OF THE TUNNEL

The most difficult part of the project and of the work was the intersection of O'Donnell with Doctor Esquerdo street, where, with the tunnel of the underground below the new underpass and the existing underpass Doctor Esquerdo street above, so it was necessary to project and construct the new tunnel with a strict height between them, almost perpendicular to both.

To solve this technological challenge, three alternatives appeared in the winning project of the bidding:

- Tunnel constructed under the protection of an umbrella of horizontal micropiles.
- Tunnel constructed by means of pushing a caisson of reinforced concrete.
- Tunnel constructed by means of the execution of in situ concrete sheet pile walls, under the existing underpass and after the construction of a top slab resting on them, further constructed by means of 11 prestressed precast beams PLN-45 and a slab of compression of 0.25 m of thickness, constituting a bridge of 10 m. of length and 14.50 m. of width.

The work was awarded in April, 1998 to the third of the three alternatives by an amount of 6.166.384 € and a term of execution of 11 months.

The works began in May 1998, with a projected completion by April, 1999.

3 GENERAL DESCRIPTION OF THE WORK

The work included the execution of an underpass in the crossing of O'Donnell with Doctor Esquerdo streets, as well as the construction of two links between the O'Donnell and Alcalde Sainz de Baranda streets and other complementary works.

The transverse section of the underpass was formed by a road of two lanes (direction M-40 outside the city) of variable width, between 3.25 m and 3.50 m, with two sidewalks of 0.75 m of width (see figure 2).

In vertical section, the axis of the underpass presents two straight alignments of 218 m and 121 m respectively, joined by a vertical circular curve of 525 m. of radius, with correspondents curves of transition (see figure 3).

The maximum slopes were 7% in the entry of Maiquez street and 6.85% in the exit to the M-30. The total length of the underpass was 547 m (see figure 3).

The work was known popularly as "the retunnel" due since it was executed between the existing Doctor Esquerdo underpass (above) and the tunnel of the underground line 6 (below). The O'Donnell tunnel project had 8,078 m^2 of in situ concrete sheet pile walls, with 0.50 m and 0.60 m of thickness.

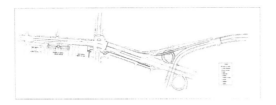

Figure 5. Plan of the works site.

Table 1. Project measurements.

Reinforced Concrete Volume	9,600 m³
Reinforcing Steel	417,510 Kg.
Prestress Steel	6,665 Kg.
Excavations	52,902 m³
Masonry Concrete	5,593 m³
Length of Soil Consolidation Injection	474 m
Road Bitumen	6,816 T

Two reinforced concrete slabs were constructed, the upper one under the surface of the street, that was supporting the traffic, was of 0.60 m thickness and 9.00 m span, and the lower one of 0.30 m thickness and equal span, leaving between both a hollow of 5.20 m, of maximum height, useful for the installation of the systems for control of gases and fires in the tunnel. The total surface of concrete slabs constructed was of 2,750 m², with an area of 39,252 m² of roads and sidewalks. In table 1, a summary of measurements of the work is presented.

The transverse section of the underpass was formed by a road of two lanes (direction M-40 outside the city) of variable width, between 3.25 m and 3.50 m, with two sidewalks of 0.75 m of width (see figure 2).

The underpass has artificial lighting, forced regular ventilation and fire and anhydride carbonic detectors.

In the intersection with the existing underpass of Doctor Esquerdo street, injections of consolidation were carried out, which fulfilled the vault of the underground line 6, executing finally the intersection with the existing underpass, by means of two "in situ" concrete sheet pile walls of 1 m. of thickness and changeable lengths between 6.00 m and 7.00 m. The toe of the sheet piles was between 2 and 3 m. over the existing vault of the tunnel line 6 of the underground.

As it has been said, the most important aspect of the work, consisted of the fact that the new underpass had to cross in it's central zone, under the existing Doctor Esquerdo underpass and at the same time above the tunnel of the line 6 of the underground, between the stations of O'Donnell and Sainz de Baranda. The available height "to thread" the new underpass in the zone of crossing, was approximately 9 m., meaning that the floor of the existing underpass constituted the roof of the new one.

Additionally there had to be solved numerous and important interferences of municipal and not municipal services existing in the zone of the crossing:

a) Dividing the existing 725 m of sewer collector in the axis of the street O'Donnell, in two collectors on both sides of the site work.

b) Supporting two existing services galleries of reinforced concrete, on both sides of Doctor Esquerdo street and the corresponding concrete sheet walls of the existing underpass.

c) Diversion of the terrestrial network of optical fibre, belonging to Retevisión.

d) Services of natural gas, telephony, water supply, sewage, electricity, etc.

The structure of the underpass was solved on both sides of Doctor Esquerdo underpass, by means of two "in situ" concrete sheet pile walls, with 0,50 m of thickness, and with two strut bracing, formed by the slab of cover to street level of 0.60 m thickness and the lower slab, that was serving as the roof, to the new step of 0,30 m of thickness, joined to the sheet pile walls.

In the zone of the crossing there was another important condition, produced by the interference of the structure of the new underpass with the existing one. Specifically, the structure of this one is formed by two concrete sheet pile walls joined by a reinforced concrete slab on top. Given the length of fixing of each one, demolition was necessary for the execution of the crossing with the new underpass.

In order to assure the stability of the sheet pile walls of the existing underpass, during the phase of cutting them, to leave the hollow for the new underpass, a "great concrete beam" was projected to hang the existing sheet pile walls. The beam had 1.50 m. of width and 3.60 m. of height, and it was supported by two piles placed on both sides of the new sheet pile walls, with 1800 mm. diameter and 19 m. length, so that was not compromising the vertical stability of the existing panels, affected by the demolition. In addition, in the lower part of them, were anchored the panels mentioned to the "structure of crossing" that was executed from the interior of the existing underpass and which is described later.

The mentioned "structure of crossing" consisted of a "bridge – slab" of 10 m. length and 14.50 m. width, constituted by 11 prestressed concrete beams in the shape of T inverted (11 PLN-45) of 45 cm. of height, arranged together, plus a reinforced concrete layer in compression with a variable height from 10 to 25 cm. The concrete beams were supported on neoprene bearings $100 \times 250 \times 19$ mm with the concrete slab fixing in the abutments constituted by a girder of tie of 0.50 m thickness and with a changeable length between 6.00 m and 7.00 m. Furthermore, this structure of crossing had a transition slab in every abutment of 0.20 m of thickness to avoid a sudden jump, on the surface road of the superior tunnel in the crossing, with the lower one.

This solution permitted continuity to the traffic of the current underpass over the new construction, constituting in addition the roof of the new underpass in the zone of crossing.

The structure of crossing was finished by a slab of reinforced concrete of 30 cm. thickness, anchored to the concrete sheet pile walls and forming a box, that

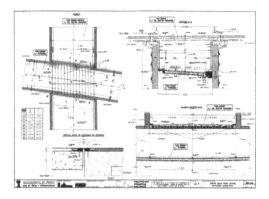

Figure 6. Structure of crossing with Doctor Esquerdo's underpass.

Figure 7. Bridge over the link to Alcalde Sainz de Baranda.

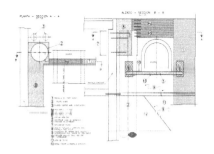

Figure 8. Supporting galleries of service.

4 THE SERVICE GALLERIES

A point of important impact in the execution of the work it was the services galleries to each side of Doctor Esquerdo street, which constituted an important barrier for the execution of the concrete sheet pile walls and whose weight was necessary to be supported in the phase of excavation of the hollow of the tunnel under them.

These galleries of reinforced concrete with an arched roof have a free distance between walls of 2.50 m and an interior maximum height in the middle of the arch of 2.65 m.

On one hand, the execution of the concrete sheet pile walls had to be interrupted on both sides of the service galleries. On the other hand, having realized the excavation of the tunnel under the gallery this one would remain exempt in the 9 m. of span of the tunnel, changing from working in compression to working in flexion, for which they were not had been designed.

Owing to this, it was necessary to support continually the total length of the galleries, from a provisional to a permanent way.

To resolve this it was necessary to construct two large beams of reinforced concrete (see figure 8) on both sides of the galleries. These beams would have a width of 0.80 m and a height of 1.50 m and would be supported by two piles of large diameter (ø = 1800 mm) on both sides of them.

In these beams there were embedded metallic profiles HEB280 every meter, placed under the galleries to support them.

Thus the services galleries would be supported continually, reproducing the initial situation of constant support on the area.

allowed the union of the concrete sheet pile walls of the existing underpass without transmitting appreciable stresses with the structure of the new the tunnel of the underground and reinforcing the monolithic response of the set formed by the new structures and the existing ones (see figure 6).

To complete the latter aspect, an injections campaign of consolidation cement with low pressure, was carried out in the surrounding areas of the vault of the underground's tunnel, to consolidate possible zones of cavities that could had been taken place in the back face of the vault and to form an "area", that would guaranteed the correct behaviour in the surrounding zones of the new work.

Besides the tunnel, the work included the connections of O'Donnell with Alcalde Sainz de Baranda streets, which were resolving with a circular link in the sense O'Donnell-Sainz de Baranda and one prestressed bridge over it with two spans of 44 m. in length. The transverse section was in a trapezium shape lightened with lateral wings where the sidewalks were placed to each side of the road, 4.50 m over the surface of soil. The abutments of the structure were joined to the retaining wall of earths of the low bow (see figures 4 and 7).

5 GENERAL DESCRIPTION OF CONSTRUCTION PROCESS

In a schematic way, first began the execution of the concrete sheet pile walls on both sides of the crossing, including the corresponding tie beams and the cover slab.

In addition the structure of the crossing from the interior of the existing underpass needed diverting the traffic for the other half of the existing underpass, excavating the concrete sheet pile walls in the zone

Figure 9. Exit west of the tunnel.

of the crossing by conventional methods, since the reduced height was insufficient to introduce the usual machinery for sheet pile walls.

Later in the adjacent zone to the concrete sheet pile walls of the existing underpass, four large diameter (1800 mm) supporting piles were constructed and the large beams for hanging the sheet pile walls of the existing underpass. In addition the structures for supporting the galleries of service were built.

Next, the area between the concrete sheet pile walls was excavated with classic systems up to the level of the slab of roof, executing this up to the zone of crossing.

The excavation of the zone of crossing, sheltered by the new concrete sheet pile walls of the structure, was realized from one of the sides for zones of 2.50 m, constructing each of them up to the floor of the underpass with a reinforced concrete slab.

For this operation, during the months of July and August, 1999, the traffic was cut in the Doctor Esquerdo's existing underpass.

Finally, once the whole excavation was finished, the paving was placed, together with the finishing and facilities, etc. of the underpass.

To construct the branches of the link with Alcalde Sainz de Baranda street, firstly the excavation of the zone for the prefabricated walls was carried out, as well as the abutments and the central support of the bridge (see figure 7).

Next it was executed the piles, the upper beams and the foundations of the retaining walls of the link with Alcalde Sainz de Baranda street.

Finally the post-tensioning of cables was carried out at the end of the superstructure of the bridge, as well as the execution of the corresponding test of dynamical load.

REFERENCES

Jurado, C. 1998. *Underpass in O'Donnell Street in the intersection with Doctor Esquerdo street and the connection of O'Donnell and Alcalde Sainz de Baranda streets.* Construction Project. Madrid

Bowles, J.E. 1996. *Foundation Analysis and Design.* Mc Graw-Hill Companies Inc.

Jiménez, J.A. et Alt. 1980. «*Geotecnia y Cimientos (I, II and III)*».Ed. Rueda.

Winter & Nilso, 1991. *Reinforced Concrete.* Mc Graw-Hill.

Shenebelli, G. *Les parois moulees dans le sol.* 1981 ED. Eyrolles.

Jiménez, J.A. et Al. 1980. «*Hormigón armado*». Gustavo Gili.

Concrete Solutions – Grantham, Majorana & Salomoni (Eds)
© 2009 Taylor & Francis Group, London, ISBN 978-0-415-55082-6

New bridge in the high velocity train to the northwest of Spain

C. Jurado
Polytechnic University of Madrid, Ingecal Ingenieros, S. L. Spain

ABSTRACT: The present article describes one of the most singular bridges in the area near to Madrid including aspects of the project and the construction of a bridge called E-8 "in pergola" over the existing railway that connects Madrid with the city of Alcobendas and with the north of Spain. The structure has 140 m length with a span of 20 m. and a vertical clearance of 7 m.

The project of the bridge was realized by means of a three-dimensional model of finite elements with the program SAP2000N that includes all the elements of the structure. The project of the bridge was realized during the year 2004 and the construction during the years 2005 and 2006.

Aspects of the model included in this paper are: the 3D finite element model of the bridge and relevant aspects of the construction.

1   INTRODUCTION

During the years 2001 to 2007 in Spain the new high speed railway line that joins Madrid with Valladolid, with a length of 179,6 kilometres has been constructed.

The time estimated to cover the distance is 50 to 55 minutes, which supposes an average speed of 215 kilometres per hour, but the railway platform is prepared to reach 350 kilometres per hour.

The cost of the works has ascended to 264.8 million Euros.

Between the most singular structures that are present in the nearest zone of the high speed line to Madrid, is the bridge E-8, called in Spanish "in pergola" type, to solve the crossing with the line of the suburban rail network in service, which joins Madrid with the near city of Alcobendas.

The adopted solution was determined by the crossing of the two railway platforms with a very acute angle of 25°. To minimize the total height of the superstructure, accomplishing the vertical clearance of 7.00 m, demanded by the Official Trains Department (RENFE) for the railway below, we came to the conclusion that the most suitable solution was a bridge called in Spanish "in pergola", constituted by prefabricated prestressed beams, perpendicular to the abutments, which cover a span of 20 m. and a slab of reinforced concrete.

The prestressed beams have a height of 1.00 m, and the thickness of the slab over them is 25 cm. These beams rest on two lintel girders of reinforced concrete of 100 cm. height and with a length of 140 meters.

The oblique angle of the intersection means a length of the abutments of the bridge in the sense of the lower railway Madrid-Alcobendas, of 140 m.

The foundation of the bridge was solved by means of piles of reinforced concrete with 1.00 m. diameter and 16.25 m. length, which are joined by a foundation beam of 1.60 m. height and 120 m. width. The

Figure 1.   Aerial view of the crossing between the new and the existing railways.

piles in the zones of "the pergola" continue upward as cylindrical pillars of 0.70 m. of diameter.

The inaugural trip from Madrid to Valladolid has been made, with satisfactory results for the minister of Civil Works on November 15, 2007.

During the trial phase the train has reached a velocity of 330 kilometres per hour.

The line Madrid – Valladolid was put in service to the public on December 23, 2007.

2   THE PROJECT

The project of the bridge was made by means of a three-dimensional finite element model using the program SAP2000N that included all the parts of the structure, such as: piles, foundation tie beams on the piles in the zone of "the pergola", abutments, prestressed beams,

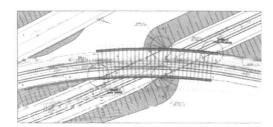

Figure 2. Aerial view of the bridge E-8 "in pérgola".

Figure 3. Plan crossing of the railway lines.

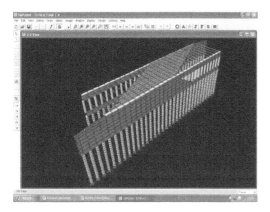

Figure 4. Three-dimensional FEM with 1664 nodes, 1323 frame and 592 shell elements.

the deck and the girders to support the prestressed beams.

The finite element model consisted of 1664 nodes, 1323 elements FRAME and 592 elements SHELL (see figure 4).

In view of oblique angle of the crossing of the new railway line of high speed over the existing railway Madrid-Alcobendas, with a very acute angle of 25°, it proved to be necessary to use very long retaining walls in the zone of the acute angles East in the abutment E-1 (side Madrid), and West in the abutment E-2 (side Valladolid).

The deck comprised 29 prefabricated prestressed beams of 1.00 m height which supports a massive slab

of reinforced concrete with 25 cm. thickness. The prestressed beams have an average separation between the axes of 2.50 m. in the centre line of the low railway from Madrid to Alcobendas, with the axes of the prestressed beams in a radial direction, according to the circle of the inferior railway.

The transversal width of the deck in plan is 15 m., with an average vertical clearance of 7,00 m under it, over the top face of the rail of the line of the Madrid-Alcobendas railway.

The distance between the supports of the prestressed beams is 20.80 m and they are supported by a lintel girder of reinforced concrete that in the zone of the abutments has a width of 1.10 m and a height of 1.00 m, where the slab of the deck does not exist, constituting both the zones of the "pergola".

The top slab of the deck connects with the abutments, so that the deck collaborates with the walls to resist the earth pressures and at the same time the back wall absorbs the displacements of the deck, which reduces the height of the neoprene bearings of the prestressed beams, constituting an integral bridge.

If the above mentioned connection were not realized between the deck and the top wall of the abutments, the movements at the top of the back wall, due to the push of the terrain and to the brakes of the train would be so important, that they would force the use of very big expansion joints between the abutment and the deck and on the other hand, the horizontal movements induced by the mobile overcharge of the great velocity train, would force the use of very big bearings, or furthermore vertical bearings in the end of the prestressed beams.

Finally, to avoid the appearance of differential settlements, between the adjacent embankment and the structure, it was infilled with a mix of granular material with cement, called in Spanish "technical blocks".

The loads considered in the project, have been those recommended by Eurocode 1 in the design of railway bridges in Spain, in agreement with the current Guidelines of railway bridges IPF-75 and with the new Code IAPF-2007. The live loads of traffic corresponded to different types of trains, with the most determining being trains of type A: UIC-71 and UIC-80 and trains type B: SW/0 and SW/2, the first for passengers trains and the second for goods trains.

A static calculation of the three-dimensional model of the bridge was performed with more than a hundred load combinations, with a dynamic modal spectral calculation.

3 DESCRIPTION OF THE STRUCTURE

The total length of the bridge in "pergola" is 140 m., with a span between the interior faces of the abutments of 20 m.

The deck comprises 29 prefabricated prestressed beams of 1.00 m of height which support a slab of reinforced concrete with 25 cm. thickness. The slopes of the deck were realized by means of mortar. The

Figure 5. Zone of the "pergola" with pillars 0.70 m. of diameter.

Figure 6. Aerial view of the railway crossing under construction.

prestressed beams have an average separation between axes of 2.50 m.

The width of the deck is 15 m., with an average vertical under-clearance of 7.0 m over the top face of the rail of the Madrid-Alcobendas railway line.

The span to calculate the prestressed beams is 20.80 m and they rest on neoprene bearings over a girder lintel of reinforced concrete, which has a width of 1.10 m and a height of 1.00 m.

The abutments in the zone of the earth pressure consisted of an 80cm thick wall of reinforced concrete and in the remaining zones, corresponding to the zones "in pergola", by circular pillars of 0.70 m of diameter, fixed in the lintel girder above and each placed each under the prefabricated beams.

The abutments have a circular disposition in plan with a slight curvature corresponding to the railway line of the train from Madrid to Alcobendas (figure 3).

The length of the piles under the foundation girder is 16.25 m. The number of piles of 1.00 m diameter is 33 for every abutment, while in the zone of the "pergola" they continue upward in pillars of 0.70 m of diameter, which support the lintel girders where the prefabricated prestressed beams rest.

Both the pillars and the wall of reinforced concrete are fixed by two girders of 1.20 m width and 1.60 m height, of reinforced concrete. These girders are used also to tie the piles of 1.00 m of diameter, being these placed in the vertical line of the pillars and neoprene bearings of prefabricated prestressed beams.

4 CONSTRUCTION PROCESS

The construction of the bridge "in pergola" in the high speed Madrid - Valladolid line, has been realized with the existing railway of the Madrid-Alcobendas line in service.

This conditioning situation imposed the solution of a prefabricated lintel with prestressed beams H-100, being installed during the nights from 12.00 a.m. to 6.00 a.m., while the trains were out of service.

The piles of the west side were compulsorily constructed with the aid of bentonite mud due to the sandy character of the ground and to the existence of groundwater with a phreatic level, which produced the falling of the walls of the perforation of the pile.

The phases of construction were organized following the modules:

– Piles
– Foundation girders
– Pillars
– Wall abutments
– Lintel Girders and neoprene bearings
– Placement of prestressed beams
– Slab of compression

Once the bridge was constructed, a test of static and dynamics loading over the bridge was carried out with satisfactory results.

After the base of the platform was finished, the rails and the catenaries were placed.

Finally the bridge was put in service the 23rd of December of 2007 with satisfactory results.

REFERENCES

MOPU. *"Instrucción relativa a las acciones a considerar en el proyecto de puentes de ferrocarril. IPF-75"*

Comisión redactora IAPF. *"Instrucción de acciones a considerar en el proyecto de puentes de ferrocarril. Borrador año 2007*

Ministerio de Fomento. *"Instrucción sobre las acciones a considerar en el proyecto de puentes de carretera. IAP-98"*

Ministerio de Fomento. *"Norma de construcción sismorresistente: parte general y edificación"*. NCSE–O2 (Real Decreto 997/2002).

Ministerio de Fomento. *Bases de calculo y diseño de pretiles en puentes de carretera*.

Eurocódigo 1. *"Bases de proyecto y acciones en estructuras"*

Eurocódigo 2. *"Proyecto de estructuras de hormigón"*

Jiménez S. & others. Eurocódigo 3. *"Proyecto de estructuras de acero"*.

Jiménez S. & others. *Geotecnia y Cimientos*. Rueda.

Ventura E. 1985. *Síntesis Geotécnica de los suelos de Madrid y su Alfoz*. Ministerio de Trasportes y Comunicaciones.

Jiménez Montoya, P., García Meseguer, A. & Morán Cabré, F. *Hormigón armado*. Ed. Gustavo Gili. 14ª edición.

Concrete Solutions – Grantham, Majorana & Salomoni (Eds)
© 2009 Taylor & Francis Group, London, ISBN 978-0-415-55082-6

Bond strength of cementitious and polymer modified repair materials

M.I. Khan, S.H. Alsayed, T.H. Almusallam, Y.A. Al-Salloum & A.A. Almosa

Department of Civil Engineering, College of Engineering, King Saud University, Saudi Arabia

ABSTRACT: A major factor that influences the durability of the repair is the bond strength between the new material and the substrate. Direct tension bond testing establishes the location of the weakest link of the composite. In this investigation, the most common repair materials, produced by international reputed manufactures, available in the local market were selected and tested for tensile bond strength. Tensile bond strength was carried out using slabs with application of the repair materials under investigation. Various surface preparation techniques were employed to assess the influence of the surface treatment. Direct tensile strength was also measured using briquettes. It has been observed that the bond strength is influenced by surface treatment and material type.

1 INTRODUCTION

Repair deterioration rates have been aggravated and accelerated by poor construction resulting from shortcomings in design, specifications, supervision, workmanship and quality control. Inadequate workmanship, procedures, or materials results in inferior repairs which ultimately fail. The need for repairs can vary from such minor imperfections to major damage resulting from structural failure. This method of repair can only prove to be successful if the new material interacts well with the parent concrete and forms a durable barrier against ingress of carbon dioxide and chlorides. The properties and durability of repair systems are governed by properties of the three phases namely, repair, existing substrate, and interface (transition zone) between them (Vaysburd et al. 2001). Properly designed, implemented, and functioning man-made systems, with a minimum number of undesirable side effects, require the application of a well-integrated systems approach as reported by Emmons & Vaysburd (1995). Smoak (2002) reported that in evaluating the causes of failures, it is essential to consistently use a systematic approach to concrete repair. The main goal of any concrete repair system is a high performance structure, regardless of what materials have been used in repairing the deteriorated structure (Poston et al. 2001).

A major factor that influences the durability of the repair is the bond strength between the new material and the substrate. The bond strength between two concrete layers is highly influenced by the roughness of the substrate surface and several factors address the adhesion between the two materials. These include: the properties of the substrate, the roughness, micro-fractures and porosity of the substrate, the properties of the repair materials, the curing procedure of the repair materials, the loading conditions

and the environmental and serviceability conditions. A durable repair may not be produced unless the effects of all such factors are considered during the design stage and when selecting the repair materials. Direct tension bond testing establishes the location of the weakest link of the composite.

Preparation of the old concrete for application of the repair material is of primary importance in the accomplishment of durable repairs. A good repair material will give unsatisfactory performance if applied to weak or deteriorated old concrete. The repair material must be able to bond to sound concrete. It is essential that all of the unsound or deteriorated concrete be removed before new repair materials are applied. The first step in preparing the old concrete for repair is to saw cut the perimeter of the repair area to a depth of 25 to 40 mm, cutting behind the steel to a depth of at least 25 mm and removing any corrosion deposits Then all deteriorated or damaged concrete must be removed from the repair area to provide sound concrete for the repair material to bond to the substrate. Reinforcing steel exposed during concrete removal requires special treatment. After the repair area has been prepared, it must be maintained in a clean condition and protected from damage until the repair materials can be placed and cured.

In Saudi Arabia, the majority of the concrete structures constructed more than three decades ago suffer because of lack of quality control and severe weather conditions. Therefore, the structures deteriorate and need urgent repair. The most common form of rehabilitation for a deteriorated structure is to remove the deteriorated concrete and replace it with a new material, therefore assessment of the tensile bond between the old concrete and the repair material is paramount for the durability of the repair. The results presented in this paper are the part of a project on performance of repair materials available in the Kingdom of Saudi Arabia (Alsayed et al. 2008).

Table 1. Details of cementitious repair mortars (CRM).

Material	Properties	Application
CRM-1	Shrinkage controlled	Hand applied Vertical & Overhead
CRM-2	Shrinkage controlled	Spraying Vertical & Overhead
CRM-3	Multi purpose For hot climates	Hand applied Vertical & Horizontal

Table 2. Details of cementitious polymer modified mortars (CPMM).

Material	Properties	Application
CPMM-1	Multi purpose For hot climates	Dry / Wet spray Vertical, Overhead & Horizontal
CPMM-2	For hot climates	Wet spray Vertical & Overhead
CPMM-3	Multi purpose	Hand applied

2 EXPERIMENTAL PROGRAM

2.1 Material selection

The selection of the repair material was based on the best selling materials manufactured by the top three companies Fosroc, Sika and MBT available in the Kingdom of Saudi Arabia. The designation, properties and application of these cementitious repair mortars and cementitious polymer modified mortars are shown in Tables 1 and 2, respectively.

2.2 Testing procedure

2.2.1 Pull-off test
This type of test measures the tensile bond or tensile strength of surface repairs and overlays. Direct tension bond testing establishes the location of the weakest link of the composite. The pull-off test involves applying a direct tensile load to a partial core advanced through the overlay material and into the underlying concrete until failure occurs. The tensile load is applied to the partial core through the use of a metal disk with a pull pin, bonded to the overlay with an epoxy resin. A loading device with a reaction frame applies the load to the pull pin. The load is applied at a constant rate, and the ultimate load is recorded. Figure 1 illustrates the principle of the pull-off test.

There are four different possible modes of failure when applying load in this manner as follows:

1. *Failure occurs at the bond surface*: the ultimate load is a direct measure of the adhesion between the overlay and the substrate concrete.
2. *Failure occurs between the disk and the overlay surface*: the tensile strength of the overlay system is greater than the failure load, and a stronger adhesive is needed.

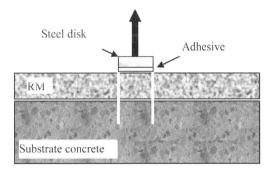

Tensile force (Pull-off)

Figure 1. Schematic diagram of pull-off test for repair material (RM).

Figure 2. View of slab prepared for application of repair material.

3. *Failure occurs in the overlay material*: the repair material is the weakest portion of the system.
4. *Failure occurs in the substrate*: the repair concrete and the bond are stronger than the existing concrete, and the repair can be considered successful provided the pull off result is adequate.

For the measurement of pull-off force, slabs having dimensions $1000 \times 1000 \times 100$ mm were cast. In order to assess the efficiency of surface preparation, three different methods of surface preparation were employed, namely: sandblasting, grinding and hammer (Fig. 2).

After preparation of the surfaces of the slabs, repair materials, as listed in Tables 1 and 2, were applied. After 28 days, pull-off force was applied as per the principle shown in Figure 1. The test was conducted at two places in each type of surface.

2.2.2 Tensile strength
To determine the direct tensile strength of a repair material, a uniaxial test is needed which is simple, economic and truly uniaxial [RILEM, 1963]. The most successful methods have been those in which the tensile loads have been applied to the concrete through friction-grip systems, through bars embedded in the concrete or through plates stuck onto the end of the

specimens (Johnston & Sidwell, 1968). High tensile strengths in the early ages (0 to 28 days) are also important so that cracking due to the restraint to shrinkage is avoided.

Mortar briquettes were used for the measurement of direct tensile strength in accordance with ASTM C190-85. The measurements were taken at 3, 7, and 28 days using three specimens for each age and the average is reported as a result.

The manufacturers' recommended mixture proportions and mixing and curing procedures were followed for each repair material. Curing was followed exactly as per manufacturers' recommendations where practicable.

3 RESULTS AND DISCUSSION

3.1 Pull-off test results

Tables 3 and 4 show the pull-off test results of cementitious repair mortars and cementitious polymer modified mortars, respectively. From this data it is apparent that the bond tensile strengths are relatively low as compared to the briquette tensile strength. In all cases except CPMM-2, sandblasted surfaces and ground surfaces showed better bond between repair material and substrate concrete. All types of materials except CPMM-2 failed at the bond line using the hammered surface. CPMM-2, failed at the bond line in all surface types which shows its lower resistance against bond with the substrate concrete. Sandblasted and ground surfaces showed the best performance. Most of the materials failed in the substrate concrete which shows that the repair is successful. Some of the failure types are shown in Figure 3.

CRM-1 and CRM-2 failed 8–10 mm below the joint with the substrate and 20–25 mm below the joint with the substrate, respectively using sandblasted and ground surfaces. Similarly, in the case of CPMM-2 and CPMM-3 the failure occurred 5–20 mm below the joint with the substrate and 5–12 mm below the joint with the substrate, respectively using sandblasted and ground surfaces. It is evident from these results that materials CRM-1, CRM-2, CPMM-2 and CPMM-3 have an acceptable bond strength for the use as repair material using proper surface preparation.

Furthermore, sandblasted surface and ground surfaces provided very similar types of results with some exceptions. The bond strength obtained from these surfaces demonstrated good correlation in both cementitious repair mortars and cementitious polymer modified mortars as shown in Figure 4. However, the results for the ground surface showed higher results at low values of bond strength which are related to cementitious polymer modified mortars.

3.2 Tensile strength (Briquettes)

Tensile strength development (Briquettes) of cementitious repair mortar at various ages is shown in Figure 5.

Table 3. Bond strength and failure location of cementitious repair mortars (CRM).

Surface condition	Strength, MPa	Location of failure
CRM-1		
Control	2.12	15 mm below bond line in substrate
Sandblasting	2.60	10 mm below bond line in substrate
Grinding	2.55	8 mm below bond line in substrate
Hammer	1.36	At bond line
CRM-2		
Control	2.38	15 mm below bond line in substrate
Sandblasting	2.66	25 mm below bond line in substrate
Grinding	2.49	20 mm below bond line in substrate
Hammer	1.42	At bond line
CRM-3		
Control	2.32	45 mm within bonding material
Sandblasting	2.15	20 mm on Substrate
Grinding	2.26	35 mm within bonding material
Hammer	1.36	At bond line

Note: Depth reported is the average of two readings.

Table 4. Bond strength and failure location of cementitious polymer modified mortar (CPMM).

Surface condition	Strength, MPa	Location of failure
CPMM-1		
Control	1.92	At bond line
Sandblasting	1.53	At bond line
Grinding	1.92	At bond line
Hammer	1.19	At bond line
CPMM-2		
Control	1.39	At bond line
Sandblasting	1.25	5 mm below bond line in substrate
Grinding	1.53	20 mm below bond line in substrate
Hammer	0.99	8 mm below bond line in substrate
CPMM-3		
Control	2.6	5 mm from top
Sandblasting	1.92	12 mm below bond line in substrate
Grinding	2.15	5 mm below bond line in substrate
Hammer	1.36	At bond line

Note: Depth reported is the average of two readings.

Figure 3. View of failure types.

CRM-1 showed higher tensile strength as compared to CRM-2 and CRM-3 at 3 and 28 days whilst at 7 days it is close to CRM-3 but higher than CRM-2. At 3 days and 7 days, CRM-3 showed a higher strength than that

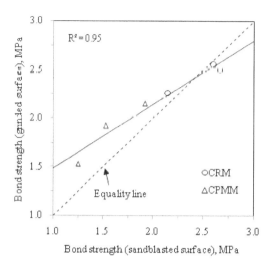

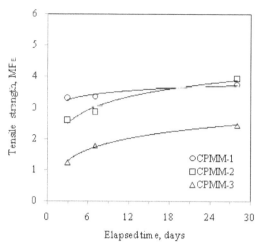

Figure 4. Comparison of bond strength of sandblasted and ground surfaces of repair materials.

Figure 6. Variation of tensile strength with age of cementitious polymer modified mortars.

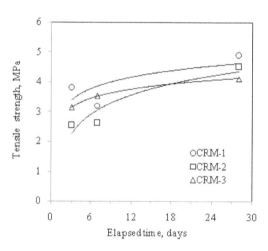

Figure 5. Variation of tensile strength with age of cementitious repair mortars.

of CRM-2, however, at 28 days it was slightly lower than CRM-2. All materials demonstrated that variation is wider at early ages whereas this variation in the tensile strength is less at 28 days. At 7 days, the rate of tensile strength development of CRM-3 was 12% and CRM-1 was about 3%. At 28 days, the tensile strength attained by CRM-3 and CRM-1 was 4.1 MPa and 4.5 MPa, respectively. The rate of increase in the tensile strength of CRM-3 from 3 to 28 day was 31% whilst for CRM-1 it was 78%.

The tensile strength development of cementitious polymer modified mortar at various ages is presented in Figure 6. The tensile strength of CPMM-1 and CPMM-2 was higher than that of CPMM-3 at all ages investigated. At 3 day, CPMM-1 showed the highest tensile strength about 3.3 MPa whereas CPMM-3 showed the lowest at about 1.2 MPa. The tensile

strength of CPMM-1 was higher than CPMM-2 at 3 days; this difference narrows down at 7 days. At 28 days both materials (CPMM-1 and CPMM-2) were almost similar. The rate of increase in tensile strength from 3-days to 28-days for CPMM-1, CPMM-2 and CPMM-3 was almost similar as can be seen from Figure 6.

4 CONCLUSIONS

Sandblasted and ground surfaces provide good and almost similar types of results with some exceptions. However, the results for the ground surface showed higher results at low values of bond strength which are related to cementitious polymer modified mortars. The hammer surface showed the lowest resistance to bond strength.

All materials investigated except CPMM-2 demonstrated good bond strength for the use as repair material provided on proper surface preparations.

CPMM-2, failed at the joint in all surface preparation types which shows its least resistance against bond with substrate concrete.

CRM-1 showed better tensile strength as compared to CRM-2 and CRM-3. However, the rate of gain of strength showed variations.

The tensile strength of CPMM-3 was lower than that of CPMM-1 and CPMM-2 at all ages investigated. Whereas the rate of gain of strength was almost similar in all three materials.

ACKNOWLEDEMENT

The authors are thankful to the Deanship of Scientific Research of King Saud University for funding this investigation.

REFERENCES

Alsayed, S.H., Al-Salloum, Y.A., Almusallam, T.H. & Khan, M.I. 2008. Classification and testing performance of concrete repair materials for durable repair under local environmental conditions. Final Report, Research project funded by Deanship of Scientific Research, King Saud University, Riyadh, Saudi Arabia.

Emmons, P.H. & Vaysburd, A.M. 1995. The total system concept – Necessary for improving the performance of repaired structures. *Concrete International*, 17(3): 31–36.

Johnston, C.D. & Sidwell, E.H. 1968. Testing concrete in tension and compression. *Magazine of Concrete Research*, 20(65): 221–228.

Poston, R.W., Kesner, K., McDonald, J.E., A.M. Vaysburd, & Emmons, P.H. 2001. Concrete repair material performance – laboratory study. *ACI Material Journal*, 98(2): 137–145.

RILEM, 1963. Direct tensile tests of concrete. *RILEM Bulletin*, No. 20, 84–90.

Smoak, W.G. 2002. Guide to concrete repair. *USBR report*, Denver Federal Center Denver, Colorado, USA.

Vaysburd, A.M., Sabnis, G.M., Emmons, P.H. & McDonald, J.E. 2001. Interfacial bond and surface preparation in concrete repair. *The Indian Concrete Journal*, 27–33.

Concrete Solutions – Grantham, Majorana & Salomoni (Eds)
© 2009 Taylor & Francis Group, London, ISBN 978-0-415-55082-6

Restoration of historic hydrotechnical concrete structures

M. Kosior-Kazberuk
Bialystok Technical University, Bialystok, Poland

M. Gawlicki
University of Science and Technology, Cracow, Poland

ABSTRACT: The Augustow Canal with 18 locks and 22 sluices was built between 1824 and 1839. The main aim of the repair and protection of the canal structures is to preserve their historical and technical values as well as the authenticity and integrity of the waterway. The investigation of the 19th century concrete structures is important because significant changes determining the applicable properties of hydraulic binders took place in the 19th century. The paper presents the methods applied for diagnostics of the state and the failure identification occurring in structural elements. The results of studies on mortar samples collected from the walls of the Canal's locks (by chemical analysis, XRD, DTA/TGA, E-SEM and SEM/EDS, etc.) are presented. The repair strategy of historic structures constructed along the Wisla – Neman Canal is described.

1 INTRODUCTION

The subjects of repair are usually those concrete structures constructed not earlier than one hundred years ago. Usually, the concrete is made of Portland cement. Concrete structures built in the 19th century are evaluated very rarely. The investigation of 19th century concrete structures is important because of the significant changes determining applicable properties of hydraulic binders which took place in the 19th century. An excellent example to assess the role of historical binders in structures is the Augustow Canal – a system of hydrotechnical structures built using water resistant artificial lime mortar.

The waterway is located on the territory of two countries: Poland (about 80 km, 14 locks) and Belarus (21,2 km, 3 locks). The eighteenth lock is situated exactly on the border. The canal which also has 22 sluices, 14 drawbridges and 65 simple road bridges, was built between 1824 and 1839 and is still operating. The working system, the route and the location of the locks have undergone no fundamental change on the Polish side since the canal was built.

The most important structures are locks, the function of which is not only to dam and maintain an appropriate level of water but also to enable vessels using the canal to manage the changing level of water (Batura 2000). They are shaped like chambers, made of broken stone with a mortar of hydraulic lime of the Vicat type, (Moropoulos et al. 2005, Landsberg 1999), similar to Portland cement.

The study of existing mortars in historic structures is an important aspect of building conservation: the choice of their components varied according to historic period and regional habits. The aim of the investigation

was to asses the condition of the main structural elements of the waterway – the lock walls. The condition of the system depends on the current state of the hydraulic lime mortar cementing the bricks and stones in the lock walls. The basic properties of the binder, the products formed as the result of chemical reactions and changes of the binders during 180 years were determined. The present paper concerns the state and the repair of structural elements of the canal objects only.

2 CANAL CONSTRUCTION AND EXPLOITATION

The Augustow system was the first complex of water constructions that was able to meet very high European standards. It was a rare phenomenon, with the exception of the Swedish Got's Canal and the English Caledonian Canal, to directly connect lakes and rivers rather than built side canals. Although the constructors lacked experience in building brick-and-stone chamber locks, the durability of those constructions is still commendable.

Among lots of problems concerning the channel construction, the difficulties in gaining suitable building materials were significant. The spending on building materials was high and some problems could not be solved using standard ways. The lack of binders resistant to long-term water influence was particularly inconvenient.

For the purpose of the Augustow Canal construction the technology of artificial hydraulic lime production was elaborated on the basis of the L. J. Vicat procedure, published in 1818. Local available raw materials were used. It was a significant technological and organizing achievement in those days. Probably, it was the first case of cement manufacturing, on such a large scale, in the world. Later, this cement was used in such important buildings as citadels, overbridges and others.

During construction of the canal more than 8300 tons of artificial hydraulic lime, derived from the burning process of a clay and lime mixture, were used (Batura 2000).

3 DIAGNOSIS OF THE CONDITION OF THE CANAL CONSTRUCTION

The selected set of samples has been characterized using a combination of chemical and microscopical techniques, (Elsen 2004, Sabbioni 2002). The samples of mortar were collected from numerous testbore-cores drilled from the stone-brick walls of lock chambers. The shafts of the walls were built of cobblestones of different dimensions, mainly granite boulders and crushed ceramics, cemented with artificial hydraulic lime mortar. The method of application of the historic mortar was as follows: the wall of cobblestones was formed and the spaces left in the pile of stones were filled with mortar of liquid consistence.

The water absorption and the density of mortar were tested on specimens devoid of coarse aggregate. The average water absorption was 17.8% and the bulk density, 1650 kg/m^3. The average value of compressive strength of the mortar tested on core samples was 17 MPa.

Chemical analysis, X-ray diffractometry (XRD), differential thermal analysis (DTA), thermogravimetry analysis (TGA) and scanning electron microscopy (E-SEM and SEM/EDS) were applied to characterize the microstructure and also to determine the hydration products of mortars.

The morphology of the mortar was investigated using the scanning electron microscope JEOL 5200 equipped with the energy dispersive X-ray analyzer EDS Link ISIS and the environmental scanning electron microscope E-SEM Philips XL 30 operated with a low vacuum in the specimen chamber. The observations were conducted with magnification from 50 to 1000 times. The selected micrographs and EDS results are presented in Figure 1.

A dense microstructure without any crystals, apart from the aggregate grains, was observed.

The pastes contained randomly oriented, irregular forms of secondary calcite, which originated in the carbonation process. The forms of CSH phase typical for present hardened binders were not found (Ramachandran & Beaudoin 2001).

The results of chemical analysis of mortar samples derived from different depths of the lock wall are presented in Table 1.

Considering the results given in Table 1, the lime used in the Augustow Canal construction was a binding material of strong hydraulicity and high content of acidic oxides: SiO_2, Al_2O_3, Fe_2O_3. The hydraulic

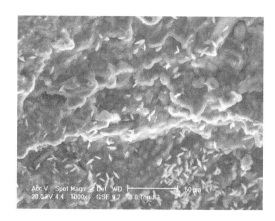

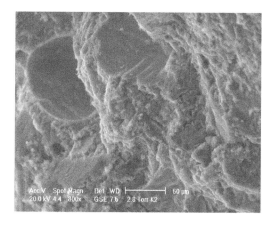

Figure 1. E-SEM micrographs of historic mortars derived from the lock's walls.

Table 1. Results (wt. %) of chemical analysis of artificial hydraulic lime mortars.

Component	K1 1.0 m from wall face	K2 3.5 m from wall face	K3 4.0 m from wall face
Insoluble matter	32.31	40.34	40.11
Loss on ignition	23.28	18.22	18.09
SiO_2	11.38	10.04	10.12
Fe_2O_3	1.75	1.67	1.70
Al_2O_3	3.80	3.45	3.58
CaO	23.24	21.40	21.98
MgO	4.11	4.06	4.07
SO_3	0.23	0.06	0.04
Cl^-	0.179	0.023	0.020
$Ca(OH)_2$	0.04	0.04	0.05

modulus (HM $= CaO[\%]/SiO_2, Al_2O_3, Fe_2O_3[\%]$) of binder was 1.7. The weight ratio of ignited binder and sand was very close to 1:1. It can be supposed that the binder tested contained relatively little amount of free CaO, and belite C2S, and gehlenite C2AS were probably its basic components, (Callebaut et al. 2001).

A significant amount of periclase, calcium aluminates CA, $C_{12}A_7$ and calcium aluminoferrite $C_2(A,F)$

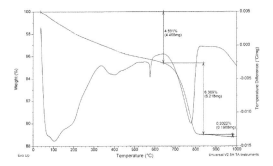

Figure 2. Thermal analysis of K4 sample.

also occurred in the hydraulic lime. The presented results do not indicate the corrosive influence of the environment on the lock walls, although the content of sulfates and chlorides in the K1 sample is significant. The long-term environment influence could cause cumulation of chloride ions; moreover the lock's walls were subjected to biological influences (e.g. plant roots).

The X-ray diffraction analysis was done using a Philips X'PERT PRO diffractometer, according to the Debye-Scherrer-Hulle powder diffraction method. CuKα radiation was used and the measuring angle range was as wide as 5 to 70° 2Θ, the tube voltage and current were 40 kV and 30 mA, respectively. The patterns obtained were compared with the PDF database (ICDD) – the component of the diffractometer software system.

The main component of the mortars was fine calcium carbonate (calcite), formed as a result of carbonation of the binder constituents' reaction with water (Callebaut et al. 2001). Strong peaks due to quartz – the basic crystalline phase of the aggregate – and feldspars (microcline and orthoclase) were observed, (Kosior et al. 2006). Other components identified in the mortar were:

Scawtite Ca$_7$[(Si$_6$O$_{18}$)(CO$_3$)] · 2H$_2$O and

Nesquehonite MgCO$_3$ · 2H$_2$O

as well as calcium hydrosilicates containing substantial amounts of aluminium and magnesium ions. There was some slight content of gypsum in the K2 and K3 samples. In any samples tested there were no lines that can be ascribed to portlandite, typical forms of the C-S-H phase, hydrogehlenite, hydrogarnets or C$_3$A · CaCO$_3$ · 12H$_2$O (Gosh 1983).

The Differential Thermal Analysis and Thermogravimetric Analysis were carried out using Universal V2.5H TA Instruments. The measurements were conducted in the range of temperature changes from 20 to 1000°C. An example thermogram of the tested mortar is presented in Figure 2.

The DTA curves showed two endothermic peaks at 500–1000°C. The first one, with a maximum at 573°C, is associated with polymorphous transformation of β-SiO$_2$ in α-SiO$_2$ (quartz is the main component of sand presented in mortars). The other one, with extreme

between 720 and 780°C, is attributed to decarbonation of CaCO$_3$, formed in the process of CO$_2$ reaction with binder hydration products (Ramachandran et al. 2003). The TGA curves, recorded at a temperature below 500°C for all samples tested, showed nearly linear weight loss with temperature increase, whereas, for the DTA curves, several endothermic peaks occurred, with maxima at 80–100°C, about 180°C, 330°C and 380–400°C.

The polymineral composition of the material tested, the lack of well-formed crystalline structure of major components, the presence of a number solid solutions and multi-stage processes occurring such as dehydration, dehydroxylation and decarbonisation make it difficult to attribute unequivocally the peaks indicated on the DTA curves.

Probably, the peaks at a temperature below 200°C should be attributed to the dehydration of unidentified semicrystalline products of hydration of gehlenite, C-S-H phase, magnesium-aluminium hydroxide and their solid solutions. The peaks observed on the DTA curve at higher temperature are mainly connected with scawtite and nesquehonite decomposition.

The studied mortars were generally in good technical condition, with no signs of chemical corrosion except advanced carbonation.

4 RANGE OF REPAIR

The canal installations were damaged greatly during World War I in 1915, as well as in battles of 1919, 1920 and 1941–1945.

Rebuilding and small modernization works were carried out during the mid-war periods. The ceramic facing on the lock's walls was replaced with concrete blocks or cut granite stones. The elements made of wood and metal were renewed. The movable bridges were replaced with fixed ones.

After World War II, the canal was rebuilt on the territory of Poland and since 1950 has been navigable over a length of 77.5 km. Ten out of 15 locks kept the original construction of the 19th Century and most of the rest kept the original brick-and-stone walls cemented with artificial hydraulic lime mortar. The current condition of the lock is presented in Figure 3. One lock only underwent thorough total rebuilding: its walls were pulled down and the lock was thoroughly modernized. However, its original historical appearance was preserved. The lock situated exactly on the border between Poland and Belarus was out of operation from 1939. It was in a ruined state, which ruled out any kind of water transport (overgrown canal and ponds, disintegration of walls and sluices and much of the machinery abandoned and dangerous). Its restoration was conducted in 2004–2007.

5 CONCLUSIONS

The application of combined test methods allowed the correct formula of compatible restoration materials to

Figure 3.   Restored lock of the Augustow Canal.

be found (Konow 2003), and provided valuable information for conservation purposes. The morphology, phase and chemical composition of products formed as the result of chemical reactions and changes of binders in contact with water and subjected to environmental influence over 180 years, were described. The results of the characterisation of samples from the lock walls clearly indicate the importance of microstructural and chemical analysis as the first step in characterisation of historic mortars because of the complexity and heterogenity of this composite material. Hydraulic lime mortar has exhibited durability through 180 years. Thanks to using the innovative cementing material during construction and suitable preservation and repair, the waterway has kept an almost unchanged form, and the majority of locks are preserved in their original construction from the 19th century. Despite the ravages of wars and the forces of nature, the Polish part of the canal has been maintained in a navigable state.

REFERENCES

Batura, W. 2000. *The Augustow Canal. The masterpiece of human hands and nature.* Torun.
Moropoulos, A., Bakolas, A., Anagnostopoulou, A. 2005. Composite materials in ancient structures. *Cement and Concrete Composites* 27: 295–300.
Landsberg, D. 1999. The history of lime production and use from early times to the industrial revolution. *Zement-Kalk-Gips* 45(6): 269–273.
Elsen, J., Brutsaert, A., Deckers, M., Brulet, R. 2004. Microscopical study of ancient mortars from Tournai (Belgium). *Materials Characterization* 53: 289–294.
Sabbioni, C., Bonazza, A., Zappia, G. 2002. Damage on hydraulic mortars: the Venice Arsenal. *Journal of Cultural Heritage*, 3: 83–88.
Ramachandran, V.S., Beaudoin, J.J. 2001. *Handbook of analytical techniques in concrete science and technology.* New York: Noyes Publications/William Andrew Publishing.
Callebaut, K., Elsen, J., Van Balen, K., Viaene, W. 2001. Nineteenth century hydraulic restoration mortars in the Saint Michael's Church (Leuven, Belgium). Natural hydraulic lime or cement? *Cement and Concrete Research* 31: 398–403.
Kosior–Kazberuk, M, Gawlicki, M., Rakowska, A. 2006, Studies of mortars produced from artificial hydraulic lime. *Structural Analysis of Historical Constructions, Proc. 5th Int. Conf., New Delhi, India, Vol. 2*: 699–705.
Ghosh, S.N. 1983. *Advances in Cement Technology.* Oxford: Pergamon Press.
Ramachandran, V.S., Paroli, R.M., Beaudoin, J.J., Delgado, A.H. 2003. *Handbook of Thermal Analysis of Construction Materials.* New York: Noyes Publications/William Andrew Publishing.
Konow, T. 2003. Restoration concrete for historical constructions – scientific studies of old concrete samples from Finland. *In ECOMAT 2003 – Materials and Conservation of Cultural Heritage, Proc. of Symp. P2*, Lausanne.

Concrete Solutions – Grantham, Majorana & Salomoni (Eds)
© 2009 Taylor & Francis Group, London, ISBN 978-0-415-55082-6

Cracking resistance of concrete overlays as predicted from the development of shrinkage stress

S.A. Kristiawan, A.M.H. Mahmudah & Sunarmasto
Civil Engineering Department, Sebelas Maret University, Indonesia

ABSTRACT: Concrete overlays have been used as a method to repair deteriorated concrete pavement in many countries. Their performance depends on various factors including their resistance to shrinkage cracking. Evaluation of their performance against shrinkage cracking may be carried out by estimating the magnitude of induced tensile stress in overlays due to restrained shrinkage. A simple method for calculation of shrinkage stress in concrete overlays has been proposed by Silfwerbrand. However, the method neglected the fact that actual bond between overlays and concrete base may vary. For fully bonded overlays, there should be a considerable degree of restraint to the shrinkage of overlays. On the other hand, partially bonded overlays will only restrain a fraction of that shrinkage. Improvement of Silfwerbrand's method is proposed to account for variation in degree of restraint provided by different type of overlays. In addition, the new method also takes into consideration the development of elastic modulus at early age. Observations of restrained shrinkage cracking on samples of concrete overlays (fully bonded and partially bonded) are used to develop and verify the proposed method. The method could be employed to set criteria in eliminating the risk of cracking in concrete overlays.

1 INTRODUCTION

Concrete pavements will deteriorate over time due to many factors including the increase of heavy traffic loads. Concrete overlay is one of among many options that may be applied to repair and rehabilitate a deteriorated concrete pavement to extend its service life. The performance of a concrete pavement after overlay will increase in term of its structural capacity, rideability, skid resistance and reflectivity characteristics.

The repaired/rehabilitated concrete pavement as a result of overlay turns into a composite structure. The new concrete overlay will shrink considerably, while the shrinkage of the base layer is negligible. The resulting differential shrinkage creates a loading case that must be accounted for. In a composite structure, differential shrinkage leads to shrinkage stresses that are mainly compressive in the base layer and tensile in the overlay. This tensile stress could cause shrinkage cracking in the overlay if its magnitude is higher than the tensile capacity of the concrete overlay.

The occurrences of shrinkage cracking on overlays have been recognised by several authors (Brown et al. 2007, Hall & Bannihatti, 2005, Schlorholtz, 2000). The problem could be reduced if the design procedure of the concrete overlay incorporates requirements for checking shrinkage stress. Such procedures may be implemented only if there is a reliable model for predicting shrinkage stress. The model should be developed to include all parameters affecting cracking performance of overlays due to restrained shrinkage. These parameters are: magnitude of shrinkage, creep, elastic modulus, degree of restraint and tensile capacity of overlays. The existing model for estimating shrinkage stress in overlays was proposed by Silfwerbrand (1997). However, the model does not recognise the fact that different types of bonded overlays will provide different degrees of restraint. This paper offers improvement to Silfwerbrand's method by taking into account variation in degree of restraint provided by different types of overlays. In addition, the proposed method also takes into consideration the development of elastic modulus at early ages. Observations of restrained shrinkage cracking on samples of concrete overlays (fully bonded and partially bonded) are used to develop and verify the proposed method. The improved method may be used as a means to evaluate whether cracking due to restrained shrinkage will occur on particular overlays or not.

2 ESTIMATION OF SHRINKAGE STRESS ON OVERLAYS

Silfwerbrand (4) developed a method to calculate shrinkage stress on overlays based on the following assumptions:

a) Concrete in base and overlay are linear elastic materials
b) Poisson's ratio ν is set to zero
c) Plane sections remain plane after bending (Bernoulli's hypothesis)
d) The shrinkage of overlay is ε_{sh} throughout
e) The shrinkage of the base is neglected

f) A good bond between overlay and base provides complete interaction

g) The movement of concrete edges are free

The shrinkage stress σ occurred in overlay due to restrained shrinkage is estimated using the following equations:

$$\sigma = \mu E^* \varepsilon_{sh} \qquad (1)$$

$$\mu = \frac{m(1-\alpha)\{m(1-\alpha)^3 + \alpha^2(3+\alpha)}{m + (m-1)\{m(1-\alpha)^4 - \alpha^4} \qquad (2)$$

$$E^* = \frac{E_1}{(1+\phi)} \qquad (3)$$

where ε_{sh}, E^* and μ are shrinkage of overlay, modified elastic modulus of overlay and degree of restraint, respectively. The degree of restraint is affected by m and α which are, respectively, ratio of elastic modulus of concrete base E_2 to E^* and ratio of depth of overlay d to total thickness of overlay plus concrete base h. Creep coefficient Φ is used to obtain E^* from elastic modulus of overlay E_1.

The method of estimating shrinkage stress as given in Equation 1 and 2 clearly shows that the degree of restraint provided by bond between the concrete base and overlay does not reflect various bonds occurred in the practice of overlay. Therefore, this paper proposes value of degree of restraint μ to be multiplied by factor τ to account for the type of bond that exists between base and overlay.

Equation 1 and 3 also shows that a single value of elastic modulus of overlay E_1 is used for the calculation of shrinkage stress. In practice, shrinkage cracking could occur at early ages (less than 10 days after application of the overlay). It means over this short period of time, that the elastic modulus of the overlay is still developing in magnitude and calculation of shrinkage stress should take into account the development of elastic modulus of the overlay with time.

Considering the above arguments, the following equation is recommended to estimate shrinkage stress in overlay:

$$\sigma_{(t)} = \tau \mu E^*_{(t)} \varepsilon_{sh(t)} \qquad (4)$$

where subscript (t) indicates that the stress may be calculated at any time by considering the changing properties of shrinkage, creep and elastic modulus over time.

3 EXPERIMENTAL PROGRAMME

A concrete mix having a compressive strength of 25 MPa was cast to produce two samples of slabs with a size of 2×2 m and a depth of 150 mm. These slabs represented concrete pavements that would eventually be overlaid at the age of 7 months with a 30 mm thickness of mortar having proportions of 1 : 2.5 by mass of cement:fine aggregate, respectively and w/c ratio of

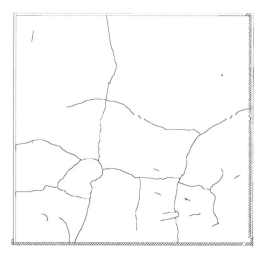

Figure 1. Cracks pattern in fully bonded overlay.

Figure 2. Cracks pattern in partially bonded overlay.

0.56. The reason for carrying out overlay at this age is that the development of shrinkage after this period is insignificant. There are two type of overlays simulated in this study. The fully bonded overlay was obtained by overlaying mortar on a rough surface of slab that had been prepared to attain good bond between concrete base and overlay. Surface preparation was carried out by hammering manually on the slab until mortar was partially removed leaving the surface of concrete slab with exposed aggregate. The partially bonded overlay was produced by overlaying mortar on the surface with no preparation. Concrete overlays were continuously monitored to identify the cracks (Figs 1–2). The time when cracks occurred was noted. It was found that cracks occurred after 7 and 10 days of drying for fully and partially bonded overlay, respectively.

Shrinkage and tensile creep of overlays were also determined by measuring their values on samples of prisms with dimensions of $100 \times 100 \times 500$ mm. Strain gauges were used for this purpose. Tensile creep was obtained at a loading equal to 30% of tensile

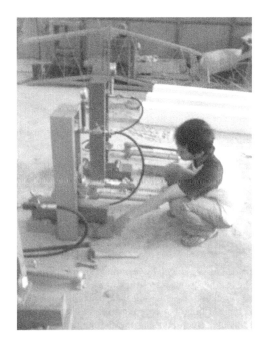

Figure 3. Loading of prism on flexural creep frame apparatus.

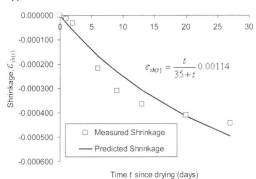

Figure 4. Shrinkage of overlay.

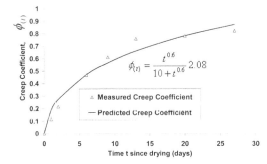

Figure 5. Creep coefficient of overlay.

strength. Arrangements of loading were similar to those of flexural tests (modulus of rupture) as seen in Figure 3. Measurement of shrinkage and tensile creep started from the age of 1 day and finished after 28 days

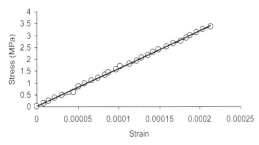

Figure 6. Stress and strain obtained during flexural tensile strength tests.

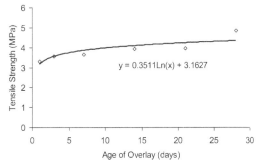

Figure 7. Tensile strength of overlay over time.

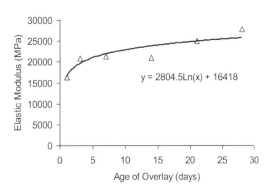

Figure 8. Elastic modulus of overlay over time.

and the results are given in Figures 4–5. It is noted that the results of tensile creep measurements are presented as creep coefficients. Equations for computed shrinkage and creep coefficient as seen in the figures were determined based on ACI 209R-92 recommendations using short-term data (28 days measurements).

Other parameters that had been measured include development of tensile strength and elastic modulus in tension up to 28 days. Tensile strength was determined according to BS1881: Part118 and at the same time strain on the tensile side of prism was measured using strain gauges. The stresses and the corresponding strains were then plotted to determine elastic modulus (Fig. 6). The development of tensile strength and elastic modulus over time are shown in Figures 7 and 8, respectively.

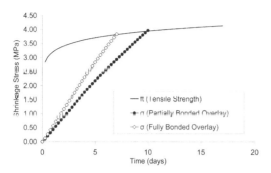

Figure 9. Shrinkage stress development in fully and partially bonded overlay.

4 RESULTS OF SHRINKAGE STRESS CALCULATION ON FULLY AND PARTIALLY BONDED OVERLAY

Equation 4 was used to calculate shrinkage stress on both fully and partially bonded overlays. The values of shrinkage, creep coefficient and elastic modulus given in Figures 4–5, 8 were utilized as input parameters. The degree of restraint was calculated using Equation 2. Since shrinkage cracking occurred after 7 and 10 days of drying for fully and partially bonded overlay, respectively, shrinkage stress for the corresponding overlay must attain the level of tensile strength (tensile capacity) at these particular times. These are achieved when the values of τ correspond to 2.35 and 1.80 for fully and partially bonded overlays, respectively. Figure 9 shows the development of shrinkage stress σ on fully and partially bonded overlays together with the tensile strength of overlay.

5 CRACKING RESISTANCE OF CONCRETE OVERLAY

Equation 4 may be rewritten as follows:
for fully bonded overlay

$$\sigma_{(t)} = 2.35 \mu E^*_{(t)} \varepsilon_{sh(t)} \qquad (5a)$$

and for partially bonded overlay

$$\sigma_{(t)} = 1.80 \mu E^*_{(t)} \varepsilon_{sh(t)} \qquad (5b)$$

The risk of cracking for particular concrete overlay is evaluated based upon the magnitude of shrinkage stress; when its magnitude attains tensile capacity, the overlay is deemed to be in a cracking condition. The magnitude of shrinkage stress itself depends on concrete overlay properties (shrinkage, creep and elastic modulus) and physical properties of overlay (relative thickness and bond type). For a particular type of concrete overlay, the values of shrinkage, creep and elastic modulus may be obtained from short-term experiments and so equations 5a–5b may be applied to determine:

a) cracking tendency
b) time of cracking likely to occur
c) the relative thickness of overlay α necessary to avoid cracking

For the current case studies, setting the value of μ less than 0.2 results in shrinkage stresses lower than tensile strength at all time. This value of μ corresponds to α equals to 0.95.

6 CONCLUDING REMARKS

Tensile stress induced by restrained shrinkage in concrete overlays can lead to cracking. A method to calculate shrinkage stress is suggested to take into account the type of bond between concrete base and overlay. Shrinkage, creep and elastic modulus are necessary input parameters that may be obtained from short-term measurements. Equations to calculate shrinkage stress may be applied to determine: cracking tendency, time of cracking likely to occur and the relative thickness of overlay necessary to avoid cracking.

ACKNOWLEDGEMENT

Financial support from the Ministry of Research and Technology, Indonesia through its Incentive Research Program for the year of 2008 is gratefully acknowledged.

REFERENCES

Brown, MD. Smith, CA. Seller, JG. Folliard, KJ. & Breen, JE. 2007 Use of Alternatives Materials to Reduce Shrinkage Cracking in Bridge Decks. *ACI Materials Journal*, Vol. 104 (6), Nov–Dec 2007: 629–637.

Hall, KD. & Banihatti, N. 2005 Structural Design of Portland Cement Concrete Overlays for Pavement. *Research Report MBTC-1052*. Dept of Civil Engineering, University of Arkansas.

Schlorholtz, S. 2000. Determine Initial Cause for Current Premature Portland Cement Concrete Pavement Deterioration. *Final Research Report*. Dept of Civil and Construction Engineering, IOWA State University.

Silfwerbrand, J. 1997. Differential Shrinkage in Normal and High Strength Concrete Overlays. *Nordic Concrete Research*, Publication No. 19: 55–68.

Concrete Solutions – Grantham, Majorana & Salomoni (Eds)
© 2009 Taylor & Francis Group, London, ISBN 978-0-415-55082-6

The importance of every detail: Performance analysis of a concrete pavement

Andreea-Terezia Mircea

Faculty of Civil Engineering, Technical University of Cluj-Napoca, Romania

ABSTRACT: The paper presents the main causes of imperfections established by an investigation performed at the request of the contractor on a concrete pavement made at the platform of an industrial company. The platform covers around $200\,000\,m^2$, and was built in 2002. Since then, numerous maintenance works have been done, especially by replacing the damaged panels. The purpose of the investigation was to identify the causes that led to the unacceptable performance of the concrete pavement claimed by the owner, and to recommend measures in order to ensure the future normal service of the pavement. Concrete is able to provide a highly durable, serviceable and attractive surface, but the quality of concrete pavements is often affected by conditions over which the designer and contractor have little control. Thus, the paper presents the performance analysis of the concrete pavement, conclusions of the investigation and recommendations made for ensuring future performance and durability.

1 INTRODUCTION

Concrete pavements can be designed for virtually any service life, from as little as 10 years to 60 years or more. The primary factors in the design life are the quality of the materials and the slab thickness. However, pavement concrete mixtures with enhanced strength and durability characteristics should be combined with a performance conception and structural design to ensure long service periods.

The platform covers around $200\,000\,m^2$, and was built in 2002 (Figure 1). Since then, a considerable amount of maintenance work has been done, especially by replacing the damaged panels.

Concrete pavements are not maintenance free. In these circumstances, frequently asked questions are: Is the design maintainable?, Are there features that may create maintenance problems in the future?, Should a longer life design be contemplated to minimize traffic disruptions for future maintenance and rehabilitation activities?

Concrete is able to provide a highly durable, serviceable and attractive surface. But the quality of concrete slabs on the ground is often affected by conditions over which the designer and contractor have little control. By keeping the causes of the imperfections in mind, it is possible to reduce the probability of unsatisfactory results.

Some curling and cracking can be expected on every project due to the inherent characteristics of Portland cement concrete such as shrinkage.

Contractors are not necessarily responsible for all imperfections. Poor design, unsatisfactory mixture proportions and improper service conditions are also implied. As presented in Figure 2, it is important to know that the pavement condition is a function of time and traffic.

Figure 1. The investigated concrete pavement.

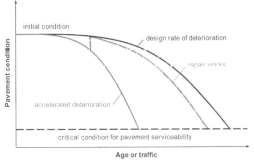

Figure 2. Pavement condition as a function of time and traffic.

Beside these general items, many more detailed items may be added to the review process. Various levels and frequencies of constructability reviews can be conducted, depending on the purpose and complexity of the project.

Briefly, the problem is related to the whole life cycle analysis, so the decision has to be taken by choosing between the following two possibilities:

• Initial low cost pavement with significant maintenance costs;
• Expensive initial investment with low maintenance costs.

The main objective of pavement design is to select pavement primary features, such as slab thickness, joint dimensions and the reinforcing system to ensure the load transfer requirements, which will economically meet the needs and conditions of a specific paving project.

The goal of all pavement design methods is to provide a pavement that performs well. That means, to provide a serviceable pavement over the design period for the given traffic and environmental loadings.

The desired pavement performance is generally described in terms of structural performance and functional performance:

• Structural performance is the ability of the pavement to support current and future traffic loadings and to withstand environmental influences. The structural performance of concrete pavements is influenced by many factors, including design, materials, and construction. The most influential design-related variables for structural performance at a given level of traffic are slab thickness, reinforcement, concrete strength and support conditions. The most prevalent type of structural distress is load-related cracking, which may appear as corner cracks, transverse cracks, or longitudinal cracks.
• Functional performance refers to the pavement's ability to provide users a comfortable ride for a specified range of speed. Most often, functional performance is thought to consist of ride quality and surface friction, although other factors such as noise and geometrics may also come into play. Functional distress is generally represented by a degradation of a pavement driving surface that reduces ride quality.

Both structural and functional distresses are considered in assessing overall pavement performance or condition.

Even well-designed and well-constructed pavements tend to degrade at an expected rate of deterioration as a function of the imposed loads and/or time. Poorly designed pavements (even if they are well-constructed) will probably experience accelerated deterioration.

With regard to the aspects presented above, the analysis of the investigated pavement was made as follows.

2 MATERIALS AND CONSTRUCTION OF THE PAVEMENT

If compressive strength is enough to sustain the mechanical actions, and the structural performance was ensured at the time of the investigation, problems refer to the rest of the properties that influence the global performance of the pavement.

Mix design is fundamental in order to ensure a good quality concrete. The contractor did not specify any reference mix, mentioning that there were several suppliers for the fresh concrete. Visual inspections revealed that the most important factor of distress for the functional performance of the pavement was excessive cracking.

Figures 3 and 4 present the short and long term evolution of the shrinkage strain for a typical C16/20 mix, with $350\,kg/m^3$ of cement, and various weather conditions at concrete placing. The evolution is shown both for Portland cement I 32.5 R and composite Portland cement II AS 32.5 R (with slag 6–20%). Peak temperatures are reached at 12–16 hours age of concrete, and the maximum temperature change and associated axial strains caused by volume contraction are shown in Table 1.

Considering the results of the analysis and the observations of visual inspections, it appears that the concrete mix that contains cement type I 32.5 R and has a high w/c ratio, with greater shrinkage potential.

As presented in Figures 3 and 4, the inevitable cracks occur within the first week from concrete placing. Analysis shows that crack widths reach about 0.4 mm in the case of cement I 32.5 R, and about 0.26 mm in the case of cement II A-S 32.5 R, both more than the allowable limit of 0.2 mm.

During repair works, the contractor replaced plain concrete with steel fiber reinforced concrete $(20\,kg/m^3)$, in order to reduce cracking.

Because steel fibers are dispersed in the entire volume of the slab, this quantity is not enough to reduce the crack widths significantly, its effect being a global increase of the ductility of concrete, a property that practically has no significance for the analyzed problem.

Tests on core samples revealed that in some areas, concrete compressive strength was beyond that prescribed by the design. Due to the uncertain age of the tested concrete, this problem could be justified also by accelerated aging of concrete. However, supplementary analyses are necessary in order to determine if some of the cracks that are currently justified by early or heavy traffic, are caused practically by the insufficient strength of concrete.

Concerning the construction, visual inspections revealed a careful construction process. No major imperfections can be attributed to poor construction practice, even if it was performed and maintained in various weather conditions.

Another important distress factor is scaling and delamination of concrete. This is caused by the

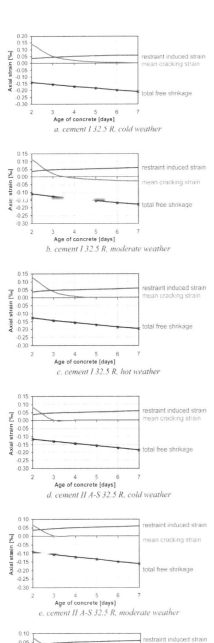

a. cement I 32.5 R, cold weather

b. cement I 32.5 R, moderate weather

c. cement I 32.5 R, hot weather

d. cement II A-S 32.5 R, cold weather

e. cement II A-S 32.5 R, moderate weather

f. cement II A-S 32.5 R, hot weather

Figure 3. Short term variation of the shrinkage strain for the reference mix C 16/20 in the concrete pavement.

sensitivity of the surface layer to freeze-thaw cycles, as a consequence of a high w/c ratio and high permeability (it should be mentioned that the small slope is not correlated with the necessary properties).

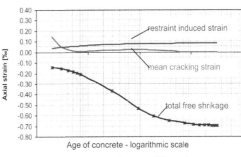

a. cement I 32.5 R, cold weather

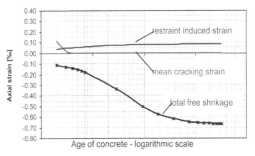

b. cement I 32.5 R, moderate weather

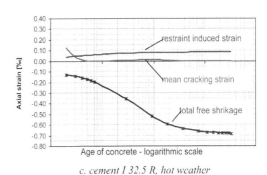

c. cement I 32.5 R, hot weather

Figure 4. Long term variation of the shrinkage strain for the reference mix C 16/20 in the concrete pavement.

Table 1. Reference temperature changes and volume contraction at the C 16/20 pavement.

Weather conditions	Cement type	Maximum effective temperature change [°C]	ε_v [‰]
cold	I 32.5 R	12.5	−0.125
moderate	I 32.5 R	9.6	−0.096
hot	I 32.5 R	11.2	−0.112
cold	II A-S 32.5 R	10.0	−0.100
moderate	II A-S 32.5 R	7.7	−0.077
hot	II A-S 32.5 R	9.0	−0.090

Due to the severe exposure conditions, besides mechanical strength, other properties of the hardened concrete are critical and have to be considered:

- Permeability: concrete with permeability classes P8 (200 mm water penetration at 28 days under a pressure of 8 bars) or P 12 (300 mm water penetration at 28 days under a pressure of 12 bars) is commonly used in pavements;
- Freeze-thaw resistance: concretes with freeze-thaw resistance from 100 cycles, up to 200 cycles are preferred for long term durability.

3 CONCLUSIONS AND RECOMMENDATIONS

The investigation made upon the concrete pavement has pointed out the following conclusions:

- At the time of investigation, the structural performance of the pavement was good;
- Functional performance presents two important distress factors:
  - Excessive cracking, due to restrained shrinkage of concrete and early and/or heavy traffic;
  - Low freeze-thaw resistance, causing scaling and delamination of the surface layers;
- The degradation mechanisms related to the above mentioned distress factors are:
  - Excessive cracking of concrete allows water to easily penetrate inside concrete. Low temperatures during winter transform the water to ice, which enlarges the initial volume and exerts pressure upon the surrounding concrete. Thus, microcracks are developed and in short periods of time initial cracks present much larger openings. This aspect affects the functional performance, and in time also reduces the structural performance beyond the acceptable limit.
  - High permeability allows water to penetrate the concrete, resulting in scaling and the occurrence of new cracks that grow further following

the above mechanism; a high w/c ratio also led to significant delamination;

- The maintenance strategy of the pavement was wrong: instead of injecting and sealing the inherent cracks with openings larger than 0.2 mm (if less, water cannot penetrate due to the concave meniscus), cracks were allowed to evolve, maintenance work consisting in simply replacing the panels reaching an unsatisfactory condition state;

In order to ensure the future performance of the pavement and a reasonable durability, the recommendations are:

- Injection of the cracks with widths more than 0.2 mm at least once per year;
- Replacement of the panels with more than 80% affected areas with a concrete based on a new mix design. Recommended concrete classes are C 28/35 or C 32/40. Recommended cement is II AS 32.5 R, and maximum water content derived from w/c ≤ 0.42. Air entraining admixtures to resist freeze-thaw and reduce permeability are also recommended;
- Because the shrinkage potential remains high, instead of injecting and sealing the excessive cracks, reinforcing of the superficial layer (at least 0.2% reinforcing ratio) with a two-directional mesh is recommended;
- In order to avoid a high consumption of reinforcing steel, checking on the need for bottom reinforcement is also recommended.

REFERENCES

A.T. Mircea, Causes of Concrete Pavements Imperfections – *International Conference Constructions 2008/C55*, Section: Civil Engineering-Architecture nr. 51, Vol. II, May 2008, Cluj-Napoca, Romania ISSN 1221-5848 pp.133–142

NE 012-1999: Cod de practică pentru executarea lucrărilor de beton, beton armat și beton precomprimat

ACI 312.1R-52: Guide for Concrete Floor and Slab Construction.

Concrete Solutions – Grantham, Majorana & Salomoni (Eds)
© 2009 Taylor & Francis Group, London, ISBN 978-0-415-55082-6

Basalt fibres: Mechanical properties and applications for concrete structures

A. Palmieri, S. Matthys & M. Tierens

Magnel Laboratory for Concrete Research, Department of Struct. Eng., Ghent University, Belgium

ABSTRACT: The use of advanced materials such as Fibre Reinforced Polymer (FRP), for reinforcing (internal reinforcement), prestressing (pre- or post-tensioning) or strengthening (externally bonded, near surface, textile reinforced mortar) of structures, have been gaining increasing interest worldwide. The effectiveness of the FRP techniques have been clearly confirmed by numerous experimental and field applications. This study focuses on the possible use of basalt fibres for FRP rebars or laminates. Basalt is a volcanic igneous rock that because of its high performance in terms of strength, corrosion resistance, temperature range, fire resistance and durability, as well as lower potential, cost may effectively replace steel, glass and carbon fibres in many applications.

The paper discusses, based on a literature review and some feasibility tests, the possible use of basalt fibres in relation to reinforcing and strengthening of concrete. Herewith, reference is made to: (1) basalt fibre composite bars or rods for internal reinforcement, (2) the possibility to use short basalt fibres, and (3) strengthening of concrete members by means of externally bonded reinforcement (EBR), near surface mounted reinforcement (NSM) or textile reinforced mortar reinforcement (TRM). In this paper, the test results in terms of tensile properties of BFRP (basalt FRP) bars and laminates, as well as confinement of plain concrete cylinders with BFRP are presented. This experimental work compares specimens made with BFRP versus GFRP (glass FRP).

1 INTRODUCTION

The effectiveness of FRP techniques as an alternative for reinforcing steel in concrete structures has been clearly confirmed by numerous experiments and field applications. As strengthening and/or reinforcing materials for reinforced concrete (RC) elements (such as beams, slabs, columns and walls) basalt fibres can be used in many applications, where conventional strengthening techniques may be less feasible and where FRP techniques with conventional fibres may imply some drawbacks.

Basalt fibres can be potentially used for different situations such as flexural strengthening; confinement; shear strengthening and internal reinforcement (rebars or short fibres). For each of these techniques, there are several material forms: basalt textile, basalt laminates; basalt sheet; short basalt fibres; basalt rods and others. The use of basalt fibres has captured the interest of the civil engineering community due to their favourable properties, such as high performance in terms of strength, corrosion resistance, temperature range, fire resistance and durability, as well as the lower potential cost with respect to other FRP materials.

Basalt is a natural, hard, dense, dark brown to black volcanic igneous rock. It is the most common rock type in the earth's crust (the outer 10 to 50 km); its origins are at a depth of hundreds of kilometres beneath the earth and it reaches the surface as molten magma. Basalt density ranges between 2700 and 2800 kg/m^3. The basic characteristics of basalt materials are high-temperature resistance, high corrosion resistance, resistance to acids and alkalis (Lisakovsky et al, 2001), high strength and thermal stability. Basalt can be formed into continuous fibres with the same technology utilized for E-Glass and AR-glass fibres but the production-process requires less energy and the raw materials are widely diffused all around the world. This justifies the lower cost of basalt fibres compared to glass fibres. Moreover, basalt fibres are environmentally safe, non-toxic, non-corrosive, non-magnetic, possess high resistance against low and high temperatures and a high thermal stability, have good heat and sound insulation properties, durability and vibration resistance. From literature review, basalt fibres present a modulus of elasticity at least 18% higher than that of E-glass fibres; they are linear elastic up to failure with ultimate strain in the order of 2%.

Some studies have already investigated fundamental properties of basalt fibres and its application as strengthening and reinforcing material. Sim (Sim et al, 2005) investigated the applicability of the basalt fibre as a strengthening material for structural concrete members through various experimental works for durability, mechanical properties, and flexural strengthening by the EBR technique. Beams strengthened with basalt fibres showed a more ductile failure than those strengthened with E-glass fibres.

Basalt fibre properties such as the higher resistance to alkalinity of the surrounding concrete, the resistance to high temperature exposure and the resistance to high moisture conditions also make these fibres a good alternative to steel, glass and carbon fibres for the Near Surface Mounted (NSM) reinforcement technique. However, it is clear that experimental investigations and some research are still necessary to validate the effectiveness of basalt fibres.

Prota (Prota et al, 2008) studied the effectiveness of confinement with basalt fibres using a cement based matrix (BRM: Basalt Reinforced Mortar) for bonding. Experimental outcomes on BRM confined concrete cylinders showed that the basalt reinforced mortar confining system could provide a substantial gain both in compressive strength and ductility of concrete members inducing a failure mode less brittle than that achieved in members wrapped with glass fibres.

Short basalt fibres (Brik, 1999) in concrete can be used to improve properties such as shrinkage and cracking, low tensile and flexural strength, poor toughness, high brittleness, and low shock resistance. Moreover the use of basalt rods for internal reinforcement (Ramakrishman et al, 2005) can offer significant advantages such as: high tensile strength; high mechanical performance/price ratio; light weight, resistance to alkali and acids; heat tolerance and fatigue resistance and compatibility with many resins.

In the following, test results in terms of tensile properties of bars and laminates, as well as confinement of plain concrete cylinders with BFRP are presented. Due to their properties being close to that of glass fibres this experimental work compares specimens made with BFRP versus GFRP.

2 EXPERIMENTAL CAMPAIGN

2.1 General approach

The current research work deals with the development of basalt fibres for civil engineering applications, namely for reinforcing or strengthening concrete structures. The experimental campaign consists of: (1) tensile tests on basalt bars; (2) tensile tests on basalt laminates; (3) compression tests on basalt cylinders.

2.2 Tensile Tests on Bars

A total of 9 bars (type Basaltoplastik Galen Ltd) were tested in this program. The bars tested were manufactured by the pultrusion technique, and had a diameter of 11, 6 and 4 mm. Two types of bar surfaces were used (sanded and ribbed). The specimens were divided into three series, depending on the bar diameter. Both ends of the rebars were anchored in the loading machine using a copper pipe with a length of 135 mm filled by epoxy resin (type PC 5800, ECC group) in order to grip the specimens without causing slippage or premature local failure during the test. Tests were carried out in displacement control with a displacement rate of the

machine head of 2 mm/min. A linear variable displacement transducer (LVDT) was mounted on the centre of the specimen test section to measure the displacement during the test.

2.2.1 Experimental results

Experimental tensile strength (f_f), tangent modulus of elasticity (F_f) and ultimate strain (ε_u) were computed from the load and displacement recorded from the test. In this work, the nominal cross-sectional area of the bar (see table 1) was used. Specimen failure in this test program occurred in two modes, namely, tensile failure and anchorage pullout failure.

Tensile failure can be described as an abrupt failure when fibres break and the bar completely ruptures around a cross-section within the test section. Anchorage pullout failure occurs when the bar slips out of the anchor. For specimens that fail with an anchorage failure, their ultimate stress and strain were not used for further analysis because the specimen did not develop full tensile strength.

All specimens behaved linearly elastic up to failure. Experimental outcomes showed that:

- In terms of tensile strength, while basalt bars with an internal diameter of 6 mm presented a value similar to that indicated by the manufacturer (1237 Mpa) the bars with an internal diameter of 4 mm presented a higher tensile strength.
- In terms of ultimate strain, experimental results were in accordance with the value indicated by the manufacturer (3%).
- The modulus of elasticity tended to decrease slightly with increasing rebar diameter.

2.3 Tensile tests on laminates

To get a clear view of differences between the mechanical properties of basalt and E-glass products, it is necessary to compare similar kinds of laminates. Therefore, roving of both materials were selected, with the same tex value and comparable sizing. In the experimental program basalt roving (type Basaltex Roving 1200 tex value and 13 μm filament diameter) and glass roving (type Owens-Corning OC SE 1500 1200 tex value and 17 μm filament diameter) were used. Two different epoxies, a tixotropic epoxy (type ECC PC 5800/BL) and a more viscous epoxy (type ECC PC 5800) were used to produce the laminates by hand lay-up. To provide appropriate anchorage during testing, rectangular tabs were glued with epoxy PC 5800/BL at both ends of each specimen in order to diffuse clamping stresses. Six specimens of each type (basalt fibres and E-glass fibres) were tested under a tensile machine of 50 kN with a displacement rate of 2 mm/min. Strain was recorded by two strain gauges glued at the centre of the specimen (1 to the front and 1 to the back).

The material properties of fibre reinforced materials are often governed by the fibres. The fibre content (V_f) of the composite is often given by the material supplier and indicates, in volume percentage, the amount of

Table 1. Tested rebar tensile properties.

Sample	Measured diameter [mm]	Nominal diameter [mm]	Surface	Q_u [kN]	f_f [MPa]	E_f [GPa]	ε_u [mm/m]	Failure aspect
B1	11.0	11.0	Sanded	59	–	37	–	Bar pullout
B6	6.5	6.0	Sanded	38	–	38	–	Bar pullout
B9	6.5	6.0	Sanded	42	1271.4	39	3.33	Tensile
B2	6.0	6.0	Ribbed	36	1273.2	–	–	Tensile
B8	6.0	6.0	Ribbed	30	–	46	–	Bar pullout
B7	4.0	3.5	Sanded	14	1478.6	45	2.59	Tensile
B4	4.0	3.5	Sanded	15	1558.7	–	–	Tensile
B5	3.5	3.5	Ribbed	17	1742.0	50	2.97	Tensile
B3	3.5	3.5	Ribbed	14	1455.1	–	–	Tensile
Average						44	2.96	

Table 2. Tensile properties of tested laminates.

Sample	Resin	t_f [mm]	Q_u [KN]	f_f [MPa]	E_f [GPa]	ε_u [mm/m]
Basalt	PC 5800/BL	1,00	2,91	138,51	19,48	11,86
E-glass	PC 5800/BL	1,00	2,70	124,91	14,06	12,25
Basalt	PC 5800	1,00	2,95	274,10	25,80	20,80
E-glass	PC 5800	1,00	2,82	218,35	18,95	18,20

fibre contained in the composite. For the tested materials produced in the lab by hand lay-up, a fibre content of 12.60% has been obtained for both materials.

2.3.1 Experimental results

The experimental results for each type of material and matrix used, in terms of maximum applied load, Q_u, nominal thickness t_f, tensile strength and ultimate strain, f_f and ε_u, as well as tangent modulus of elasticity, E_f, are reported in table 2. The results in this table are the average values of 6 laminates and are referred to a fibre volume fraction of 12,6%. The average experimental stress-strain relationship is reported in figure 1.

Experimental results showed that the tensile strength of basalt laminates was 10% and 25% higher than that of E-glass laminates, respectively for laminates made with epoxy PC 5800/BL and PC 5800. The modulus of elasticity was 38% and 36% higher than that of glass laminates. The ultimate strain of basalt and E-glass laminates was almost the same using epoxy PC 5800/BL, while basalt laminates presented a larger elongation at failure using epoxy PC 5800. Moreover, experimental results showed that the epoxy resin had a great influence on laminates. Laminates made with epoxy PC 5800 showed a tensile strength almost twice as high as that of laminates made with epoxy PC 5800/BL. This can be explained by the insufficient wetting and impregnation capabilities of epoxy PC 5800/BL versus epoxy PC 5800. The modulus of elasticity of basalt laminates and E-glass laminates made with epoxy PC 5800 was respectively 15% and 24%

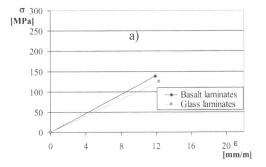

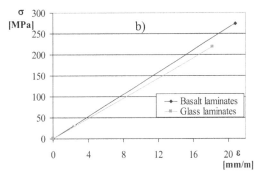

Figure 1. Stress-strain relationship laminates impregnated with a) PC5800/BL b) PC 5800.

higher than that of laminates made with PC 5800/BL. Also in terms of ultimate strain, the effectiveness of laminates made with epoxy PC 5800 was higher than those made with epoxy PC 5800/BL.

2.4 Confined cylinders

An investigation has been carried out on 10 FRP confined concrete cylinders (5 cylinders made by basalt roving and 5 made by E-glass roving). The cylinders consisted of an FRP tube filled with concrete. The FRP tubes, made with a filament winding machine, were composed of three layers and were identical for both materials. A first layer was wound at an angle of

Table 3. Confinement effects.

Sample	f_{cc} [MPa]	$\xi = f_{cc}/f_{co}$ [–]
Basalt	91.0	1.84
E-glass	84.1	1.70
Concrete	49.5	1.00

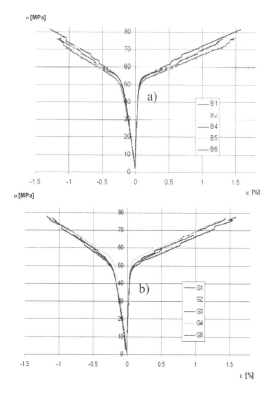

Figure 2. Stress-strain relationship of a) BFRP and b) GFRP confined cylinders.

of 88 degrees with the rotation axis, the second at 56 degrees and the third was identical to the first. Impregnation with epoxy was conducted (type PC 5800) on the mould and manually. For production reasons, the FRP tubes were made slightly cone-shaped (height of 300 mm and diameter of 160 mm for the upper side and 150 mm for the lower side). After hardening, the fibre reinforced cylinders were filled with concrete ($f_c = 50$ MPa W/C = 0,5). The cylinders were tested in a compression machine of 5000 kN through monotonically applied axial loading in displacement control with a rate of 0.005 mm/s. On the outside surface of the tubes, six strain gauges (three for the axial strain and three for transversal strain) were installed at the middle height of the specimens to measure both longitudinal and lateral strain during loading.

2.4.1 Experimental results

The mean experimental results on 5 cylinders for each FRP type and on 3 cylinders of unconfined concrete used as control specimens are reported in Table 3 in terms of confined strength f_{cc} and strengthening factor (the ratio between the average concrete strength of reinforced cylinders and unconfined ones).

For both materials, the curves are bilinear in nature with a small transition zone. In the first linear zone, concrete primarily takes the axial load; the slope of the confined concrete is the same as the slope for unconfined concrete.

Fig. 2 shows the axial stress plotted as a function of axial strain (drawn on the left side) and of lateral strain (drawn on the right side) for respectively basalt and glass cylinders.

At stress levels near to the ultimate stress of the unconfined concrete, a transition zone to the second portion of the bilinear curve starts. This region signified that concrete starts cracking and radially expanding, so that the FRP tube starts to show its full confining characteristics. As can be seen in table 2 the experiments showed that basalt fibre reinforced cylinders provided substantial gain in strength: 84% of average compressive strength increase was recorded in comparison to the control specimens. Making a comparison with E-glass reinforced cylinders ($\xi = 1,70$), it is concluded that basalt fibre cylinders are more effective than E-glass reinforced cylinders. The failure aspects of all the composite specimens were characterized by fracture of the FRP tube with concrete bursting at half height of the cylinders. One can therefore state that the slight cone shape of the cylinders didn't significantly influence the tests.

3 CONCLUSIONS

The presented experimental work, in terms of tensile properties of BFRP bars and laminates as well as confinement of plain concrete cylinders with BFRP, has been conducted as a feasibly study to investigate the possible use of basalt fibres in relation to reinforcing and strengthening of concrete. Tensile tests on basalt bars showed a tensile strength higher than that suggested by the supplier. From tensile tests on laminates, experimental results have shown that basalt laminates were stronger and stiffer than glass laminates, while the choice of epoxy viscosity significantly influenced the laminate properties both for BFRP and GFRP. Confinement with BFRP was 14% more effective than that with GFRP in terms of compressive strength. Based on these data and further results found in the literature, basalt fibres seems to be a promising solution for reinforcing and strengthening of concrete. However, it is clear that further research is necessary to validate the effectiveness of basalt fibres.

ACKNOWLEDGEMENT

The authors wish to acknowledge the companies Galen Ltd, Basaltex, Owens-Corning and ECC group for providing testing materials.

168

REFERENCES

Brik, V.B., "Performance Evaluation of Basalt Fibres and Composites rebars as Concrete Reinfocement", Tech Res submitted to NCHRP-IDEA, Project 45, 1999.

Lisakovski, A.N., Tsybulya, Y.L., and Medvedyev, A.A., "Yarns of Basalt Continuous Fibres," in The Fibre Society Spring 2001 Meeting, Raleigh, NC, May 23–25 (2001).

Prota A., Di Ludovico, M. and Manfredi, G. "Concrete confinement with BRM systems: experimental investigation", Proceedings of the 4th International Conference on FRP Composites in Civil Engineering, July 2008, (from CD).

Ramakrishman, V and Panchalan, R.K. "A New Construction Material-Non-Corrosive Basalt Bar Reinforced Concrete" ACI Journal Special publication, Vol. 229, September 1 2005, pp. 253–270.

Sim J., Park C. and Moon D.Y., "Characteristics of Basalt Fibre as a Strengthening Material for Concrete Structures," Composites. Part B Engineering, Vol. 36, 2005, pp. 504–512.

Tierens, Matthias, "Verkenned onderzoek naar het gebruik van basaltvezels voor de versterking van betonconstructies"(in Dutch) master dissertation (supervisor S. Matthys), Gent University May, 2007.

Concrete Solutions – Grantham, Majorana & Salomoni (Eds)
© 2009 Taylor & Francis Group, London, ISBN 978-0-415-55082-6

Overview upon restrained shrinkage cracking of RC structures

C. Mircea

National Building Research Institute [INCERC], Cluj-Napoca, Branch, Cluj-Napoca, Romania
Technical University of Cluj-Napoca, Cluj-Napoca, Romania

ABSTRACT: Cracking of concrete is a common complaint. Due to the non-linear behaviour of concrete, cracking, tension stiffening, creep and shrinkage calculations related to the Serviceability Limit States are relatively complicated, and often neglected in current practice. Shrinkage is the most challenging among these phenomena. External and/or internal restraint to shrinkage generates parasite tensile stresses, inducing time-dependent cracking, widening existing cracks, all these resulting in a lower structural rigidity and corrosion of reinforcement.

Aspects of over 5 years of investigation concerning cracking states induced by shrinkage of concrete are presented. Several reinforced concrete structural types are concerned: structural walls, mass elements (e.g., piers, abutments and portals), slab systems, slabs on ground etc. Considering the specifics of each structural type, and based on the data acquired from the technical documents, inventory and monitoring of the cracks, analyses, interpretations and predictions are presented in a synthetic form. The conclusions point out some of the code recommendations in the context given by the actual common practice.

1 INTRODUCTION

Control of degradation of reinforced concrete structures is a difficult task, with many objective or subjective (i.e., related to the human factor) parameters involved, as shown in Figure 1. Actual performance based structural design, considers only the objective parameters. For example, cracking of concrete is a common complaint. Even if Ultimate Limit States design covers all objectives parameters, due to the non-linear behaviour of concrete, cracking, tension stiffening, creep and shrinkage, calculations related to the Serviceability Limit States are complicated, less understood and often neglected in current practice. Thus, its safety may become uncertain in time, but durability of the structure is almost sure to be compromised, and consequently its sustainability. Moreover, concrete mix design is related mainly to the strength properties, and less to other properties of equal importance (e.g., shrinkage properties, permeability). A sustainable structural design should take account of all objective and subjective parameters in a holistic approach, offering for the future a more consistent practice.

Free deformation of concrete is mainly related to two local parameters: temperature and water content. As with other materials, concrete expands when heated and shrinks when cooled. Even if it has a porous structure, unlike other materials, concrete presents a great difference in deformation when drying and absorbing water. Thus, moisture deformation that occurs during initial drying is only slightly compensated by any subsequent reversible deformation.

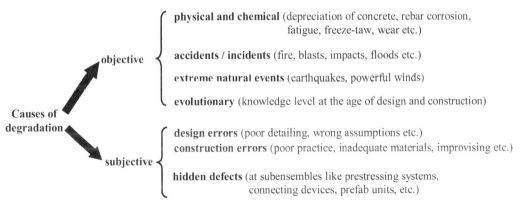

Figure 1. Causes for the degradation of reinforced concrete structures.

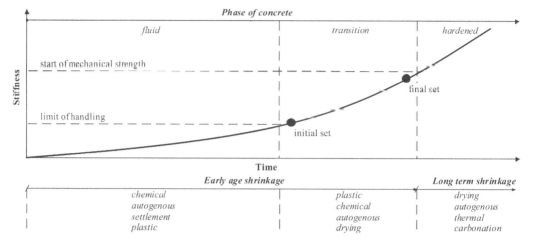

Figure 2. Shrinkage stages and types.

External and/or internal restraint to shrinkage generates parasite tensile stresses inducing time-dependent cracking and widening of load induced cracks, all these resulting in a lower structural rigidity and corrosion of reinforcement. Whether cracking due to shrinkage initiates or not depends on: concrete mix composition, curing of concrete, degree of restraint to shrinkage provided by external supports, internal connections and embedded bonded reinforcement, size of the structure, size of the exposed faces to drying, quality of concrete through its stiffness and strength properties, and level of the tensile stresses induced by the external load. Nevertheless, even if very complex, concrete remains a strong sustainable option for the built environment. Next, several examples will be briefly presented, in an attempt to share recent experience acquired in respect of restrained shrinkage of concrete.

2 BASIC SHRINKAGE MECHANISMS OF CONCRETE

Concrete is a complex material, which up to service changes its properties. Figure 2 shows the basic types of shrinkage, related to the phases of concrete and time. Their mechanism is briefly presented forward.

Thermal contraction: Temperature gradients generated by the heat developed during hydration of cement lead to a general high temperature of the element. On cooling (Fig. 3.a), the element reduces its volume, while the supports provide an external restraint to it. Temperature varies also within the depth of concrete and leads to an internal restraint, as shown in Figures 3.b and 3.c. Other significant thermal contractions may rise in the elements subjected to important temperature gradients in service.

Autogenous shrinkage: Appears due to the chemical reactions (i.e., hydration of cement) within concrete. It is associated with the loss of moisture from the capillary pores during hydration, and initiates

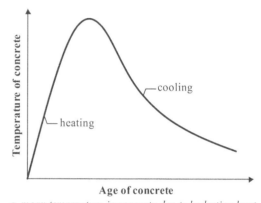

a. mean temperature in concrete due to hydration heat

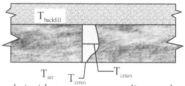

b. inside temperature gradient at element with one side backfill

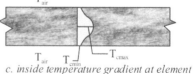

c. inside temperature gradient at element with both faces exposed

Figure 3. Temperature distribution due to hydration of cement.

at the beginning of concrete setting. It lasts several years during hardening of concrete (Fig. 4), but only the first month has practical significance.

Drying shrinkage: Is the result of the loss of moisture from the cement paste constituent, which reduces its volume. Primary factors of influence

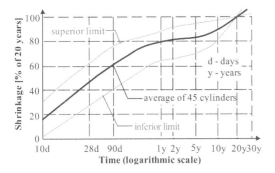

Figure 4. Time evolution of autogenous and drying shrinkage.

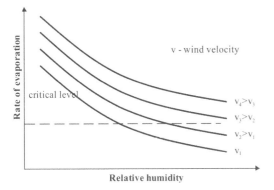

Figure 5. Rate of moisture evaporation related to humidity and wind velocity.

are w/c, dosage and nature of the aggregates. Other significant factors may be the admixtures that influence the water content of the fresh concrete. The major progress of this type of shrinkage lasts about 3 years (Fig. 4), but continues during the entire lifetime of the structure.

Settlement shrinkage: Occurs before the initial set of concrete, due to gravity, when the constituents are consolidating. Autogenous shrinkage, air voids rising to the surface, and bleeding, all contribute to settlement shrinkage. Concrete that is not thoroughly consolidated during placement will experience a greater degree of shrinkage.

Plastic shrinkage: Results from the very rapid loss of moisture at the surface of plastic concrete, caused by a combination of factors like concrete and surroundings temperature, relative humidity and high wind velocity at the surface of concrete (Fig. 5). When moisture from the surface of fresh concrete evaporates faster than it is replaced by bleed water, the surface of concrete reduces its volume. Thus, within the superficial layer appear cracks with variable length, width and spacing.

Carbonation shrinkage: Carbonation occurs when cement paste in the hardened concrete reacts with moisture and carbon dioxide in the air, as shown in Equation 1. This leads to a slight shrinkage and a decrease of the pH factor in concrete. The amount of carbonation is dependent on the concrete density

and quality, but is usually limited to less than 20 mm of depth on the exposed surface. It also depends on the age and environmental conditions.

$$H_2CO_3 + Ca(OH)_2 \rightarrow CaCO_3 + 2H_2O \qquad (1)$$

3 MASS ELEMENTS

The volume change of mass concrete elements has three major interdependent components. In cold weather, volume contraction due to initial heating caused by hydration of cement lasts approximately one week. Obviously, while continuous base support and air temperatures balance the thermal gradient, the restraint contribution has a key relevance for each particular case. On the other hand, concrete also suffers autogenous shrinkage and drying shrinkage. Because steel and concrete have comparable values for the linear thermal coefficient, the reinforcement embedded within the mass element will generate internal restraint only against autogenous and drying shrinkage. On this background of restrained volume contraction, tensile creep of concrete compensates partially the volume reduction, and concrete continues to improve its strength properties at various rates. Hence, it is obvious that a complete evaluation must consider time as the fourth dimension and the weather parameters as variables.

Figure 6 presents the typical cracking pattern and sequence propagation that results from base restrained volume change, as described in ACI 207.2R 1995. The first crack (i.e. crack 1) initiates around the middle of the restrained edge, and progresses up to the top. If L/H > 2.0 the crack extends to about 0.20H–0.30H, the crack becomes unstable it will propagate up the entire height of the element. Consequent to the redistribution of the initial base restraint, a new pair of cracks (i.e. cracks 2) occur at nearly half of the uncracked regions at the base, and progress towards up in the same conditions like the first one if L/2H > 1.0. All successive cracking groups develop in a similar mode, until the sum of all crack widths compensates the volume change. The maximum width for each crack is reached above the top of the previous initiated cracks. While ACI Committee 207 considers a constant base restraint factor of the axial strain, which is a conservative approach in design, Mircea et al 2007 proposed, after numerous investigations, a decreasing base restraint factor with each redistribution.

The basic assumptions of both approaches are shown in Figure 7. When a crack reaches the critical height, its propagation becomes unstable. As Figure 8 shows, the maximum shear force of the internal stress block corresponds to the crack initiation (i.e. stage 2), while the maximum base related moment of internal stresses corresponds to the critical height of the crack (i.e. stage 4). Beyond this stage, both steel and concrete cannot sustain the maximum restraining moment at the base, and parasite stress concentrations at the crack ends will make the crack unstable and free to propagate to the top of the element.

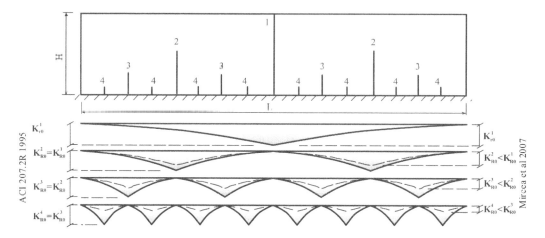

Figure 6. Typical cracking pattern and sequence propagation and scheme for the base restraint redistribution.

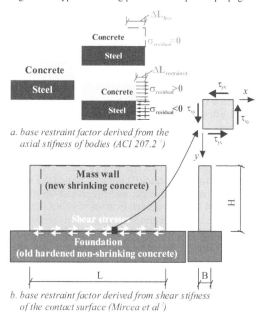

a. base restraint factor derived from the axial stifness of bodies (ACI 207.2 [1])

b. base restraint factor derived from shear stifness of the contact surface (Mircea et al [2])

Figure 7. Basic mechanisms of the base restraint.

4 SHEAR WALLS

The same types of shrinkage have relevance like in the case of mass elements, but with different proportions in magnitude. Initial heating is much less than in the case of mass elements, but severe cracking states can also occur in shear walls. Mircea et al 2008 reported severe and chaotic cracking patterns in the shear walls of an industrial building. Horizontal cracks were developed mainly in the 40 cm thick boundary walls of the basement (see Fig. 9.a), maximum crack widths reaching 0.5 mm. However, numerous vertical cracks also occurred in the 25 cm thick inside walls (see Fig. 9.b), both from the basement and first floor. Crack spacing varied in general between 2.0 m and 3.0 m, crack widths varied in general between 0.2–1.5 mm, but crack openings up to 3.5 mm were found too.

In this case, poor practice in construction (i.e., inadequate treatment of the technological joints) led to horizontal discontinuities, revealed as horizontal cracks in Figure 9. A similar computational approach can be applied like as in the case of mass concrete members, but more attention should be given to the restraint on several edges, where over posing of effects can be successfully applied. Particular attention should also be given to the contraction joints of the structure. If no joints are intended, code provisions do not cover the amount of reinforcement needed to control the shrinkage induced cracks, and special analyses are mandatory in this respect. Supplementary costs may also rise from this approach.

5 SLABS ON GRADE

Figure 10 presents typical cracking patterns caused by shrinkage at slabs on ground. Superficial cracks (Fig. 10.a) may occur on the conditions for plastic shrinkage are met (see Fig. 5). Concrete expands and shrinks with changes in moisture and temperature. The overall shrink tendency can cause cracking due to the external restraint provided by the subgrade. When the tensile stresses induced by the restraint to volume reduction exceed the tensile strength of concrete, the element cracks, as shown in Figures 10.b and 11.

Map cracking (crazing) is a network of fine fissures on the concrete surface that enclose small (12 to 20 mm) and irregular areas (Fig. 10.c). Shallow; often only 3 mm deep cracks occur. This is the result of restrained drying shrinkage of the surface layer after set (visible the day subsequent to placement, but not later than the end of the first week.), and is associated with the following poor practices: overfinishing the new surface or finishing while there is bleed water on the surface, use of mixes with high water-cementitious materials ratios, late or inadequate curing, spraying water on the surface during finishing and sprinkling cement on the surface to dry bleed water.

Severe cracks were also found due to the same base restraint to thermal contraction and drying shrinkage

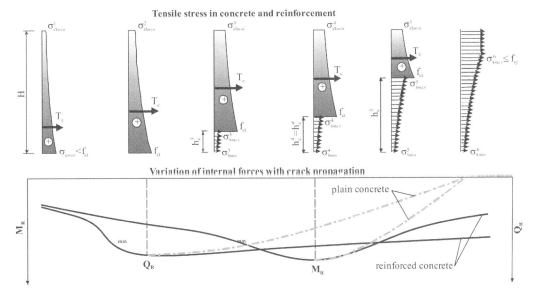

Tensile stress in concrete and reinforcement

Variation of internal forces with crack propagation

Figure 8. Efforts evolution at a single crack initiation at a wall with $L/2H \geq 1.0$.

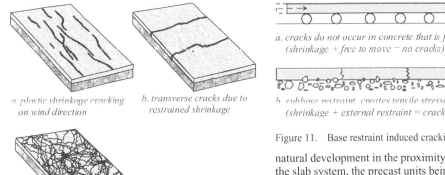

a. basement boundary wall

b. inside wall

Figure 9. Cracking patterns at shear walls.

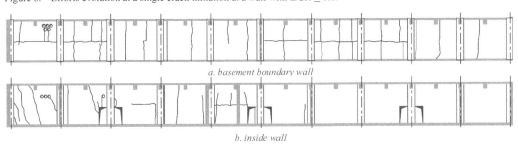

a. plastic shrinkage cracking on wind direction

b. transverse cracks due to restrained shrinkage

c. map cracking (crazing) due to shallow drying shrinkage

Figure 10. Typical shrinkage cracking patterns at slabs on ground.

in slabs systems made of precast units and a cast in situ layer of concrete, as shown in Figure 12. In this particular case, the lack of a contraction joint led to its

a. cracks do not occur in concrete that is free to shrink (shrinkage + free to move − no cracks)

b. subbase restraint creates tensile stresses and cracks (shrinkage + external restraint = cracks)

Figure 11. Base restraint induced cracking.

natural development in the proximity of the middle of the slab system, the precast units being also subjected to disarrangement. Discrete restraint provided by the columns has also to be considered in this situation, the columns also being the subject of parasite stress states.

6 DIRECT TENSION MEMBERS

Work performed by Base & Murray 1982 and by Gilbert 1992, lead to generally accepted analytical approaches for restrained direct tension members (see Fig. 13).

Once that concrete reduces its volume, a gradually increasing reaction force N(t) occurs until the

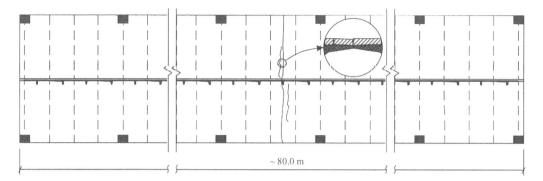

Figure 12. Severe shrinkage induced cracks at mixed slab system.

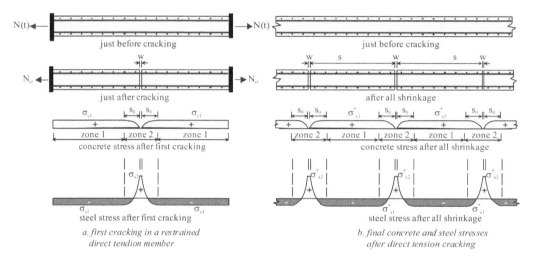

*a. first cracking in a restrained
direct tendion member*

*b. final concrete and steel stresses
after direct tension cracking*

Figure 13. Cracking in a restrained direct tension member.

first crack initiates (Fig. 13.a). Prior to the first crack occurrence, usually within the first two weeks from starting of drying shrinkage, the entire tension is sustained mainly by concrete. Just after that, boundary reaction forces decrease to the value N_{cr}, when these are sustained in the cracked cross-section just by reinforcement. Stress in concrete near the crack reduces well below the tensile strength of concrete f_t, where concrete shortens elastically and the crack width becomes w. In sections adjacent to the crack, stresses in steel and concrete vary significantly, bond being partially destroyed. Influence of the crack extends to a distance s_0 on both sides of the crack. Outside this zone, stresses in steel and concrete are not influenced by the crack any more. In zones 1, at distances more than s_0 from the crack, the stresses in steel and concrete are σ_{s1} and σ_{c1}. Since the steel stress in the crack is tension and the total elongation of the reinforcement is null, σ_{s1} must be a compressive stress. Thus, due to equilibrium conditions, tensile force sustained by concrete $A_c\sigma_{c1}$ is greater than the cracking reaction force N_{cr}. In zones 2, where the distance to the crack is less than s_0, concrete stress varies from 0 within the crack to σ_{c1} at s_0. Steel stress varies from σ_{s2} (tension) in the crack, to σ_{s1} (compression) at s_0.

The prediction equations proposed by Base & Murray 1982 are based on numerical analyses of restrained members using the finite difference method. Gilbert (Gilbert,1992) on the other hand, used the basic principles of equilibrium and compatibility to derive a series of expressions to calculate the final stresses in the concrete and steel, the number of cracks, and the average crack width. The equations developed by Gilbert imply that concrete strength did not change with time; this assumption is reasonable because the analysis considers behaviour after the formation of the first crack, which occurs usually at a late age.

The effects of creep are taken into account by using the effective modulus of elasticity of concrete. The analytical model developed first by Gilbert was modified and recalibrated by Nejadi & Gilbert (Nejadi & Gilbert, 2004) on the grounds of further comparison of the experimental results and analytical predictions. Disregarding some discrepancies, it can be seen that both methods result in similar predictions of crack width, as a function of reinforcement ratio and shrinkage strain.

Tam & Scanlon 1986 carried out a parametric study to investigate the effects of the following factors: span length, steel area, bar size, and the relationship used to

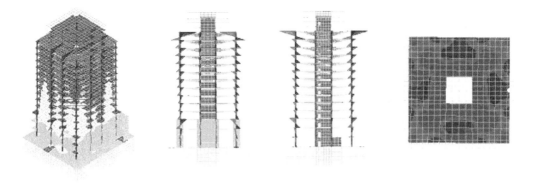

Figure 14. Creep and Shrinkage Effects – study for the Transylvanian Business Centre – Sibiu.

calculate the slip length. Based on their analyses, they reported that the number of cracks increases with span length. Therefore, crack spacing is essentially independent of span length. They also noticed that the number of cracks increases with the amount of reinforcement, and larger bar sizes lead to fewer cracks, but the cracks would be wider.

7 COMPLEX STRUCTURES

Concrete structures are generally designed based on FE linear structural models, with or without redistribution of the internal forces according to the specifics of each design situation. However, as shown above, increased sizes of the structure require consideration of long term behaviour of the structure, sometimes even sequential behaviour, in order to obtain accurate predictions. Therefore, long-term and sequential analysis procedures must be included and several scenarios considered.

Figure 14 shows the structural behaviour of the Transylvanian Business Centre from Sibiu, obtained by implementing long term specific algorithms in the FE analysis by the using the initial strain concept.

8 CONCLUSIONS

The above paragraphs present a few cases of inadequate results obtained due to poor structural design and/or poor construction practice, but most of all due to the lack of correlation between structural design and material's design. Increasing the concrete class just by increasing the Portland cement content, often results in a vulnerable a low quality construction

Considering several exposure conditions of concrete members, EN 1992 recommends superior classes of concrete, sometimes much more than needed for an adequate structural performance. This leads to the conclusion that durability based design requires superior concrete quality. However, this indirect approach needs a very close cooperation of the structural engineer with the concrete manufacturer. Indirect design methods, based on constructive provisions, presume a more refined concrete mix design and production

technology, instead of just producing standardised compositions. Moreover, insufficient research was done for the majority of admixtures from the market. Therefore, other options should be also considered (e.g. the use of non-corroding reinforcement, direct approach in structural design through specific calculations, development of protective systems etc.).

A sustainable built environment can be achieved only with more consistent work, based on a holistic thinking, by coordinating and adjusting at every step all aspects of the design. Finally, the author wishes to emphasize that even if shrinkage of concrete is inevitable, shrinkage is controllable and concrete still deserves the first option in choosing the primary structural material.

REFERENCES

ACI Committee 207, 1995, *Effect of Restraint, Volume Change, and Reinforcement on Cracking of Mass Concrete (ACI 207.2R-95)*, American Concrete Institute, Farmington Hills, MI, 26 p.

Mircea, C., Filip, M. & Ioani, A. 2007. Investigation of Cracking of Mass Concrete Members Induced by Restrained Contraction. *ACI Special Publication SP-246CD: Structural Implications of Shrinkage and Creep of Concrete*, 2007, Farmington Hills, 229–244.

Mircea, C., Păstrav, M. & Filip, M. 2008 Repair of an Industrial Building with Glass Fibre Sheets. *Concrete Durability: Achievement and Enhancement, Proceedings of the International Conference held at University of Dundee*, Dundee (UK), 8–9 of July 2008, 561–570.

Base, G.D. & Murray, M.H. 1982. *New Look at Shrinkage Cracking*. Civil Engineering Transactions, IEAust, V.CE24, No. 2, May 1982, 171 p.

Gilbert, R.I. 1992. *Shrinkage Cracking in Fully Restrained Concrete Members, ACI Structural Journal, Vol. 89, No. 2,* March–April 1992, 141–149.

Nejadi, S. & Gilbert, I. 2004. Shrinkage Cracking and Crack Control in Restrained Reinforced Concrete Members. *ACI Structural Journal, Vol. 101, No. 6,* November–December 2004, 840–845.

Tam, K.S.S. & Scanlon, A. 1986. Analysis of Cracking Due to Restrained Volume Change in Reinforced Concrete Members, *ACI Journal, Vol. 83, No. 4,* July–August 1986, 658–667.

EN 1992-1-1, *Eurocode 2: Design of concrete structures – Part 1: General rules and rules for buildings*, 226 p.

Concrete Solutions – Grantham, Majorana & Salomoni (Eds)
© 2009 Taylor & Francis Group, London, ISBN 978-0-415-55082-6

Repair and restoration of limestone blocks on Agulhas lighthouse – A world heritage structure

S.J. Pinker & D. Kruger

Department of Civil Engineering Science, University of Johannesburg, South Africa

ABSTRACT: A lighthouse situated at the southernmost tip of Africa and constructed from limestone blocks has been exposed to massive swells, treacherous seas and relentless winds since 1849. The small town of Agulhas thrives on tourism and at the centre of it all, lays the lighthouse which is in desperate need of rehabilitation of the limestone blocks that have kept the structure sturdy for the past 159 years. Limestone weathering and degradation due to intense winds, rain and pollution in the air have caused cracked masonry, spalling of surface material and loss of structural integrity in some areas, which are among some of the problems requiring attention. Limestone decay and inadequate maintenance have lead to serious degradation of lower sections of the lighthouse structure creating concern for its sustainability as an operational structure and a world heritage tourist destination. Extensive research on limestone repair techniques such as re-pointing, patching and stone replacement are discussed in this paper taking key performance factors such as porosity, hardness, density and colour testing into consideration. Application of breathable water and acid rain repellents are possible solutions to prolong and slow future degradation. The conclusion reached is that applying the best repair techniques along with the latest technology in limestone repair, patching and cleaning will restore a once magnificent heritage structure close to its original state.

1 INTRODUCTION

Limestone blocks have been subjected to decomposition and weathering resulting in severe degradation to a once scenic and structurally sound structure and tourist attraction. The structure is vulnerable to further rapid degradation and thus in desperate need of repair.

The Cape Agulhas Lighthouse, a world heritage site situated at the very southern tip of Africa, has been exposed to the elements for the past 159 years without any major maintenance or repair during these years. The solid limestone blocks used to construct the structure in 1848 now present signs of extensive decomposition, severe weathering and in some cases, loss of structural integrity.

The limestone blocks used in this structure are relatively soft and pervious to water and air which allow them to be susceptible to the penetration and attack of acid rain, air pollutants and salt crystallization all of which lead to the degradation and decomposition of the limestone blocks.

The urgent need for a sustainable repair technique utilizing a durable patch repair material in conjunction with water repellents specifically suited to the limestone of this particular area is discussed and analyzed in this paper. Taking a number of key performance factors such as density, porosity and colour matching into account when selecting a methodology for restoring the structure is crucial to a sustainable and durable limestone repair project.

2 LIMESTONE WEATHERING

Limestone composition consists mainly of calcium carbonate, this allows it to be vulnerable to attack from acid rain, which is a weak carbonic acid that slowly but surely causes degradation and weathering of the stone or structure. Weathering usually occurs along the joints and bedding planes which are present due to the sedimentary nature of the rock. This causes the surface to become uneven, leading to grooves along the surface called lapis or karren. Over time, as the limestone is subjected to further weathering, the lapis or karren become known as grykes which separate raised sections called clints. (www.geographypages.co.uk, July 2008).

In cold climates, however, a porous limestone can suffer rapid degradation due to freeze-thaw cycling. Water may be contained in the small pores of the rock which freezes at sub-zero temperatures resulting in a volumetric increase of about 9 percent. Due to the pore being completely filled with water, this freezing causes rupturing of the pore and can then lead to physical cracking. Once the crack has opened, water is allowed to ingress further once it has thawed, thus allowing it to again freeze under sub-zero conditions in pores deeper in the stone, thus propagating the cycle. (www.findstone.com, June 2008).

There is constant wetting and drying of the surface of a limestone wall over the years due to rain and varying levels of moisture in the air. Gradually more of the exposed surfaces begin to erode leading to natural weathering. A decade of extremely wet weather interspersed with unusually low temperatures has accelerated this natural weathering. Atmospheric pollutants such as sulphur dioxide, combines with the calcium carbonate in the stone and creates a hard gypsum layer covering the surface of the stone. As a result, moisture inside the stone cannot escape and salts may crystallise behind this hard layer and eventually cause spalling, which leaves a weak exposed surface, more vulnerable to natural weathering. Salt weathering and salt crystallization are major factors that affect the weathering of limestones, especially in coastal regions such as Agulhas lighthouse which lies approximately 200 m from the water's edge. Salt weathering is the physical growth of salt crystals within the pores of the limestone. These crystals grow and cause internal pressures on the pores which eventually cause cracks and physical disintegration. (www.buildingconservation.com, June 2008).

Limestone also tends to weather more dramatically in harsh environments. In such environments, strong winds, sand blasting and heavy rains mechanically attack a structure, which results in accelerated degradation and weathering. Another aspect inherent in most limestone materials is micro-structures called stylolites which are pressure-solution features formed during the compaction and lithification of limestones. These stylolites form natural planes of weakness and can often transmit fluids because they are not fully closed. Any expanding clays in the limestone can react to such fluids and physically weaken the limestone. (www.geographypages.co.uk, July 2008).

Failures of buildings made of stone are commonly related to water intrusion and contaminants, such as soluble salts carried in the water as well as combinations of the above mentioned aspects. Water migration can cause corrosion of the anchoring system, failure of the mortar joints, and exfoliation of the stone, cracking, spalling and bowing. Typical waterproofing repairs as well as typical structural repairs to stone would include: installation of new flashings, re-pointing of mortar joints, repair of corroded anchoring systems, patching of cracks and spalls, application of water repellents and replacement of badly degraded blocks, some of which will be discussed in detail in this paper (Goldstein, 2008).

3 LIMESTONE REPAIR

3.1 *Limestone cleaning*

Limestone cleaning is a relatively simple technique but special attention has to be taken on the choice of cleaning agent. Chlorine, bleach and any other harsh cleaners should be avoided since they will further degrade the limestone. Limestone should be treated and cleaned with agents that contain two ingredients; a non-ionic detergent and an alkali, such as ammonia and baking soda.

3.2 *Limestone block replacement*

There is the option of removing severely damaged blocks and replacing them with a new, equi-dimensional block that can easily be found in nearby quarries of the area and cut to the right shape. This is needed in a few areas of the lighthouse. These new blocks that are installed can be bonded with a mortar or epoxy resin to secure them in place. It must be noted that the blocks that are used to replace the old chiselled out sections should be the same material as previously used, or alternatively one that is more porous and less dense than the original rock to ensure that a seal is not formed and that water already trapped inside the original rock can evaporate and not cause it to rot or degrade further.

3.3 *Surface preparation for patch material*

Surfaces to receive a patching material must be sound and free of all dust, dirt, grease, laitance and/or any other coating or foreign substance which may prevent proper adhesion. All loose and deteriorated masonry from the repair area must be removed using manual or pneumatic cutting tools. The area to be repaired should be cut to provide a minimum depth of 7 mm. It is important to bear in mind that incorrect installation will cause repairs to fail prematurely.

3.4 *Limestone patch*

It is necessary to perform various tests on the insitu limestone as well as the replacement limestone before commencing any repair. It is imperative to know exactly what type of limestone was used during the original construction in order to adopt and create the right patch material to suit the structure. Factors to be considered are the limestone density, the degree of hardness and the porosity of the stone. If patches and pointing mortars are too dense or impermeable, the moisture in the stones will be forced back through the stone which will accelerate degradation and deterioration (Berkowitz, 2008).

3.5 *Limestone consolidation*

Limestone consolidation should not be confused with limestone preservation. Preservation is the aim of trying to protect and preserve the structure from the elements and to stop weathering altogether. Limestone consolidation, on the other hand, has the aim of trying to stabilise the friable material, but at the same time permitting natural weathering to take place. The primary aim of consolidation is thus trying to slow the rate of decay. The most important aspects to consider when trying to consolidate limestone are strength, porosity, permeability, thermal dilation and colour. The most widely used polymers as consolidation material are silanes, although they may not be suitable in every situation. The theoretical end product of polymerisation of the simplest silanes is silica. It is necessary for the consolidation material to penetrate at least 25 mm into the stone to reach all areas of friable material. (www.buildingconservation.com, June 2008).

3.6 *Re-pointing repair*

Re-pointing requires cutting out failed joint mortar and applying new mortar that is finished to replicate the original mortar style. Two important points to take into account when re-pointing are that the new mortar should be of a weaker compressive strength than the surrounding masonry to avoid spalls at the joints, and joint sealant is not recommended, as it prevents the migration of damaging moisture from within the wall system. All historic re-pointing projects should conform to ASTM Standard Guide for Re-pointing. (www.buildingconservation.com, June 2008).

3.7 *Water and acid rain repellents*

These products protect and strengthen deteriorating carbonate building stones such as marble, limestone, dolomite and travertine by forming a stable, well-adhered, hydroxylated, conversion layer on carbonate mineral grains. This conversion layer dramatically increases the resistance of marble and limestone surfaces to acid attack and improves the ability of a variety of chemical compositions to react with or bond to surfaces such as those found on the Lighthouse.

Limestone treatment water repellents provide long-lasting repellence without altering the natural appearance of the substrate. These water repellent products are used as a surface treatment on natural stone and masonry surfaces to impart a water-repellent surface and reduce water absorption into the substrate.

Before finalizing the selection of the product that will be used to reduce water absorption and acid attack, it should be shown by insitu tests that these products will not seal the surface of the substrate since water vapour within the limestone must be allowed to evaporate.

4 MATERIALS AND TEST METHODS

Limestone specimens collected from the Cape Agulhas lighthouse were cut into manageable sizes and labelled specimens 1 through 3 as shown in Figure 1.

These samples were trimmed to size using a diamond cutting machine (Metkon Metacut-A 250). A Mini-scan® apparatus from Hunterlab was used to record accurate colour readings of both limestone samples from the lighthouse, as well as the patch materials made in the laboratories. The colour readings were made using the mini-scan in similar environments to reduce the risk of deviations in readings due to human error. The colour testing method used was the CIELAB colour system. Varying fields of this colour system exist, but in this system, geometrically equal distances in all colour sectors more or less correspond to the differences perceived visually. Due to this reason, this

Figure 1. Trimmed Limestone specimens.

Figure 2. CIE colour space.

system can be referred to as an almost naturally sensitive system. Figure 2 shows the CIE colour space coordinates distinguishing between colours (Zhang, 2005). The system uses three perpendicular axes of equal scale made up of 'L', 'a' and 'b'. This means that every colour has an 'L', 'a' and 'b' value assigned to it, which would fall somewhere on the 3-dimensional set of axes. Presented as dimensionless numbers, 'L' indicates brightness, where 'L = 100 means white and 'L = 0 means black. On one of the horizontal axes, '+a' means red and '−a' means green, the scale of 'a' usually ranges from 'a' = −270 to 'a' = +270. The second horizontal axes represents the letter 'b', where '+b' means yellow and '−b' means blue and the range in from 'b' = −1487 to 'b' = +200 (Zhang, 2005).

When dealing with colour testing and matching, it can often be useful to assign a value of variance to each sample so that columns can be compared to other columns. In this case the colourimetric calculation formula of CIELAB is expressed by the value ΔE, where the 'Δ' means discrepancy and 'E' is the first letter of the word 'Empfindung' which means 'feeling'. The formula is as follows:

$$\Delta E = \{(\Delta L)^2 + (\Delta a)^2 + (\Delta b)^2\}^{0.5} \qquad (1)$$

Table 1. Limestone specimen colour variance results.

Reading No	Sample 1			Sample 2			Sample 3			ΔE
	L	a	b	L	a	b	L	a	b	
1	77.1	3.6	15	79	3.1	15	76	4.8	19	5.3
2	71.9	4.1	16	71	2.8	16	79	4.9	20	8.7
3	72.8	3.4	14	67	3.1	15	77	4.8	20	11.3
4	74.2	3.4	15	79	2.2	14	78	5	20	8.0
5	73.3	3.4	16	79	3.2	16	78	5.4	21	7.8
6	77.2	3.8	16	76	3.6	16	76	4.9	20	4.5
7	77.2	3.7	16	75	2.8	15	78	5.2	20	6.9
8	73.8	3.9	16	78	3.4	16	76	5.2	21	6.6
9	77.2	3.7	16	78	3	15	71	4.4	16	7.1
10	77.9	3.7	17	75	2.2	14	71	4.3	17	7.4
Average	75.3	3.7	15.6	75.6	2.9	15.2	76.0	4.9	19.3	4.6

with ΔL, Δa and Δb the maximum discrepancies measured on the white/black, red/green and yellow/blue columns for various specimens.

Density is another key-performance factor of the limestone that has to be calculated for the reason that the density of the patch samples and replaced limestone blocks, whether it be partial or full replaced blocks, has to be lower than that of the original block. The reason is so that water and air can still move uninhibited through the blocks to ensure no water or vapour gets trapped inside the original stone. If a seal is formed and water is trapped, it will cause the original stone to decompose at a faster rate. A very simple explanation of this concept is to imagine placing a plastic glove on a person's hand. Moisture is not allowed to escape and water and water vapour will be trapped inside the glove.

In order to calculate densities accurately, specimens were weighed and placed in an oven to dry and then weighed at intervals between drying times until the difference between weight measurements was negligible. The porosity of the specimen was calculated as the ratio of volume of water absorbed by the specimen to the volume of the trimmed specimen. Porosity is the ratio of the volume of the pore spaces or voids in a material to the total bulk volume. Porosity is important for developing the right patch material or for selecting the right replacement block material since porosity will greatly affect the breathability and characteristics of the replacement material to be suitable as a repair material.

5 RESULTS

Using the Hunterlab Mini-scan, 10 colour readings each were taken of specimens 1, 2 and 3. Using these averages in equation 1, the colour variance (ΔE) value was calculated for the three insitu limestone specimens in order to give an indication of typical colour variances in the natural stone. These values measured are presented in Table 1 which shows the averages of the 10 measurements of 'L', 'a' and 'b' for each specimen.

Table 2. Colour variance in specimens.

ΔE value	Visibility of colour difference
Up to 0.2	Not visible
0.2–0.5	Very slight
0.5–1.5	Slight
1.5–3.0	Obvious
3.0–6.0	Very obvious
6.0–12.0	Large

Table 3. Limestone specimen density and porosity.

Specimen No	1	2	3
Dry Weight (g)	218.9	80.06	158.7
Wet weight (g)	303	98.85	201.7
Block Volume(cm^3)	160	50	100
Porosity (%)	38	23	27
Density(kg/m^3)	1368.1	1601.2	1587.0

These values can be used as a guide when designing the patch material or choosing a suitable replacement limestone block. From these values and referring to Table 2 it can be observed that the colour difference between the averages of the specimens fall into the 'Very obvious' category. Replacement limestone or patch material should thus have a similar ΔE value to make it colour compatible.

Density and porosity was then calculated from weight and volume measurements done on the trimmed specimens taken from the lighthouse.

These results are presented in Table 3 and provide base values for comparison when selecting patch material or replacement limestone.

6 DISCUSSION

Limestone in most cases may be impure, i.e. having a lower percentage of calcium carbonate. Limestone is a sedimentary rock and can contain a large number and variety of impurities. Because of these factors, limestones vary to a large degree in their physical and visual appearance. This has been found to be the case for the limestone used to construct the Cape Agulhas lighthouse. This structure exhibits a wide variety of limestone blocks each with its own colour and set of physical properties such as porosity and density. For this reason, there is more freedom of choice when selecting repair patch materials or replacement blocks, and less need for an exact colour, density and porosity match to be achieved.

When producing colour patches consistent with the limestone blocks found on the lighthouse, Table 1 can be used as a guide to check adequate colour performance from patching repair solutions. As can be seen from Table 1, specimen colour indicators vary to a degree and thus allow some room for colour experimentation to take place.

Physical performance factors such as achieving adequate density and porosity can often prove to be a more significant problem. While it may be simpler to achieve the optimum solution for only one of these properties at a time, it is often more complex to achieve both the optimum density and porosity simultaneously.

As mentioned earlier, the limestone performance factors found on the Cape Agulhas lighthouse vary to a significant degree. Because of the large spread of colour variance on the existing limestone blocks, correct colour matching of patch repair materials or replacement blocks is not as crucial as the matching of the porosity and density properties. It is clear that for a sustainable repair and restoration project, further testing needs to be performed on a wider range of blocks found on the limestone to ensure that the patching material produced will be suitable for most, if not all of the structure.

Using and applying the best repair techniques along with the latest technology to ensure that colour, texture and physical properties are suitably matched will help to ensure that this once beautiful world heritage structure will be rehabilitated and restored to last another 100 years.

REFERENCES

Berkowitz, J.C. (2008) *Director of Conservation*; Superstructures Engineers & Architects, Private e-mail to author.

Goldstein, K. (2008). President – Everest Waterproof ing and Restoration Inc, Private e-mail to author

http://www.buildingconservation.com/articles/stoneconsol/stoneconsul.htm (June 2008)

http://www.findstone.com/gelime.htm (June 2008)

http://www.geographypages.co.uk/weathering.htm (July 2008)

Zhang, Y. (2005). *Methodology for Aesthetic Repair and Rehabilitation of Architectural concrete.* Unpublished Masters Dissertation. University of Johannesburg, Johannesburg, South Africa.

Concrete Solutions – Grantham, Majorana & Salomoni (Eds)
© 2009 Taylor & Francis Group, London, ISBN 978-0-415-55082-6

Bacterial concrete – An ideal concrete for historical structures

Darshak B. Raijiwala, Prashant S. Hingwe & Vijay K. Babhor

Department of Applied Mechanics, S V National Institute of Technology, Surat, Gujarat, India

ABSTRACT: The application of a novel approach in concrete crack remediation by using microbiologically induced calcite precipitation ($CaCO_3$) is presented in this paper. A common soil bacterium called 'Bacillus Pasterii was used to induce $CaCO_3$ precipitation. As a metabolic by product of Bacillus Pasteurii, inorganic $CaCO_3$ precipitates outside the cell and can persist in environment for a much longer period of time. After application of this type of Biosealant, the concrete would be expected to be self – remediating because Bacillus Pasteurii has the ability to produce endospores so as to endure extreme environmental conditions. The efficiency of the Microbiologically Enhanced Crack Remediatior (MECR) was evaluated by comparing the compressive strength of treated mortar cubes with those of control specimens. Environmental Scanning Electron Microscope (ESEM) analysis was used to check the involvement of Bacillus Pasteurii in $CaCO_3$ precipitation. Energy – Dispersive X-ray diffraction (XRD) analysis was used to quantify $CaCO_3$ distribution in the regions of treated cubes. Based on observations made in this study, it is clear that MECR increases compressive strength of treated cube specimens by 12–13% and has remarkable potential as a sealant for concrete cracks. This technique is highly recommended for monumental & historic (heritage) structures.

1 INTRODUCTION

Cracking of concrete is a common phenomenon. Without immediate and proper treatment, cracks in concrete structures tend to expand further and eventually require costly repairs. Even though it is possible to reduce the extent of cracking by available modern technology, remediation of cracks in concrete has been the subject of research for many years. There are a large number of products available commercially for repairing cracks in concrete structures: epoxy, resins, epoxy mortar and other synthetic mixtures. Currently, these types of synthetic filler agents are extensively used in concrete crack repairs. Because cracking in concrete structures continues over a long period of time, this type of onetime, quick remedy should also be applied repeatedly as expansion of cracks continues.

Cracks and fissures are a common problem in building structures, pavements, and historic monuments. We have introduced a novel technique in repairing cracks with environmentally friendly biological processes that are continuously self-remediating. In the study, *Bacillus pasteurii* that is abundant in soil has been used to induce $CaCO_3$ precipitation. It is therefore vital to understand the fundamentals of microbial participation in crack remediation.

Definition: "Bacterial Concrete" is a concrete which can be made by embedding bacteria in the concrete that are able to constantly precipitate calcite. This phenomenon is called microbiologically induced calcite precipitation. It has been shown that under favorable conditions, for instance, Bacillus Pasteruii, a common soil bacterium, can continuously precipitate a new highly impermeable calcite layer over the surface of an already existing concrete layer. The favorable conditions do not directly exist in a concrete but have to be created.

2 CHEMISTRY OF THE PROCESS

Microbiologically enhanced crack remediation (MECR) utilizes a biological byproduct, $CaCO_3$ which has shown a wide range of application potential as a sealant (Ramachandran et al, 2001). Its prospective applications include remediation of surface cracks and fissures in various structural formations, in-base and sub-base stabilization, and surface soil consolidation. In principle, MECR continues as microbial metabolic activities go on. This inorganic sealant not only is environmentally innocuous but also persists in environments for a prolonged period, (Bang et al, 2001).

Microbiologically induced calcium carbonate precipitation (MICCP) is comprised of a series of complex biochemical reactions, (Stocks-Fischer et al, 1999) including concomitant participations of *Bacillus pasteurii*, urease (urea amidohydrolase), and high pH. In this process, an alkalophilic soil microorganism, *Bacillus pasteurii*, plays a key role by producing urease that hydrolyzes urea to ammonia and carbon dioxide (Bachmeier et al, 2002). The ammonia increases the pH in the surroundings, which in turn induces precipitation of $CaCO_3$, mainly as a form of calcite. In aqueous environments, the overall chemical equilibrium reaction of calcite precipitation can be described as:

$$Ca^{2+} + CO_3^{2-} \rightarrow Ca\,CO_3\downarrow \qquad \text{----- Equation 1}$$

The solubility of $CaCO_3$ is a function of pH and is affected by ionic strength in the aqueous medium. In a Urea-CaCl$_2$ medium that supports microbial growth, additional ions including NH^{4+}, Cl^-, Na^+, OH^-, and H^+, may affect chemically induced $CaCO_3$ precipitation (CICCP) at different pHs. MICCP occurs via far more complicated processes than chemically induced precipitation. The bacterial cell surface with a variety of ions can nonspecifically induce mineral deposition by providing a nucleation site. Ca^{2+} is probably not utilized by microbial metabolic processes; rather it accumulates outside the cell. In medium, it is possible that individual microorganisms produce ammonia as a result of enzymatic urea hydrolysis to create an alkaline microenvironment around the cell. The high pH of these localized areas, without an initial increase in pH in the entire medium, commences the growth of $CaCO_3$ crystals around the cell.

Possible biochemical reactions in Urea-CaCl$_2$ medium to precipitate $CaCO_3$ at the cell surface can be summarized as follows:

$$Ca^{2+} + Cell \rightarrow Cell\text{-}Ca^{2+} \qquad \text{Equation 2}$$

$$Cl^- + HCO_3^- + NH_3 \rightarrow NH_4Cl + CO_3^{2-} \qquad \text{Eqn. 3}$$

$$Cell\text{-}Ca^{2+} + CO_3^{2-} \rightarrow Cell\text{-}Ca\,CO_3 \qquad \text{Equation 4}$$

3 IMMOBILIZATION OF BACTERIA

To protect the cells from the high pH of concrete and high metabolic activity of the bacterium, the microorganisms are immobilized in polymer, lime, silica fume, and fly ash, and then applied in concrete crack remediation. Aliphatic Polyurethane (HYDROTHANE-330) has been used as a vehicle for immobilization of bacterium whole cells because of its mechanically strong and biochemically inert characteristics (Klein et al, 1981, Wang et al 1993). Typically, porous matrices of HYDROTHANE-330 and other materials not only increase the surface area but also minimize diffusional limitations for substrates and products. A variety of hydrophilic prepolymers can be used for the immobilization of microbial cells, which allow enzymatic and metabolic activities to remain intact for extended periods.

Use of the immobilization technique:-

- It is the technique in which microorganisms are encapsulated in different porous materials to maintain high metabolic activities and protection from adverse environments.
- For immobilization, different materials like polyurethane (PU) polymer, lime, silica, fly ash can be used.

- PU can be used widely because of its mechanically strong and biochemically inert characteristics.
- PU mixes produce open cell foam as a result of condensation of polycyanates (R-CNO) and polyols (R-OH).

4 EVIDENCE OF CALCITE PRECIPITATION INDUCED BACILLUS PASTEURII

Upon polymerization, PU foam is pliable and elastic with an open-cell structure of matrices (Fig. 1.a). Micrographs (Figs. 1.b-1.c) showing cell-laden PU matrices indicate that immobilization caused no apparent morphological damage to the cells and microorganisms are entrapped throughout the polymer matrices where cells are adhered or embedded with some clumping. As shown in Figs. 1.d and 1.e, calcite precipitation occurred throughout the entire matrices, including the inside of pores as well as the surface areas. It is also apparent that calcite crystals grow around the microorganisms and PU matrices (Fig. 1.f).

5 STRENGTH AND DURABILITY PERFORMANCE OF BACTERIAL CONCRETE:

- The performance of MICCP in concrete remediation was examined using hairline-cracked cement mortar beams remediated in the medium with B. pasteurii. Various levels of performance enhancement were observed in the treated specimens;
- Reduction of the mean expansion due to the alkali aggregate reactivity by 20%.
- Reduction of sulfate effects by 38%; reduction of the mean expansion by 45% after freeze-thaw cycle; and higher retaining rates (30% more) of the original weight.

From X-ray diffraction (XRD) analysis studies, $CaCO_3$ crystals were determined as calcite as shown in Figure 2.

- The microbiological enhancement of concrete was further supported by SEM analysis evidencing that a new layer of calcite deposit was present providing an impermeable sealing layer, and increasing the durability of concrete against the freeze-thaw cycles and chemicals with extreme pH.

5.1 *Efficiency of bacteria as biosealent:*

The results suggest that PU provides cells with protection from a high pH of concrete and further supports the growth of bacteria more efficiently than other filling materials.

6 EXPERIMENTAL PROCEDURES

6.1 *Micro-organisms and growth condition*

A stock culture of B.pasteurii is generally maintained in a solid medium containing : 10 g trypcase; 5 g yeast

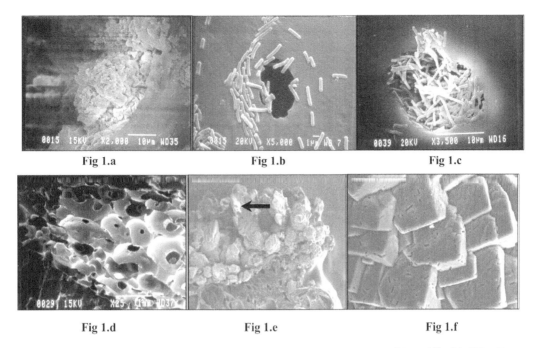

Fig 1.a   Fig 1.b   Fig 1.c

Fig 1.d   Fig 1.e   Fig 1.f

Figure 1. Scanning Electron Micrographs of Calcite Precipitation Induced By B. Pasteurii Immobilized In PU. **a.** Porous PU matrix without microbial cells showing open-cell structures. Bar, 1 mm; **b.** Distribution of microorganisms on the PU surface. Bar 1 μm; **c.** Microorganisms densely packed in a pore of the PU matrix. Bar 10 μm; **d.** Calcite crystals grown in the pore (shown in c) of the PU matrix. Bar 10 μm; **e.** Calcite crystals grown extensively over the PU polymer. Bar 500 μm; **f.** Magnified section pointed with an arrow in e shows crystals embedded with microorganisms. Bar 20 μm.

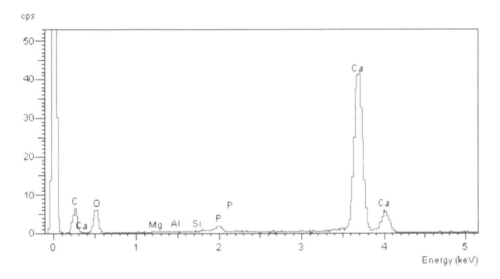

Figure 2. Energy Dispersive X-Ray Spectrum Shows the Abundant Presence of Ca.

extract; 4.5 g tricine; 5 g (NH$_4$)$_2$SO$_4$; 2 g glutamic acid and final concentration of 1.6% agar, which is auto-claved separately and added after-wards. For quantity culture use in MICCP *B. Pasteurii* is cultured in yeast extract (YE) broth medium with 30°C for 24–30 h with aeration facilities.

6.2 *Transfer of bacteria:*

After sufficient growth of bacteria in the laboratory the culture is transferred to cracked mortar cubes by mixing with sand and the required amount of bacteria culture.

187

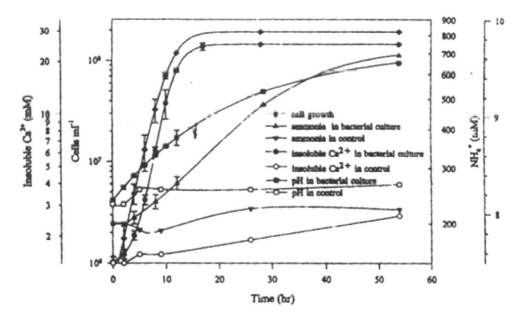

Figure 3. Microbiologically Induced CaCO₃ Precipitation In The Presence of B. Pasteurii at 25.2 μm of Cacl₂ at 25°C.

6.3 Strength characteristics:

After successful transfer of bacteria, trials for compressive strength with different concentrations and with immobilization of bacteria were carried out. And samples were taken for Environment Scanning Electron Microscope (ESEM) and X-Ray diffraction analysis for visualize and quantified microbial growth in cracked region of mortar cubes.

7 CONCLUSIONS

The positive potential of MICCP that has been demonstrated in our study offers an interesting concept of the crack remediation technique in various structures. The following summarizes our preliminary findings on MICCP.

- The rod shape impression (Figure 4 – Calcite Crystals Developed at Higher Magnification, Bar 10 μm) shows that rod shaped impressions are very widely dispersed on calcite crystals, which are of the same size as B. Pasteurii and confirms that the bacteria serve as a nucleation site for calcite crystals for precipitation.
- From Figures 5 & 6 it is seen that using an immobilized cell concentration of 9.0×10^8 with different depths of cut like 10.0 mm, 20.0 mm, 25 mm, an increase in compressive strength was observed of 15.0%, 8.0% and 8.0% respectively. The maximum compressive strength was for a depth of cut 10.0 mm, which indicates that bacteria near the surface of mortar cubes are more active than bacteria away from the surface. From Figure 6b it is also clear that the compressive strength is around 5%

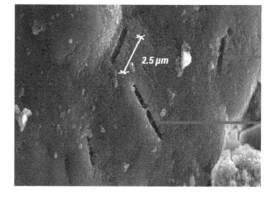

Figure 4. Rod shape impression.

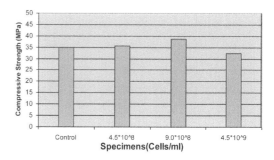

Figure 5. Comp. Strength at Different Concentration.

higher than cells without immobilization. This also supports that immobilization of cells increases cell retentivity and provides space to bacteria for more bacteriogenic activity.

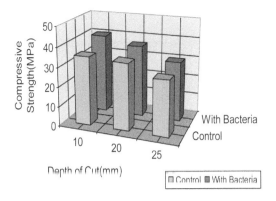

Figure 6a. Comparison of Compressive Strengths of Cubes with Various Crack Depth.

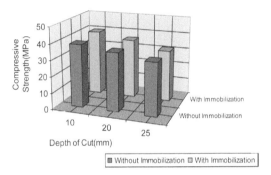

Figure 6b. Comparison of Compressive Strengths of Cubes with Various Crack Depth.

- A concentration of 9.0×10^8 cells per ml is most suitable for the maximum compressive strength. Specimens with higher concentration do not give higher compressive strength values probably because the greater population of bacteria does not have enough nutrients to multiply.
- Comparisons of compressive strength of specimens with various crack depths with a single immobilized cell concentration, MICCP in concrete using Hydrothane-330 matrix immobilized with B. Pasteurii 9.0×10^8, was most effective when compared to MICCP with a mixture of Sand and B. Pasteurii keeping the same concentration of bac-

teria. The degree of enhancement in compressive strength by Hydrothane-330 immobilized bacteria is between 4 to 5% as compared to crack remediation with sand–mixed B. Pasteurii and 14% increase as compared to control specimens.
- Energy Dispersive X-ray (EDX) analyzer which indicated the abundant presence of Ca. This was enough proof to conclude that the precipitation was indeed calcite ($CaCO_3$) crystals
- B. Pasteurii however, has an ability to produce endospores, a dormant form of the cell, to endure extreme environments, so if in future cracks treated with this technique widen then again B. pasteurii starts metabolic activity, which leads to accumulation of insoluble $CaCO_3$
- The optimum concentration of B. Pasteurii should be decided considering parameters like crack size, frequency of reaction mix application, length of microbial treatment, remediation temperature and material for immobilization, environmental condition etc.

Based on the observations made in this study, it is clear that MICCP has excellent potential in cementing concrete as well as several other types of structural and non-structural cracks.

REFERENCES

Bachmeier K., Williams A. E., Warmington J., and Bang S. S., Urease Activity in Microbiologically-Induced Calcite Precipitation. *Journal of Biotechnology*, 2002, *93*:171–181.

Bang S. S. and Ramakrishnan V., Microbiologically-enhanced crack remediation (MECR). *Proceedings of the International Symposium on Industrial Application of Microbial Genomes*, 2001, Daegu, Korea, 3–13.

Klein J., and Kluge M., Immobilization of microbial cells in polyurethane matrices. *Biotechnology Letters*, 1981, *3*, 65–70.

Ramachandran S. K., Ramakrishnan V. and Bang, S.S., Remediation of Concrete Using Microorganisms. *American Concrete Institute Materials Journal*, 2001, *98*:3–9.

Stocks-Fischer S., Galinat J. K., and Bang S. S., Microbiological precipitation of $CaCO_3$. *Soil Biology and Biochemistry*, 1999, *31*, 1563–1571.

Wang X. and Ruchenstein E. Preparation of porous polyurethane particles and their use of enzyme immobilization. *Biotechnology Progress*, 1993, *9*, 661–665.

Concrete Solutions – Grantham, Majorana & Salomoni (Eds)
© 2009 Taylor & Francis Group, London, ISBN 978-0-415-55082-6

Microbial precipitation for repairs of concrete structures

D.D. Sarode
Indian Institute of Technology Bombay, India
University Institute of Chemical Technology, Mumbai, India

Abhijit Mukherjee
Thapar University, Patiala, India

ABSTRACT: Large numbers of reinforced concrete structures deteriorate or become unsafe before their design service life. The presence of voids, pores in concrete and subsequent cracks are responsible for early deterioration. Natural cementation of geological formation occurs over geological time due to physical, chemical and biological reactions. *Bacillus pasteurii* a microorganism from soil (ATCC 11859), was used for inducing calcium carbonate precipitation in voids, pores and microcracks of concrete. These microbiologically induced substances are pollution free and naturally ubiquitous in all environments. Quantification and characterization was done using titration, X-ray diffraction, X-ray fluorescence and microscopic examination. Reinforced concrete samples have been corroded with accelerated initial corrosion and subsequently remediated with microbial deposition. They have been subjected to an aggressive environment by keeping them in a salt mist chamber. Considerable reduction in mass loss of the reinforcing bar and an increase in pullout strength was observed. A reduction in water and chloride permeability was observed.

1   INTRODUCTION

Structures, natural and manmade, often deteriorate from weathering, natural disasters such as earthquakes, cyclones and floods. Although natural structures such as anthills, coral reefs etc. can heal themselves and grow, manmade structures lack such ability. Hundreds of thousands successful structures are annually constructed worldwide with artificial materials such as reinforced concrete which deteriorate and become unsafe much before their design service life. The deterioration process of structures with cement composites has two stages: initially aggressive fluids penetrate or are transported through the pore structure of concrete, which is followed by actual chemical or physical deterioration reactions. The rate at which these environmental agents ingress into the body of concrete depends on the porosity in the concrete cover region. Earlier research has proved that low permeability concrete lasts longer without exhibiting signs of distress and deterioration. Ingress of moisture and chlorides may lead to early onset of deterioration in the cement mortar finishes and concrete cover to the reinforcement, which leads to early onset of corrosion in reinforcement. Once the corrosion starts further distress takes place at a much faster pace. *Bacillus pasteurii* a soil bacterium was successfully used for biomineralisation in the concrete by deposition of microbiologically induced calcite precipitation (MICP). The calcite deposition reduces the porosity of concrete in the top few millimeter in the cover region. This reduces the rate of ingress of water and chlorides, improving the durability of concrete. If such deposition can be done in the micro cracks we have progressed one step in self healing concrete structures.

During the process of biomineralization, the enzymatic hydrolysis of urea takes place forming ammonia and carbon dioxide. Urease which is provided by bacteria catalyzes hydrolysis of urea (Equation 1), resulting in increase of pH in the surrounding medium. This condition helps to deposit calcium carbonate in the presence of mineral ions (Ca^{+2} and CO_3^{-2}).

$$H_2N\text{-}CO\text{-}NH_2 + H_2O \rightarrow 2NH_3 + CO_2 \uparrow \qquad (1)$$

MICP is a complex mechanism and is a function of cell concentration, ionic strength, nutrient and pH of the medium. The possible biochemical reaction in a microbial medium along with Urea-CaCl$_2$ precipitates calcium carbonate at the cell surface as given by following reactions. (Stocks et al. 1999)

$$Ca^{2+} + Cell \rightarrow Cell\text{-}Ca^{2+} \qquad (2)$$

$$Cl^- + HCO_3^- + NH_3 \rightarrow NH_4Cl + CO_3^{2-} \qquad (3)$$

$$Cell\text{-}Ca^{2+} + CO_3^{2-} \rightarrow Cell\text{-}CaCO_3 \qquad (4)$$

Researchers have studied the microbiological precipitation using different microbes (Stocks et al. 1999, Bang et al. 2000, Morita 1980, Lee 2003). The effect of strain, pH and calcium metabolism has been studied (Hammes et al. 2002, Hammes et al. 2003) Researchers have successfully done MICP in sand

(Stocks et al. 1999), stones,(Castanier et al. 1999, Disck et al. 2006) ornamental stones (Carlos et al. 2003) and monumental stones.(Tiano et al. 1999) Deposition reduces the water absorption rate of limestone. A reduction of water absorption by about 60% by the samples treated with MICP was observed. (Dick et al 2006, Tiano et al. 1999). Cracks in granite were effectively remediated by MICP using microorganisms mixed with a filling material such as silica fume and sand. (Gollapudi et al. 1991) The potential for bioclogging in micro fractured aquifers was demonstrated by stimulation of ground water microbes.(Ross et al. 2001). To reduce the harmful effects of the environment on the microbial metabolism some researchers used polyurethane to immobilize the bacterium or the enzyme. (Klein & Kluge 1981, Wang & Ruchenstein 1993, Bang et al 2001)

The strength improvement in cement mortar due to MICP was studied using *Bacillus pasteurii* (Ramachandran et al. 2001) and Shenwalla in a tropical environment. (Ghosh et al 2005) Strength as well as durability properties of concrete with MICP were studied.(Ramakrishnan et al. 2003) They not only observed improvement in strength, modulus of rupture and stiffness but also observed better resistance to alkali aggregate reaction, resistance to sulfate and freeze thaw. MICP on cementitious materials using *Bacillus Sphaericus* was attempted. (Muynck et al. 2008) They observed that the surface deposition of calcium carbonate crystals decreased the water absorption by 65 to 90% depending on the porosity of the specimens.

2 MATERIALS AND METHODS

2.1 Microorganisms and growth conditions

Bacillus *pasteurii* (ATCC 11859) obtained from the National Collection of Industrial Microorganisms (NCIM) from the National Chemical Laboratory Pune, India was used. The characteristics were observed by growing them in the Petri dish. For detailed understanding of the growth characteristics, the growth in static and shaking conditions, at different pH was studied.

2.2 Microbial media, cell concentration and chemicals

Initially 3 different media Yeast Extract Broth (YEB) which was prepared by Yeast extract and peptone, Brain Heart Infusion (BHI), Nutrient Broth (NB) were used. All nutrient media were of the Himedia make. In the NB medium, 5 different concentrations of cells as given in Table 1 were used. Hydrochloric acid, urea, calcium chloride, salt (NaCl), agar slants, ethylenediamino tetra acetic acid disodium salt, NaOH and Hydroxy Naphthol.

2.3 Concrete mix proportion and sampling

Ordinary Portland cement of nominal strength 53MPa, fine aggregate (medium-sized natural/river sand), grit

Table 1. Media, cell concentration and percent purity of calcium carbonate in precipitate.

Sample	Cell concentration cells/ml	%$CaCO_3$ purity
NB1	2.6923×10^4	96.53
NB2	4.3846×10^4	90.97
NB3	6.8461×10^4	91.48
NB4	8.1153×10^4	95.52
NB5	11.5384×10^4	89.96
BHI	4.3846×10^4	78.34
YEB_1	6.8461×10^4	88.45
YEB_2	6.8461×10^4	90.97

Table 2. Quantification of precipitate.

Medium	NB1	NB2	NB3	NB4	NB5
Weight of precipitate gms	98.48	100.21	108.2	109.3	175.6

and crushed stone coarse aggregate with a maximum size of 20 mm was used. Two sets of cement mortar cubes of side 25 mm were cast using cement and sand in the proportions 1:3 (by mass). One set of cubes was cast by adding 143 ml water (CMC_W). The other set of cubes were cast using 143 ml 18 hr old cell suspension from a YEB medium (CMC_M).

For concrete, the water-cement ratio was 0.52 and aggregate-cement ratio 4.9. The resulting strength of concrete was 25 MPa. For samples from media instead of water respective media were used. The media used contained cells concentration as per Table 1. Control samples were cast and cured in potable water.

Samples were cast using the above mix. Three types of samples were cast, cube samples of 150 mm sides, small cylindrical samples 100 mm diameter and 50 mm height and samples for corrosion studies in the form of a cylinder with 100 mm diameter and 230 mm height with sand blasted 12mm diameter steel reinforcing bars centrally placed so as to have uniform cover from all directions. The bottom edge of the reinforcement and the area near the top surface of the concrete cylinder were covered with teflon tape before casting. To avoid corrosion of the exposed part of the reinforcement, it was coated with epoxy paint. Cube samples were used for water permeability tests (DIN 1048) and small cylindrical samples were used for conducing chloride penetration tests (ASTM C 1202).

3 EXPERIMENTAL PROCEDURE

3.1 Preparation of media

Different media (NB, BHI and YEB) were prepared using distilled water. Dissolved material was sterilized using an autoclave at 120°C for 20 mins. The media was allowed to cool and the same used for preparation of seed culture, media for casting and curing.

Seed culture was prepared by taking a saline suspension of 18 hrs old culture from nutrient agar slants. 2% of this was added as inoculum in a known quantity of NB medium and incubated for 18–24 hrs at room temperature under shaking conditions (180 rpm). This seed culture was used for preparation of media required in casting and curing of concrete.

2% of urea and 0.3% of $CaCl_2$ solutions were prepared in distilled water separately. The solutions were filter sterilized through $0.2\,\mu$ filter paper in laminar flow.

3.2 Procedure for microbial precipitation

Samples were kept in the media and seed culture was added. After 24 hrs urea and calcium chloride solution were added. Every 7 days 10% of the media was replenished so as to supply nutrients to the microbes. At the same time, the urea- $CaCl_2$ solutions were also added so as to maintain the concentration of urea in the total curing medium at 2% and that of $CaCl_2$ 0.3%. After 28 days of curing, samples were removed from the curing tub and white deposits of carbonates were observed on the surface of samples. The mortar samples with microbial precipitation kept for air drying cemented together. Precipitation was observed not only on the surface of the samples but also into voids and pores of the samples.

The supernatant liquid and precipitate were analysed for calcium content. The quantification of the precipitate in the curing tub was done. The precipitate was further characterized and morphological studys were done using X-ray diffraction (XRD), X-ray fluorescence (XRF) and Scanning Electron Microscope (SEM) examination. Those samples with MICP were further used for testing of their water and chloride penetration to assess improvement. Corrosion studies were done on bigger cylindrical samples to see the effect of MICP.

4 CHARACTERIZATION OF DEPOSITS

4.1 Titrimetric analysis of Deposits

Microbial deposits obtained in all the trials were analysed using USP method for calcium assay. From these results (Table 1) it was clear that in the precipitate from the BHI medium the percentage purity of $CaCO_3$ was quite low. This may be due to production of more cell mass during biomineralisation.

For precipitate from the YEB medium, the average percentage purity of $CaCO_3$ was observed to be 89.71. In the precipitate from the NB medium, when the cell concentration was lowest (2.693×10^4 cells/ml) the percent purity of $CaCO_3$ found to be maximum in the precipitate.

4.2 XRD analysis

Chemical characterization was done using an energy dispersive X ray analyzer that indicated the abundant

Table 3. Elemental analysis by X-ray fluorescence spectroscopy.

Analyte	% Concentration in Lab grade $CaCO_3$	% Concentration in MICP
O	76.921	74.885
Ca	35.24	31.893

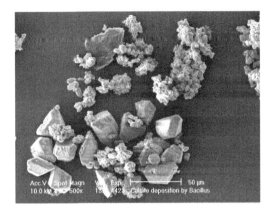

Figure 1. SEM Image of deposits at 500 X.

presence of Ca. Most of the major peaks in the XRD trace of the microbial precipitate matched those in the XRD trace of laboratory grade calcium carbonate. This confirmed that the microbial precipitate was calcium carbonate. To know the exact form, the peaks were compared with XRD traces of three mineral forms Argonite, Vaterite and Calcite of calcium carbonate. It was observed that all the major peaks occurring at the same 2θ even the d spacing were matching with that of calcite. Hence we can conclude that the precipitate obtained in different trials of MICP was the calcite form of calcium carbonate.

4.3 X-ray florescence Spectroscopy

The X-ray florescence results for laboratory grade calcium carbonate and microbial precipitate obtained in a trial of MICP are given in Table 3. The analysis gives the concentration of different elements present. Comparison clearly shows that concentration of calcium and oxygen in the precipitate are quite close to the concentrations in calcium carbonate. Hence we can conclude that the precipitate was the calcium carbonate.

4.4 Optical and Scanning Electron Microscopy

The presence of crystalline calcite associated with bacteria indicates that bacteria served as nucleation sites during the mineralization process. Fig. 1 shows a scanning electron micrograph of the precipitate from thye trial at flask level, while standardizing the media.

From the image it is clear that there are rhombohedral crystals of calcite present in the precipitate.

Some SEM images of precipitate showed that rod shaped objects were found dispersed in the crystals. These objects measured 1–3 μm in length and 0.5 μm or less across, consistent with the dimensions of *B. pasteurii*. These objects showed that bacteria were abundant and in intimate contact with the calcite crystals during the process of precipitation.

5 EVALUATION OF PROPERTIES

5.1 *Evaluation of water Permeability*

Microbial precipitation into pores and voids in concrete blocks the passage of water and air. To see the improvement due to this, a water permeability test as per DIN 1048 was done.

For conducting the water permeability test the cube samples were firmly secured in position and water nozzles with adjustable pressure were connected on the surface of the cubes. 1, 3 and 7 bar pressure was applied sequentially for 24 hours each. The samples were removed and split into two halves using a wedge and a compression testing machine. The penetration depth of water into the concrete was measured and marked with paint. The depth of penetration in mm indicated a measure of permeability.

5.2 *Evaluation of chloride permeability*

Chloride ingress into the concrete particularly in coastal and marine structures initiates corrosion of embedded steel at an early age. As the corroded volume is greater it exerts pressure on the cover concrete leading to cracking. Microbial precipitation into pores and voids will definitely slow down the ingress of chlorides. To assess the same, a Rapid Chloride Penetration test was done as per ASTM 1202. Specimens were conditioned using resin, vacuum pump and desiccator. The specimen was installed into the cell. Suitable sealant along with a rubber gasket was used to make it water tight. Two halves of the test cells were sealed together. One side of the cell containing the top surface of the specimen was filled with 3% NaCl solution, which was connected to the negative terminal of the power supply. The other side of the cell was filled with 0.3 N NaOH solution, which was connected to the positive terminal of the power supply. Electrical connections to a voltage application device and data reading apparatus were done. A constant voltage of 60 V was applied and the initial current reading after 1 minute was recorded. During the test, the air temperature around the specimen was maintained between 20° to 25°C. The current was recorded every 15 mins,. It was ensured that each half of the test cell was filled with the appropriate solution for the entire period of 6 hrs of the test. The test was terminated after 6 hrs and the specimen was removed. A graph was plotted for current versus time. The area under the curve was measured using a trapezoidal rule. The area under the curve gives the charge

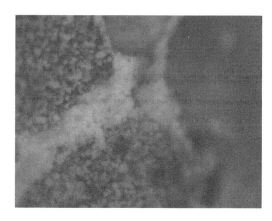

Figure 2. MICP in crack portion at 40X.

passed during the test period, measuring indirectly the chloride permeability. If the sample used is not of 95 mm, the charge passed was adjusted by multiplying the value obtained by the ratio of the cross sectional areas of the standard and the actual specimens so as to give the charge passed through the standard sample of 95mm. The charge passed was compared with the standard range given in the ASTM standard to get the class of the permeability.

5.3 *Evaluation of Corrosion Resistance*

Due to reduction in water and chloride permeability there will be delay in initiation and a slower rate of corrosion. To assess the same, corrosion studies were done on samples with microbial precipitation. The samples were cured in respective media with a predetermined concentration of cells as given in Table 1. The corrosion of reinforcement would take a number of years to show signs of deterioration of the concrete. To simulate the condition of a corroded reinforced concrete section. The process of corrosion was accelerated by impressing anodic current of 100 mA to all the samples.(Tamer et al. 2003) Owing to this, corrosion of reinforcement will take place at much faster rate and the corrosion products will exert a pressure on the surrounding concrete. When a small crack of about less than 1 mm was observed, the sample was removed.

5.4 *Potential for crack sealing*

To see the potential for crack sealing due to MICP, the cracked samples were again kept in microbial media with cell mass for 28days to remediate the crack. Regular replenishment of media and urea and calcium chloride was done. The samples after crack remediation were removed from the media. Whitish deposits were noticed at many places along the crack, sealing the same due to MICP.

Fig. 2 shows the deposited carbonates in a cracked portion seen under optical microscope (LEICA make). It was clear that the deposition had taken place

Table 4. Water and Chloride Permeability for NB1 and control.

Sample	Avg. permeability (mm)	Avg. charge passed (Coulomb)	Avg. after current 1 min. (mA)
Control	32.67	7432.16	59.67
NB1	9.67	6126.54	49

Table 5. Pullout force and mass loss in bar.

Sample	Force in kN	Loss in mass gms	Percent loss
Control 1	20	6.8	2.2
BHI	26.46	7.97	2.58
NB1	27.3	3.97	1.23
NB2	26.22	3.97	1.28
NB3	24.23	5.91	1.91
NB4	20.84	3.44	1.11
NB5	27.49	3.31	1.07

throughout the crack. It was found that full-grown calcite crystals with distinct and sharp edges had grown all over the surface of the cracked section thus acting as an agent that eventually plugged and remediated the cracks. The precipitate obtained during crack remediation was observed under an optical microscope (Fig. 2). The image clearly shows the deposited calcite crystals in the cracked portion. A closer view of the deposits clearly showed the rhombohedral crystals of calcite. It was found that full-grown calcite crystals with distinct and sharp edges had grown all over the surface of the cracked section thus acting as an agent that eventually plugged and remediated the cracks.

5.5 Effect of aggressive environment

After the crack sealing, to know the extent of corrosion in all the samples, potentiodynamic scanning was done to get the corrosion current at the initial stage. Now to simulate the field conditions i.e. to expose the concrete to an aggressive (corroding) environment, the samples were kept in a salt mist chamber which was as per ASTM B117 (CO.FO.ME.GRA Milano, Italy make). Every day half cell potential was measured using a Ag/AgCl electrode. Potentiodynamic scanning was done at regular intervals to get the corrosion current. The corrosion current gives an indication of the extent of corrosion in the reinforcement embedded in the concrete.

This procedure was repeated 4 times so that the total duration for which the remediated cylinders were kept in salt spray conditions would be about 2 months. Monitoring of crack width was also done regularly at the time of potential scanning. This gave the trend of reinforcement corrosion in the concrete samples in which MICP had taken place. After 2 months in the salt spray environment, pullout tests were done on the samples using a universal testing machine (Autograph AG-10TG Shimadzu).

6 RESULTS AND DISCUSSION

MICP on cement mortar indicated that the precipitation takes place on the surface of the mortar. The cubes with precipitation kept for air drying cemented together, indicating the deposits had some cementing characteristics.

The results of the water and chloride permeability tests are given in Table 4. From these results, we can see that in the sample with MICP using NB medium with a cell concentration of 2.69×10^4 cells/ml, water permeability reduced to 9.67 mm from 32.67 mm observed in control samples. The chloride permeability was measured as charge passed in 6 hours in Coulombs. For the sample with MICP, the charge passed reduced to 6126.54 Coulombs from 7432.16 Coulombs observed in the control sample. The current after 1 min. also reduced to average 49 mA in samples with MICP from an average of 59.67 mA observed in control samples. Trials with remaining other cell concentrations and media were in progress at the time of writing.

The pullout test results along with percent mass loss for different samples are given in Table 5. The percent loss of mass was calculated which was a good indicator of the extent of corrosion.

It was observed that, for all samples with MICP, the percent mass loss was less as compared to control samples. The maximum mass loss was observed in the reinforcement from the control sample Therefore we can conclude that the maximum extent of corrosion in the reinforcement was in the control samples. This had weakened the bond between the reinforcement and surrounding concrete to a greater extent. The effect of the same was observed by less pullout force in pullout tests on control samples. It was clear that a good correlation existed between pullout force and percent mass loss.

The visual inspection of split surfaces after pullout, showed the corrosion products oozing out from the crack region covering the concrete from the crack portion. On the crack side surface of the cross section the MICP was clearly seen throughout the depth of the crack. Its concentration went on reducing as we went in from the cylinder's surface. The maximum whitish deposits were observed in the case of BHI. It was observed that more corrosion products oozed out from the control sample as compared to the sample with MICP. For samples from the NB medium, as the concentration of cells increased the extent (quantum) of corrosion products oozing was reduced. Maximum pullout force and minimum percent mass loss were observed when the concentration was 11.5384×10^4 cells/ml.

Hence we can conclude that in the samples with MICP, the corrosion resistance was much better as compared to the control. This was due to precipitation of calcite in the pores and voids of concrete which prevented the salty fog present in the salt mist chamber from penetrating into the concrete so as to reach the reinforcement and accelerate the corrosion process. Due to the absence of MICP in the control, the ingress of salty moisture into the body of concrete until it reached the reinforcement through the continuous channel of voids, pores accelerated the corrosion.

6.1 *Visual observations of the reinforcement*

The maximum extent of corrosion for the reinforcement was observed from the control. Ribs at the lower end were corroded to the maximum extent. There was localized corrosion occurred in the reinforcement from cylinders in BHI, NB1, NB2, NB4 and NB5. However the corrosion on the bar from the NB3 was greater which is clear from the increased percent mass loss. The reinforcement from BHI showed maximum loss in weight, this may be due to more content of cell mass in the precipitate indicated by less purity of calcium carbonate.

7 CONCLUSIONS

Microbial action precipitates calcite into voids and pores in mortar and concrete. This blocks the passage of water and air. Owing to blockage of passage for water, a reduction in water and chloride permeability was observed. This technique of calcite deposition can be used to remediate cracks in concrete. The corrosion current in samples with MICP was always less than the control sample. This indicates that the extent of corrosion in the reinforcement from control samples was more. This was clear from the reduction in the pullout force required for the reinforcement in the control samples as compared to the pullout force in samples with MICP. It was clear that a good correlation existed between pullout force and percent mass loss. As the pullout force decreased percent mass loss increased. It was clear from the corrosion studies on samples with MICP on the surface and top few millimeters in the cover portion of the concrete that MICP is very useful. This deposition occupied the void space in the concrete and restricted the ingress of moisture and other corroding agents, thereby reducing the rate of corrosion of the embedded reinforcement. It was observed that as the cell concentration increased, the percent loss of mass for the steel reinforcement reduced. This indicates that there was reduction in the extent of corrosion in the samples with a higher concentration of microbes due to higher precipitation in voids and pores of the concrete. The control sample split surface showed more oozing of corrosion products as compared to samples with MICP due to a greater extent of corrosion.

The successful remediation of cracks in cement composites due to MICP will be useful in nuclear structures such as impermeability of spent fuel storage facilities, and nuclear reactor structures. Some of the nuclear plants are aging and micro cracks in them can be remediated, extending the service life of existing plants. Use of spore forming type microbes which can be activated as and when required by a supply of nutrients will help us to develop a technology that heals the structures automatically. This will be of great help in repairs of structures having access problems, particularly in nuclear structures.

ACKNOWLEDGEMENTS

The authors sincerely acknowledge the financial support by the Atomic Energy Regulatory Board, Government of India, for this work. Thanks to junior research fellow microbiologist on this project Mr. Huzaifa Choonia and Miss Ravleen Kaur for their help in microbiological aspect of the work.

REFERENCES

ASTM C 1202. 1997 "Standard Test Method for Electrical Indication of Concrete's ability to Resist Chloride Ion Penetration. American Society for Testing of Materials.

Bang, S. S.; Galinat, J. K.; Ramakrishnan, V., 2000. "Carbonate biomineralization induced by soil bacterium *B. megaterium,*" *Enzyme and Microbial Technology* 28: 404–409.

Bang, S. S.; Galinat, J. K.; Ramakrishnan, V., 2001. "Calcite precipitataion induced by Polyurethane-immobilized Bacillus Pasteurii," *Enzyme and Microbial Technology* 28(4-5):404–409.

Carlos, R. N.; Manuel, R. G.; Chekroun, K. B.; Maria, T. G., 2003. "Conservation of ornamental stone by *Myrococcus xanthus* induced carbonate Biomineralisation," *Applied and Environmental Microbiology,* 69(4):2182–2193.

Castanier, S.; Leverel, Le M. G.; Perthuisot, J. P., 1999. "Calcium carbonate precipitation and limestone genesis – the microbiologist point of view," *Sediment Geology* 126:9–23.

DIN 1048. Regulation for testing concrete used during erection of concrete and reinforced concrete structures Regulations. The German Reinforced Concrete Association.

Dick, J.; Windt, W. D.; Graef, B. D.; Saveyn, H.; Meeren, P. V.; Belie, N. De; Verstraete, W., 2006. "Biodeposition of a Calcium Carbonate layer on degraded limestone by Bacillus species," *Biodegradation* 17:357–367.

Ghosh, P.; Mandal, S.; Chattopadhyay, B. D.; Pal, S., 2005. "Use of microorganisms to improve the strength of cement mortar," *Cement and Concrete Research*, 35:1980–1983.

Gollapudi, U. K.; Knutson, C. L.; Bang, S. S.; Islam, M. R., 1991. "A new method for controlling leaching through permeable channels," *Chemosphere*, 30:695–705.

Hammes, F.; Verstraete, W., 2002. "Key roles of pH and calcium metabolism in microbial carbonate precipitation,". *Environmental Science and Bio-Technology* 1:3–7.

Hammes, F.; Boon, N.; Villers, J. D.; Verstraete, W.; Douglas, S. S., 2003. "Strain specific Ureolytic microbes calcium carbonate precipitation," *Applied and Environmental Microbiology* 69(8):4901–4909.

Klein, J; Kluge, M., 1981. "Immobilisation of microbial cells in polyurethane matrices," *Biotechnology letter* 3:65–70.

Lee, Y. N., 2003. "Calcite production by Bacillus amyloliquefaciens CMB01," *The journal of Microbiology* 12:345 348.

Morita, 1980. "Calcium carbonate precipitation by marine bacteria," *Geomicrobiol Journa,* 2:63–82.

Muynck, W. D.; Debrouwer, D.; Belie, N. D.; Verstraete, W., 2008, "Bacterial carbonate precipitation improves the durability of cementitious materials," *Cement and Concrete Research*, 38:1005–1014.

Ramachandran, S. K.; Ramakrishnan, V.; Bang, S. S., 2001. "Remediation of concrete using microorganisms," *American Concrete Institute Materials Journal* 98:3–9.

Ramakrishnan, V.; Panchalan, R. K.; Bang, S. S., 2003. "Bacterial Concrete – A Concrete for the future. Special Presentation." in: Innovative World of Concrete conference, Pune, Maharashtra, India, December.

Ross, N.; Villemur, R.; Deschenes, L.; Samson, R., 2001. "Clogging of a Limestone Fracture by Stimulating Groundwater Microbes," *Water Resources* 35(8):2029–2037.

Stocks Fischer, S.; Galinat, J. K.; Bang, S. S., 1999. "Microbiological precipitation of CaCO$_3$," *Soil Biology and Biochemistry* 31(11):1563–1571.

Tamer A., Maaddawy E. and Soudki K. A., 2003. Effectiveness of Impressed Current Technique to Simulate Corrosion of Steel Reinforcement in Concrete, Journal of Materials in Civil Engineering, 15(1): 41–47.

Tiano, P. L.; Biogiotti; Mastromei, R. G., 1999, "Bacterially bio-mediated calcite precipitation of monumental stones conservation: Methods of evaluation," *Journal of Microbiological Methods,* 36:139–145.

United States Pharmacopoeia, USP 29 NF 24, calcium carbonate, The United States pharmacopeial convention 2005.

Wang, X.; Ruchenstein, E., 1993. "Preparation of porous polyurethane particles and their use of enzyme mobilization". *Biotechnology Progress* 9:661–665.

Concrete Solutions – Grantham, Majorana & Salomoni (Eds)
© 2009 Taylor & Francis Group, London, ISBN 978-0-415-55082-6

Rejuvenating our aging structures – Cumberland street parking garage rehabilitation

Philip Sarvinis

Read Jones Christoffersen Ltd., Toronto, Canada

ABSTRACT: Built in 1968, this multi-level concrete parking facility has been subjected to years of moisture and chloride ion attack causing it to experience corrosion-related deterioration to the point that its structural integrity was being compromised. Over the years, moisture and chloride contamination has caused both the embedded mild steel and paper wrapped "Button-Head" post-tensioning to experience corrosion-related deterioration, which in turn, has resulted in the failure of the post-tensioning system. After years of monitoring and structural capacity calculations, sections of the structure were deemed to be no longer able to safely support the applied loading. This paper will illustrate the procedures implemented in the design of the new post-tensioning system, which enabled us to rejuvenate this 40-year old structure and extend its overall effective service life. This paper will also outline how the new system was installed while the parking facility remained operational.

1 BRIEF DESCRIPTION OF THE FACILITY

1.1 General Description

The Cumberland Street Parking Garage is located in the heart of Downtown Toronto's High Fashion District in an area where vehicle parking is at a premium. As the area continues to develop with the introduction of high rise condominiums, parking is becoming an even bigger problem for local shoppers and workers in the Yorkville/Cumberland corridor.

The parking structure, which was built in 1968, is typically above ground and consists of five parking levels, four of which are suspended parking levels. 3½ of the parking levels are above ground and 1½ are below ground. The structure has a rectangular footprint with approximate plan dimensions of 260 feet (79.3 m) in the north/south direction and 100 feet (30.5 m) in the east/west direction. The garage has the capacity to hold approximately 300 vehicles with approximately 120,000 sq. ft. (11,150 m²) of parking (excluding occupied space, mechanical rooms, parking offices, stairwells, etc…) of which 24,000 sq. ft. (2,230 m²) is on grade and 96,000 sq. ft. (8,920 m²) is suspended slab.

The parking structure is bound on the east and west sites by adjacent buildings with minimal separation between the structures. Retail space occupies a portion of the north and south elevation of the structure and a retail car rental company and an auto carwash and detailing company also occupies a portion of the basement level of the garage.

The functional layout of the garage provided two-way traffic flow to all levels of the split-level garage, which act in a helical fashion, and has parking stalls which are perpendicular to the flow of traffic.

Figure 1. South Elevation of Structure.

1.2 Structural Description

The suspended slabs of this parking garage are typically 8" (200 mm) thick cast-in-place post-tensioned concrete one-way spanning slabs. The slabs are supported by 30" (750 mm) deep by 18" (450 mm) wide post-tensioned concrete beams spaced at 28 foot (8.53 m) centres and are also typically one-way spanning. The post-tensioning system within the suspended slabs is an unbonded post-tensioning system known as a "Button-Head System". The post-tensioning system in the beams is also the "Button-Head System" however it is of the bonded type (i.e. placed in solid grout filled ducts). The suspended slabs have a nominal amount of mild reinforcing steel embedded within them (i.e. # 3 bars at 16" (400 mm) centres each way) to control cracking due to temperature fluctuations. (Figure 2).

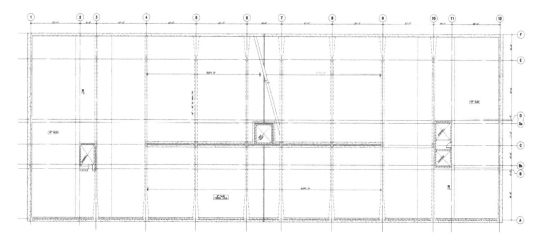

Figure 2. Typical Floor Plan.

Figure 3. Overview of Garage Interior.

The columns, foundation walls and stairwell cores are all of normally reinforced cast-in-place concrete construction. The foundations are a combination of reinforced concrete strip and pad footings.

As noted above, the parking slab arrangement is one of a split level design dividing the structure in half in the east/west direction. There is also one expansion joint which travels in the east/west direction dividing the structure once again into two halves in the north/south direction. These division points define the various quadrants of this garage.

1.3 Moisture Protection Systems

Up until 1998, the suspended slabs of this facility were not waterproofed to protect them against the ingress of moisture and harmful chlorides. At the time of construction, a surface sealer was applied to the top surface of the suspended parking slabs. In 1998, a staged installation of waterproofing (thin traffic deck coating) began and today all but the exposed roof level remains to be protected with a waterproofing system.

2 BUTTON-HEAD POST-TENSIONING SYSTEM

2.1 Post-Tensioning Basics

Post-tensioning is a method of introducing internal forces into a concrete element after the concrete element has been cast in order to counteract the external loads that will be applied to the element when it is put into use. When the concrete element is post-tensioned, the concrete portion of the element is put into compression and the steel portion of the element is put into tension. Putting the components of the element into this state prior to applying a service load puts the building materials in a stronger state resulting in a stiffer element that has more capacity to resist tensile forces.

The first successful use of unbonded post-tensioning systems dates back to 1920–1930 in Europe and it was not until the 1960's that it became popular in North America.

2.2 Button-Head System

The term "Button-Head" refers to the shape of the wire's end after it is anchored.

The "Button-Head System" installed in this facility's parking slabs consists of ¼" (6 mm) diameter high strength steel (240,000 psi or 11,500 kPa), cold drawn, stress relieved wires which are bundled together into tendons and wrapped with protective greased paper. Tendon sizes ranged from 7 wires to 19 wires in each bundle and were typically spaced at 4 foot (1.2 m) centres. The Button-Head System is an early type of post-tensioning technology and possesses very little protection against moisture and chloride attack. The post-tensioning system in the suspended slabs of this parking garage is an unbonded system meaning it is anchored to the concrete slab at the end anchor locations only. The post-tensioning in the concrete beams is a grouted or bonded system and thus not as susceptible to moisture deterioration as the unbonded slab system.

Figure 4. Typical Button-Head Anchor Configuration.

Figure 6. Typical Button-Head Strand Configuration in Slab.

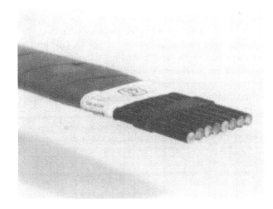

Figure 5. Typical Button-Head Wire Layout.

All wires of the post-tensioning strand are cut to the same length and then pass through pre-drilled holes in the anchor head. To secure and seat the wires against the anchor head, a small button like configuration was cold formed onto the end of each wire. The dead end sides of the strands were cast into the structure at the time of concrete placement. At the live end, the tendon wire first passed through a bearing end plate before they were threaded through the live end anchor head and seated with Button-Heads. During the concrete placement the bearing end plate was also cast into the concrete leaving the ends of the wires and anchor head at the live end exposed for stressing. After stressing, the wires elongate creating a gap between the anchor head plate and the embedded bearing plate. In order to maintain this tension force and transfer it into the bearing plate, transfer plates were installed in the gap between the anchor and bearing plates. The live anchor was then cast in concrete.

Unfortunately, the cross selection of the tendon was typically found to be irregular shaped primarily due to the manual paper wrapping operation. This cross section makes removal of the Button-Head tendon impossible as compared to present day post-tensioning systems.

Because the unbonded Button-Head post-tensioning system is susceptible to moisture and chloride attack, it was found that as the structure continued to age and experience this ongoing moisture and chloride ingress, the embedded post-tensioning strands were deteriorating and becoming de-stressed, resulting in a reduced load carrying capacity of the suspended parking decks.

3 HISTORY OF PAST EVALUATIONS

Read Jones Christoffersen's involvement with this structure dates back to 1985. Over the years, the structure and in particular the post-tensioning system was evaluated to identify signs of corrosion related deterioration, tension deficiencies in the post-tensioning and reductions in load carrying capacity.

The parking slab evaluation included traditional condition survey techniques such as visual reviews, acoustical sounding (chain drag, hammer tap surveys), chloride ion content testing, comprehensive strength surveys, deflection surveys and load tests. The review of the post-tensioning expanded the traditional survey to include the introduction of inspection recesses in the slabs to exposed short sections of the embedded post-tensioning system. The inspection recesses allowed the condition of the paper wrapping and protective grease to be reviewed and analyzed, allowed the exposed section of wires to be reviewed for corrosion and allowed us to test for tension deficiencies in the post-tensioning system.

Over the years, as deterioration was detected with respect to embedded mild reinforcing steel, repair programs were implemented to locally repair any concrete deterioration as well as protect the suspended slabs with a traffic bearing waterproofing system. These programs also included the replacement of expansion joint seals.

However, with respect to the post-tensioning system, it was analyzed on a quadrant by quadrant basis and once it was found to have lost load carrying capacity to a point which was deemed to be below the design loads for which they currently support, external reinforcing was installed. Being unable to physically remove and replace the defective "Button-Head" post-tensing system, an external post-tensioning system was designed and installed through the existing structural slabs to replace the original "Button-Head" system. The external post-tensioning was done on a quadrant by quadrant basis.

It should be noted that the deterioration of the post-tensioning system to date has been limited to the

Figure 7. Overview of Slab Surface Illustrating Excavation for New Post-Tensioning Strands.

unbonded post-tensioning system within the slabs. The bonded post-tensioning system in the beams appears to be functioning well.

4 EXTERNAL POST-TENSIONING REHABILITATION SOLUTION

4.1 General System Description

Once a quadrant was deemed to be in need of external reinforcement, calculations were performed based on the extent of mild reinforcing and transverse post-tensioning to determine the maximum spacing between the proposed external post-tensioning system (i.e. 8'–0" or 2.4 m centres). (Refer to Figure 7). The new external post-tensioning consists of 7-wire high strength (270,000 psi or 12,900 kPa) low relaxation steel strands in extruded plastic sheathing. The sheathing was high-density polyethylene with a minimum wall thickness of 0.06 inches (1.5 mm). The tendons were placed at 8-foot centres (2.4 m) and at each location there were 3 tendons which ran the full length of the quadrant. At end bays, each line of external post-tensioning had an extra two tendons (5 total) to accommodate the extra span and flexural moment due to the end conditions.

At each external post-tensioning location, the tendon groups were threaded through the existing

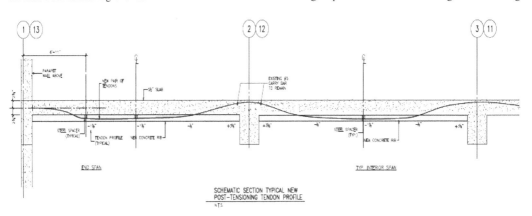

Figure 8. Typical External Post-Tensioning Strand Profile.

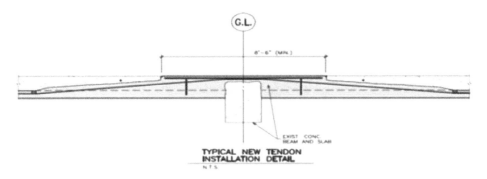

Figure 9. Tendon Profile at Support Beam.

suspended slab and anchored at each end. The dead ends were typically located at the expansion joint whereas the live ends (stressing end) were at the building exterior elevation to accommodate the stressing. The tendons had a draped profile with high points overtop of beam support lines and low points at mid-bay between beams. (Refer to Figure 8).

At high points, troughs were chipped in the slab surface to allow the new tendons to obtain the appropriate concrete cover. (Refer to Figures 9 and 11).

In order to achieve the appropriate drape, through slab openings were chipped in the suspended slabs to allow the tendons to be threaded down to the underside of the slab and back up at next beam line. (Refer to Figure 12).

The exposed portion of the tendons below the existing slab were then encapsulated in reinforced concrete beams or ribs to provide the required slab-strand integration and overall fire protection to the system. (Refer to Figures 10 and 13).

It should be noted that during the process, the existing post-tensioning system remained in place and was de-stressed simultaneously with the stressing of the new external system to ensure the existing suspended slabs were not overstressed at any time during construction. The new tendons were stressed to a final effective stress of 173 ksi (1193 MPa).

4.2 Advantages of the Solution

The primary advantages of the external post-tensioning solution were as follows:

1. It allowed the majority of the existing suspended slabs to be reused.
2. It permitted the work to be undertaken in phases which allowed traffic flow to continue through the garage during the construction and minimize lost parking. (Refer to Figure 14)
3. It minimized the need for shoring and re-shoring as the existing post-tensioning system was maintained during construction until the new external system was ready to be engaged.
4. The solution was approximately half the construction cost of the complete slab replacement option which does not include the lost revenue component associated with complete replacement.

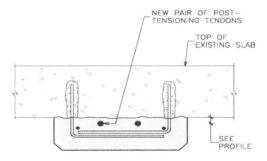

Figure 10. Tendon Layout on Underside of Slab.

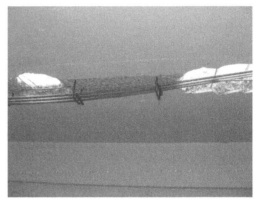

Figure 12. Tendon Layout on Underside of Slab.

Figure 11. Tendon Layout Overtop of Support Beams.

Figure 13. Overview of Concrete Ribs Encapsulating New Tendons on the Underside of the Slab.

Figure 14.   Overview of Typical Phasing of Work.

Figure 15.   Overview of the Finished Product.

5   CONSTRUCTION COST AND SERVICE LIFE

The construction cost for this external post-tensioning solution was in the order of $28.00 per square foot of suspended slab (Canadian Dollars). We are of the opinion that the suspended slabs of this facility will remain in a serviceable condition, if properly maintained, for another 40 years. The maintenance will involve repair and renewal of the waterproofing as well as localized repair to corrosion related deterioration as it develops.

6   CLOSING REMARKS

Deteriorated parking structures can be repaired and/or rehabilitated to extend their effective service life and maximize economic return. Through innovative thinking, we can all achieve practical results.

REFERENCES

Newman, Alexander, P.E. 2001. *Structural Renovation of Buildings*: McGraw Hill, USA
ASTM Standard A414, ASTM International
ACI Standard 421–16, American Concrete Institute.
PTI Acceptance Standard for Post-Tensioning Systems, Post-Tensioning Institute

AUTHORS BIO

Philip Sarvinis, P.Eng., graduated with honours from the University of Toronto in 1989 and is currently the Managing Principal of Read Jones Christoffersen's Building Science and Restoration Team in Toronto, Ontario Canada. He has 19 years experienced in the evaluation, rehabilitation and protection of concrete structures and is currently licensed to practice engineering in the Provinces of Ontario, New Brunswick, Newfoundland and Nova Scotia. He is currently the President of the Building and Concrete Restoration Association of Ontario, and a member of the following organizations: the Canadian Parking Association, the International Concrete Repair Institute, the American Concrete Institute, the Canadian Society of Civil Engineers, the American Society of Civil Engineers and Construction Specifications Canada (Toronto Chapter).

Concrete Solutions – Grantham, Majorana & Salomoni (Eds)
© 2009 Taylor & Francis Group, London, ISBN 978-0-415-55082-6

Approach for the safe use of coatings for stopping corrosion in chloride bearing concrete: Experiences from long term surveys with installed electrodes

U. Schneck

CITec Concrete Improvement Technologies GmbH, Dresden OT Cossebaude, Germany

ABSTRACT: In preparation of concrete repairs in an underground car park a coating was evaluated on a chloride bearing concrete, which should stop further corrosion activity by the reduction of water content. In a trial application, an extended corrosion survey was done on a representative concrete area: first the top side and soffit of the concrete slab were surveyed by potential- and resistivity measurements, accompanied by chloride- and moisture measurements. Then reference- and counter electrodes were placed close to the reinforcement permanently at 6 locations selected according to the previously obtained measurement results. After an initial polarization- and impedance measurement, the concrete repair was done and an epoxy based surface protection coating was applied. With follow-up measurements on the built-in electrodes, the development of the concrete impedance and the corrosion currents was monitored, which could verify the drying out of the concrete and the limitation of corrosion activity.

1 APPROACH FOR THE MONITORING TASK

1.1 The repair principle W-Cl

One option for controlling chloride induced corrosion activity in reinforced concrete slabs is to reduce the water content and hence to slow down the cathodic reaction. This can be done in a very cost effective way by applying a coating on the concrete surface and can be considered if the concrete doesn't show damage and if the chloride content doesn't exceed about 1% by cement mass.

On the other hand, the decision for such a repair must be made with great responsibility because the corrosion activity will not stop in every case, and new or continuing corrosion elements may form and some concrete repair may be needed, especially when the water content of the concrete is quite high.

So it is advised to verify the feasibility of a coating of a chloride bearing concrete in a trial application first, as is recommended in the present German guideline Richtlinie für Schutz und Instandsetzung von Beton-Bauteilen of the DafStB (German Board for Reinforced Concrete)

1.2 The underground car park

The underground car park, built ca. 1985, is a two-storey structure with 4 half levels underneath a business building comprising in-situ concrete elements. During the repair of the existing coating concrete deterioration after reinforcement corrosion was observed (see fig 1).

Figure 1. View on the concrete surface during the repair works.

Chloride contents of about 1% by cement mass were measured, and it was to be decided whether only the loose concrete was to be replaced before coating the concrete again, or if all chloride containing concrete was to be removed first.

1.3 The approach for measuring the feasibility of a new coating on top of chloride bearing concrete

With a condition survey the whole concrete surface (ca 300 m^2) was investigated first: rest potentials, the surface resistivity and the concrete cover were measured on the top side of the concrete slab, then drilled

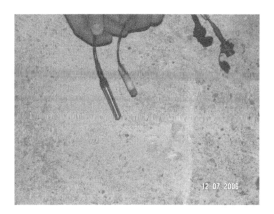

Figure 2. Reference electrodes and stainless steel counter electrodes for the permanent mounting in the concrete and further polarization and impedance measurements.

Figure 3. Measurement setup with electrode sockets on the soffit of the ceiling/concrete slab and potentiostat.

concrete samples for the analysis of chloride and humidity content on interesting locations according to the non-destructive measurements and selection of 6 locations for mounting of reference- and counter electrodes (see fig 2). These were taken adjacent to the reinforcement with an on-site interpretation of the rest potentials (2× each for "passive state", "active state" and "intermediate state" of corrosion activity).

Across the sub area of the permanently mounted electrodes, rest potentials and resistivity were measured also on the soffit of the concrete slab (where also the sockets of the electrodes were placed). The setup for the polarization- and impedance measurements can be seen in fig 3; the equipment used and the methods applied are shown in table 1 and table 2.

The extended electrochemical measurements (polarization- and impedance measurements) were done initially before continuing with the concrete repair as seen in fig 1 and then every 4 months over 1 year and – as the data of one location had to be checked one more time – again after 9 months. Since the climate in this section of the structure was

Table 1. Devices used for the extended electrochemical measurements.

Method	Devices
Polarization measurement	Ecochemie Autolab PGSTAT100 with FRA module
Impedance spectroscopy	GPES and FRA 4.9 software CITec miniature-MnO₂-reference electrodes CITec Cu/CuSO₄ reference electrode stainless steel electrode

Table 2. Listing of the measurement projects during the extended electrochemical measurements.

Step	Projects
01	open circuit potential (240 s)
02	impedance spectroscopy (100 kHz to 10 mHz in 4 steps/decade)
03	open circuit potential (300 s)
04	polarization measurement from ($E_{rest} - 825$ mV) to ($E_{rest} + 225$ mV) with 2,0 mV/s at 0,76 mV/step
05	open circuit potential (180 s)

Table 3. Numbering, classification and detailed data of the measurement locations.

Location	E_{rest} vs. CSE on concrete surface [mV]	Classification according to rest potentials on concrete surface	E_{rest} vs. MnO₂, related to CSE, at rebar [mV]	Concrete cover [mm]	Chloride content at rebar [% cement mass]
A1	−409	"active"	−357	38	0.15
I1	−290	"intermediate"	−460	31	1.76
P1	−144	"passive"	−100	29	
A2	−410	"active"	−471	17	1.34
I2	−318	"intermediate"	−203	13	0.39
P2	−237	"passive"	−177	29	

reasonably constant, temperature effects were not taken into consideration.

2 RESULTS OF THE EXTENDED ELECTROCHEMICAL MEASUREMENTS

2.1 Description of the measurement locations

Table 3 shows detailed data from the chosen locations for the extended electrochemical measurements: the rest potentials measured on the concrete surface vs. CSE (which were the criteria for their selection and classification), the rest potentials measured on the built-in reference electrodes close to the reinforcement (measured vs. MnO₂, related to CSE), the concrete cover and the determined free chloride content in the reinforcement vicinity.

Table 4. Results from the extended electrochemical measurements.

Date	A1	I1	P1	A2	I2	P2
E_{rest} [mV vs. MnO$_2$]						
30.01.2007	−405	−476	−158	−482	−204	−217
30.05.2007	−406	−250	−201	−405	−170	−162
24.09.2007	−381	−386	−165	−383	−158	−213
28.01.2008	−341	−372	−130	−399	−172	−204
16.10.2008	−323	−364	−101	−394	−207	
Change over the last 9 months	▶	▶	▲	▶	▼	
I_{corr} [µA]						
30.01.2007	35	48	13	130	179	13
30.05.2007	55	40	13	1,171	100	12
24.09.2007	45	47	18	726	145	21
28.01.2008	21	55	14	503	95	26
16.10.2008	18	128	46	637	113	
Change over the last 9 months	▶	▲	▲	▶	▶	
$R_{electrolyte}$ [Ω]						
30.01.2007	1,135	867	1,127	18	201	1,545
30.05.2007	1,000	602	812	157	241	1,259
24.09.2007	1,321	931	1,172	811	608	2,748
28.01.2008	1,629	2,443	3,864	900	2,985	3,902
16.10.2008	3,213	3,767	2,741	425	1,836	
Change over the last 9 months	▲	▲	▼	▼	▼	

All locations except A2 were situated in areas without concrete replacement, and the electrodes of A2 were placed in a small rib of old concrete where the surrounding concrete had been removed by sand blasting.

It can be seen in table 3 that the rest potentials measured on the concrete surface don't correlate with the chloride content necessarily. If negative potentials were found at quite low chloride content, an increased water content could be measured at the same time. The water contents ranged in total between 1.74% and 3.54%.

2.2 Results from the polarization- and impedance measurements

In table 4 all calculated results from the electrochemical measurements are to be seen. The corrosion currents have been I-R corrected. Since there was no information about the respective reinforcement area, the currents could not be transformed into corrosion density values. The electrolyte resistances of the concrete were calculated from the impedance measurement results at ca 1 kHz.

A very interesting development of the corrosion currents can be seen at A2, which increased from 130 µA during the initial measurement to almost 1,200 µA on the first follow-up measurement. Later, the currents were decreasing, but did not fall down to the initial value.

All other locations showed much lower changes in their corrosion currents, but an overall tendency of increasing electrolyte resistances can be stated – at least to a significantly increased level, compared to the results from the initial measurement. This was an essential result, because the increasing electrolyte resistance can be interpreted as a decrease of water content – which was to be tested by the monitoring project.

The rest potentials show some changes without a clear tendency, and no simple conclusions can be made from these data.

However, some uncertainties especially in the rest potentials and corrosion currents must be taken into account because of the varying electrolyte resistances as these can have a considerable influence on the measurement conditions, so a combined data assessment is advised.

2.3 Evaluation of the results

Obviously in the area of A2 a new corrosion element has been formed between chloride containing original concrete and fresh alkaline repair concrete that boosted the corrosion current to a much higher level. In addition, the repair concrete introduced even more water to the concrete slab that had to evaporate later. With the reduction of water content, the corrosion current dropped and the electrolyte resistance increased.

The expected further decrease of the corrosion current in A2 during the last measurement date could not be found, but at the same time the electrolyte resistance had dropped by 50%. It was recommended to check the new coating on possible fresh cracks in the area of A2, which would explain this behaviour thus avoiding the necessary conclusion that the coating is an unsafe approach in that case.

The corrosion currents of the locations A1, I1, P1, I2 and P2 show some deviation, but all stayed on their initial level – regardless how much chloride was in rebar vicinity (note A1 and I1). At I1, where even 1,7% chloride were present, the same increase of electrolyte resistance of 2,000 to 3,000 Ω could be observed as on the other locations (except on A2).

3 CONCLUSIONS

With extended corrosion measurements, the feasibility of an epoxy based coating for stopping corrosion activity in chloride containing concrete could be verified.

However, great care has to be taken if such corrosion protection is being done in combination with patch repairs, because new and very intensive corrosion elements may be triggered in the interface between old and fresh concrete even at chloride contents less than 1%, due to the incipient anode effect.

The need for an individual verification of this inexpensive corrosion protection has been experienced in other projects too (Schneck, 2007)

REFERENCES

DAfStb-Richtlinie. 2001. Schutz und Instandsetzung von Beton-Bauteilen, Teile 1-4. *Beuth-Verlag, Berlin*

Schneck, U. 2007. Effiziente und sparsame Überwachung des Instandsetzungsprinzips W-Cl – Kontrolle der Wirksamkeit von Beschichtungen auf chloridbelasteten Stahlbetonplatten. *Proc. 6th Colloquium Industrial Floors, Esslingen*

Sulphate resistance of cements evaluated according to ASTM C1012 and AFNOR P18-837 standards

K.K. Sideris

Laboratory of Building Materials, School of Civil Engineering, Democritus University of Thrace, Greece

P. Manita

Department of Civil Engineering, Technological Educational Institute of Serres, Greece

ABSTRACT: Sulphate resistance of cements and concretes is of great concern in cases where concrete structures are in contact with underground water. In this paper, the sulphate resistance of different types and strength of cements (CEM II, CEM IV and CEM I (sulphate resistance cement)) were experimentally investigated. A total of four different cements were tested. Their sulphate resistance was assessed against different aggressive environments (Na_2SO_4, $MgSO_4$ and mixed (NA_2SO_4 and $MgSO_4$ solutions) according to two different Standards: the ASTM C1012 Standard and the AFNOR P18-837 standard (which is proposed to work as the reference method for the coming European Standard prEN TC51). According to the results of this paper the sulphate resistance of the cements evaluated with the AFNOR P18-837 Standard are in good accordance with the results obtained with the ASTM C1012 Standard.

1 INTRODUCTION

Sulphate attack is one of the mechanisms minimizing the effective life of concrete structures. Sulphates usually exist in significant concentrations in underground water and attack mainly underground concrete structures. The attack mechanism mainly depends on the cation of the salt (Al Amoudi, 1999).

In the case of Na_2SO_4, the cementitious matrix of the material is corroded through the formation of crystalline compounds of ettringite $C_6AS_3H_{32}$ and gypsum CSH_2. The mechanism providing the above results is:

$$Ca(OH)_2 + Na_2SO_4 + 2H_2O \rightarrow CaSO_4 \cdot 2H_2O + 2NaOH \quad (Eq\ 1)$$

$$C_4AH_{13} + 3CaSO_4 \cdot 2H_2O + 14H_2O \rightarrow$$
$$6\ CaO \cdot Al_2O_3 \cdot 3H_2SO_4 \cdot 32H_2O + Ca(OH)_2 \quad (Eq\ 2)$$

$$4Ca(OH)_2 \cdot Al_2O_3 \cdot H_2SO_4 \cdot 12H_2O +$$
$$2Ca(OH)_2 \cdot H_2SO_4 \cdot 2H_2O + 16H_2O \rightarrow$$
$$6CaO \cdot Al_2O_3 \cdot 3H_2SO_4 \cdot 32H_2O \quad (Eq\ 3)$$

$$C_3A + 3CaSO_4 \cdot 2H_2O + 26H_2O \rightarrow$$
$$6\ CaO \cdot Al_2O_3 \cdot 3H_2SO_4 \cdot 32H_2O \quad (Eq\ 4)$$

Since ettringite is expansive in nature (it has a density of $1.73\ g/cm^2$ compared with an averaged density of $2.50\ g/cm^2$ for the other products of hydration), expansion and cracking of concrete are the results of sulphate attack caused by Na_2SO_4 (Lea, 1970) The principal method available to limit sulphate attack caused by Na_2SO4 is the use of cements with lesser amounts of C_3A, such as ASTM Type II or Type V Portland

cements instead of Type I Portland cement. An alternative method is the use of additives – such as natural pozzolans, fly ashes or silica fume. These materials decrease the total amount of C_3A available in the mixture, when they replace a portion of the cement. Additionally they reduce the total $Ca(OH)_2$ content available in the mixture, as following:

$$Silica_{(from\ Pozzolan)} + Ca(OH)_{2\ (From\ Cement)} \rightarrow Secondary\ CSH \quad (Eq\ 5)$$

(In the presence of water and at normal temperature)

The attack mechanism is different in the presence of MgSO4:

$$Ca(OH)_2 + 2MgSO_4 + 4H_2O \rightarrow 2CaSO_4 2H_2O + 2MgOH$$
$$C_xS_yH_z + xMgOH + (3x-0,5y-z)H_2O \rightarrow xCa(OH)_2.H_2SO_4.2H_2O + xMgOH + 0,5y2Si\ O_2.H_2O$$
$$4MgOH + SH_n \rightarrow M_4SH_{8,5} + (n-4,5)H_2O \quad (Eq\ 6)$$

Magnesium hydroxide formed in Equation (6) is insoluble and its saturated solution has a pH value equal to 10.5 instead of 12.4 and 13.5 which are the pH values of saturated solutions of $Ca(OH)_2$ and NaOH respectively. The consequences of this low pH value can be summarized as follows (Rasheeduzzafar et al, 1994; Al Amoudi et al, 1995; Cohen & Bentur, 1988).

(i) Secondary ettringite will not form (not stable in low pH values),

(ii) $MgSO_4$ reacts with CSH to produce gypsum, brucite (MgOH) and silica gel (S_2H) (Equation 7),

(iii) CSH tends to liberate lime to raise the decrease pH.

Liberated lime reacts further with $MgSO_4$ producing more $MgOH$ (Equation 6), iv) concentration of gypsum and brucite will increase while CSH is destabilizing and becomes less cementitious, v) magnesium silicate hydrate formed by the reaction of brucite with hydrosilicates (Equation 8) is non-cementitious. The main cause of deterioration due to $MgSO_4$ is therefore the conversion of CSH in a fibrous, amorphous material with no binding properties (Rasheeduzzafar et al, 1994).

Protection solutions against Na_2SO_4 mentioned above are less effective in the case of $MgSO_4$ environments simply because there are different mechanisms to be faced: C_3A, C_4AF and their hydrates are not involved in equations 6 to 8 whereas $Ca(OH)_2$, which is the primary reactant constituent in $NaSO_4$ environments (Equation 1), constitutes in $MgSO_4$ environments the first line of defence by protecting CSH. It is well documented (Cohen & Bentur, 1988) that mixtures having sufficient resistance against Na_2SO_4 attack (such as blended cements and especially silica fume blended cements) suffer extended deterioration in $MgSO_4$ environments.

Sulphate resistance is usually measured according to the ASTM C1012:2001 standard, which describes the use of mortar bar specimens. Specimens are pre-cured to a compressive strength of 20 MPa and then are stored in a solution of 5% Na_2SO_4. Expansion measurements are taken for a period of six months. In order to reduce the time period needed for extracting results, mortar bar specimens of smaller size are used in the AFNOR-P18-837 French Standard. In this Standard the bar specimens are of dimensions $20 \times 20 \times 160$ mm. They are stored under water for 28 days and then they are immersed in a 1.6% Na_2SO_4, which is much less aggressive. Expansion is measured for a time period of three months. This national Standard is proposed to work as the reference method for the coming European Standard prEN TC51.

In this work the performance of four different cements against different sulphate solutions is investigated according to ASTM C1012 and AFNOR P18-837 Standards.

2 MATERIALS AND METHODS USED

The materials used were four Greek cements of different strength categories according to EN-197/1: CEM I42.5N, CEM II42.5N(B-M), CEM IV32.5N(B_M) and CEM I52.5R with low alkali content (sulphate resistant cement, SR). The chemical composition of these cements is presented in Table 1.

All mixtures were produced according to DIN 1164 using EN-196/1 normal sand. Water to cementitious material ratio was equal to 0.50 in all cases.

The sulphate resistance was determined according to two different standards: ASTM C-1012 and AFNOR P18-837.

Table 1. Chemical analysis of the cements tested.

	SR	I42.5N	II42.5N	IV32.5N(A)
SiO_2	20.78	19.64	22.71	29.46
Al_2O_3	3.97	4.62	6.06	7.80
Fe_2O_3	4.38	3.27	3.43	3.60
CaO	64.47	63.59	58.87	47.15
MgO	1.60	1.91	1.67	2.96
SO_3	1.99	3.03	2.65	2.85
K_2O	0.41	0.62	1.18	0.75
Na_2O	0.28	0.38	0.43	0.75
L.I.	0.87	—	—	4.70

According to ASTM C-1012, the length change should be evaluated on cement prisms. Their dimensions were $40 \times 40 \times 160$ mm. The compressive strength was evaluated on 50mm mortar cubes. According to the standard all specimens were cured in a curing room at a temperature of 35–38°C until a compressive strength of 20MPa was achieved. After this happened, specimens were immersed in the aggressive solutions. These were a 5% Na_2SO_4, a 8.7% $MgSO_4$ and a mixed (2.5% Na_2SO_4 and 4.35% $MgSO_4$) solution.

The temperature of the solutions was 18–20°C. Length change measurements were performed at different ages up to 6 months, whereas compressive strength measurements were performed at the ages of 1, 4 and six months. Expansion of all mixtures is plotted in Figure 1.

The specimens produced for the evaluation of sulphate resistance according to AFNOR P18-837 were prisms with dimensions $20 \times 20 \times 160$ m. These specimens were used for measuring the expansions of the mixtures. The compressive strength was measured on $40 \times 40 \times 160$ mm specimens, according to the procedure defined in EN196-1. All specimens were cured for 28 days in a curing chamber. Thereafter they were immersed in the tested solutions: 1.6% Na_2SO_4, 2.8% MgSO4 and a mixed (0.8%NA_2SO4 +1.4%MgSO4) solution. According to the procedure defined in this Standard, the sulphate resistance should be evaluated by measuring the linear expansion up to the age of three months. In this paper the expansion was measured up to the age of six months, in order to totally compare the results obtained according to the two different Standards. These measurements are plotted on Figure 2.

The compressive strength measured after 90 and 180 days of immersion in the tested solutions is presented for all mixtures in Table 2.

3 RESULTS AND DISCUSSION

According to ASTM C1157/02, specific cement is characterized as *"cement with high sulphate resistance"* only if the expansion measured after six months of immersion in the aggressive solution is less than

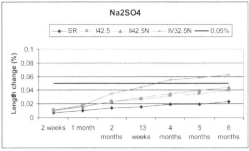

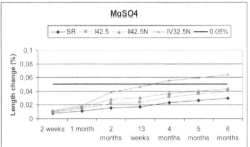

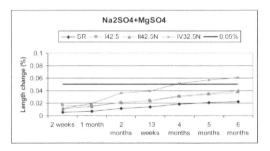

Figure 1. Length change measurements in different aggressive solutions according to ASTM C-1012.

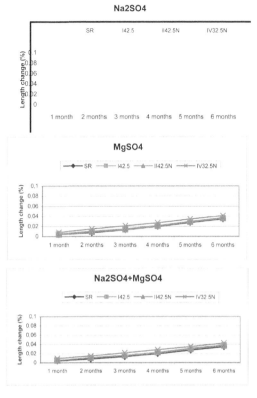

Figure 2. Length change measurements in different aggressive solutions according to AFNOR P18-837.

Where : $fs_{(T)}$ = the compressive strength of the mixture immersed in the sulphate solution tested at the age (T),
$f_{(T)}$ = the compressive strength at the same age of the mixture cured in the curing chamber.

The calculated values of compressive strength reductions were calculated using the measured values presented in Table 2 and are plotted for all mixtures on Figure 3.

All the tested cements are in general in the same rank as when evaluated using the criterion of their length change. When tested according to ASTM C1012, the three cements of the upper strength classes – S.R., CEM I42.5N and CEM II42.5N- performed residual compressive strength after six months of immersion in the Na_2SO_4 solution greater than 70%. In this case the CEM IV32.5N cement performed residual compressive strength equal to 65,63%.

Sulphate Resistant cement is the only cement that performed residual strength greater than the limit value in all three aggressive solutions. The rest of the tested cements performed lower residual compressive strength values after six months of immersion in the $MgSO_4$ and the mixed solution.

The length change of all cement mixtures when tested according to AFNOR P18-837 was in general in good accordance with the one obtained according to ASTM C1012. The S.R. cement still performed the

0.05%. If the value of the expansion measured at the same age is at the range of 0.05%–0.1%, the cement is characterized as *"cement with moderate sulphate resistance"*. The length change measured in all cement mortars according to ASTM- C1012 is plotted in Figure 1. It is clearly observed that three of the four cements tested – S.R., I42.5N and II42.5N- are well classified as cements with high sulphate resistance against all aggressive solutions tested in this research. However sulphate resistant cement –S.R.- performed the best in any case. Cement CEM IV 32.5N is classified as cement with moderate sulphate resistance.

Sulphate resistance of cements may also be evaluated using the criterion of the compressive strength reduction. According to Al-Amoudi (1995), reduction of compressive strength down to 30% is the limit value for a mixture with high sulphate resistance. However, such limitations do not exist in any of the Standards used in this research.

The reduction of the compressive strength was calculated according to the following formula:

Compressive strength reduction = $fs_{(T)}/f_{(T)}$,

Table 2. Compressive strength (MPa) of mixtures at different ages.

	AFNOR P18-837		ASTM C1012	
	90 days	180 days	90 days	180 days
Sulphate Resistant Cement (S.R.-CEM I52.5R)				
Curing chamber	63.1	68.1	79	87
$MgSO_4$	55.7	59.5	71	65
Na_2SO_4	61.4	64.7	76	70
Mixed	56.0	58.0	68	60
CEM I42.5N				
Curing chamber	49.4	50.3	62	68
$MgSO_4$	42.3	43.3	52	42
Na_2SO_4	39.9	41.4	56	48
Mixed	40.5	42.2	51	40
CEM II42.5N (B-M)				
Curing chamber	53.8	55.0	57.5	59.5
$MgSO_4$	45.9	47.1	51.5	33.0
Na_2SO_4	44.3	45.7	56.0	43.4
Mixed	44.9	46.3	48.5	32.4
CEM IV32.5N (B-M)				
Curing chamber	40.6	45.6	49	54
$MgSO_4$	28.4	32.8	35	27
Na_2SO_4	30.1	35.8	50	37
Mixed	29.0	34.3	36	29

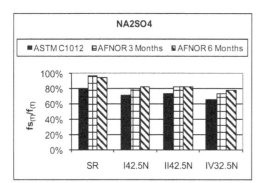

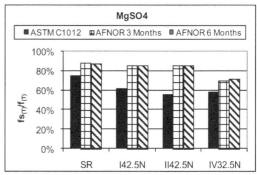

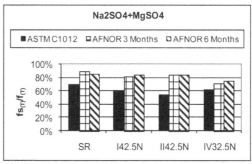

Figure 3. Compressive strength reduction after immersion in the aggressive solutions.

smaller expansion in all three tested solutions and the CEM IV32.5N cement performed the worst. CEM I and CEMII cements suffered expansions between those mentioned above. However due to the weaker solutions the expansion values measured were very close to each other and is difficult to distinguish the behaviour of cements in high and moderate sulphate resistance. Since the relative ranking of cements remained the same at the age of three and six months, it is proposed to use the length change value of 0.02% at the age of three months as the limit for sulphate resistant cements. This seems to be in good accordance with the limitation value of 0.05% used by ASTM C1157/02.

Residual compressive strength of cements tested according to AFNOR P18-837 was not influenced from the immersion period and remained almost the same after three and six months of immersion. Relative ranking of cements remained the same as when tested according to the criterion of their length change as well as when tested according to ASTM C1012. A compressive strength reduction of 20% could be proposed in this case as the limit between sulphate resistant and not sulphate resistant cements.

4   CONCLUSIONS

Sulphate resistance of cements may be well assessed according to AFNOR P18-837 Standard. The results obtained after three months of immersion in the defined solutions coincide well with the results obtained according to ASTM C1012 Standard after six months of immersion.

Working with AFNOR P18-837 reliable results may be available after three months of immersion in the aggressive solutions. The residual compressive strength values measured at this age are almost the same with the values measured after six months of immersion.

The length change value of 0.02% after three months of immersion in the aggressive solutions may be proposed as the limit among cements with high and moderate sulphate resistance.

REFERENCES

AFNOR P18-837 1993: "Produits Spéciaux destinés aux constructions en béton hydraulique-Produits de calage et/ou scellement á base de liants hydrauliques – Essai de tenue á l' eau de mer et/ou á l' eau á haute teneur en sulfates", Boutique AFNOR.

Al-Amoudi, O.S.B., Maslehuddin, M. and Saadi, M. M. 1995: Effect of Magnesium Sulfate and sodium sulfate on the durability performance of plain and blended cements. ACI Materials Journal, Vol. 92, No 1, pp. 15–24.

Al-Amoudi, O.S.B., 1995: Performance of fifteen reinforced concretes in magnesium-sodium sulfate environments. Construction and Building Materials, 1995, Vol. 9, No 1, pp. 25–33.

Al-Amoudi, O.S.B.:, 1999 "Mechanisms of sulfate attack in plain and blended cements: A review." Proceedings of the conference Extending Performance of Concrete Structures, pp. 247-260, International Congress "Creating with Concrete", Dundee, 6-10 September 1999.

ASTM C 1012-01,2001: "Standard Test Method for Length change of Hydraulic Cement Mortars Exposed to a Sulfate Solution", ASTM, West Conshohocken, Pa.

ASTM C 1157-02,2002 "Standard Performance Specification for Hydraulic Cement", ASTM, West Conshohocken, Pa.

Cohen, M.D. & Bentur, A., 1988: Durability of portland cement – silica fume pastes in magnesium sulfate and sodium sulfate solutions. ACI materials Journal, Vol. 85, No 3, May-June, pp.148–157.

Lea F.M., 1970: The chemistry of Cement and Concrete, 3rd edition, Edward Arnold, London.

Rasheeduzzafar, Al-Amoudi, O.S.B., Abduljauwad, S. N. and Maslehuddin, M, 1994.: Magnesium-sodium sulfate attack in plain and blended cements. ASCE Journal of Materials in Civil Engineering, Vol. 6, No 2, May, pp201–222.

Concrete Solutions – Grantham, Majorana & Salomoni (Eds)
© 2009 Taylor & Francis Group, London, ISBN 978-0-415-55082-6

Performance of "waterless concrete"

H.A. Toutanji

Civil and Environmental Engineering Department, the University of Alabama in Huntsville, AL, USA

R.N. Grugel

Marshall Space Flight Center, Huntsville, AL, USA

ABSTRACT: Waterless concrete consists of molten elementary sulfur and aggregate. The aggregates in a lunar environment will be lunar rocks and soil. Sulfur is present on the Moon in Troilite soil (FeS) and, by oxidation of the soil, iron and sulfur can be produced. Sulfur concrete specimens were cycled between liquid nitrogen ($\sim -191°C$) and room temperature ($\sim 21°C$) to simulate exposure to a lunar environment. Cycled and control specimens were subsequently tested in compression at room temperatures ($\sim 21°C$) and $\sim -101°C$. Test results showed that due to temperature cycling, the compressive strength of cycled specimens was 20% of those non-cycled. This reduction in strength can be attributed to the large differences in thermal coefficients of expansion of the materials constituting the concrete which promoted cracking. Similar sulfur concrete mixtures were strengthened with short and long glass fibres. The lunar regolith simulant was melted in a 25 cc Pt-Rh crucible in a Sybron Thermoline high temperature MoSi2 furnace at melting temperatures of 1450 to 1600°C for times of 30 min to 1 hour. Glass fibres and small rods were pulled from the melt. The glass fibres were used to reinforce sulfur concrete plated to improve the flexural strength of the sulfur concrete. Beams strengthened with glass fibres showed to exhibit an increase in the flexural strength by as much as 45%.

1 SULFUR/REGOLITH CONCRETE

Sulfur, a thermoplastic material, is melted and mixed with an aggregate after which the mixture is poured, moulded and allowed to harden. Sulfur concrete is not concrete in the traditional sense because little chemical reaction happens between the components. It is considered well established as a building material to resist corrosive environments, or in areas where there is high acid or salt content. Sulfur concrete usually contains 12–22 weight % sulfur and 78–88 weight % aggregate in its composition. The sulfur could consist of 5% plasticizers. The aggregate may include coarse and fine particles. Sulfur is generally expected to melt at about 119°C and stiffen above 148°C; therefore, the sulfur and aggregate must be mixed and heated at a temperature between 130°C and 140°C. Thus, the environment in which sulfur concrete is used must not have a temperature greater than the melting point of sulfur (Vaniman et al. 1992).

Commercial use of sulfur "concrete" on Earth is well established, particularly in corrosive, e.g., acid and salt, environments. Having found troilite (FeS) on the Moon raises the question of using extracted sulfur as a lunar construction material, an attractive alternative to conventional concrete as it does not require water. Table 1 is a record of the amount of sulfur found during some of those missions (Gibson & Moore 1973).

Sulfur can be extracted from lunar soil by heating the lunar soil to over 1100°C in a vacuum environment

Table 1. Sulfur abundance in Apollo soil samples.

Apollo mission	Sulfur abundance $\mu gS/g$
Apollo 14	706–778
Apollo 15	517–712
Apollo 16	474–794

resulting in the release of sulfur in the form of SO_2 and H_2S. Then by using the Claus process, pure elemental sulfur and water are created, as shown in Equation (1). The Claus process takes the hydrogen sulfide and the sulfur dioxide and passes them through a heated catalyst bed at a temperature of 323°C. The catalyst that is usually used is bauxite (Al_2O_3), which has been found in the Apollo lunar soil samples (Lunar Sourcebook 1991)

$$SO_2 + 2H_2S \rightarrow 2H_2O + 3S \tag{1}$$

Test results have shown that the compressive strength of sulfur concrete is higher than that of hydraulic concrete. The addition of silica to sulfur concrete increases the compressive strength by as much as 26% (Toutanji et al. 2006). Mechanically, silica is very similar to the silicate minerals in the lunar regolith. This addition of silica to the sulfur concrete decreases the required sulfur content but also improves the mechanical properties of the system.

Figure 1. Waterless lunar simulant specimen.

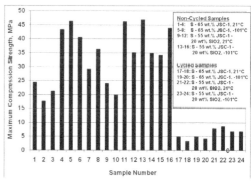

Figure 2. The max compressive strength exhibited by the cycled and non-cycled samples.

1.1 Specimen preparation

Cubes measuring 50.8 mm (2 in) were cast made of sulfur concrete (Fig. 1). Two different sulfur concrete mixtures were made: a) 35% purified sulfur and 65% JSC-1 aggregate by mass and b) 20% silica, 25% purified sulfur and 55% JSC-1 aggregate by mass. To examine the effect of severe environmental conditions on the concrete mixtures, the effect of the addition of silica, which is the main element of the lunar regolith composition, was also studied.

The specimens were subjected to light freeze-thaw exposure (room temp. to $\sim -2°C$) and severe freeze-thaw exposure (room temp. to $\sim -191°C$). The specimens were tested in compression after each exposure using a Universal Hydraulic Testing Machine. The size of the light exposed specimens were cubes measuring $50.8 \times 50.8 \times 50.8$ mm and were subjected to 50 cycles. For the severe freeze-thaw specimens, the cubes measured $25.4 \times 25.4 \times 25.4$ mm and were subjected to 80 cycles. A minimum of four specimens were tested for each test.

1.2 Results

The cubic blocks that were 50.8 cm on each side were cut into eight 25.4 cm cubes. Cubes were packaged with a k-type thermocouple in sets of eight and put into a Styrofoam container, into which liquid nitrogen (LN2) was poured and allowed to cool and evaporate. LN2 decreased the temperature from room temperature (RT), at about $21°C$ to about $-191°C$. The temperature cycle was repeated 80 times. Compression testing took place at a constant crosshead speed of 0.127 cm/minute. Samples of both compositions that were cycled and non-cycled underwent compression testing, one set happening at room temperature, about $21°C$, and the other set at about $-101°C$.

The average maximum compression strength for the non-cycled samples was 35 MPa and 7 MPa for the cycled samples, a 5 times difference. An explanation for the cycled samples' failure at a load 5 times less than the non-cycled samples could be that the de-bonding of the particles of aggregate with sulfur that occurred during the cycling left cracks which weakened the sample before compression testing. There was no difference in behavior between samples tested at $-101°C$ and at $21°C$. Samples cycled to $-191°C$ were weaker than at other temperatures, as shown in Figure 2. The concrete did in fact show a transition when it was cycled to $-191°C$ but what is not known is whether $-191°C$ is the definitive temperature or if the transition could have occurred at different temperatures. Although, it is not definitively determined, the rate of de-bonding may have been related to the number of cycles in the plastic range, as well as the heating and cooling rates. Sulfur and aggregate have vastly different coefficients of expansion that causes de-bonding at unknown temperatures, unknown because many properties of sulfur are still unknown such as exactly where sulfur transitions from elastic to plastic behavior.

2 GLASS FIBRE-REINFORCEMENT

Glass fibre and glass rebar reinforcement are used in a large number of terrestrial civil infrastructure applications. Glass fibres and glass rods are also ideal candidates for use in lunar construction.

Utilization of the lunar regolith for production of fibreglass and glass rods will require knowledge of the glass forming capability of this material. Some work has been performed with lunar simulants. Tucker & Ethridge (1998) studied the fibre forming characteristics of Minnesota Lunar Simulant-1 (MLS-1) and Minnesota Lunar Simulant-2 (MLS-2). MLS-1 simulated the mare composition and MLS-2 the lunar highlands. It was found that MLS-2 was easier to draw into a fibre. However, both simulants led to "fragile" glasses which tended to crystallize quite readily. Addition of boric oxide (8 wt. %) extended the viscosity such that continuous fibres were easily drawn using a fibre pulling apparatus. Glass formation appears to be easier with JSC-1 simulant.

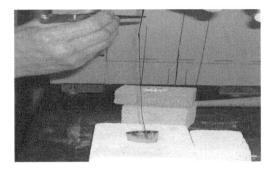

Figure 3. Hand-drawing glass fibres.

Figure 5. Specimens strengthened with long glass fibres.

Figure 4. Glass fibres drawn from lunar regolith simulant.

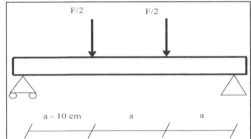

Figure 6. Four-point bend test setup.

2.1 *Specimen preparation*

JSC-1 lunar soil simulant was melted in a high temperature furnace at temperatures between 1450°C and 1600°C for times of 30 minutes to 1 hour using a 25cc platinum crucible. The crucible containing the melt was placed on a refractory brick and glass fibre was hand drawn directly from the melt using an alumina rod. The fibre was drawn through two felt pads containing a polyamide solution to provide a protective coating for the glass fibre. The coated fibres were then placed in a low temperature furnace and the polymer coating was cured at 200°C for 12 hours. The coating acts as a barrier to atmospheric moisture, which is known to degrade glass strength (Tucker et al. 2006). The drawing process and drawn fibres are shown in Figures 3 and 4.

To aid in drawing continuous glass fibre, a KC135 fibre-drawing apparatus has been refurbished (Tucker et al. 2006). The perform mechanisms and low temperature furnace were replaced with a high temperature furnace. The furnace windings are Pt/Rh wire, which gives the furnace a capability of reaching 1600°C. The fibre diameter is controlled by the furnace temperature and/or the take-up reel rotation rate which determines the draw rate. Fibre diameter is controlled primarily through take-up speed, although temperature changes of the glass viscosity can also be used. Coating is achieved by running the fibre through a small cup containing a polymer solution. The polymer is then UV cured before winding.

Plates measuring 101.6 mm × 254 mm × 12.7 mm were cast, as shown in Figure 5, consisting of sulfur powder and JSC-1. By mass the mixtures were 35% sulfur and 65% JSC-1. Long and short glass fibres produced from lunar regolith simulant, were used to reinforce the sulfur concrete. The percentage of the glass fibre in the mix was about 1% by mass. The diameters of the short fibres ranged from 3 to 20 micrometers and those of the long fibres were between 0.50 mm and 1 mm. Every plate was divided into three small beams, measuring 33.8 mm × 254 mm × 12.7 mm. Plain sulfur-regolith concrete and the sulfur-concrete reinforced with glass fibres were tested for load and deflection, using a four-point bending test setup shown in Figure 6.

2.2 *Results*

Adding glass fibre, derived from lunar regolith simulant, significantly increased the overall strength of the concrete as shown in Figure 7. As compared to the control specimens (SC), specimens strengthened with long hand drawn glass fibres (SCLGF) and specimens strengthened with short glass fibres (SCSGF) have exhibited an increase of more than 40%. This is a preliminary data and more tests are currently conducted to study the effect of the glass fibres on the ductility and strain energy capacity of the sulfur concrete.

3 CONCLUSIONS

To determine the structural integrity of sulfur concrete, sulfur concrete specimens were subjected to cycling between room temperature to −191°C. Results

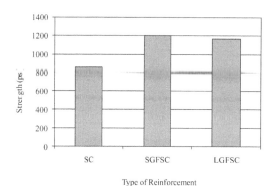

Figure 7. Ultimate strength values of sulfur/JSC-1 simulant concrete with and without glass fibres.

showed a significant reduction in strength, compared to non-cycled samples. This reduction in strength can be attributed to the fact that sulfur and aggregate have vastly different coefficients of expansion that cause de-bonding at different temperatures. Regolith derived glass fibres can also be used with sulfur concrete, also made with in-situ regolith. The addition of fibre showed a significant improvement in the strength. Both short glass and long glass fibres were shown to increase the strength of sulfur concrete by as much as 40%. More tests are currently being conducted to study the effect of the glass fibres on the ductility and strain energy capacity of the sulfur concrete.

ACKNOWLEDGEMENTS

The authors are grateful to the Marshall Space Flight Center and the In-Space Fabrication and Repair Element for their support of this work and the financial support of NASA grant NNM05AA22A. The authors also would like to thank Drs. Ed. Ethridge and Dennis Tucker of NASA-Marshall Space Flight Center for providing the glass fibres.

REFERENCES

Gibson, E.K. Jr, Moore, G.W. 1973. Carbon and sulfur distributions and abundances in lunar fines. *Proceedings of the Fourth Lunar Science Conference*, Vol 2: 1577–1586.
Lunar Sourcebook, 1991. A user's guide to the Moon. (ed). Heiken G.H, Vaniman, D.T. & French, B.M. (eds), Cambridge University Press, New York.
Toutanji, H. et al. 2006. Glass fiber reinforced sulfur concrete for habitat structures. *Proceedings: ASCE Habitation 2006: Conference on Habitation Research and Technology Development*. Orlando, FL.
Tucker, D.S. and Ethridge, E.C. 1998. Processing glass fiber from Moon/Mars resources. *Space 98*.
Tucker, D. et al. 2006. Production of glass fibers for reinforcement of lunar concrete. *Proceedings: 43rd American Institute of Aeronautics and Astronautics (AIAA)*. Reno, NV.
Vaniman, D. et al. 1992. Uses of lunar sulfur. *Second Conference on Lunar Bases and Space Activities of the 21st Century*, Wendell Mendell (ed), Lunar and Planetary Institute: 429–435.

Concrete Solutions – Grantham, Majorana & Salomoni (Eds)
© *2009 Taylor & Francis Group, London, ISBN 978-0-415-55082-6*

Repair of the reinforced concrete structure of the "Prokop" railway station in Belgrade

D. Vasović
Faculty of Architecture, Belgrade University, Belgrade, Serbia

T. Vasović
RELIT co., Belgrade, Serbia

M. Glišić
Faculty of Architecture, Belgrade University, Belgrade, Serbia

ABSTRACT: This paper will present an example of successful repair of the reinforced concrete structure of the new railway station "Prokop" in Belgrade, applying polymer-modified materials. The long span ribbed slab structure was damaged due to low temperatures and extensive ice formation, during concrete casting. Huge parts of the ribs and beams reinforcement were exposed to the environment, so durability and bearing capacity of the structure was threatened. During laboratory research and testing a repairing material adequate for field application was designed. Treatment of the reinforced concrete structure elements depended on the scale of the damage, and the concrete mix used for the repair included a polymer-modified cement mix.

1   INTRODUCTION

This paper presents an example of the successful repair of a structure in-between floors above the platform of the new Prokop railway station in Belgrade. The structure is a reinforced concrete ribbed slab, with two spans of 9.6 m, roughly 4000 m^2 of concreted area.

The repair of the damaged reinforced concrete slab was designed to address the following:

- The damage occurred during casting of the concrete;
- The degree of damage was estimated to be considerable by a visual examination, posing a threat to the bearing capacity of the structure;
- The repair of the structure was estimated to be very expensive, and the Client ordered the Contractor to find an appropriate solution or remove the entire structure.

Owing to the large extent of the damage and the expense that would emanate from the partial removal of the structure, it was required that all consequences of the damage be reviewed once again, as well as the potential possibility of a repair, even if only partial.

2   DESCRIPTION OF THE DAMAGE

Following a meticulous examination of all available data, it was established that the damage had occurred in a rather extraordinary way. The structure was cast in place. The casting had been performed in wintertime conditions, with outside temperatures dropping to subzero levels. Just before casting, the snow had covered the formwork and reinforcement. The site engineer ordered the snow to be melted with burners in order to remove it from the reinforcement. The melted snow turned to water and dripped to the bottom of the formwork. The high quality work of the formwork did not allow the water to drain, but retained it in the lower part, around the reinforcement. The casting process lasted deep into the night due to the size of the slab. During the night, the temperatures dropped to several degrees below zero. The water (melted snow) froze and formed ice layers around the reinforcement. It was therefore impossible to cast concrete in the space already occupied by the ice. A few days later, temperatures rose above zero, so concrete continued to harden. When the desired hardness was attained, the formwork was taken off and the sight shown on Figures 1 and 2 appeared. No one could explain how such damage had occurred. An intentional bad casting was even suspected.

Examination of concrete cast into the slab showed that concrete satisfied, even surpassed the requested 40 MPa, achieving more than 50 MPa and the required characteristics (hardness, frost resistance). It was determined that the quality and the compaction of concrete had not been an issue, but the fact that it had not been cast into the entire cross-section of the formwork. These tests and calculation simulations showed that the structure had the required degree of safety, that structural integrity of the structure had not been jeopardised and that it was solely necessary to repair the protective layer of the reinforcement.

Figure 1. Example of concrete beam damage.

Figure 4. Repaired ribs and beams.

Figure 2. Example of concrete rib damage.

Figure 5. Repaired ribs and beams.

Figure 3. Repair procedure.

3 REPAIR PROCEDURE

Because of the significant height and available manpower, it was decided that the reinforcement be protected with a new layer of protective concrete by mortaring. Such simple work technology required a high quality repair material. It was suggested that the new protective layer be made with a high-efficiency fine concrete of high adhesive quality. For the purposes of this repair process, new materials with high mechanical characteristics were developed, based on polymers

and cement. In order to successfully repair concrete, several years' experience on repairs demonstrated that it was indispensable to establish good adhesion to the existing concrete. Particular attention had therefore been given to realisation of the highest possible quality link between the existing and new concrete. By application of polymer-based materials, the adhesion to the existing base was achieved up to 2.0–2.5 MPa.

The surface of the ribs and beams was carefully prepared. After the removal of the deteriorated concrete layer, the surface was protected with a polymer-cement based coating and sprayed with the cement based binder composed of the polymer/cement/crushed aggregate (size 2–4 mm) mixture in the 1:1:1 ratio. The binder was applied on the prepared surface 24 hours before the rendering using a special tool called a "hedgehog." After 24 hours, the surface of the prepared concrete and steel was rendered with high-performance small size aggregate polymer modified concrete.

The repair procedure showed, in this example, a high degree of applicability. The extent and the manner of the repair required a procedure easily applicable on large fragmented ceiling surfaces (ribbed slabs and beams), at height, from a scaffold, and which could be easily applied even by a large number of averagely trained workmen (rendering by high-performance small size aggregate polymer modified

Table 1. Applied concrete mixture.

Material	Per m^3	Units
Portland cement CEM I	617	kg
Crushed stone aggregate 0–2 mm	1234	kg
Styrene-butadiene-latex*	108	kg
Accelerating admixture	20	kg
Added water	134	kg
Total water content (W/C = 0.31)	193	kg

*45% water solution SBR latex.

Table 2. Mechanical Properties of the Mixture.

Properties	Values	Units
Initial slump (small mortar cone d = 60/70 mm,	173	mm
H = 100 mm) Unit weight (wet density)	2172	kg/m^3
Compressive strength		
2 days	31.0	MPa
7 days	47.0	MPa
28 days	63.0	MPa
Tensile strength		
2 days	7.4	MPa
7 days	8.7	MPa
28 days	9.7	MPa
Adhesion to concrete surface – 28 days	4.3	MPa
Abrasion resistance*	6.94	mm^3/50 mm^2

*According to JUS B.B8.015, Wear Test. Grinding wheel method.

concrete with a bricklayer's trowel from a receptacle). The speed and the quality of repair have confirmed in full the applied procedure.

4 MATERIAL

In order to predict behavior on the site, laboratory testing of the proposed mixture was carried out. As described before, a high-performance small size aggregate polymer modified concrete was applied. The composition of applied mixture is given in Table 1.

Mechanical properties of the applied mixture are given in the Table 2.

5 CONCLUSIONS

This case study is an example of successful repair of the reinforced concrete ribbed slab of the new railway station "Prokop" in Belgrade, applying polymer-modified materials. Although the initial appearance of the damaged building gave the hopeless impression that almost everything had to be removed, careful examination of the severely damaged structure showed that cost-effective and time-saving procedures could be used. High quality materials and an innovative technique enabled a great part of the reinforced concrete structure to remain in use, which brought huge savings. Extensive laboratory testing and experience from previous repairs were needed to make the materials which best suited the complex requirements of the repair procedure.

REFERENCES

Neville, A. M., 1995, *Properties of Concrete*, London, Pearson Education Limited.

Aïtcin, P.-C., 1997, The Art and Science of High-performance Concrete, *Proceedings of Mario Collepardi Symposium on Advances in Concrete Science and Technology*, Roma.

Aitcin, P.-C., 1998, *High-Perfomance Concrete*, London, E & FN Spoon.

Skalny, J.N. ed., 1989, *Materials Science of Concrete*, Westerville, American Ceramic Society.

Vasović, T., Vasović, D., 1995, Cement Stone Density – Key Factor for High-Performance Concrete, *Second Canmet/Aci International Symposium on Advances in Concrete Technology*, Las Vegas, June 11–14, 1995.

Vasović, D., Vasović, T., 1996, Quality of Concrete Protecting Cover next to Reinforcement in Concrete Construction, *7th International Conference on the Durability of Building Materials and Components*, Stockholm, May 19–23, 1996.

Concrete Solutions – Grantham, Majorana & Salomoni (Eds)
© 2009 Taylor & Francis Group, London, ISBN 978-0-415-55082-6

Structural assessment of agricultural buildings

E. Rodum
Norwegia Public Roads Administration, Trondheim, Norway

Ø. Vennesland
Department of Structural Engineering, Norwegian University of Science and Engineering, Trondheim, Norway

H. Justnes & H. Stemland
SINTEF Building and Infrastructure, Trondheim, Norway

ABSTRACT: Norwegian agricultural buildings are different from most of the European types – especially is that valid for the cowsheds that normally are standing on a manure cellar below. Nowadays almost all floors between the cowshed and the manure cellar are made of reinforced concrete with openings in the floor below the cowshed where the droppings from the cattle are shoveled down. Lately, however, accidents have happened causing the mentioned floor between the shed and the cellar to collapse. Because of this type of accidents many cows or oxen has died and it is just by pure luck that no human beings have been killed.

Most of the material in the presented paper is from a project where much of Norwegian agriculture – together with NTNU and SINTEF – was involved and where the intention was to present guide lines for structural assessment of the concrete structures and to present measures that might be taken in order to minimize the problem.

It has been found that the type of attack depends on whether there is an open or closed connection between the cellar and cowshed. If the connection is open the cause for damage is mainly corrosion of the embedded steels due to chlorides from the liquid manure, and if the cellar is closed the main reason for damage is disintegration of the concrete due to sulphate attack.

1 INTRODUCTION

The project "Concrete in agriculture" was initiated in September 1998. In the beginning it was focused on the durability situation for manure cellars and the main intention was to investigate the extent of damage. Two extensive field investigations – one in Middle Norway (Report – Damage to Concrete . . . ., 1999) and one in the Eastern Part of Norway (Berge et al., 2000) were crucial in this work. In addition a field investigation was started in the county of Rogaland (Vernhardsson et al., 2001)

In Middle Norway 281 manure cellars were investigated. Mainly two types of damage were found: corrosion of embedded steel and disintegration of the concrete. In about 10% of the cellars the damage was so severe that strengthening of the concrete structures was recommended. The repair costs were estimated to be 33 million NOK. Based on these numbers, the repair costs of manure cellars for the whole country were estimated to be 5 billion NOK. This first part of the project was made mainly to get an estimate of the visual damage and the costs. It was therefore mainly based upon visual investigation without any systematic selection of test material.

Later on – in order to get realistic recommendations both for repair of existing structures and building of new ones – a more detailed study was made. In Middle Norway this meant a thorough structural assessment of four manure cellars – including both a visual inspection in the field and laboratory tests on collected concrete core samples. More information is given in the report STF A04618 (Rodum et al, 2004).

2 FIELD TESTS

The investigation was made on manure cellars in the counties of Møre og Romsdal and Sør-Trøndelag, located on four different farms (1, 2, 3 and 4), built in the years 1971–83. The cellars studied were selected in cooperation with the agricultural departments of the county governors in both counties based on earlier results. All cellars had been exposed to liquid manure from cows all the time. Three of the cellars (2, 3 and 4) were open between cowshed and the manure cellar through grates while the last one (1) was closed with scrape grooves below the manure grate.

The manure level at the time of inspection varied from 0 to about 0.7 meters. The concrete surfaces had not been cleaned prior to inspection.

Visual inspection was mainly made from floor level. More detailed inspection as well as field measurements and core drilling were made from scaffold.

The field tests included:

- Visual inspection of all (in some cases selected) structural parts
- Measurements of cover thickness and carbonation depth
- Drilling cores for laboratory tests
- Drilling dust for chloride analyses

The selection of field tests was based on the existing documentation of the environment and possible causes for deterioration in manure cellars (Berge et al, 2000; Rodum et al, 2004). To some extent also the laboratory information given in an earlier report was included (Vernhardsson et al, 2001).

3 ENVIRONMENTAL CONDITIONS

The choice of tests was based on information about what one might expect to find in the cellars and what could possibly happen with time. The chemical composition of manures in the Netherlands is shown in Table 1. (Hoeksma, 1988)

Table 1 Chemical composition of manure, in g/l

3.1 Possible detrimental reactions

Hydrogen sulphide (H_2S) is a gas that might be formed in a septic environment. Dissolved in water H_2S is a weak acid that might react with $Ca(OH)_2$ forming calcium sulphide:

$$Ca(OH)_2 + H_2S \text{ (aq)} = CaS + 2H_2O \qquad (1)$$

S^{2-} might oxidise because of aerobic bacteria in the cellar and/or by oxygen forming SO_4^{2-}:

$$S^{2-} + 2O_2 = SO_4^{2-} \qquad (2)$$

In addition the manure itself may contains sulphate SO_4^{2-} reacts with $Ca(OH)_2$ forming gypsum ($CaSO_4 \cdot 2H_2O$) that acts expansively:

$$2H_2O + Ca(OH)_2 + SO_4^{2-} = CaSO_4 \cdot 2H_2O + 2OH^- \quad (3)$$

Table 1. Chemical composition of manure, in g/l.

Chemical component	Cattle	Fattening pigs	Laying hens
Nitrogen	3.8–7.6	4.3–11.5	5.9–15.7
Ammonium	0.2–4.4	1.3–5.5	2.6–9.2
Phosphate	1.3–3.1	3.6–6.6	0.3–12.0
K_2O	3.3–11	2.0–6.1	0.3–11.5
CaO	1.6–3.3	2.4–4.4	0.9–19.6
MgO	0.8–1.6	0.6–2.0	0.1–2.4
Cl^-	1.8–4.2	0.6–3.3	0.1–3.2
SO_4^{2-}	2.0–3.0	1.0–2.0	2.0–4.0
Acetic Acid	5.5–7.0	3.2–11.0	11.0–22.0
Propionic acid	1.6–2.0	0.7–3.0	4.0–7.5
pH	7.0–8.8	7.3–8.6	6.7–8.3
Dry matter (%)	7–9	4–11	11–18

The gypsum might react further and form ettringite:

$$Ca_3Al_2O_6 \cdot 6H_2O + 3CaSO_4 \cdot 2H_2O + 20H_2O =$$
$$3CaO \cdot Al_2O_3 \cdot 3CaSO_4 \cdot 32 H_2O \qquad (4)$$

Any possible diffusion of water soluble sulphates into the concrete has the same detrimental effects as shown (expansion due to formation of gypsum or ettringite).

A dangerous form of sulphate attack is the formation of thaumasite. This demands available carbonate. The silicate comes from the C-S-H-phase and the concrete is destroyed. The carbonate might come from any calcareous filler, from carbonation or from urea in the liquid manure.

The liquid manure always contains much urea (($NH_2)_2CO$). This might react with calcium hydroxide forming calcium carbonate and ammonia that evaporates:

$$Ca(OH)_2 + (NH_2)_2CO = CaCO_3 + 2 NH_3 \qquad (5)$$

Carbon dioxide (CO_2) is a gas that might be found in high concentration in the air of manure cellars. CO_2 reacts with $Ca(OH)_2$ in the concrete, forming calcium carbonate that again causes a lowering of the passivating high pH of the concrete:

$$Ca(OH)_2 + CO_2 = CaCO_3 + H_2O \qquad (6)$$

For continuous supply of CO_2-containing water the calcium carbonate will react and form calcium hydrogen carbonate:

$$CaCO_3 + H_2O + CO_2 = Ca(HCO_3)_2 \qquad (7)$$

This compound is water soluble and might leach out of the concrete.

4 LABORATORY TESTS

Based on the possible mechanisms of damage the following laboratory methods were chosen:

- Determination of the strength of concrete
- Phenolphthalein-test to determine depth of neutralization (often called carbonation depth)
- Determination of penetration depth of chlorides
- Determination of penetration depth of sulphates/sulphides
- Determination of penetration depth of nitrates/nitrites
- Scanning Electron Microscope (SEM) – analyses

The concrete dust for chemical analyses were made by milling concrete from the cover (distance from surface) at layers 0–1, 1–5, 5–10, 10–15, 15–20 and 20–25 mm. In some cases also 30–35 and 35–40 mm was included. Where the chlorides had penetrated deep into the concrete also layers of 5 mm were milled into 60 mm or 70 mm.

From behind, a slice of ca 10 mm thickness was made by dry sawing. This slice was crushed to dust.

The chloride content was determined by a spectrophotometric method; Sulphate content by a gravimetric method and nitrate content by ionic chromatography as water soluble NO_3^-.

The main results from the mechanical and chemical analyses are shown in the Figures 1–5.

The main finding from the SEM-analyses (Justnes et al, 2006) was that both ettringite and thaumasite

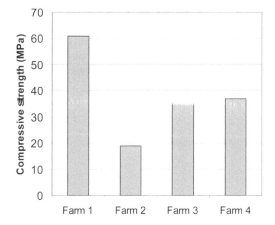

Figure 1. Compressive strengths of cores drilled from the beams of farms 1–4.

formations were documented in the concrete from farm 1 (with closed cellar).

Figure 2 Chloride profiles from drilled cores in beams. Farm 1 has a closed connection between cellar and cowshed while 2, 3 and 4 have open connections. For farm 2–4 the cores are drilled below and between the grates, respectively.

5 CONCLUSIONS

All four investigated manure cellars were built in the years 1971 to 1983. Three of the cellars were open between cowshed and manure cellar through grates while the last one was closed with scrape grooves below the manure grate. The investigation revealed that the damage situation was related to the connection (open or closed) between cowshed and manure cellar.

The main reason for damage in the three cellars that were open to the cowshed was corrosion of the reinforcement due to chlorides. The damage was mainly localized to structural parts in direct contact with droppings of urine or manure through the manure grates. Other parts of the structure had a limited

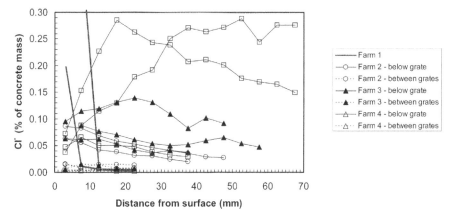

Figure 2. Chloride profiles from drilled cores in beams. Farm 1 has a closed connection between cellar and cowshed while 2, 3 and 4 have open connections. For farm 2 4 the cores are drilled below and between the grates, respectively.

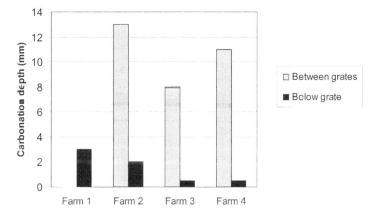

Figure 3. Mean values of neutralization (carbonation) depths from cores drilled from beams, between and below grates, respectively. Farm 1 has a closed connection and the farms 2–4 have open connections between cowshed and manure cellar.

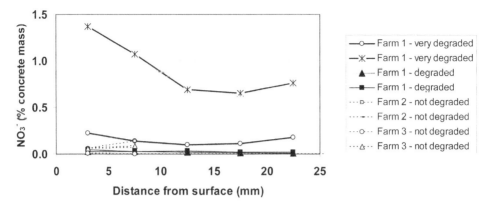

Figure 4.   Nitrate profiles from farms 1 (closed connection), 2 and 3 (open connections).

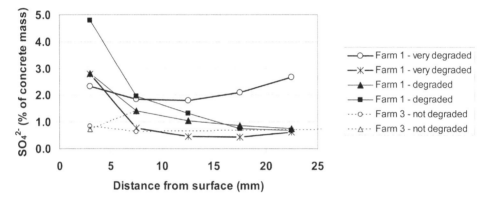

Figure 5.   Sulphate profiles from farms 1 (closed connection) and 3 (open connection).

development of damage. No disintegration of the concrete was observed in any of the three cellars with open connection.

The main reason for damage in the cellar that was closed to the cowshed was disintegration of the concrete due to sulphate attack, both ettringite and thaumasite formation. The damage was mainly observed on the beams, and especially along edges and in connection to casting faults. The attack was most pronounced in areas around the hatches for manure droppings.

Most of the used methods for concrete repair (electrochemical methods and mechanical repair) are not relevant for manure cellars. All these methods are technically demanding, costly and very uncertain regarding their long term effect in manure cellars. The exception is mechanical repair if the corrosion should be initiated by carbonation and for damage to the concrete, especially if the scale is limited. Mechanical repair combined with sacrificial anodes should be further evaluated.

Periodic supervision, strengthening (temporary or permanent) and replacement are the most relevant measures. Preventive measures should also be taken in order to 1) protect the concrete beneath the grates from being directly exposed to the droppings and 2) ventilate the closed cellars.

REFERENCES

Berge, E and Vernhardsson, C: "Damage to concrete in manure cellars", ITF report 111/2000, Norwegian University of Life Sciences (earlier Norges Landbrukshøgskole), May 2000 (In Norwegian)

Hoeksma, P: "De sammenstelling van drijfmest die naar akkerbouwbedrijven wordt afgezet" (The composition of slurry marketed to arable farms), in Dutch, IMAG-DLO, Wageningen, 141 p, 1988.

Justnes, H. and Rodum, E.: "Case Studies of Thaumasite Formation", The Seventh CANMET/ACI International Conference on Durability of Concrete, Montreal, Canada, May 28 - June 3, 2006.

Rodum, E et al: "Concrete in agriculture, Investigation of four manure cellars in Sør-Trøndelag and Møre og Romsdal", Sintef report STF A04618, Trondheim, 2004 (In Norwegian).

The agricultural departments of the county governors in Sør-Trøndelag and Møre og Romsdal: "Report from the project Damage to concrete in manure cellars – investigation in Sør-Trøndelag and Møre og Romsdal", 1999 (In Norwegian).

Vernhardsson, C and Berge E: "Investigation of manure cellars in Rogaland", ITF report 118/2001, Norwegian University of Life Science (earlier Norges Landbrukshøgskale), April 2001 (In Norwegian).

Concrete Solutions – Grantham, Majorana & Salomoni (Eds)
© 2009 Taylor & Francis Group, London, ISBN 978-0-415-55082-6

Long-term investigation of alternative reinforcement materials for concrete

H.G. Wheat

Mechanical Engineering Department, Texas Materials Institute, University of Texas at Austin, Austin, TX

ABSTRACT: One of the techniques to reduce the corrosion of steel in concrete involves changing the reinforcement material. In this investigation, this strategy has been implemented by using coated steel or a more corrosion resistant material than black steel. This paper describes an investigation in which the corrosion behavior of a number of materials has been compared with that of black steel for more than five years. The materials include bars made of epoxy coated steel, galvanized steel, nylon-coated steel, polyvinyl chloride-coated steel, stainless-clad and stainless steel.

The materials are being tested using a procedure similar to ASTM G109, a test procedure in which current is measured between steel in chloride-free and chloride-contaminated concrete. The test procedure is being complemented with electrochemical tests and tests involving acoustic emission (AE) techniques. A good correlation was obtained based on the average number of pulses between AE sensors placed on the sides of the concrete macrocells.

1 INTRODUCTION

When reinforced concrete is exposed to corrosive environments, its structural integrity can be severely compromised (Andrade et al, 1990; Broomfield, 1997; Chen & Wheat, 1996). The process is usually associated with chloride and/or carbon dioxide penetration through the concrete, followed by rust staining, concrete cracking and spalling.

In many cases, the level of damage in reinforced concrete is determined based on the corrosion activity of the steel inside (Clifton et al., 1975; Dunn et al., 1984; Hurley & Scully, 2005; Ing et al., 2003). The linear polarization resistance (LPR) technique is often used to measure corrosion rate, where the corrosion rate is inversely related to the polarization resistance, Rp. Another procedure that is used involves concrete macrocells and measurement of current between reinforcing bars in chloride-contaminated and chloride-free concrete. Electrochemical tests can also be carried out on macrocells when the reference electrode is placed in the reservoir containing the salt solution and connection is made to the reinforcement. Based on these tests, the condition of the steel can be assessed. The information may then be used to determine whether the structure is in need of repair.

The techniques that are used to determine corrosion rates are based on electrochemical techniques and derived from principles applicable to uniform corrosion processes (Andrade et al., 1990). In the case of chloride-induced corrosion, there is usually an initiation period when chlorides begin to penetrate through the concrete. Subsequently, if a chloride threshold is achieved, there is a decrease in the corrosion potential,

Ecorr, and an increase in the corrosion rate, which is related to 1/Rp. An example of this type of behavior is shown in Figure 1. This plot was determined based on cyclical exposure of cylindrical reinforced concrete lollipop specimens in which the water: cement ratio was approximately 0.68 (Zhang et al., 1994). In this case, the steel was black steel, the cover was 2.5 cm, and the time to corrosion was approximately 16 weeks.

While the electrochemical techniques are based on uniform corrosion, it is known that the corrosion processes are often localized, especially initially. In addition, alternative materials that are being developed and used as reinforcement are often non-metallic or coated, they are usually inherently more corrosion resistant, and thus the time required for testing will be extended. This is especially true when the quality of the concrete is improved. For example, when Type F fly ash was added to replace cement at a level of 35% in specimens similar to those in Figure 1, corrosion was not observed even after more than 104 weeks of cyclical saltwater exposure (Zhang et al., 1994). When concrete macrocells containing alternative reinforcement materials were exposed to cyclical saltwater ponding, corrosion was not observed even after 198 weeks (Wheat & Liu, 2007). This was for concrete with a water:cementitious material ratio of 0.32.

It is also true that electrochemical techniques that work reasonably well for metallic materials may provide limited information for some of the alternative materials (Kahhaleh et al., 1993; Kahhaleh et al., 1994; McDonald et al., 1995; Chen & Wheat, 1996; Kahrs et al., 2001; Hurley & Scully, 2005; Lysogorski et al., 2005, Wheat, 2006). This is because the chloride

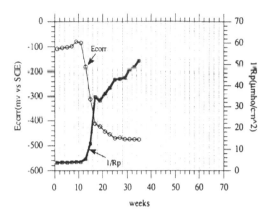

Figure 1. Corrosion potential and 1/Rp vs time (Zhang et al., 1994).

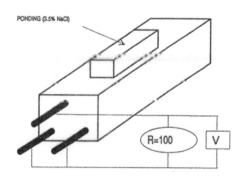

Figure 2. Schematic of macrocell specimens.

thresholds are usually different and the changes in potential may be different.

Therefore, to gain additional information about the performance of alternative reinforcement materials, it would be desirable to have a non-destructive technique that focuses on assessing concrete damage; in this case concrete damage that has resulted from corrosion. Thus, one of the overall goals of this research is to investigate the usefulness of a non-destructive technique such as acoustic emission, as a complement to electrochemical techniques.

Acoustic Emission (AE) is a technique that can be used to examine the behavior of materials deforming under stress. This is because the technique is particularly sensitive to cracks and crack propagation. Previous investigations have concentrated on utilizing AE techniques to study the ongoing process of rebar corrosion in concrete during corrosion experiments (Dung et al., 1984; Yuyama et al., 1990; Li et al., 1998; Matsuyama et al., 1993; Muravin et al., 1996; Yoon et al., 2000; Ing et al, 2003; Ohtsu & Tomoda, 2006). In the present research, the goal is to investigate the possibility of extending the application of AE to intermittent or one-time tests associated with rebar corrosion (Wheat & Thakar, 2005).

Results from AE monitoring of steel-reinforced concrete specimens showed that there can be a relationship between AE activity and corrosion behavior (Li et al., 1998, Ohtsu & Tomoda, 2006). The investigators concluded that continuous AE monitoring could detect the onset of rebar corrosion earlier than other methods such as galvanic corrosion and half-cell potential measurements. The results were based on tests carried out during corrosion experiments in which reinforced concrete specimens, typical of those used in ASTM G109 were used.

Yoon and coworkers (Yoon et al., 2000) obtained similar results when they evaluated concrete beams that were prepared using unreinforced, notched-unreinforced, reinforced, and corroded-reinforced specimens.

While continuous AE monitoring has definite potential as a complement to electrochemical tests, there are still several questions that remain. For example, it is not clear whether this technique can be used to detect and differentiate different levels of concrete damage in concretes made with (1) different mix designs or (2) different reinforcement materials, such as epoxy-coated, galvanized, or stainless steel bars.

2 BACKGROUND

To address the issue of different mix designs, selected macrocell specimens from a previous project were examined using AE techniques (Zhang et al., 1994, Wheat & Thakar, 2005). AE data for specimens exhibiting different levels of corrosion activity and electrochemical data were compared. These specimens were made in accordance with ASTM G109 and a schematic representation of the specimens is shown in Figure 2.

The plastic dikes that initially held the sodium chloride solution for ponding were removed from the macrocell specimens. MISTRAS 2001 software and the equipment developed by the Physical Acoustics Corporation were used to collect the AE signals. R61 piezoelectric sensors were attached to the exposed areas of embedded rebar and signals were manually generated. The standard lead-break test generally used for calibration was used to generate the AE signals. The test involves breaking the point of a 0.3 mm lead on the surface of the specimens. The breaking of the lead generates sound waves that are reflected off the surface of the specimens and are consequently picked up by the sensors. The lead-break test was carried out at fixed points on the exposed rebar edges and at 2 in. (51 mm) intervals along the top side of the macrocell specimen at fixed intervals of time of 10 seconds.

Twenty macrocell specimens with varying degrees of rebar corrosion were used. The corrosion results found in the previous investigation for four of the specimens after one year of exposure to cyclical ponding are shown in Table 1. The signals generated by the tests were collected in the form of plots such as Energy vs time, Hits vs time, Amplitude vs time, etc; however,

Table 1. Data from previous investigation, (Zhang et al., 1994)

Sample Number	Water: Cementitious Material	E_{corr} (mV vs. SCE)	1/Rp (μS/cm^2)	Current (μA)
2	0.46	−230	5	4
4	0.66	−520	52	78
5	0.69	−475		50
7-CFA	0.49	−260		15

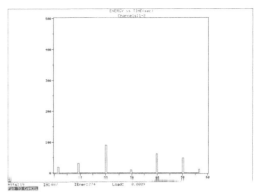

Figure 3b. Energy vs time curve for specimens with high corrosion activity.

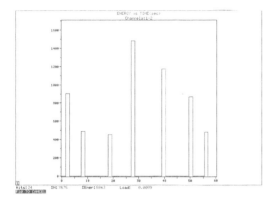

Figure 3a. Energy vs time curve for specimens with low corrosion activity.

the plots of Energy vs time were the most useful. The signals generated by the specimens that showed the highest level of corrosion activity, based on corrosion tests, were the weakest. On the other hand, signals from the specimens that showed the lowest level of corrosion activity, based on corrosion tests, were much stronger. This is consistent with the work done by Yoon and coworkers, who found that in their corrosion tests, fewer AE events were recorded as the degree of corrosion increased (Yoon et al., 2000). This seems to suggest that AE techniques may be used intermittently or on a one-time basis to characterize or differentiate corrosion activity.

The AE signals vs time plots for specimens showing low and high corrosion activity are shown in Figures 3a and 3b, respectively. The designation for specimen 7-CFA indicates that Type C fly was added to the mix.

3 EXPERIMENTAL PROCEDURE

Concrete macrocells were made in accordance with ASTM G 109 (Figure 2), except they contained a bent bar at the top. Candidate materials were cast in concrete having a width of 254 mm, a depth of 229 mm, and a height of 203 mm. Two types of concrete mixes were used. One type (Mix 1265) had a water: cement ratio of 0.57 and it was used for epoxy coated reinforcement only. The reinforcement could contain a maximum of 2% defects. The other mix (Mix

4904) contained fly ash; it had a water:cement ratio of 0.39, and a water:cementitious material ratio of 0.32. The latter concrete mix resulted in extremely good concrete; a type that might be used with inherently corrosion resistant materials. It was used for all of the types of alternative reinforcement: epoxy coated steel (bendable, E or nonbendable, NBE); polyvinyl chloride coated steel (PVC); nylon coated steel (N); galvanized steel, types A and B (GA and GB); stainless steel (SS); and stainless clad steel (SSC). The as-received bars for Mix 4904 had few defects and no attempt was made to induce additional damage. Black steel (B) was used as the control in both cases. More than one hundred specimens were made using Mix 1265 and 10 are being used in this investigation. One hundred and seventy six specimens were made using Mix 4904 and 20 are being used in this investigation.

The specimens made using Mix 4904 have been exposed to 3.5% NaCl on a cyclical basis of two weeks wet and two weeks dry for more than five years. Electrochemical tests are performed monthly, on the last day of the wet cycle. The specimens made using Mix 1265 were exposed for two years.

The specimens were also tested on a one-time basis using AE techniques. This is because of the encouraging results shown in Figures 3a and 3b, which indicated a correlation between electrochemical and AE results. The AE tests were carried out by a company that specializes in infrastructure inspection, TechCorr Inspection & Engineering. The author proposed loading the specimens using 22.5 N or 225 N. However, representatives from TechCorr suggested loading by pushing the bent bar, pulling the bent bar, or counting the number of pulses between 2 sensors placed on either side of the concrete macrocells at a separation distance of 254 mm, about 38 mm from the top surface. It should be noted that in several cases with top and bottom bars, four sensors were used; two about 38 mm from the top surface and two about 38 mm from the bottom surface. The lead-break test was used for calibration.

4 RESULTS AND DISCUSSION

The specimens were exposed to saltwater conditions and tested for at least two and as many as five years. In the specimens from Mix 1265, corrosion initiated after about one year and cracking is readily observable on the macrocells. In the specimens from Mix 4904, the electrochemical test results have changed very little over the past five years. The test results for selected specimens are shown in Tables 2 and 3. The time of exposure is given in weeks.

When other specimens from Mix 4904 were examined after two years of exposure, the values of corrosion potential and current were very similar to their present values. Those specimens were autopsied and no corrosion was evident, even on the black steel that was the control. The same was true when similar specimens containing the coated bars were examined using electrochemical impedance spectroscopy after fours years of exposure (Wheat & Liu, 2007). It is interesting to note that the corrosion potentials and current values have remained virtually the same even after five years of exposure. It should be pointed out that the bars had few defects and the concrete quality is excellent, with a water:cementious material ratio of 0.32.

When AE stimulation was attempted using 22.5 N and 225 N, there was no response. In addition, when the bent bars were pushed together or pulled apart, the results were not consistent or repeatable. However, the results based on average number of pulses from sensors placed on either side of the macrocell were consistent and repeatable. In addition, they correlated well with corrosion behavior. The AE results are also shown in Tables 2 and 3.

The specimens from the two mixes provided an excellent opportunity to compare electrochemical

Table 2. Energy vs time curve for specimens with low corrosion activity.

Sample	Ecorr (mV vs SCE)	Current (μA)	Avg No. of Pulses (based on AE)	Weeks
EB410	−47	5	70	268
NBE10	−180	7	79	268
PVC12	−44	4	75	268
N9	−51	4	77	268
GA9	−368	9	74	268
B5	−281	10	69	268
SS12	−95	5	70	268
SSC12	−192	15	53	268

Table 3. Information for Selected Concrete Macrocells (Mix 1265).

Sample	Ecorr (mV vs SCE)	Current (μA)	Avg No. of Pulses (based on AE)	Weeks
B11	−545	50	47	104
E5	−195	57	53	104
E9	−178	59	49	104

results and AE results determined in the pulse mode. Specimens from Mix 1265 showed significant corrosion activity while specimens from Mix 4904 showed very little corrosion activity. As before, corrosion activity resulted in higher currents and lower AE activity and low corrosion activity resulted in lower currents and higher AE activity. This seems to be consistent with the likelihood that the more corroded specimens have resulted in more concrete damage, more cracks, and a more open structure. It should also be noted that in selected macrocells that contained top and bottom bars, the pulses were lower in number for the sensors placed 38 mm from the top than for those placed 38 mm from the bottom. The only questionable result is the one from SSC12. The current value is still relatively low, but the average number of pulses is also low. This may in some way be related to galvanic interaction between the black rebar core and the stainless cladding.

5 CONCLUSIONS

Macrocell specimens made from two concrete mixes were used in this study. One mix (Mix 1265) was a standard mix with a water:cement ratio of 0.57. Epoxy coated reinforcement was cast in concrete made with that mix and the reinforcement could contain a maximum of 2% defect area. The other mix (Mix 4904) was made with fly ash as partial replacement for cement and had a water:cementitious material ratio of 0.32. Macrocells containing all of the alternative reinforcement materials were made using this mix of excellent concrete quality. In addition, the reinforcement had little or no damage. The specimens from Mix 4904 have been tested for at least five years.

Specimens made from Mix 1265 have undergone significant corrosion, as indicated by the high current values, and they have experienced cracking. Specimens made from Mix 4904 have shown little change in electrochemical parameters over the five-year period.

Current values are still relatively low and the potentials have remained virtually the same. This would indicate that corrosion has not occurred and that the condition of the reinforcement is still good.

When AE tests were carried out on similar concrete specimens reinforced with black steel, it was consistently observed that specimens with significant corrosion activity (and presumably extensive concrete damage) had very little acoustic emission activity while those with much less corrosion activity had substantial AE activity.

Because of these encouraging results, specimens from Mixes 1265 and 4904 were also subjected to testing using AE techniques. AE results using the pulse mode proved to be consistent and repeatable. In general, they correlated well with the results from electrochemical tests in that low corrosion activity resulted in a higher number of pulses and high corrosion activity resulted in a lower number of pulses. The issue with SSC type specimens will be investigated further.

The use of intermittent or one-time AE in the pulse mode may have the potential to be used as a complement to electrochemical tests.

ACKNOWLEDGMENTS

The author is grateful to Stan Botten from Physical Acoustics Corporation and to Wen-Yuan Yong, Mirage Thakar, and Juan Maldonado for their help on this project. The author is also grateful to Wyatt Hunter from TechCorr, who performed the AE tests.

REFERENCES

Andrade, C., Alonso, M.C., and Gonzalez, J.A.(1990), "An initial effort to use corrosion rate measurements for estimating rebar durability," in Berke, N.S., Chaker, V. and Whiting, D. (eds) Corrosion Rates of Steel in Concrete, American Society for Testing and Materials, STP 1065, Philadelphia, PA, pp 29–37.

Broomfield, J.P. (1997), Corrosion of Steel in Concrete, E. & F. Spon, London, U.K.

Chen, H. and Wheat, H.G. (1996), "Evaluation of Selected Epoxy-Coated Reinforcing Steels," Paper No. 329, Corrosion '96, National Association of Corrosion Engineers, Houston, TX.

Clifton, J.R., Beeghly, H.F., and Mathey, R.G. (1975), in Corrosion of Metals in Concrete, SP 49, American Concrete Institute, Detroit, 1975, p 115.

Dunn, S.E., Young, J.D., Hartt, W.H., and R.P. Brown (1984), "Acoustic Emission Characterization of Corrosion Induced Damage in Reinforced Concrete," Corrosion, Vol. 7, No. 7, pp. 339–343, July 1984.

Hurley, M.K., and Scully, J.R. (2005), "Threshold Chloride Concentrations of Selected Corrosion Resistant Rebar Materials Compared to Carbon Steel," Corrosion 2005, Paper No. 05259, NACE International, Houston, TX.

Ing, M., Austin, S. A, and Lyons, R. (2003), "Development of an Acoustic Evaluation Technique to Diagnose Steel Corrosion in Concrete," Research in Progress Symposium, Corrosion 2003, National Association of Corrosion Engineers (NACE) International, San Diego, March 16–20, 2003.

Kahhaleh, K.Z., Chao, H.Y., Jirsa, J.O., Carrasquillo, R.L., and Wheat, H.G (1993)., "Inspection and Determination of Damage to Epoxy-Coated Reinforcement," Research Report 1265-1, January 1993.

Kahhaleh, K.Z., Jirsa, J.O., Carrasquillo, R.L., and Wheat, H.G. (1994), "Macrocell Corrosion Study of Fabricated Epoxy-Coated Reinforcement," in Corrosion and Corrosion Protection of Steel in Concrete, Vol. 2, R.N. Swamy, Ed., Sheffield Academic Press, Sheffield, U.K., 1244–1253.

Kahns, J.T., Darwin, D., and Locke, Jr C.E. (2001), "Evaluation of Corrosion Resistance of Type 304 Stainless Steel Clad Reinforcing Bars," Report on Research sponsored by The National Science Foundation, Research Grant No. CMS-9812716, Kansas Department of Transportation Contract No. C1131, and Structural Metals, Inc., Structural Engineering and Engineering Materials SM Report No. 65, August 2001.

Li, Z., Li, F., Zdunek, A., Landis, E., and Shah, S.P. (1998), "Application of Acoustic Emission Technique to Detection of Reinforcing Steel Corrosion in Concrete," ACI Materials Journal, pp. 68–76, Jan-Feb. 1998.

Lysogorski, D.K., Cros, P., and Hartt, W.H. (2005), "Performance of Corrosion Resistant Reinforcement as Assessed by Accelerated Testing and Long Term Exposure in Chloride Contaminated Concrete," Corrosion 2005, Paper No. 05258, NACE International, Houston, TX.

Matsuyama, K., Fujiwara, T., Ishibashi, A. and Ohtsu, M. (1993), "Field Application of Acoustic Emission for Diagnosis of Structural Deterioration of Concrete," Journal of Acoustic Emission, Vol. 11, No. 4, pp. S65–S73.

McCarter, W.J., and Vennesland, O. (2004), "Sensor systems for use in reinforced concrete structures," Construction and Building Materials, Vol. 8, Number 6, 2004.

McDonald, D.B., Sherman, M.R., and Pfeifer, D.W. (1995), "The Performance of Bendable and Nonbendable Organic Coatings for Reinforcing Bars in Solution and Cathodic Debonding Tests," Report No. FHWA-RD-94-103, Federal Highway Administration, Washington, D.C., January 1995.

Muravin, G., Lezvinsky, L., and Muravin, B. (1996), "Nondestructive Testing and Diagnostics of Reinforced Concrete Structures by the Acoustic Emission Method," in Concrete in the Service of Mankind; Concrete Repair, Rehabilitation and Protection, R.K. Dhir and M.R. Jones, Eds., E & FN Spon, London, U.K.

Ohtsu, M. and Tomoda, Y. (2006), "Quantitative NDE of Corrosion in Reinforced Concrete by Acoustic Emission," Proceedings of Concrete Solutions 2006 (Second International Conference on Concrete Repair), St. Malo, France. BRE IHS Press, 2006.

Tuutti, K. (1982) "Corrosion of Steel in Concrete," Technical Report, Swedish Cement and Concrete Research Institute, Stockholm, Sweden, 1982.

Virmani, Y.P., Clear, K.C., and Pasko, Jr., T.J. (1983), "Time-To-Corrosion of Reinforcing Steel in Concrete Slabs, V.5: Calcium Nitrite Admixture or Epoxy-Coated Reinforcing Bars as Corrosion Protection Systems," Report No. FHWA/RD083/012, Federal Highway Administration, Washington, D.C., September, 1983.

Wheat, H.G. and Thakar (2005), "Efforts to Determine the Relationship Between Corrosion Rate and Concrete Damage Using Acoustic Emission," Journal of ASTM International, Volume 2, No. 8, September, 2005.

Wheat, H.G. (2006), "Laboratory and Field Observations of Corrosion of Alternative Reinforcement Materials," National Association of Corrosion Engineers (NACE) International, Corrosion "06, Paper 355, San Diego, CA, March 2006.

Yoon, D.-J, Weiss, W.J., and Shah, S.P. (2000), "Detecting the Extent of Corrosion with Acoustic Emission," Transportation Research Record 1698, Paper No. 00–1425.

Yuyama, S., Nagataki, S., Okamota, T., and Soga, T. (1990), "Several AE Sources Observed During Fracture of Repaired Reinforced Concrete Beams," in Progress in Acoustic Emission V Proceedings of the 10th International Acoustic Emission Symposium, Sendai, Japan, October 22–25, 1990, Yumaguchi, K, Takahashi, H. and Nitsuma, H, eds., The Japanese Society for Non-Destructive Inspection Publishers.

Zhang, H., Wheat, H.G., Sennour, M.L., and Carrasquillo, R.L. (1994), "Electrochemical Testing of Reinforced Concrete," Corrosion and Corrosion Protection of Steel in Concrete, International Conference on Corrosion Protection of Steel in Concrete, University of Sheffield, Sheffield, England, edited by R. Narayan, published by Sheffield Academic Press, Vol. 2, pp 721–730, July 24–29, 1994.

Patch repairs

Concrete Solutions – Grantham, Majorana & Salomoni (Eds)
© 2009 Taylor & Francis Group, London, ISBN 978-0-415-55082-6

Improving the bond cohesion of patch repair material and substrate with nano-fibres

P. van Tonder, D. Kruger & J.J. Duvenage
Department of Civil Engineering Science, University of Johannesburg, Johannesburg, South Africa

ABSTRACT: Nanotechnology can be seen as the next industrial revolution. Great breakthroughs have been made in the last 25 years, in the chemical, physical and biological sectors. The effectiveness of nano-technology in construction is a recent area of interest.

This paper deals with the research of adding carbon nano-fibres to the substrate and patch repair material. The research was conducted to see what effect the introduction of carbon nano-fibres had on the bonding between the concrete substrate and the patch repair material. Furthermore, it was investigated whether the properties of the nano-fibres could be utilized to increase the longevity and durability of the patched area. The carbon nano-fibres were added to the bonding surface in different ways, and a simple tensile test was performed to determine the tensile loads reached, and compared to those specimens not having carbon nano-fibres. This paper summarises all the results obtained and presents the conclusions of this research.

1 INTRODUCTION

The repair of concrete surfaces has become an everyday occurrence in today's lifestyle. Both new and old concrete structures show exhibit fatigue that minimizes the service life of the structure. Carbon nanofibres have excellent mechanical properties that may be used in concrete to make it an excellent material of tomorrow (Balaguru, 2006). The addition of the nanofibres to a concrete mixture may have the potential to enhance the general strength of the concrete and prevent crack propagation in cement composites (Nanoscience And Nanotechnology In Cementitious Materials, 2004; Makar et al, 2005). If the concrete can be reinforced by using nanofibres, theoretically the formation of cracks can be interrupted as soon as they start to initialize.

If the strength of concrete can be improved by the addition of carbon nanofibres, the addition of the nanofibres in the patch repair process may yield the same increase of strength as the patch repair.

The aim of this study was to determine if carbon nanofibres and graphite fibres can improve the bond cohesion between the substrate and the patch repair material.

2 EXPERIMENTAL PLANNING

2.1 *Fluorination*

Untreated carbon nanofibres and graphite fibres have dispersion problems when added to water. To improve the fibre/matrix adhesion of the fibres, the surface polarity of the fibres needs to be increased and more sites for hydrogen bonding are needed (Bismark et al,

Sept 1997). This is done in order to accomplish good shear transfer from the matrix material and the filling fibre material.

Fluorination is a process in which fluorine is introduced to the nanofibres. Fluorination can be seen as an alternative technique to increase the polarity of the carbon. After fluorination the carbon nanofibres and graphite fibres can disperse in water.

2.2 *Mixture design*

Table 1 shows the mixture design of the "old" concrete which served as the concrete requiring patch repair. The same mixture design was used for the concrete with carbon and graphite fibres. The details of the mixes are given in Tables 1 and 2.

Cubes were cast in "dog bone" moulds. Each specimen was cast using the mixture design in table 1. One-half of the specimen was cast first. The following day the samples were removed and cured in water at a constant 22°C. After seven days the specimen was removed and the bonding surface was roughened by using a chisel and hammer. This simulated a dry joint between the new and previously cast concrete. Five

Table 1. Mixture design.

Material	Quantity	
	kg/m³	l/m³
PPC 42.5R Cement Top	450.0	
Sand	1406.2	
Stone	0.0	
Water		225.0

Table 2. Mixture design.

Percentage	Quantity ml
1%	3.2
3%	9.6
5%	16.1

specimens were cast. This sample served as the benchmark for the rest of the tests with carbon nanofibres (CNF) and graphite fibres added.

Table 2 shows the quantity of carbon nanofibres and graphite fibres for 1 litre of concrete mix.

2.3 Casting of the cubes

For the carbon nanofibre and graphite fibre additions the same procedure was used as described for the normal specimens. The only difference was the second half of the mould contained the concrete with carbon nanofibres and graphite fibres added in 1%, 3% and 5% of the cement volume. Five specimens were cast for each carbon nanofibre percentage.

The final specimens cast consisted of nanofibres added directly to the bonding plane. The same procedure was used as described for the normal specimens. The difference was that, after the bonding plane was roughened, a thin layer of fibres was added to the plane. The specimens were then placed back into the moulds and the other half was cast by using the mixture design in table 1. This was done both for the carbon nanofibres and graphite. Five blocks for each were cast.

After casting, the specimens were left for seven days to cure in a bath at 22°C. After this period the specimens were removed from the bath and the tensile strength between the concrete and the substrate was determined.

2.4 Testing procedure

A direct tensile test was used to perform tests on the specimens. This was done by using a tensile testing machine. The machine pulled the vertically-orientated specimens apart at a constant rate of 0.25 mm/s.

The SABS standard specifies the following procedure as described in the SANS 6253:2006 code[5]:

- The specimen must first be removed from the bath and all the water and grit from the surface removed. The specimen needs to be weighed to determine its mass and tested directly after it has been removed from the bath and cleaned.
- The steel loading fixtures of the tensile testing machine are cleaned to ensure no inaccurate readings.
- The specimen is placed centrally in the tensile testing machine making sure the testing plates are parallel to each other.
- A constant load increase of 0.03 MPa/s ± 0.01 Mpa/s needs to be applied to the specimen until failure of the specimen occurs. The maximum load applied to the specimen is then recorded.

Table 3. Statistical analysis.

Specimen Type	Mean	Specimen Standard Deviation
Normal	1.595	0.575
CNF 1%	1.236	0.239
CNF 3%	1.345	0.344
CNF 5%	1.436	0.375
CNF Surface	1.103	0.160
Graphite 1%	1.151	0.205
Graphite 3%	1.635	0.434
Graphite 5%	1.285	0.639
Graphite Surface	1.780	0.168

Table 4. Simplified Tensile Test Results (MPa).

Specimen Type	1	2	3	Max	Average
Normal	1.813	1.825	2.064	2.064	1.901
CNF 1%	1.112	1.291	1.368	1.368	1.257
CNF 3%	1.082	1.397	1.637	1.637	1.372
CNF 5%	1.433	1.558	1.641	1.641	1.544
CNF Surface	1.026	1.075	1.214	1.214	1.105
Graphite 1%	1.029	1.100	1.351	1.351	1.160
Graphite 3%	1.246	1.323	1.435	1.435	1.335
Graphite 5%	0.825	1.441	1.339	1.441	1.201
Graphite Surface	1.706	1.739	1.783	1.783	1.743

3 RESULTS

3.1 Statistical analysis

The variance and standard deviation for each series was determined. The following results where obtained showed in Table 3:

Using the values obtained in Table 3, the data were simplified to only three samples per series. These results can be seen in Table 4. All the outliers were removed.

The results were plotted for each separate series and compared this against the benchmark (normal) sample results.

3.2 CNF results

The normal values in Table 4 were used as the benchmark values against which the separate percentage of carbon nanofibre addition was tested.

These values were plotted and shown in Figure 1, it can bee seen that the addition of the fibres make the bonding strength between the substrate and the patch repair material weaker, but the strength values increase upwards from 1% to 5%. Unfortunately, the specimen was restricted to 5%. It would have been desirable to have extended the range up to 10% carbon nanofibre addition. Further testing is recommended.

Due to the reinforcement addition limit of 5%, a trend line was extrapolated to determine at which

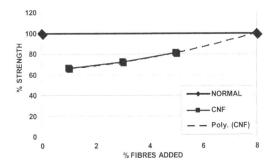

Figure 1. CNF% vs. Normal%.

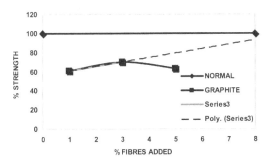

Figure 2. Graphite% vs. Normal%.

percentage the fibres will become effective in increasing the bond cohesion. The trend line in Figure 1 indicated that at approximately 8% addition of carbon nanofibres, the CNF results will reach the benchmark results for the unreinforced specimen.

3.3 Graphite results

The normal values in Table 4 were used as the benchmark values against which the separate percentage of graphite addition was tested. The values for the addition of graphite nanofibres in table 4 are plotted on as shown in Figure 2. The highest strength was achieved for 3% addition of fibres.

At 5% the strength seemed to drop again. A trend line was drawn, but these results need to be verified.

3.4 Surface application results

The following results were obtained for the addition of graphite and carbon nanofibres directly to the bonding plane as seen in Table 4. The values for the different surface applications in Table 4 are plotted and shown in Figure 3.

The values in figure 3 show that both the graphite surface specimens and the CNF surface specimens application did not reach the benchmark (normal) value. This could have been due to too much or too little fibre added to the bonding plane. The surface

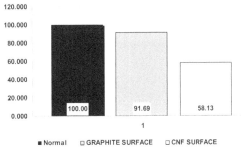

Figure 3. Surface application% vs. Normal%.

roughness of the samples may also have played a role. Further testing is needed to determine if the application of nanofibres directly to the bonding plane can increase bond cohesion.

4 CONCLUSIONS

In the test done for the 1%, 3% and 5% of nanofibres added, as well as the addition of fibres directly to the bonding plane, the benchmark value wasn't reached. The results show that the addition of nanofibres to the mix made the bonding strength between the substrate and the patch repair material weaker, however the trend line suggests that the addition of nanofibres can be effective, but only at larger percentage addition of nanofibres.

The results obtained for the addition of fibres directly to the bonding plane showed that both the graphite and CNF didn't reach the benchmark value. This may be due to too much or too little fibre being added to the bonding plane. Further research is necessary to confirm if adding more or less fibre at the bonding plane will increase the bond between the patch repair material and substrate.

REFERENCES

Balaguru, P.N. Nanotechnology And Concrete. Background Opportunity and Challenges. Rugters the State University of New Jersey, United States of America, Proceedings of the Nanotechnology in Construction, 2006

Bismarck et al; Influence Of Fluorination On The Properties Of Carbon Fibres; Journal of Fluorine Chemistry 84 (IYY7), pp127–134, Sept 1997

Makar, J.; Margeson, J.; Luh, J.; Performance, Innovations and Structural Implications 3rd International Conference on Construction Materials:, Vancouver, B.C., Aug. 22–24, 2005, pp. 1–10

Nanoscience And Nanotechnology In Cementitious Materials; Cement International, 4/2004, vol4, pp 81–85

SANS 6253:2006; Concrete tests – Tensile splitting strength of concrete; South African National Standards

Concrete Solutions – Grantham, Majorana & Salomoni (Eds)
© 2009 Taylor & Francis Group, London, ISBN 978-0-415-55082-6

Long term evaluation of concrete patch repairs

J. Pretorius
Department of Mechanical Engineering Science, University of Johannesburg, South Africa

D. Kruger
Department of Civil Engineering Science, University of Johannesburg, South Africa

ABSTRACT: The vast majority of concrete structures have performed satisfactorily for many years and concrete was, until recently, seen as a maintenance free building material. This perception together with the expectation that university buildings should last for hundreds of years resulted in the decision to use off-shutter concrete as the principle building material and finish for the buildings of the University of Johannesburg. Terms such as concrete durability, concrete carbonation, acid rain and related concrete construction problems were unknown when the construction of the University of Johannesburg was in full swing during the early 1970's. Environmental attack is currently receiving a lot of attention as it is substantially reducing the lives of many concrete structures around the world as is the case with the buildings of the University of Johannesburg. As the exposed concrete was considered to be maintenance free, very little time and money were spent on the maintenance of the external facade of the building during these years.

Different combinations of concrete patch repair, corrosion inhibitors and coatings were applied during a long term pilot project addressing concrete patch repair at the University of Johannesburg. Pre-repair as well as post repair concrete conditions were carefully noted. Ten years later another assessment was made regarding the performance of the various combinations used during the initial repair. This allows the researcher to evaluate the long term performance of the various concrete repair combinations providing some interesting insights.

1 INTRODUCTION

1.1 General remarks

During the last fifty years enormous numbers of structures have been built worldwide with ever increasing speed of design and construction, which has had a detrimental effect on the durability of concrete. Population growth throughout the world with the passage of time after World War II, directly coupled with the energy and consumer needs of the people, have led to the creation of a much more aggressive environment due to pollution. The combined effect of poor concrete durability and a more aggressive environment has resulted in an increase in the rate of concrete degradation. Although many concrete structures have performed well and have shown excellent durability over many years, many more are showing signs of deterioration caused primarily by reinforcement corrosion. (St John et al. 1998)

A lack of understanding of the corrosion mechanism in reinforced concrete and inadequate design and construction provisions for durability has resulted in an increasing incidence of concrete durability problems in recent decades. In many parts of the world infrastructure maintenance takes preference to infrastructure provision through new construction. This is proved by the five hundred million Pound estimated annual concrete repair bill for the United Kingdom while in the United States of America ten million Dollars of research has been undertaken by the Strategic Highway Research Programme to address bridge deck repair and rehabilitation. (Neville. 1995 & Mays. 1992)

South Africa currently produces some thirty million cubic meters of concrete and mortar each year making it probably the most widely used construction material locally and world wide when considering both volume and diversity. (Addis. 1998) The vast majority of concrete structures have performed satisfactorily for many years and concrete was until recently seen as a maintenance free building material. This perception together with the expectation that university buildings should last for hundreds of years resulted in the decision to use off-shutter concrete as the principle building material and finish for the buildings of the University of Johannesburg. (Kruger & Pretorius. 1996)

Terms such as concrete durability, concrete carbonation, acid rain and related concrete construction problems were unknown when the construction of the University was in full swing during the early 1970's. At the time of construction, the campus was a dream project for any consultant or contractor. The capital worth, in terms of today's values, was in the order of one billion Rand. Design and construction of the project was completed over a period of five years using novel rapid construction techniques, which included a number of design and construction innovations. Time constraints, under which the project was planned, designed and constructed, resulted in very little attention being given to aspects such as ensuring durable concrete and adequate cover over reinforcing steel. Environmental attack is currently receiving a lot of attention as it is substantially reducing the lives of many concrete structures around the world as is the case with the buildings of the University of Johannesburg. (Kruger & Pretorius. 1996)

Over the following twenty-five year period, the inadequate concrete cover has exposed itself through cracking, the leaching of corrosion products and spalling. As the exposed concrete was considered to be maintenance free, very little time and money were spend on the maintenance of the external facade of the building during these years. Ongoing evaluation and diagnostic studies undertaken on the University buildings indicated that carbonation of concrete and inadequate cover of the reinforcement steel could be identified as the main causes of the concrete degradation and reinforcing steel corrosion, taking place on the campus buildings. Spalling of concrete on high locations and the aesthetically unacceptable appearance of cracks and leaching gave rise to concern.

Concrete repair and rehabilitation measures are extremely costly and the difficulty in fully determining the extent of the problem is well known. The decision on a suitable repair strategy is also a very difficult one and therefore it was decided to initiate a pilot project on a certain portion of the campus buildings in the hope to gain answers to these questions.

1.2 *Concrete as construction material*

Concrete has generally been regarded as chemically and dimensionally stable in most environments leading to the belief that concrete is an inherently durable and versatile material. It has a high compressive strength, abrasion resistance and fire resistance; it protects embedded steel from corrosion, and it is durable and economic. Concrete's strength lies in the compression forces it can withstand while it's weakness in tension could only be improved with the addition of embedded steel in the concrete matrix, as was done towards the end of the nineteenth century. This perception of concrete being an inherently durable material is also associated with reinforced concrete structures, which are expected to be relatively maintenance-free during their service life. It is because of this belief that many buildings in South Africa including the University of Johannesburg, have been built using reinforced concrete as the primary building material. (Kruger & Pretorius. 1996)

Textbooks as recent as 1976 on the design of reinforced concrete structures contained remarks such as 'compared with other building materials reinforced concrete is distinguished for its very long service life. This arises from the fact that the strength of concrete does not decrease with time but on the contrary it increases with time, and the steel embedded in it is protected against corrosion.' Today such statements appear somewhat naïve although they are in fact essentially true but need to be qualified. Concern was also expressed in the 1980's about the implication of the gradual increase of strength of Portland cement with a possible effect on the durability of the concrete if the concrete mix was based on a strength only specification. This would imply that the required strength could be achieved with considerably lower cement content and a higher water-cement ratio than normally used having serious implications for the durability of the concrete. Concrete should therefore be properly specified to ensure durability to the environment to which it will be exposed. This means specifying minimum cement contents, maximum water-cement ratios and concrete strength, which are compatible with the former two characteristics. (Mays. 1992)

Until very recently, little if any attention was given to the maintenance of concrete structures, as it was believed that these structures were maintenance free and built to last indefinitely. Although many concrete structures have shown excellent durability performance throughout the years, there is an increasing number that have deteriorated rapidly after construction. Many examples of early damage to reinforced concrete exist, primarily as a result of premature corrosion of the steel reinforcing bars, which led to investigations concerning the deterioration of reinforced concrete.

A survey concluded in 1954 showed that even on a moderately exposed site, the trouble free life of reinforced concrete is likely to be short if the cover to the reinforcement is meager. This also holds true if the binding wire projects towards the surface or if for any reason the concrete is pervious to moisture. (Mays. 1992) In the rapid expansion of the construction industry during the 1960's such lessons were often forgotten and today many modern concrete structures need substantial repairs and maintenance during their service life with the resultant costs to the economy reaching 3–5% of the Gross National Product in some countries. (Hewlett. 1998) The growing backlog of unserviceable structures is draining the economies of many countries world-wide.

Actual deterioration of concrete has only recently been accurately quantified and the extent of the problems, regarding the premature corrosion of reinforced steel bars, realised. This premature corrosion of steel reinforcing can be attributed to a number of factors influencing the durability and degradation of the concrete. Assuming concrete to be forever durable has proven to be extremely foolish, as it is known today that maintenance needs to be done on concrete structures in order to ensure the longevity of the structure.

2 PATCH REPAIR PILOT PROJECT

2.1 *Location of pilot project*

The location for pilot repair project was carefully chosen as concrete repair work needed to continue

Figure 1. Eastern and western facing facades of laboratory buildings.

Figure 2. Nine panels per eastern and western facing facades of laboratory buildings.

while students were attending classes. Inconvenience to building users was assessed during the project while results obtained from the study would be used to evaluate complete concrete repair work to be undertaken on all of the University of Johannesburg buildings.

It was decided to use one of the three story laboratory buildings situated on the western side of the campus. This decision was based on the fact that the laboratory buildings are comprised of an eastern and western facing façade as can be seen from Figure 1.

Each side consisted of nine panels per floor level at three different levels as can be seen from Figure 2. Panels were sized at nine meters long by two meters high.

The panels are clearly separated by construction joints which end in nibs as can be seen from Figure 3.

This relates to fifty four similar sized panels which could be used for the evaluation of different repair strategies using various repair materials, corrosion inhibitors, and concrete coatings. Evaluation could be conducted directly after completing the concrete repair as well as over a longer time period. An additional advantage of selecting this specific location for the concrete repair pilot project was that student movement through the building would not be impaired.

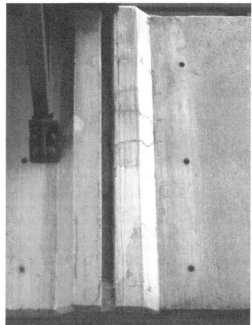

Figure 3. Concrete degradation on nibs clearly separating panels in the laboratory buildings.

2.2 Pre repair evaluation

Pre repair evaluation started by high pressure water cleaning of all the facades in order to facilitate a visual inspection. All facades were visually inspected and twenty seven different irregularities such as efflorescence, honeycombing, discoloration, shutter lines and the positions of core samples were identified and recorded on data sheets for each panel.

Six core samples per panel were taken to determine depth of carbonation. These samples consisted of three in the top third and three in the bottom third. The samples were also spaced at thirds across the length of the panel. Results obtained varied greatly with carbonation depths starting at as little as 5 mm right through to 30 mm. Variations not only occurred from panel to panel but also within measurements from the same panel.

A cover-meter survey was conducted on each panel with node points at 500 mm centre to centre throughout the length and height of each panel. The results indicated that cover depth of reinforcing steel also varied greatly from as little as zero cover to 45 mm cover.

In addition to the cover-meter survey, the core samples and the visual inspection an electro-potential contour plot for each panel was done to try and determine the distribution and probability of reinforcement corrosion taking place as can be seen from Figure 4.

2.3 Patch repair methodology

All areas where spalling of concrete occurred and where cores for carbonation depth determination were

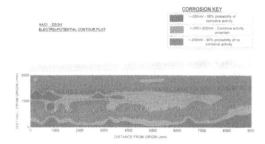

CORROSION KEY

Figure 4. Typical electro-potential map.

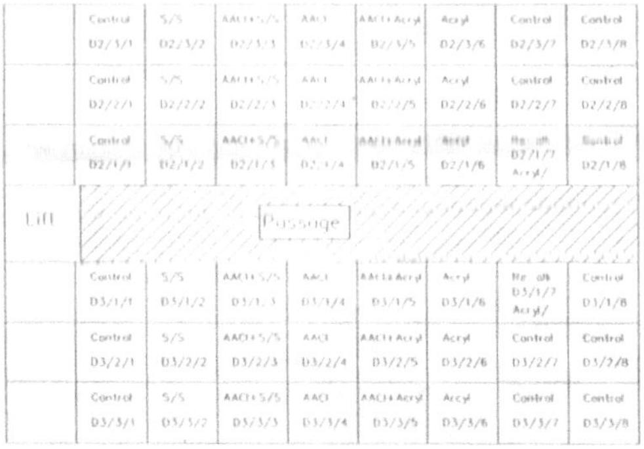

Figure 5. Layout of coatings and applications.

taken were identified as patch repair areas. These patch repair areas were limited in number and extent on the panels, however much repair work was identified on the nibs separating the panels.

Patch repair areas were demarcated and the edges were cut using an angle grinder to eliminate feather edges. An electrical concrete breaker was used to remove the affected concrete in the area of the corroded reinforcing steel. Concrete was removed to a depth of 20 mm behind the reinforcing steel in order to enable access to completely clean the reinforcing steel.

Shot blasting was used to clean corrosion residue from the reinforcing steel as well as to remove any loose concrete debris due to micro-cracking.

Due to the high daytime temperatures and dry windy conditions in Johannesburg during the spring months, all patch repair areas were thoroughly soaked with water before patch repairs commenced. This was required to ensure adequate water availability and adhesion when patch repairs were conducted.

2.4 Patch repair materials

Three different repair materials were used as patch repair material during the pilot repair project. For purposes of this report the repair materials are labeled type A, B and C. To ensure consistence only one repair material (Type A) was used throughout the pilot repair project with the other two materials (Type B and C) being used as control samples.

Repair material A is defined as a non-shrink, multi purpose patch repair material with rapid strength development characteristics. It is packaged as a one pack easy to use product which has the appearance of concrete and is waterproof. Reinforcing steel within the patch repair area needed to be treated with a zinc rich primer while a slurry coat should be applied to the concrete substrate directly before applying the patch repair material A.

A patch repair material of a different manufacturer was chosen to represent repair material B. This non-shrink repair material is classified as a single component medium-weight concrete and masonry reinstatement mortar for use in hot climates. This repair material is said to be suitable for a wide range of concrete and masonry repair, particularly in vertical

and overhead locations without the use of formwork. Repair mortar B is said to be suitable where superior chloride and carbon dioxide resistance is required.

Reinforcing steel within the patch repair area needed to be treated with a zinc rich primer while a primer coat was applied to the thoroughly soaked substrate before applying patch repair material B.

A third single component polymer modified, fibre reinforced, cement based repair mortar was used and labeled repair material C. This repair mortar was classified as being free of chlorides, non-shrink with good bonding strength to concrete. As with material A and B this repair mortar manufacturer specifies their own primer on exposes reinforcing steel and wetting of the concrete substrate at least twelve hours before application while their own primer compound is recommended before applying the repair mortar.

2.5 Concrete repair methodology and coatings

The concrete repair methodology applied in the repair project was consistent with concrete repair methodology used throughout the world. However, many different concrete repair materials as well as concrete protective coatings and applications are available on the market at the time of the concrete repair pilot project. It was therefore decided to use different combinations of repair materials, protective applications and protective coatings and to evaluate the different combinations on a long term basis. As the university buildings primarily consist of off shutter concrete, the ideal situation would be to move away from acrylic based coatings which would introduce a maintenance component.

Figure 5 indicates the layout of the laboratory buildings together with the numbering system used as well as the coatings used. The very first column of blocks, three above the lift and the three blocks below the lift were not numbered as no repairs were done to these panels at all. They were left as is in order to compare with repaired sections on a long term basis.

Each panel was numbered starting by identifying the position of the panel in either the D2 Lab or D3 lab block. Panels in the D2 block face in a westerly direction while panels in the D3 block face in an easterly

Figure 6. View of D2/1/1 after completion of concrete repair.

direction. The second number indicates the vertical position of the panel, either on the first, second or third floor. Panel position away from the lift shaft is indicated by the last digit starting with 1 right through to the eighth panel away from the lift.

The first set of panels from floor one to three on both sides of the laboratory building was viewed as concrete repair control panels. These panels were cleaned using high pressure water and repaired using repair material C. No protective substances or coatings were applied to these panels. The final result can be seen in Figure 6.

Concrete repair using material A was done on both sides of the laboratory. The extent of the use of repair material A ranged from the second set of panels right through to the sixth set as well as the eighth set, from floors one to three on both sides of the laboratory. In addition to the concrete repair, a catalysed silane/siloxane mixture was applied by flooding the surface with two coats wet-to-wet. The silane/siloxane mixture is a water repellent treatment which repels water from the concrete resulting in keeping the concrete dry.

The third numbered set of panels were treated with a combination of a Migrating Corrosion Inhibitor (MCI) and a silane/siloxane. The MCI was applied after the concrete repair had been completed. Before the silane/siloxane combination was applied the MCI application was conducted according to the manufacturer's specification.

A Migrating Corrosion Inhibitor had been applied exclusively on the fourth numbered set of panels once the concrete repair work had been completed. No change of the natural concrete colour had occurred up to this point.

The next set of panels was treated with both the MCI as well as a pure acrylic coating after concrete repair had been concluded. The acrylic coating system has good adhesion properties to most substrates and possesses excellent weathering properties especially with regard to resistance to ultra violet attack. Colour change from the original concrete colour is unavoidable in addition to the aspect of maintaining the acrylic coating.

After concrete repair had been completed on the sixth set of panels an acrylic coating was applied exclusively. This set of panels was treated with exactly the same acrylic coating as used in the set of panels where the MCI was used.

Panel set numbers seven and eight were designated to be repaired using two different repair materials without using any additional coatings or treatments. Concrete repair on panel seven was conducted using repair material B while repair material A was used on panel eight. The panels on the first floor on both sides of the laboratory of panel set seven were identified to be treated by making use of the process of re-alkalisation. Panel set number one, seven and eight were viewed as control panels in order to evaluate the performance of different concrete repair materials without coatings or treatments.

3 LONG TERM EVALUATION

3.1 *Initial considerations*

The concrete repair pilot project was initiated by the then Rand Afrikaans University. In 2005 the Rand Afrikaans University merged with the Technicon Witwatersrand resulting in the birth of the University of Johannesburg. Financing of the pilot repair project was channeled through the operations department of the Rand Afrikaans University.

The pilot project was initiated in 1996 with the repair work being completed by the end of 1997. One of the reasons for conducting the concrete repair pilot study was that many external facades of the University buildings showed signs of concrete degradation. A major concern to the university operations department was that there was as no definite answer as to which method of repair or which repair material would provide the best results.

3.2 *Results*

A long term evaluation was conducted ten years after the pilot project had been completed. As the institutional powers that be had changed, the evaluation was done from a research perspective instead of from an operations perspective. This therefore impacted on the financial means available to do the long term evaluation.

Aesthetics of the concrete repair was an important consideration. As can be seen from Figure 7 some of the repairs blended in extremely well with the substrate colouring.

However some of the repairs stood out quite prominently as can be seen from the repairs on the right hand side of Figure 8.

Figures 7 and 8 indicate the visual effect of the repaired area using repair material C. All panels in this series did not show any signs of new concrete degradation in the form of spalling.

Panels treated with the silane / siloxane combination showed no signs of recurring concrete degradation as

Figure 7. View of D2/1/1 ten years after completion of concrete repair.

Figure 8. View of D2/1/2 ten years after completion of concrete repair.

Figure 9. View of D2/2/1 during long term evaluation.

can be seen from Figures 9 and 10. These panels were repaired using repair material A and after 10 years repairs have blended in well with the substrate.

In panels where the migrating corrosion inhibitor and the silane / siloxane were applied, no evidence could be found of recurring concrete degradation.

Figure 10. View of D2/2/2 during long term evaluation.

Figure 11. Close up view of interface between D2/4/2 and D2/5/2 during long term evaluation.

As can be seen from Figure 11, some concrete degradation had started on the nib interface and spalling is imminent. This panel had only been treated with a migrating corrosion inhibitor. The onset of spalling was also evident on the adjacent panel where an MCI and an acrylic coating had been used. Investigation however showed that a water supply had steadily been dripping water onto the interface of the nib from the floor above. Apart from this apparent degradation as a result of abundant water ingress, no other signs of further concrete degradation could be found on the panels covered with an acrylic covering except for the one small patch indicated in Figure 12.

Repair material A was used to conduct the repairs on the eighth set of panels. The western facing panel on the second floor showed some signs of new concrete degradation as can be seen in Figure 13.

The panel on the top floor of the eighth set of panels repaired with material A showed some of the patches starting to delaminate together with some evidence of new delamination starting as can be seen in Figure 14. Patches showing delamination were close to the top of the panel which could indicate that water ingress played an important role.

It was found that in areas where panels were subjected to water, a black algae growth had returned as

Figure 12. View of D2/2/8 during long term evaluation.

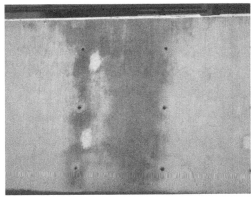

Figure 15. View of D2/4/2 during long term evaluation.

Figure 13. View of D2/2/8 during long term evaluation.

Figure 16. View of un-numbered control panel which had no repair done to it.

The onset of spalling could be seen in numerous places all over the panel shown in Figure 16. Exactly the same could be said for the other un-numbered control panels.

Visible concrete degradation results indicated in the photographs outlined in this report were the only evidence found of continued long term degradation after a period of 10 years since completion of the pilot concrete repair project.

4 CONCLUSIONS

It was clear that panels that had been used as control panels and were left untouched definitely showed accelerated signs of degradation. Long term evaluation of these panels indicates the typical state in which the panels that were repaired would have degraded to further if concrete repair had not been initiated.

Ten years after the concrete repair project had been completed by making use of three different repair materials as well as various concrete protection applications, it can be concluded that no real difference existed at this stage between the different repair methodologies applied.

Figure 14. View of D2/8/3 during long term evaluation.

can be seen from Figure 15. The extent of the original black algae can be seen from Figure 15 where only the panels numbered 1 had been cleaned. The rest of the panels still required cleaning before concrete repair work commenced.

The unnumbered panels which were left uncleaned, untreated with no concrete repairs showed definite signs of further degradation as can be seen in Figure 16.

It is therefore recommended that degradation evaluations be continued and since signs of the start of degradation on some of the repaired panels is apparent, the frequency of inspection and reporting be increased.

REFERENCES

St John, DA; Poole, A.W; Sims, I; Concrete Petrography – A Handbook of Investigative Techniques. Arnold. London. 1998.

Neville, A.M; Properties of Concrete. Fourth Edition. Longman. 1995.

Mays, G; Durability of Concrete Structures – Investigation, Repair and Protection. Edited by G. Mays. E & FN Spon. London. 1992.

Addis, B; Fundamentals of Concrete. Published by the Cement and Concrete Institute. 1998

Kruger, D; Pretorius P.C; Rehabilitation of Concrete: The Owners Dilemma. Guest Lecture Cement and Concrete Institute. March. 1996.

Hewlett, P.C; Lea's Chemistry of Cement and Concrete. Fourth Edition. Arnold. London. 1998.

Concrete Solutions – Grantham, Majorana & Salomoni (Eds)
© *2009 Taylor & Francis Group, London, ISBN 978-0-415-55082-6*

Controlling cracks in concrete repair materials by using nanofibres as reinforcement

P. van Tonder, D. Kruger & J.H. Strydom
Department of Civil Engineering Science, University of Johannesburg, Johannesburg, South Africa

ABSTRACT: The combined effect of inadequate concrete durability and a more aggressive environment has resulted in an increase in the rate of concrete degradation. Much time and money is allocated to infrastructure maintenance and repair which could have been better utilized for infrastructure provision. Unfortunately, repair materials have in most cases also posed additional problems because of inadequate durability.

This paper deals with research using carbon nano-fibres to reinforce and enhance concrete repair materials. The critical problem encountered with cement-based repair materials is often early age plastic shrinkage. In an experimental project, substrate bases were cast and cured until fully hardened. An overlay of carbon nano-fibre reinforced concrete repair material was placed on this substrate base with protuberances and subjected to controlled drying conditions. Cracks appearing on the overlay were measured after two days and classified accordingly. Bases with unreinforced concrete repair material overlays were compared to bases with standard fibre reinforced concrete repair materials as well as bases with fluorinated carbon nano-fibre reinforced overlays. This paper presents these results and draws a conclusion as to the application of nano-fibres as reinforcement in repair materials.

1 INTRODUCTION

Most civil engineering structures utilize concrete as main construction material due to the advantages of ease of casting concrete in different shapes and sizes as well as concrete's excellent performance under compression forces. Due to inadequate design, unreliable contractors and lack of proper construction control, the specified cover over reinforcement is not always provided and as a result such concrete structures degrade due to corrosion of the reinforcing steel leading to cracking and spalling of the surface concrete over time. This process is accelerated when the structure is exposed to aggressive environments such as marine, high humidity, chemical or other corrosive conditions, In many countries, a large portion of the construction budget is spent on rehabilitation and repair of such concrete structures.

Cement based composite materials such as fibre reinforced mortars (FRM) or fibre-reinforced concretes (FRC) are increasingly being used as repair materials and are often applied as an overlay (Mindess, 1990). The aim of such an overlay is to provide additional cover over the reinforcing steel to prevent corrosion. To be effective, such overlay materials should prevent ingress of corrosive agents and should thus be dense with a high resistance to micro cracking. There are a variety of fibres available to reinforce these repair materials in order to prevent or retard cracking. Such fibre materials were already added to concrete in the early 1900's and the technology has now advanced to a state of sophistication using micro- and nanofibres.

The advantages of using fibres as additional micro-reinforcing in concrete when compared to conventionally reinforced concrete, are the higher post cracking loading capacities and improved toughness with higher energy absorption. It also results in a large reduction of shrinkage cracking of concrete in both the plastic and drying stages (Turatsinze, 2007)

Nanofibres are defined as a fibre having at least one dimension of 100 nanometer (nm) or less. The name is derived from the scientific measurement unit representing a billionth of a meter, or three to four atoms wide, a nanometer. Nanofibres generally have a diameter of less than one micron. Nanofibres typically used in concrete include carbon nanotubes and graphite fibres. Carbon nanotubes are a form of carbon which are cylindrical in shape and graphite fibres are platelets which are perfectly arranged in various orientations with respect to the fibre axis.

2 EXPERIMENTAL SETUP

To effectively study shrinkage induced cracking in cement based repair materials, it is essential for the tests to represents stress conditions similar to those experienced in the field. A technique producing realistic shrinkage conditions was recently developed by researchers at the University of British Columbia (UBC) (Banthia, 1996, 2000). In this method a

Removable Spacers

Figure 1. Plexiglass mould (400 × 100 × 100 mm).

Table 1. Substrate base mix design.

Mix design (bases)	Substrate (kg/m^3)
Cement (CEM 142.5R)	402
Water	225
Sand	486
Aggregate (14 mm)	1270
Plasticizer (L per kg Cement)	3

concrete substrate layer is firstly cast and left to dry until fully hardened. A layer of repair material is then placed on this substrate base. It is ensured that the substrate has surface proportions which enhance its roughness to impose a uniform restraint on the shrinking overlay. The whole assembly is now subjected to a drying environment to induce shrinking in the overlay.

2.1 Shrinkage test

Shrinkage tests were performed in four main stages. During the first stage, substrate bases were cast according to the UNC method. In the second stage, repair material overlays were placed over the substrate bases. The third stages involved curing and for this the overlay base combinations were placed in an environmental chamber to cure under controlled conditions. In the final stage, the specimens were carefully examined and cracks were identified and measured.

2.1.1 Substrate bases

The moulds used to cast the substrate bases are shown in Figure 1. Spacers were used at each end of the mould to make it possible to use the same mould to cast both the repair overlays and the substrate bases. When casting the bases, the spacers were inserted which reduced the length of the moulds by 80 mm. During placement of the repair overlay, the spacers were removed in order to allow the overlay to also cover the ends of the substrate bases.

Concrete with a compressive strength of 60 MPa using a mix design as per Table 1 was used for the substrate bases. On the surface of these bases, 17 mm aggregates were placed such that half of the aggregates protruded from the bases in order to fully restrain the subsequent overlay. Two 280 mm Y10 steel reinforcing bars were positioned in the longitudinal direction of the substrate base to prevent cracking or breakage of the bases during handling. After 24 hours, the moulds were removed and the bases were placed in a curing bath with a temperature of 24°C and left to cure for 28 days.

2.1.2 Placement of overlay

Both the carbon nanotubes (CNT) and the graphite nanofibres (GNF) were fluorinated prior to its placement in the repair mortar. This was done in order to improve the bonding properties and dispersibility of these fibres. Once the substrate bases were fully hardened, a 30 MPa repair overlay was placed over each base. The first overlay was a no fibre concrete repair material, while for the second, third and fourth overlays, fluorinated CNT were added in proportions of 1, 3 and 5% of volume cement respectively. GNF were added in the same proportions as the CNT's in a set of separate overlay specimens. The overlays were 40 mm thick and overlapped with the substrate bases on the top and bottom sides to prevent curling up.

The entire assembly was placed in an environmental chamber and left to harden for 48 hours. After the first two hours, the moulds were carefully removed in order to significantly increase the total surface exposed to the drying environment. The time of demoulding was thus critical and affected both the amount and rate of shrinkage cracking.

After the total of 48 hours had passed, the specimens were removed from the chamber and cracks were marked and measured. Data recorded included; the number of cracks, crack lengths and crack widths. The measurements were taken to an accuracy of 0.01 mm using a Vanier and a Feeler gauge.

The temperature in the environmental chamber for the test was maintained at 40°C while the relative humidity was maintained as close to 5% as possible. Both the temperature and humidity were logged with a Lutron HT-3009 portable data logger. The data was closely monitored and the necessary adjustments were made to keep the temperature and humidity constant.

2.1.3 Mix design

The mix design, tabulated in Table 1, was used to cast the 60 Mpa high strength substrate bases.

The mix design for the overlays is tabulated in Table 2. This mixture was a 30 Mpa mixture.

3 RESULTS

The total crack area versus the percentage of fibres added by volume of cement can be seen in Figure 2.

It is estimated from Figure 2 that the total crack area will near 0 mm^2 when 17% CNT is added.

The total crack area versus the percentage of Graphite fibres added is shown Figure 3.

Table 2. Overlay mix design.

Mix design (overlays)	Substrate (kg/m³)
Cement (CEM IV 32.5R)	250
Water	225
Sand	585
Aggregate (14 mm)	600
Nanofibres	1%, 3%, 5%

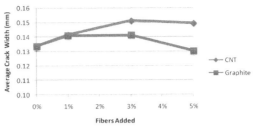

Figure 5. Total crack width of both specimens with CNT and Graphite.

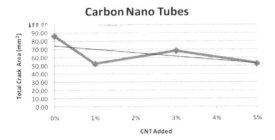

Figure 2. Total crack area of specimens with CNT.

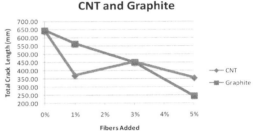

Figure 6. Total crack length of both specimens with CNT and graphite.

Both the CNT and the Graphite fibre results are plotted in this graph.

Figure 6 is a graph of the total crack length for both fibre types versus the percentage of fibres added to the overlay mixture.

The evaporation rate of the environmental chamber was calculated to be 0.61 kg/m²/hour.

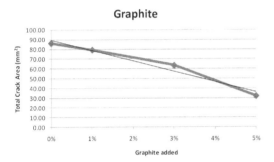

Figure 3. Total crack area of specimens with graphite fibres.

3.1 Carbon nanotube results

Figure 2 shows a decrease in total crack area from 86.16 mm² to 52.37 mm² by adding 1% CNT to the cement mixture. When 3% CNT fibre was added to the overlay material, the total crack area decreased to 86.0 mm² which is higher than the 52.37 mm² which was measured when only 1% fibre was added. When 5% CNT was added, the total crack area was measured as 52.80 mm². Thus overall the total crack area reduced form 86.16 mm² to 52.80 mm² when 5% CNT was added.

The average crack widths for both fibres are plotted in Figure 5. When 1% CNT fibre was added, the average crack width decreased from 0.14 mm to 0.13 mm. When 3% CNT was added the average crack width increased to 0.15 mm. In the 5% CNT specimens, the average crack width was also measured to be 0.15 mm. Thus overall the average crack width increased from 0.13 mm to 0.15 mm when 5% CNT was added.

Figure 6 shows the behaviour of the specimens with regards to crack length. The total crack length of the CNT specimens reduced from 645 mm to 370 mm when 1% CNT was added. The crack length decreased to 450 mm when 3% CNT was added and when 5%

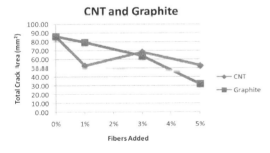

Figure 4. Average crack area of specimens with CNT and graphite.

It is estimated from Figure 3 that there will near 0 mm² when 8% Graphite fibres is added.

The total crack area versus the percentage of fibres added is plotted in Figure 4. Both fibres types are plotted on the same graph to indicate the difference in performance between the fibres.

Figure 5 is a graph of the average crack width versus the percentage of fibres added to the overlay mixture.

CNT was added, the total crack length decreased to 354 mm. The crack length thus reduced form 645 mm to 354 mm when 5% CNT was added.

3.2 Graphite fibre results

As can be seen in Figure 3, the total crack area decreased from 86 mm^2 to 79 mm^2 when adding 1% graphite fibres to the cement repair mixture. When 3% GNF was added the total crack area was measured to be 64 mm^2. The total crack area was reduced to only 32 mm^2 when adding 5% GNF. Thus overall the total crack area reduced from 86 mm^2 to 32 mm^2 by adding 5% GNF.

The average crack widths are plotted in Figure 5. When 1% GNF was added, the average crack width increased form 0.13 mm to 0.14 mm. When 3% GNF was added the average crack width was again measured to be 0.14 mm. In the 5% specimens the average crack width of the Graphite samples decreased to 0.13 mm. Thus overall the crack width remained unchanged at 0.13 mm when 5% GNF was added.

Figure 6 shows the behaviour of the specimens with regards to total crack length. The crack length of the specimens was reduced from 645 mm to 563 mm when 1% GNF was added. The total crack length was reduced to 450 mm when 3% GNF was added and when 5% GNF was added, the total crack length reduced to 245 mm. In summary, the total crack length reduced from 645 mm to 245 mm when 5% GNF was added.

4 CONCLUSION

The aim of the test was achieved by noting that by the addition of nano fibres to the overlay material resulted in lower total crack areas and thus a more durable and dense repair material. It is important to note that the total crack area is in essence the sum of all the crack widths multiplied by the sum of all the crack lengths. The total crack area is thus an indication of the crack lengths, the crack widths and the number of cracks.

The CNT and the GNF fibres reduced the crack area significantly. The crack width was however not significantly reduced by the fibres. This is an interesting result, which certainly was not expected and might be due to the nano size of the fibres. The fibres might thus not be large enough to restrain large cracks, but it was found that both fibre types are capable of reducing crack lengths and thus reducing the total crack area. The number of cracks was also substantially reduced by the fibres.

The use of Carbon Nanotubes and Graphite fibres was thus found to be useful to control crack widths in concrete repair materials and thus ensuring a more durable and sustainable repair.

REFERENCES

Banthia, N; Yan, C; and Mindess, S; Restrained shrinkage cracking in fibre reinforced concrete: A novel test technique. Cement and Concrete Research, 26(1)1996: 9–14.

Banthia, N; and Yan, C; Shrinkage Cracking in Polyolefin Fibre-Reinforced Concrete. ACI Materials Journal 97(4)2000: 432–437.

Mindess, A. B; Fibre Reinforced Cementitous Composites. Taylor and Francis, 1990

Tu, L; Development of Surface Fluorinated Polypropylene Fibres For use in Concrete. Johannesburg: RAU.

Turatsinze, Q. T; Modelling of Debonding Between Old Concrete and Overlay. Materials and Structures, 1045–1059. 2007.

Repair strategy

Concrete Solutions – Grantham, Majorana & Salomoni (Eds)
© 2009 Taylor & Francis Group, London, ISBN 978-0-415-55082-6

The new VIAP way of monitoring and planning in the Iraqi culture

Ali Al-Khatib

Versar International Assistance Projects (VIAP), Baghdad

ABSTRACT: Construction quality management involves an industry accepted series of processes designed to ensure that construction of a definable feature of work is built as per the plans and specifications, on time, within a defined budget and in safe working environment. However, what if the environment is one of the most violent places in the world, is a very hot country on the eastern periphery of Arab countries and is affected by a lack of experienced craftsmen? Dictatorial centralized state control and weak professional identities have been influential. However, the VIAP construction management style has resulted in the development of specific quality and collaboration practices in Iraq reconstruction.

Improvements of construction quality in Iraq require cultural change with a greater component of individual responsibility for work. A complication is that successive improvements in information communications technologies, leading to greater opportunities for rapid coordination in projects, could aggravate individual stress in a culture characterized by lack of clearly assigned responsibilities. However, it has become customary to use newly graduated Iraqi engineers when introducing new quality practices and methodologies as in some instances the relevant skill-set and experience might not be available within the projects' area.

1 INTRODUCTION

Every conflict generates risks to human health and to the environment. The post-conflict situation in Iraq compounds a range of chronic environmental issues, and presents immediate challenges in the fields of reconstruction and administration (USAID 2007).

Infrastructure is a broad concept that encompasses many of the obvious, physical features of civilization, such as roads, bridges and highways; transport and ports; basic utilities such power, water and sanitation; and also schools, health care facilities and public buildings (Jahan and McLeary 2005).

Rebuilding Iraq is an enormous task. Iraq is a large country with historic divisions, exacerbated by a brutal and corrupt regime. The country's 27 million people and its infrastructure and service delivery mechanisms have suffered decades of severe degradation and under-investment. Elements of the old regime engage in a campaign of sabotage and ongoing resistance, greatly magnifying the natural challenges of rebuilding Iraq (Hamre, Mendelson and Orr 2003).

VIAP (Versar International Assistance Projects) has established itself as the leading quality assurance, construction management and engineering services firm in conflict/post-conflict regions. Leveraging technology with a global reach, VIAP provides a cost-effective means of ensuring contract compliance, construction quality and delivery of projects under the toughest of conditions. VIAP today consists of 600 engineering, construction management and academic professionals in Iraq, supporting both the US Army Corps of Engineers and U.S. Air Force Reconstruction Programs.

As part of VIAP's ongoing commitment to provide these services, it also recognized the critical and essential needed to assist in the capacity development of countries such as Iraq and Afghanistan. Capacity development helps to build the individual, organization and nation in order for a government to provide for the essential services of their people.

To support the nation's challenges, the American Concrete Institute (ACI), a world leader in concrete research, specification and code publication, selected to team with VIAP and begin filling the most critical void in the Iraq reconstruction community- internationally recognized, codified in accredited certifications (Watkins and Al Khatib 2007).

Materials' testing in conflict environments is subject to graft and corruption. That reality has plagued the rebuilding efforts in Iraq. It has potentially cost millions of dollars and may prove fatal (UNEP 2003)

Too many qualified engineer professionals, superintendents, project managers, entrepreneurs and business owners, engineering and construction educators, technicians and laboratory specialists have died or fled the country.

2 TECHNICAL PROBLEMS IN THE RECONSTRUCTION PROGRAM

Deterioration of concrete caused by hot weather, inferior materials and poor workmanship are the main causes of concrete failure. Alkali-aggregate reaction within the concrete has resulted in cracking and white surface staining. Aggregates were not always properly graded and produced a poorly consolidated and therefore weaker concrete.

Other problems caused by poor workmanship are not unknown in later concrete work. If the first layer of concrete is allowed to harden before the next one is poured next to or on top of it, joints can form at the interface of the layers. In some cases, these cold joints permit water to infiltrate, and subsequent freeze-thaw action causes the joints to move. Dirt packed in the joints allows weeds to grow, further providing paths for water to enter. Inadequate curing can also lead to problems. If moisture leaves newly poured concrete too rapidly because of low humidity, excessive exposure to sun or wind, or use of too porous a substrate, the concrete can develop shrinkage cracks and will not reach its full potential strength.

3 MAJOR SIGNS OF CONCRETE DETERIORATION

Cracking occurs over time in virtually all concrete. Cracks may vary in depth, width, direction, pattern, location, and cause. Cracks can be either active or inactive. Active cracks widen, deepen, or migrate through concrete. Dormant cracks remain unchanged.

Diagonal cracks appear in the plaster on both sides of the wall and typically follow the mortar joint lines. The cracking is more prevalent in walls constructed of Cement Masonry Units (CMU) block as opposed to the more traditional fire brick.

Photographs of cracked load bearing walls from many projects in Iraq are presented. Those photos represent just a sampling of the more recent projects with cracking issues and do not fully encompass all of the projects on which cracking has been a quality issue during life of the program (VIAP-IRAQ, 2008).

Structural cracks can result from temporary or continued overloads, uneven foundation settling, or original design inadequacies. Structural cracks are active if the overload is continued or if settlement is ongoing; they are dormant if the temporary overloads have been removed, or if differential settlement has stabilized. Thermally-induced cracks result from stresses produced by temperature changes.

It is common construction in Iraq to construct load bearing masonry walls without using any reinforcement. It is possible to design and construct load bearing masonry walls that can function properly in non-seismic zones without the use of steel reinforcement. The IBC code requirements do not specifically call out the size, spacing, and location of horizontal and vertical steel reinforcement. The codes reference ACI design criteria and the contractor's design engineer is responsible for determining the size and placement of the reinforcement that is required. The contractors have submitted designs for non-reinforced load bearing masonry walls that have been reviewed and accepted based on the assumption that the walls would be constructed with sufficient quality to perform as

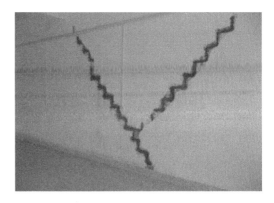

Figure 1. Typical diagonal cracking in masonry.

Figure 2. Typical diagonal cracking in masonry.

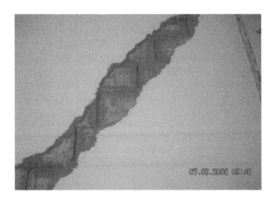

Figure 3. Cracking along mortar joints.

designed. This assumption has been proved to be wrong the majority of the time.

The mortar used is generally batched with poor quality control, over proportioning and water content. Mortar is mixed on the ground using a traditional ponding technique with no attempt made to measure the proportions of sand, cement, and water. No quality control test samples are made for compressive strength testing of the mortar to determine if it meets design strength. The high variability in the mix method would make any such test samples highly unreliable as a measure of the actual in-situ strength of the mortar.

Figure 4. Repairs to cracked masonry.

Weak mortar reduces the wall's ability to resist the compressive and shear loads on the wall.

Traditional Iraqi masonry construction techniques do not include placing mortar in the head joints (vertical joints) between the brick or block units. The lack of mortar in the head joints greatly reduces the ability of the loads to be transferred between units, especially in CMU construction where the head joint is a much larger percentage (±50%) of the total contact area than it is for brick (±25%). Filling of the head joints after the units are placed is prohibited by IBC and AICS Standards[7]. Walls are typically constructed with no mortar in the head joints, and the masons will only go back and fill in the head joints later if forced. The most common practice is to rely on the plaster coat to fill in these joints. This poor head joint construction reduces the wall's ability to evenly distribute loads and resist shear forces.

4 RECOMMENDATIONS

The addition of vertical steel reinforcement and fully grouted cells would greatly increase the compressive and shear load carrying capacity of the walls. The addition of horizontal course reinforcement would significantly increase the ability of the wall to carry shear loads. These measures are standard in most masonry design guidelines, including the US Military are Facilities Criteria (UFC), and would add an extra factor of safety to prevent poor workmanship issues from leading to wall failures.

VIAP's recommendation to the contractors requiring vertical and horizontal reinforcement in all load bearing masonry walls, and require a minimum of horizontal reinforcement in all non load bearing masonry walls[8]. The use of vertical and horizontal reinforcement in the masonry walls will help to eliminate these problems and increase the overall quality and safety of the buildings being constructed in the Iraq reconstruction Program.

REFERENCES

American Concrete Institute, 2007, ACI 530/ASCE 5/ TMS 402 (2.3).
American Concrete Institute, 2007, ACI 530.1/ASCE 6/ TMS 602.
Hamre, J. Mendelson, G and Orr, R., July 17, 2003, Iraq Reconstruction Assessment Mission, *www.cfr.org/content/publications/attachments/iraq-Trip_Report.pdf.*
Jahan, S and McCleary, R., 2005, Making Infrastructure Work for the Poor, *www.undp.org/poverty/docs/fpage/synthesisreport.pdf.*
UNEP, 2003, Desk Study on the Environment i.n Iraq, postconflict.*unep.ch/publications/Iraq_DS.pdf.*
USAID, 2007, Community-Based Development in Conflict-Affected Areas, *usa-id.gov/our_work/.../publications/docs/CMM_CBD_Guide_May_2007.pdf.*
VIAP-IRAQ, 2008, Load Bearing Masonry Wall Recommendations, *internal newsletter.*
Watkins, S. and Al Khatib, A., Dec.2007, Sustainable Development: VIAP Lessons in Continual Education in Iraq, *Proceedings of Erbil International Education Conference, Iraq, 1:32–35.*

Concrete Solutions – Grantham, Majorana & Salomoni (Eds)
© 2009 Taylor & Francis Group, London, ISBN 978-0-415-55082-6

Application of the new European concrete repair standards BS EN 1504 parts 1 to 10 to a range of reinforced concrete structures

John P. Broomfield

Broomfield Consultants, East Molesey Surrey, UK

ABSTRACT: This paper describes how the principles of the new European Concrete Repair Standards EN 1504 parts 1 to 10 can be applied to a wide range of concrete repair procedures including cathodic protection, patch repairs, coatings and waterproofing to extend the life of building facades, walkways and bridges.

The paper discusses the investigation process, the use of life cycle costing, the design and specification of concrete repairs and corrosion protection systems and the execution of the works within the framework of EN1504 Parts 1 to 10 "Products and systems for the protection and repair of concrete structures"

1 THE STRUCTURES

The University of East Anglia was founded in the 1960s and the main campus buildings, including the "Teaching Wall" and the well known "Ziggurat" residential blocks (featured on the English Heritage website) were laid out by Sir Denys Lasdun. The University is proud of its architecture which as been supplemented by other famous architects.

The "teaching wall" consists of a shallow "W" of reinforced concrete buildings approximately 500 m long, five and six storeys high (see Figure 1). The runs of offices, laboratories and lecture rooms are interrupted by lift and stair "towers" at intervals along its length with water tanks and plant rooms above the main building roof level. The exposed concrete facades are a feature of the teaching wall and various parts of the campus which were given grade II* and grade II listing during the process of the works. Various sections of the teaching wall and other campus buildings are linked with elevated walkways.

2 THE CONDITION AND SITUATION OF THE STRUCTURES

Broomfield Consultants were appointed as corrosion specialist consultants to Jacobs Babtie Consultants to help them to conduct "Forensic Structural Engineering" initially to the "Biotower" (Phase 1) and then to all of the reinforced concrete structures with exposed concrete façades on the campus. Work was conducted in close collaboration with the university departments affected, as well as the Estates Department who ran the project, English Heritage and the Norwich City Planning office who gave the planning consents for the work.

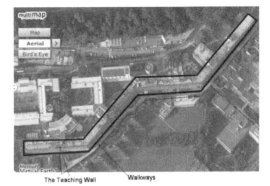

The Teaching Wall   Walkways

Figure 1. Showing the teaching wall and adjacent walk-ways.

Phase I work was on the "Biotower", a lift and stair tower with air conditioning plant room and a water tower above. Detailed investigation showed low cover and carbonation to be prevalent with some admixed calcium chloride in some batches or "lifts" of concrete, probably used to speed up setting during winter construction. These problems were all leading to reinforcement corrosion. A number of options for repair were investigated, including the possibility of cladding the façade and "air conditioning" it to remove moisture and stop reinforcement corrosion according to DD ENV1504 part 9 Principle 8 (see section 3). However, this was untried technology and it was considered that no contractor would offer any warranties on such an installation. For that reason, impressed current cathodic protection was applied according to DD ENV1504-9 Principle 10 (cathodic protection). The specification was according to BS EN 12696:2000 "Cathodic protection of steel in concrete" and used

Figure 2. Walkway showing teaching wall behind. Deicing salts and leachate run down can be seen where the water-proofing and drainage has failed, allowing corrosion of the slim pier supports.

probe anodes installed and wired up inside the building to minimise the effect on the façades.

The Phase 2 works were on the library walkway, shown in Figure 2, a concrete stairway to another walkway showing severe corrosion damage and two further stair/lift towers in the teaching wall. A detailed quantitative condition survey revealed areas of concrete damage due to corrosion from carbonation. This was principally due to low cover, indifferent quality concrete and the age of the structure. Other areas were deteriorating due to deicing salt ingress, particularly on the elevated walkways and access stairways. Using the survey data, calculations were made of ongoing chloride and carbonation ingress on a 30 year life projection (see Broomfield 2006).

Corrosion modelling was done by using Fick's law of diffusion calculations on cover depth measurements combined with carbonation depths and chloride dept profiles (Broomfield 2007). This showed that other than the areas showing immediate damage, few areas were found to be susceptible to future corrosion.

3 APPLYING THE PRINCIPLES OF EN1504 TO THE REHABILITATION PROCESS

Under DD ENV1504 Part 9 Section 5.2 the following options are given:

a. Do nothing for a certain time
b. Re-analysis of structural capacity
c. Prevention or reduction of further deterioration without improvement of the concrete structure
d. Improvement, strengthening or refurbishment
e. Reconstruction of all or part of the structure
f. Demolition

Given that the structures are part of a listed site, that further deterioration could lead to health and safety problems in some areas and that the university has set

aside a budget for its "Concrete Preservation Plan" options c) and d) were relevant.

The standard options for intervention on a reinforced concrete structure suffering from reinforcement corrosion are given in Table 2 of DD ENV 1504-9 as:

Principle 7 – Preserving or restoring passivity
7.1 Increase cover
7.2 Replace contaminated concrete
7.3 Electrochemical realkalization
7.4 Realkalization by diffusion
7.5 Electrochemical chloride extraction

Principle 8 – Increasing Resistivity
8.1 Limiting moisture content by surface treatments, coatings or sheltering

Principle 9 – Cathodic control
9.1 Limiting oxygen content (at the cathode) by saturation or surface coating

Principle 10 – Cathodic protection
10.1 Applying electrical potential

Principle 11 – Control of anodic areas
11.1 Painting reinforcement with coatings containing active pigments
11.2 Painting reinforcement with barrier coatings
11.3 Applying inhibitors to the concrete

These can be expressed in practical terms as:

a) Do nothing for a certain time
b) Complete or partial demolition and rebuild DD ENV 1504-9 Principle 3.4
c) Patch repair of local damaged areas DD ENV 1504-9 Principle 3 and 7.2
d) Ingress control via coatings, membranes, sealers, water stops, enclosures or other barriers DD ENV 1504-9 Principles 1, 2 and 8
e) Impressed current cathodic protection DD ENV 1504-9 Principle 10 (BS EN 12696)
f) Galvanic cathodic protection DD ENV 1504-9 Principle 10
g) Electrochemical realkalization DD ENV 1504-9 Principle 7.3 (CEN/TS 14038-1)
h) Electrochemical chloride extraction EN 1504-9 Principle 7.5 (CEN/TS 14038-2 – In Draft)
i) Corrosion inhibitors. DD ENV 1504-9 Principle 11.3.

Being part of a listed building and suffering from corrosion damage, options a) and b) were not feasible. Option c) was required in some areas. Option d) was used but in some areas its use was constrained by the requirement to retain the board marked finish to the concrete on the listed façades. However, control of ingress of carbon dioxide (CO_2) chloride ions and water was required.

To this end, a proprietary architectural coating was trialed, for approval by the university and by the local authority conservation officer. This coating "tones down" changes in concrete colour and finish and was considered ideal for minimising the visual impact

Figure 3. Incipient anode formation around an old repair on the Biotower plant room prior to Phase I repair and impressed current cathodic protection.

of patch repairs on the board marked finish on the concrete façades. The selected coating had anticarbonation properties and was also compatible with a silane for control of moisture and chloride ingress.

In this phase of the works, impressed current cathodic protection, e) in the list above, was not required on a large enough area to be cost effective. However, given the presence of active chloride induced corrosion, an alternative was to use galvanic anodes installed in the patch repairs to minimise incipient anodes (see Broomfield 2007 and Figure 3 below). The other electrochemical treatment techniques f), g), h) were not considered suitable for this project.

4 DESIGN AND SPECIFICATION OF THE WORK

Techniques selected therefore included localised galvanic cathodic protection to minimise the incipient anode effect around patches in areas of high chloride (DD ENV 1504-9 Principle 10 CP). An example of such an effect is shown in Figure 3. Penetrating sealers were required to keep out further chloride according to principles 1.1 and 6.1 and to reduce moisture (Principle 8). Anticarbonation coatings were required to reduce the rate of carbonation (Principles 1.3c and 6) and a renewal of the waterproofing membrane on the walkway decks was specified to keep moisture and chlorides out of the deck concrete (Principle 1.1). The membrane and improvements of drainage provided reduction in water leakage, sheltering the walkway substructure from deicing salt run down. These techniques were used along with conventional patch repair where required (Principles 3 and 7.2).

Detailed analysis of the condition survey results allowed the determination of treatments to different elements of the structures as shown in Table 1.

The following specifications were written for the job:

1. A concrete repair specification based on:

    a. Materials according to DD ENV 1504 Part 9 and BS EN 1504 Part 3 (Class R4 structural grade repair mortar)

    b. Patch Repair Preparation according to BS EN 1504 part 10, section 7 and appendix A7

    c. Material Application, BS EN 1504 part 10 section 8 on application, Appendix A8.

    d. Special section on galvanic anode application from manufacturer's data sheets.

    e. Testing on site and of site samples using test methods and values in Part 10 Appendices A7, A8 and A9

2. A coating specification for silicate coating and silane impregnation – based on

    a. Materials according to DD ENV 1504 Part 9 and EN 1504 Part 2 (1.3C for anticarbonation coating and

    b. Surface Preparation according to BS EN1504 part 10, sections 7 and 8 and appendix A8

    c. 1.1(H) and 1.2(I) for silane impregnation for moisture/chloride ingress control)

    d. Manufacturer's literature for application

    e. Site Testing according to Part 10 Appendices A8 and A9

3. An application specification for a waterproofing membrane

    a. lifting and retaining the original paving slabs,

    b. conducting concrete repairs

    c. repairing and improving drainage and falls

    d. applying new waterproofing system

    e. replacing paving slabs

5 SITE TESTS

After applying coatings, cores were taken and sent for testing. Carbon dioxide permeability tests (BS EN 1062-6:2002) gave far better than the 50 m minimum values recommended in the specifications, but starting at values of 23 m and 30 m before the coating was applied.

The water permeability test results were:
    Coated 0.03 and 0.04 kg/m^2.h$^{1/2}$
    Partial coated 0.05 kg/m^2.h$^{1/2}$
    Uncoated 0.11 and 0.12 kg/m^2.h$^{1/2}$

BS EN 1062-3:1998 Table 1 states:
    I High >0.5 kg/m^2.h$^{1/2}$
    II Medium 0.1 to 0.5 kg/m^2.h$^{1/2}$
    III Low <0.1 kg/m^2.h$^{1/2}$

Given the requirements for a coating with architectural properties, and the fact that most areas of low cover were repaired, the coated values falling in

Table 1. EN1504-9 Principles, methods, relevant standard and treatments applied.

EN1504-9 Principle	Method/Principle	EN Standard	Elements treated	Materials used
1 Protection against ingress and 8 Increasing Concrete Resistivity	Hydrophobic Impregnation, EN 1504 Part 2 Principle 1.1 and 2.1	EN1504 Part 2 EN 1062-3 maximum value $w = 0.035 \, \text{kg/m}^2.\text{h}^{0.5}$	Walkways below deck level where deicing salts were applied and chloride level at rebar is below the threshold for corrosion	Silane compatible with cosmetic coating used to "tone down" repairs
1 Protection against ingress	Anticarbonation Coating, EN 1504 Part 2 Principle 1.3c	EN 1504 Part 2 EN1062-6 Permeability to CO_2 $S_D > 50 \, \text{m}$	Parapets on walkways above deicing salts where chloride levels are very low	Cosmetic coating with anticarbonation properties
1 Protection against ingress	Waterproofing membrane EN1504 membrane EN1504 Part 9 Principle 1.7	EN1504 Part 9 Principle 1.7 No membrane tests listed in 1504	Walkway decks	Waterproofing system
3 Concrete Restoration 7 Preserving or Restoring Passivity	Hand Applied Mortar, Principle 3.1, 7.2	EN1504 Part 3 Class R4 Compressive Strength $> 40 \, \text{MPa}$ Adhesive Bond $> 2 \, \text{MPa}$	All Damaged elements	Pre bagged Patch Repair Material
10 Cathodic Protection	Local galvanic Anodes, Principle 10	Galvanic anodes not covered in 1504 or yet in EN12696	Patch repairs with chloride levels in excess of the corrosion threshold	Zinc anodes encapsulated in a proprietary activating mortar

the II medium range was judged to be an acceptable performance. Also, renewal of the waterproofing and the drainage would reduce the amount of water run down on the substructure, reducing further the rate of chloride ingress.

Pull off tests on concrete patch repairs can be conducted according to EN 1542, ISO 4624 and BS 1881 Part 201 and 207 as described in EN 1504-10 A9.2 Test or observation No 35,. Recommended values are given in EN1504-10 Table A2. In this project, pull off tests achieved 0.8 MPa or better.

6 CONCLUSIONS

Work was successfully completed in 2007. There was minimal disruption to campus activities and both the University and the listing officer were pleased with the final finishes on the listed elevations.

It should be noted that the specifications were written prior to full publication of all parts of EN1504 and the associated test methods. Not all testing on this project was fully compliant with the specific CEN test mentioned but used equivalent British Standards or other tests in use at the time.

In conclusion, it can be seen that concrete repair systems can be designed, performance specified and repair products applied using the EN 1504 set of documents along with their associated test methods following the principles of corrosion engineering to ensure corrosion prevention before it initiates and corrosion control once damage has initiated.

The first critical part of any repair and refurbishment project is a condition survey which quantifies the type and extent of damage to ensure that:

- only areas in need of treatment are treated;
- appropriate treatments are selected;
- the current and future requirements of the structure are fully considered in the repair design process.

Appropriate repair systems and materials can then be selected based on the principles of EN1504 and repair designs and specifications prepared using the product characteristics specified in EN1504 and the associated test methods. Principles 7 to 11 of DD ENV1504 Part 9 are not the most obvious breakdown of repair and corrosion control options and Principle 1 must also be included in any analysis. However, using them and the other relevant parts of EN1504 along with the specified test methods for laboratory and site testing does ensure that a comprehensive approach is taken that considers all practical options for achieving a repair. It also ensures that the repairs meet the requirements for the structure, its condition, its environment and its future management. Once the full range of concrete repair products is available with CE marking based on EN 1504 criteria, it should become possible to write performance specifications and know that products used will meet the required standards.

ACKNOWLEDGEMENTS

The author would like to acknowledge Martin Lovatt, Project Manager for The University of East Anglia

Estates Department and Andrew Brown, the Engineer for Jacobs for their contributions and for permission to publish this article.

REFERENCES

Broomfield, J.P. 2006. "A web based tool for selecting repair options and life cycle costing of corrosion damaged reinforced concrete structures, *Concrete Solutions, Proc. 2nd Intl Conf.*, St Malo, France, Ed. M.G. Grantham, R Jaubertie, C. Lanos, Watford. BRE Press, pp 505–513.

Broomfield, J.P. 2007 *Corrosion of Steel in Concrete – Understanding, Investigation and Repair*, London, Taylor & Francis

BIBLIOGRAPHY

BS EN 1504-1:2005, *Products and systems for the protection and repair of concrete structures. Definitions, requirements, quality control and evaluation of conformity. Definitions*

BS EN 1504-2:2004, *Products and systems for the protection and repair of concrete structures. Definitions, requirements, quality control and evaluation of conformity. Surface protection systems for concrete*

BS EN 1504-3:2005, *Products and systems for the protection and repair of concrete structures. Definitions, requirements, quality control and evaluation of conformity. Structural and non-structural repair*

BS EN 1504-4:2004, *Products and systems for the protection and repair of concrete structures. Definitions, requirements, quality control and evaluation of conformity. Structural bonding*

BS EN 1504-5:2004, *Products and systems for the protection and repair of concrete structures. Definitions, requirements, quality control and evaluation of conformity. Concrete injection*

BS EN 1504-6:2006, *Products and systems for the protection and repair of concrete structures. Definitions, requirements, quality control and evaluation of conformity. Anchoring of reinforcing steel bar*

BS EN 1504-7:2006, *Products and systems for the protection and repair of concrete structures. Definitions, requirements, quality control and evaluation of conformity. Reinforcement corrosion protection*

BS EN 1504-8:2004, *Products and systems for the protection and repair of concrete structures. Definitions, requirements, quality control and evaluation of conformity. Quality control and evaluation of conformity*

DD ENV 1504-9:1997, *Products and systems for the protection and repair of concrete structures. Definitions, requirements, quality control and evaluation of conformity. General principles for the use of products and systems*

BS EN 1504-10:2003, *Products and systems for the protection and repair of concrete structures. Definitions. Requirements. Quality control and evaluation of conformity. Site application of products and systems and quality control of the works*

BS EN 14630:2006, *Products and systems for the protection and repair of concrete structures. Test methods. Determination of carbonation depth in hardened concrete by the phenolphthalein method*

BS EN 14629:2007, *Products and systems for the protection and repair of concrete structures. Test methods. Determination of chloride content in hardened concrete*

BS EN 12696:2000, *Cathodic Protection of Steel in Concrete*

CEN/TS 14038-1:2004, *Electrochemical realkalization and chloride extraction treatments for reinforced concrete – Part 1: Realkalization*

Concrete Solutions – Grantham, Majorana & Salomoni (Eds)
© 2009 Taylor & Francis Group, London, ISBN 978-0-415-55082-6

The schematic program of bridge inspection and condition assessment

P. Chupanit
Bureau of Road Research and Development, Department of Highways, Bangkok, Thailand

T. Pinkaew & P. Pheinsusom
Department of Civil Engineering, Chulalongkorn University, Bangkok, Thailand

ABSTRACT: In Thailand, there are over 17,000 bridges, which can be categorized into slab type, plank girder type, multi-beam type, I-girder type and box-girder type. The ages of bridges are varied and the maximum is about 60 years. Owing to the bridge collapses in many countries, Thailand Department of Highways (DOH) started to develop a program to evaluate bridge condition. In the program, the visual inspection is performed with the load testing. From the visual inspection, the bridge condition can be described into 6 service condition levels, which are very good, good, fair, poor, critical and failed. The load testing is performed following *National Cooperative Highway Research Program by ASSHTO* and the load testing result is represented by operating and inventory rating factors. During load testing, the trucks with allowable configuration and weight limit are used. With obtained results, a maintenance or rehabilitation policy is provided in order to guarantee safety for road users.

1   INTRODUCTION

Owing to many bridge collapses around the world, the Thailand Department of Highways (DOH) has started a strategic program for bridge inspection and condition assessment. In the program, all bridge information is gathered and recorded on the webpage based inventory system. All bridges are visually inspected and their service conditions are indicated by a parameter named overall condition index (OCI). The developed OCI is adapted from the overall condition rating (OCR) by AASHTO (AASHTO 1994), in which the condition level is changed from 0–9 to 0–5. In addition to visual inspection, bridge load testing is conducted for those bridges that are in poor condition. Then the results from both visual inspection and load testing are used to specify the rehabilitation approach.

2   BRIDGE INVENTORY SYSTEM

All bridge information is gathered including bridge name, bridge location, bridge type, bridge dimension, bridge barrier, pier type and number of traffic channels. In addition to the collection of physical information, data regarding to the bridge usage is also surveyed, which includes traffic volume (AADT and truck volume). In addition, the bridge history is determined, such as year of construction, year of rehabilitation and rehabilitation method. All information is recorded on the webpage based inventory system, which was developed according to the DOH's policy. The developed system is not a commercial one.

The advantage of the webpage based system is that it can be accessed by authorized staff from many places. According to the updated information (year 2008), there are approximately 17,500 bridges over the whole country, in which 45% are slab type bridges, 25% plank girder bridges, 10 % multi-beam bridges, 15% I-girder bridges and 5% box-girder bridges. The age of all bridges varies between 0–60 years.

3   BRIDGE VISUAL INSPECTION

Thailand Department of Highways (DOH) developed a procedure for bridge visual inspection. All bridges are visually inspected and their photos are systematically taken, so that the locations of damage can be identified from the pictures. The photography process starts from the superstructure and then thesubstructure. The photos are taken span by span, which starts from the lowest kilometer span to the highest kilometer span. The bridge condition is rated by using a parameter called overall condition index (OCI), which is illustrated in Table 1. The OCI is adapted from overall condition rating (OCR) by AASHTO[1], in which the number of service condition levels has been reduced, because the nine service condition levels in AASHTO are difficult to differentiate. From the project, 40% of all bridges were visually inspected in year 2008. The results revealed that 20% of bridges were in very good condition, 45% were in good condition, 24% were in fair condition, 8% were in poor condition, 3% were in critical condition and none were in a failed condition.

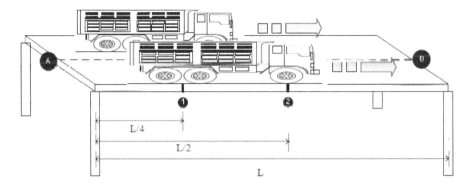

Figure 1. Loading configuration on tested bridge.

Table 1. Overall condition Index (OCI).

OCI	Condition	Description
5	Very Good	No problem noted.
4	Good	A few minor problems. Hairline cracking but no spalling is noted.
3	Fair	Some minor problems without effect on overall strength. All primary structural elements are sound with minor section loss. Cracks with width less than 2.0 mm or spalling less than 5% is noted.
2	Poor	Advanced section loss or spalling are found. Crack with width 2.0-4.0mm or spalling greater than 5% is noted. Bridge needs local repair and additional support.
1	Critical	Loss of section or spalling have seriously affected one or more structural components. Advanced damage present in structural component(s), or obvious movement affecting structural stability is noted. Fatigue crack or shear crack is noted. Crack with width greater than 4.0 mm is noted. Bridge needs temporary support and needs major rehabilitation and strengthening.
0	Failed	Out of service and has to be reconstructed.

4 BRIDGE LOAD TESTING

The bridge load testing was performed according to a report from National Cooperative Highway Research Program (NCHRP 1998) and according to AASHTO (AASHTO 1989). For load testing, three types of electronic instruments were installed on each bridge; strain gauges, displacement transducers and accelerometers. Strain gauges were used to measure expansion and contraction at the structure surface, while displacement transducers were used to measure vertical displacement or deflection of bridge structures. The accelerometers were used to measure the vibration of bridge during the load tests. Two 25-ton Thai trucks were used to load the bridge during the test as shown in Figure 1. After load testing, all the test data collected were used for bridge evaluation.

The load carrying capacity of a bridge was calculated and presented by a parameter called "Rating Factor" (RF) as follows.

$$Operating\ RF = \frac{Rs\text{-}Fd}{Fl+I} \qquad (Equation\ 1)$$

Where
Rs = Reduced Strength
Fd = Factored Dead Load Effects
Fl = Factored Live Load Effects
I = Impact

$$And\ Inventory\ RF = \frac{3}{5}\ Operating\ RF \qquad (Equation\ 2)$$

Both rating factors are calculated by taking into account measured data from a load test. The inventory rating factor represents a loading capacity for long term usage, whereas the operating rating factor represents a loading capacity for short term usage. If the inventory rating factor is less than 1.0, that bridge is at risk for carrying the required vehicle traffic in the long term. If the operating rating factor is less than 1.0, that bridge is at risk for carrying the required vehicle traffic in the short term.

In the project, 12 slab-type bridges were sampled and tested in order to determine their safe load carrying capacities. The load testing results are used to determine the relationship between service conditions and safe load capacity, which are shown in Table 2. In Table 2, some bridges have inventory rating factors less than 1.0. These are really in need of special consideration. The project is still ongoing and more load test results will be obtained.

5 REHABILITATION STRATEGY

With results from visual inspection and load testing, the rehabilitation strategy can be developed by following Table 3, Table 4 and Table 5. It starts with specifying the class of load carrying capacity in Table 3, which depends on operating and inventory rating factors. Then, the rehabilitation strategy can be

Table 2. Rating factors and service conditions of 12 Slab-Type bridges.

Bridge No.	Service Condition	Inventory RF	Operating RF	Rehabilitation
B1	Very Good	1.21	2.01	R6
B2	Poor	0.66	1.11	R3
B3	Fair	0.85	1.42	R4
B4	Critical	0.48	0.80	R2
B5	Poor	0.79	1.31	R3
B6	Fair	1.11	1.84	R5
B7	Critical	0.88	1.46	R3
B8	Fair	1.00	1.67	R5
B9	Poor	0.82	1.36	R3
B10	Fair	0.65	1.09	R4
B11	Fair	1.10	1.80	R5
B12	Good	0.94	1.56	R4

Table 3. Class of bridge load carrying capacity.

Class	Operating RF	Inventory RF	Remarks
Class A	≥ 1.0	≥ 1.0	Good for long term and short term usage.
Class B	≥ 1.0	< 1.0	Risk for long term usage. Short term is serviceable.
Class C	< 1.0	< 1.0	Risk for long term and short term usages.

Table 4. Rehabilitation strategy.

Service condition	Class		
	Class A	Class B	Class C
Very Good	R6	R4	R3
Good	R5	R4	R2
Fair	R5	R4	R2
Poor	R5	R3	R2
Critical	R5	R3	R2
Failed	R1	R1	R1

Table 5. Description of rehabilitation methodology.

No.	Methodology	Rehabilitation Plan					
		R1	R2	R3	R4	R5	R6
1	Usage Control						
	1.1 Limit weight/Limit vehicle speed		X	X			
	1.2 Detour vehicle/Decrease traffic lane		X	X			
2	Routine Maintenance						
	2.1 Visual Inspection		X	X	X	X	X
	2.2 Corrosion inspection/Chloride and carbonation test		X	X	X	X	X
	2.3 Repair local damage (Spot)		X	X	X	X	X
	2.4 Protection		X	X	X	X	X
3	Structural Repair						
	3.1 Partial repair (Concrete, Joint, Bearing)					X	X
	3.2 Full repair (Concrete, Joint, Bearing)		X	X			
4	Structural Strengthening						
	4.1 Partial Strengthening				X		
	4.2 Full Strengthening		X	X			
5	Replacement						
	5.1 Partial Replacement		X				
	5.2 Total Replacement	X					

developed by integrating the class of load carrying capacity with service condition from visual inspection. The rehabilitation strategy ranges from R1 to R6 in Table 4. Finally, the maintenance or rehabilitation methods can be obtained from Table 5. The rehabilitation methods could be usage control, routine maintenance, structural repair, structural strengthening or bridge replacement. Examples of rehabilitation methods are shown in Table 2.

6 CONCLUSIONS

Thailand Department of Highways (DOH) is developing a schematic program for bridge inspection and condition assessment. In the program, a bridge inventory system is developed, which is a webpage based application. In the program, visual inspection to assess service condition is required with bridge load testing. However, bridge load test is performed only if a bridge is in poor, critical or failed condition. With both assessments, the maintenance or rehabilitation plan can be established.

REFERENCES

AASHTO, Guide Specifications for Strength Evaluation of Existing Steel and Concrete Bridges," Washington, D.C., 1989.
AASHTO, Manual for the Condition Evaluation of Bridges, Washington, D.C., 1994, Revised by 1995, 1996, 1998 and 2000 Interim Revisions.
"National Cooperative Highway Research program," Research Results Digest No. 234, November 1998.

Concrete Solutions – Grantham, Majorana & Salomoni (Eds)
© 2009 Taylor & Francis Group, London, ISBN 978-0-415-55082-6

Application of a statistical procedure to evaluate the results from potential mapping on a parking garage

J.J.W. Gulikers

Rijkswaterstaat Bouwdienst, Utrecht, The Netherlands

ABSTRACT: The commonly used potential criteria suggested by ASTM C-876 to indicate the probability of reinforcement corrosion are considered too strict to cover a wide range of structures exposed to a variety of environmental conditions. This problem can be solved by employing a statistical procedure which eventually results in more reliable potential criteria for a specific concrete structure or component. Moreover, a quantification of the relative amount of corroding steel is obtained. This procedure is applied to the data obtained from a potential survey on a parking garage. This case study demonstrates the possibilities of the method to obtain relevant output. In addition, some of the pitfalls and limitations are discussed in more detail.

1 INTRODUCTION

Half-cell potential mapping is a widely employed measurement technique which allows for a fast and essentially non-destructive assessment of concrete structures with respect to reinforcement corrosion. This method normally involves measuring the potential difference between a reference electrode placed on the concrete surface and the embedded reinforcing steel. In practice, most often a copper/copper sulfate (CSE) or silver/silver chloride (SSC) half-cell is used as the reference electrode. Although half-cell potential mapping neither quantifies the actual rate of corrosion nor the residual cross sectional steel area, past experience has demonstrated that the half-cell potential method has proved to be very effective for locating corroding reinforcement in concrete structures.

Half-cell potential surveys are frequently undertaken on small patch areas of concrete: however, areas of corrosion may then be missed. Therefore it is recommended to apply half-cell mapping to very large areas of structure, preferably the complete elements of a structure, when signs of corrosion are visible on the concrete surface. For such large scale surveys it is common to use a portable "wheel" and "multicell" devices with computerized data recording and presentation.

The interpretation of the potential data thus obtained is largely based on experience. On the basis of a large number of observations in North America, criteria have been developed that have proved to be reliable when used on bridges made from Portland cement concrete suffering from chloride ingress. These empirically derived criteria, sometimes referred to as the Van Daveer criteria, have been adopted by ASTM C 876-91. Despite the fact that these criteria are not universally applicable, it is common to find that this so-called Numeric Magnitude is employed on a wide range of concrete elements exposed to a variety of exposure environments.

According to Gu & Beaudoin (1998) and the RILEM Recommendation (Elsener (2003)) several factors have a significant influence on the readings and consequently it is recognized that the use of fixed boundary limits to distinguish between active and passive zones as proposed in ASTM C 876-91 (1999) can be misleading. In this respect variations in cover thickness, temperature, moisture conditions, type of cement and surface carbonation may complicate the interpretation.

In view of these inherent difficulties in interpretation, the RILEM recommendation advocates the use of a statistical analysis of the potential readings. Ideally, such an analysis will result in potential criteria corresponding to any desired probability of finding active corrosion. Thus for each concrete component of a structure specific potential criteria can be derived which will reflect the exposure conditions at the time the survey was made. In addition, the analysis will yield the relative amount of corroding steel present. A detailed description of the theoretical backgrounds and assumptions made has been given by Gulikers (2007). In this paper the application of the statistical analysis on a case study will be reported.

2 PRESENTATION OF MEASUREMENT DATA OF POTENTIAL MAPPING

2.1 *Equipotential contour plot*

The most common way of presenting measurement data is given by drawing a potential map of the area surveyed. The readings are plotted automatically or manually on the map and lines are drawn between points of equipotential. These contour lines delineate areas where corrosion is likely to occur (anodic areas) and areas where corrosion is less likely to occur (cathodic areas). The greater the potential difference between neighbouring measurement points, and the steeper the gradient of potential lines then the greater the possibility of significant corrosion in the anodic areas. While these equipotential contour plots are not as quantitative as simply following the ASTM criteria, such mapping of anodic areas is valid over a wider range of structures and conditions (Broomfield (1997)). According to Elsener (2003) contour plots react very sensitively to single erroneous readings (e.g. very negative values) because fictitious contour lines are plotted around the erroneous point.

2.2 *Histograms or frequency distributions*

In order to allow for a statistical interpretation the data can be presented in histograms or frequency distributions. For a sound analysis it is then required that a regular grid size is employed over the complete concrete surface as the change to a smaller grid size in certain areas would erroneously cause a shift in the relative amount of corroding steel. However, in a histogram or frequency distribution the information on the position of the reading is lost. Such a representation of data is only of interest when a significant number of readings is available pertaining to a concrete surface exposed to similar environmental conditions, particularly moisture.

In a histogram or a frequency distribution all potential readings are grouped in classes of potential and the number of readings falling in a class is counted. If a significant number of readings are available a small class width can be chosen and the readings can be presented in a frequency distribution by a continuous line. From such a graph, a first impression can be obtained on the mean value, variation, and skewness. Based on certain assumptions, the relative amount of actively corroding steel can be estimated from a frequency distribution. In Figure 1 an example characteristic for an ideal situation is given, in which 20% of the data pertains to actively corroding steel. Such a graph shows the individual contributions of passive and corroding steel in the frequency distribution, as well as the overall frequency distribution, assuming normal (Gaussian) distributions for the potentials. The "shoulder" in the overall distribution resulting from the contribution of the corroding steel is clearly noticeable. For smaller relative amounts of corroding steel the "shoulder" will become less pronounced. However, it should be noted that in practice only the overall frequency distribution can be plotted, reflecting the overall results of a potential survey. Consequently, the contribution of the corroding steel has to be estimated based on experience or derived by calculation.

2.3 *Cumulative frequency distributions*

An alternative way of presenting a large amount of potential readings is by a cumulative frequency

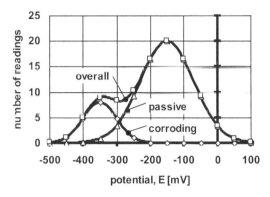

Figure 1. Frequency distribution of potential mapping for an ideal situation characterized by $\mu(E_{corr}) = -350\,mV$, $\sigma(E_{corr}) = 50\,mV$, $\mu(E_{pas}) = -50\,mV$, $\sigma(E_{pas}) = 80\,mV$; 20% corroding steel, 1000 readings, class width 5 mV (assuming Gaussian distributions).

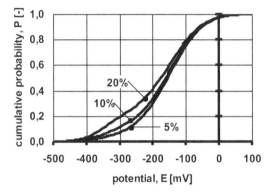

Figure 2. Cumulative frequency distribution of potential mapping for an ideal situation characterized by $\mu(E_{corr}) = -350\,mV$, $\sigma(E_{corr}) = 50\,mV$, $\mu(E_{pas}) = -50\,mV$, $\sigma(E_{pas}) = 80\,mV$; 5, 10 and 20% corroding steel (assuming Gaussian distributions).

distribution. On the vertical axis the cumulative probability is indicated, in a range from 0 to 1 (or 0 to 100%). Figure 2 shows such a distribution when a linear scale for the vertical axis is used for the same ideal situation as depicted in Figure 1, however now comprising 3 values for the relative contribution of corroding steel ($p_{corr} = 0.05$, 0.10 and 0.20).

In such a graph, using a linear scale for the vertical axis, the contribution of corroding steel will show up less pronounced than in a frequency distribution. However, this can be improved by presenting the cumulative frequency distribution on so-called normal probability paper. In the ideal situation that the potentials of both corroding and passive steel conform to a normal density distribution, a probability plot will comprise 2 linear portions: see Figure 3. In the extreme case that only passive steel is present, i.e. for $p_{pas} = 1.0$ ($p_{corr} = 0.0$), one straight line will appear, and likewise for $p_{corr} = 1.0$ ($p_{pas} = 0.0$). Then the intersection of the straight line with P = 50% corresponds to the mean potential of either corroding or passive steel, $\mu(E_{corr})$

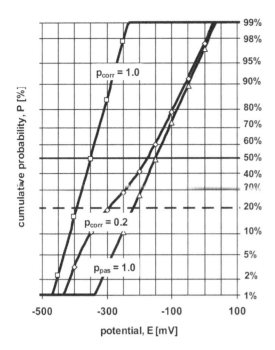

Figure 3. Cumulative frequency distribution of potential mapping plotted on normal probability paper (ideal situation).

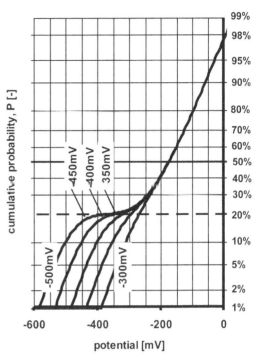

Figure 4. Cumulative frequency distribution of potential mapping plotted on normal probability paper (ideal situation), indicating the influence of $\mu(E_{corr})$.

and $\mu(E_{pas})$, respectively. The slope of the straight line reflects the standard deviation: for $\sigma(E) = 0$ a vertical line will ensue and the higher the standard deviation σ the flatter the slope will be. However, mostly the measured potentials include both actively corroding and passive steel. As $\mu(E_{corr})$ and $\mu(E_{pas})$ usually will differ by at least 100 mV, in practice 2 distinct straight lines can usually be identified. In the example presented in Figure 3, 20% of the potentials pertain to corroding steel and such a situation creates an interjacent portion between the corroding and the passive portion. From such a cumulative frequency graph it can be deduced that the measured potentials comprise a certain amount of corroding steel, however, it is not always possible to obtain a reliable estimate of p_{corr}. This is dependent on the level of p_{corr}, on the difference between $\mu(E_{corr})$ and $\mu(E_{pas})$ as well as on the levels of $\sigma(E_{corr})$ and $\sigma(E_{pas})$. An improved estimate of p_{corr} can be obtained when e.g. the difference between $\mu(E_{corr})$ and $\mu(E_{pas})$ increases, as depicted in Figure 4. Then a distinct shoulder appears at a cumulative probability level $P = 20\%$ for $\mu(E_{corr}) = -500$ mV.

3 INTERPRETATION OF POTENTIAL MAPPING

Provided that a significant amount of measurement data is available, this information can be used for a statistical analysis. The result of such an analysis will be a more reliable estimate of the relative amount

of corroding steel, p_{corr}, as well as improved potential values corresponding to any arbitrary level of the probability of reinforcement corrosion occurring. The underlying calculation procedure has been described in detail by Gulikers (2007). In essence, the procedure includes the decomposition of either the overall frequency density distribution or the overall cumulative probability plot, into a corroding and a passive component by regression. The primary assumption in this statistical analysis is that the potentials of both corroding and passive steel conform to a normal probability distribution. Eventually, this statistical treatment will result in values for p_{corr}, $\mu(E_{corr})$, $\sigma(E_{corr})$, $\mu(E_{pas})$ and $\sigma(E_{pas})$. The quality of the output of such an analysis is dependent on the quality of the input. A case study using data obtained from a tunnel wall was elaborated by Gulikers and Elsener (2008) and this study resulted in satisfactory results. However, it is anticipated that in a number of situations major problems may be encountered which would render the calculation results less realistic. In this article some of these problems are addressed for data obtained from a parking garage.

4 CASE STUDY – PARKING GARAGE

4.1 Measurement results of potential mapping

In order to evaluate the condition of the floors of a multi-storey parking garage regarding reinforcement corrosion, an extensive non-destructive survey

Table 1. General overview of the results of potential mapping on floor A.

Section	A_I	A_II	A_III	A_IV
Number [—]	5057	4153	520	1054
$\mu(E)$ [mV]	−266.8	−257.9	−365.8	−282.7
$\sigma(E)$ [mV]	89.7	70.9	105.4	104.6
E_{max} [mV]	+55	+5	−175	40
E_{min} [mV]	−575	−610	−620	−605

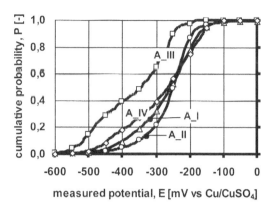

Figure 5. Cumulative frequency distribution of the potential readings for floor sections A_I to A_IV.

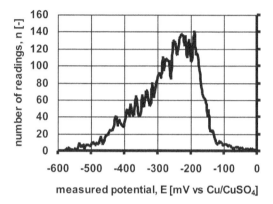

Figure 6a. Frequency distribution of the potential readings for floor section A_I for a class width of 5 mV.

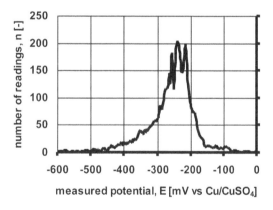

Figure 6b. Frequency distribution of the potential readings for floor section A_II for a class width of 5 mV.

was performed using potential mapping. Preliminary investigations had indicated that excessive amounts of chloride had penetrated towards the embedded reinforcing steel. In the outer layer, the concrete had been severely contaminated by chlorides with contents ranging from 2 to 7% by mass of cement. Potential mapping was performed using a so-called potential wheel according to a 0.25 m × 0.50 m grid. In view of the presence of chlorides, the concrete floors were suspect of localised reinforcement corrosion and with these grid dimensions it was anticipated that all corroding sites would be detected. For practical reasons each of the four concrete floors was subdivided in a number of sections. Prior to execution, the concrete surface to be measured was pre-wetted to ensure good electrolytic contact between the reference electrode (Cu/CuSO₄; CSE) and the concrete. Unfortunately, no further details on cover thickness, type of cement, concrete quality, presence of patch repairs, and other relevant data were available to support the interpretation of the measurement results. In this paper, the statistical analysis will be demonstrated and evaluated for the potential readings obtained for floor A.

Floor A is subdivided into 4 sections referred to by A_I to A_IV. The overall characterisation of the potential readings for each floor section is presented in Table 1 regarding number of readings, the mean, maximum and minimum potential (with respect to a Cu/CuSO₄ reference electrode), as well as the standard deviation.

The cumulative frequency distribution of the 4 sections is presented in Figure 5, whereas the frequency density distributions for the individual floor sections are shown in Figures 6a to 6d (based on a potential class width of 5 mV).

From these results it can be deduced that approximately 19% of the readings of floor section A_I, 11% of section A_II, 48% of section A_III, and 28% of section A_IV demonstrate potential values less than −350 mV vs Cu/CuSO₄. According to ASTM C-876 then there will be more than 90% probability that at these areas corrosion of the reinforcement steel is occurring at the time of measurement.

However, the cumulative frequency distribution as well as the frequency density distributions indicates distinct differences between the individual floor sections. These differences also come into clear view when the cumulative frequency distribution is presented on normal probability paper, as shown in Figure 7.

4.2 Statistical analysis of the results

In order to obtain an improved estimate of the relative amount of corroding steel in the floor sections the potential readings were subjected to a regression analysis. In addition, the output comprised a quantification of the mean value and the standard deviation of the potentials of the passive and the corroding steel

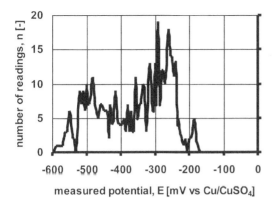

Figure 6c. Frequency distribution of the potential readings for floor section A_III for a class width of 5 mV.

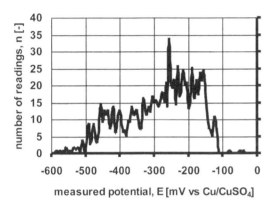

Figure 6d. Frequency distribution of the potential readings for floor section A_IV for a class width of 5 mV.

within a particular floor section, i.e. μE_{corr}, $\sigma(E_{corr})$, $\mu(E_{pas})$, and $\sigma(E_{pas})$. This regression analysis was performed on both the cumulative frequency distribution (presented in Figure 5) as well as the frequency density distributions as presented in Figures 6a to 6d. The overall results corresponding to the best fit, are presented in Table 2a. In addition, the regression was also performed by using a fixed value of $\sigma(E_{corr}) = 50\,mV$ as advocated by Gulikers and Elsener (2008). These results are summarized in Table 2b.

The calculated results indicate that for all sections a significant area of the floor contains corroding steel. According to the regression analysis in which all parameters are optimised as to achieve the overall best fit (Table 2a), the relative number of potential readings pertaining to corroding steel, p_{corr}, ranges from 32.4% (section A_IV) to 51.4% (section A_I). These levels seem unrealistically high but due to the lack of reliable information on the actual condition of the floor these results cannot be validated. When it is assumed that $\sigma(E_{corr}) = 50\,mV$ the relative number is significantly less with 3 sections demonstrating a p_{corr} of approximately 23%, whereas for Section A_III the regression analysis results into $p_{corr} = 40\%$.

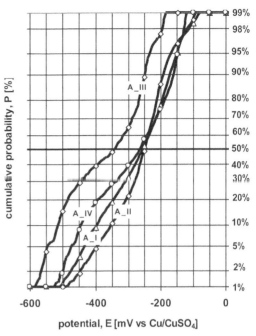

Figure 7. Cumulative frequency distribution of the potential readings for floor sections A_I to A_IV (normal probability paper).

Table 2a. General overview of the results of regression analysis performed on the half cell potential data of floor A.

Section	A_I	A_II	A_III	A_IV
p_{corr} [−]	0.514	0.481	0.456	0.324
$\mu(E_{corr})$ [mV]	−317.8	−282.1	−469.9	−404.0
$\sigma(E_{corr})$ [mV]	87.1	83.1	63.8	70.6
$\mu(E_{pas})$ [mV]	−214.6	−235.0	−280.6	−219.4
$\sigma(E_{pas})$ [mV]	47.4	31.0	44.7	64.4

Table 2b. General overview of the results of regression analysis performed on the half cell potential data of floor A; $\sigma(E_{corr}) = 50\,mV$.

Section	A_I	A_II	A_III	A_IV
p_{corr} [−]	0.226	0.225	0.400	0.236
$\mu(E_{corr})$ [mV]	−377.7	−342.0	−477.6	−425.1
$\sigma(E_{corr})$ [mV]	50.0	50.0	50.0	50.0
$\mu(E_{pas})$ [mV]	−229.4	−235.9	−285.1	−229.0
$\sigma(E_{pas})$ [mV]	50.1	36.7	50.5	72.1

Furthermore the results can be translated in potential criteria related to any arbitrary probability of reinforcement corrosion. These criteria can be regarded as an improvement compared to the strict criteria suggested by ASTM C-876. In Table 3 the criteria are presented corresponding to 10% and 90% probability

Table 3. Calculated potential criteria corresponding to probabilities of 10% and 90% of reinforcement corrosion.

Section	A_I	A_II	A_III	A_IV
E (90%) [mV]	−378	−336	−416	−407
E (10%) [mV]	−288	−269	−358	−320

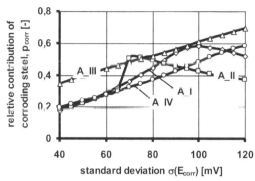

Figure 10. Relative contribution of corroding steel, p_{corr}, as a function of the standard deviation of potentials for corroding steel, $\sigma(E_{corr})$ for each floor section.

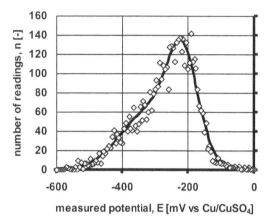

Figure 8. Frequency distribution of the potential readings for floor sections A_I; original data and regression curve.

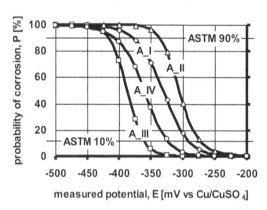

Figure 9. Potential criterion as a function of probability of steel corrosion for each section.

of steel corrosion, whereas in Figure 9 for ach floor section the relationship between probability of steel corrosion and the potential criterion is shown.

5 DISCUSSION

Given a significant amount of potential readings, the data can be used for a statistical analysis to provide an improved quantification of the relative amount of corrosion and as well as potential criteria related to a probability of steel corrosion. For the case study addressed in this paper, the regression analysis revealed that the outcome of such a regression analysis in terms of relative amount of corroding steel could be unrealistically high. This can be improved by adjusting the magnitude of some parameters, e.g. the standard deviation of potentials for corroding steel to $\sigma(E_{corr}) = 50$ mV. However, care should be exercised in doing this as it can easily lead to arbitrary choices as to achieve a more realistic result. The mathematical treatment of this case study highlighted the fact that the outcome is very prone to the chosen level of $\sigma(E_{corr})$, as shown in Figure 10.

A validation of the outcome by visual inspection of the electrochemical condition of the embedded steel is urgently required to allow for a sound evaluation of this analysis. In addition, it should be emphasized that the case study appeared to be more complicated to arrive at an unambiguous interpretation.

6 CONCLUDING REMARKS

A regression analysis performed on the results of potential mapping could prove to be of help to support the interpretation. In addition to the relative amount of corroding steel such a statistical procedure results in improved potential criteria with respect to the probability of steel corrosion for any arbitrary level. Moreover, this procedure takes into account the specific nature of a structure and the exposure conditions prevailing during the measurements. However, care should be exercised with the calculated outcome as in certain situations this could turn out to be unrealistic. Moreover, the consultant should be aware that the outcome of a regression analysis is dependent on the implicit assumptions made. The availability of additional information on the concrete structure, e.g. age, exposure conditions, geometry, cover depth, presence of surface cracking and spalling, concrete composition, type of cement, temperature and humidity at the time of measurement, is required in order to arrive at a reliable and realistic interpretation. Such an interpretation of non-destructive measurements should always be validated by visual inspection of the condition of the reinforcement steel at selected spots identified by the analysis to contain corroding or passive steel.

ACKNOWLEDGEMENTS

The author expresses his appreciation to Dr. Andreas Burkert of BAM (Berlin) for providing the data of the potential survey.

REFERENCES

ASTM C 876-91. 1999. *Standard test method for half-cell potentials of uncoated reinforcing steel in concrete.* West Conshohocken: ASTM.

Broomfield, J. 1997. *Corrosion of steel in concrete.* London: E&FN Spon.

Elsener, B. 2003. Half-cell potentials measurements – Potential mapping on reinforced concrete structures, *Materials and Structures* 36(8): 461–471.

Gu, P. & Beaudoin, J.J. 1998. Obtaining effective half-cell potential measurements in reinforced concrete structures. *Construction Technology Update No. 18.* Ottawa: IRC

Gulikers, J. 2007. *Half-cell potential mapping on reinforced concrete structures – Development of a procedure for the statistical analysis of measurement results.* Utrecht: Rijkswaterstaat.

Gulikers, J. & Elsener, B. 2008. Statistical interpretation of results of potential mapping on reinforced concrete structures. In L. Binda, L., M. di Prisco, & R. Felicetti, (eds), *On site assessment of concrete, masonry and timber structures SACoMaTiS: 221–230; Proc. intern. RILEM symp., Varenna, 1–2 September 2008.* Bagneaux: RILEM.

Concrete Solutions – Grantham, Majorana & Salomoni (Eds)
© 2009 Taylor & Francis Group, London, ISBN 978-0-415-55082-6

Durability and serviceability problems in the Kuwait National Assembly structure and ongoing investigations and repairs

M.N. Haque
Civil Engineering Department, Kuwait University, Kuwait

M. Al-Busaily
Kuwait National Assembly

ABSTRACT: The Kuwait National Assembly (KNA) which is situated on the sea-shore was completed in 1985. Accordingly, the structure is subjected to aggressive marine conditions in hot-dry and hot-humid exposure conditions. The KNA structure was also attacked during the Iraqi invasion which burned and ruined some portions of the structure. Most of the structural elements of the KNA are of precast and prestressed concrete. The fire caused extensive cracking and spalling of the concrete. In addition, there have also been several water leakage, joint leakage, cracking, efflorescence and corrosion problems in the basement of the Assembly and the multi-storey car park. Several investigations are in progress to limit further damage to the facilities. The paper includes the results of investigations undertaken and some remedial measures and repairs recommended.

1 INTRODUCTION

Serviceability and durability of a concrete building require studies of the environmental exposure and loading conditions, in order to decide the best maintenance methods to protect the structural members. Kuwait is located in an arid desert area which has hot-dry weather in the inland areas and hot-humid-salty weather in the coastal areas. Dusty winds blow aggressive salts which precipitate on the concrete buildings especially at high humidity in the coastal structures, as is the case with the Kuwait National Assembly.

The State of Kuwait lies at the North-West corner of the Arabian Gulf, with which it shares its East border. To the South and South-West, it shares borders with the Kingdom of Saudi Arabia, and to the North and West it shares borders with the Republic of Iraq (see Figure 1).

Most of Kuwait's mainland is a flat sandy desert, gradually sloping down from the extreme west of Shigaya and Salmi (300 m high) to reach sea level on the East Coast of the Gulf (Kuwait Facts).

Owing to the location of Kuwait in the Sahara geographical region, its weather is characterized by long, hot and dry summers and short, warm and sometimes rainy winters. Dust storms almost always occur during the year and humidity rises during summer. The highest temperature ever recorded in Kuwait was 51°C, recorded in July 1978, whilst the lowest was −4°C, recorded in January 1964. The average temperatures range from 45°C in summer to an average of 8°C in winter. Such climate fluctuation is often accompanied by a change in the annual averages of rainfall which ranges from 22 mm to 352 mm.

Figure 1. Kuwait Geological Map.

Arabian water records high levels of salinity, where chloride content is nearly (1.6 − 2) times higher than water of the Mediterranean and the Atlantic sea (Haque et al. 2006).

2 KUWAIT NATIONAL ASSEMBLY (KNA) AT A GLANCE

KNA is located in Kuwait City, in the Al-Qeblah area facing the Gulf road (see Figures 2 and 3). Jorn

Figure 2. A Satellite picture shows the location of the KNA at the Gulf Road.

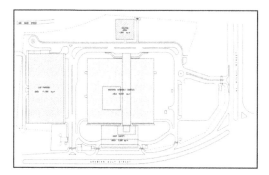

Figure 3. The KNA layout.

Figure 4. The KNA Under construction.

Figure 5. The main canopy under construction.

Figure 6. The KNA after construction.

Figure 7. Proximity of the KNA to the Gulf Coast.

Utzon, best known as the architect who designed the Sydney Opera House, was invited to design the Kuwait National Assembly. The project lasted for 7 years (1978–1985) and cost about 25,869,703 KD (KNA).

Figures 3, 4, 5 and 6 show the KNA during and after the construction phase. It is clear that most of the structural elements are pre-cast and pre-stressed concrete, specially the monumental main canopy which covers the VIP entrance, and the secondary canopy which covers the Assembly Hall.

The main building is composed of a basement, ground floor, first floor and a car park.

2.1 Aggressive environment of KNA

The KNA is located in very harsh and aggressive environmental conditions. According to Euro codes, such a monumental building is designed to serve for 100 years, however, the exposure conditions where the KNA is located would adversely affect the service life and durability of the concrete construction.

Referring to the classification established by Haque, et al. (2006) based on proximity to the sea, it is clear that the KNA is classified as a coastal structure, since it is only about 200 m away from the Gulf coast. (see Figure 7). The concrete would be attacked by several corrosive agents such as:

a) Dampness attracting salts and fungal growth
b) Chloride build-up from salt spray, soils and ground water
c) Carbonation due to moderate relative humidity (55–75%)
d) Sulphate-rich coastal soils induce sulphate attack
e) During Iraqi invasions, the main hall and the western part of the building was set on fire resulting in significant damage.

276

Figure 8. Basement wall leakage (2006).

Figure 9. Basement wall expansion joint (2007).

2.2 *General Repair of KNA*

In July 1991, reconstruction processes were undertaken to repair the damaged parts of the building, which was finally completed in October 1992 with a cost of 19,394,705 KD.

General methods for repairing fire damaged structural reinforced concrete consisted of the following:

a) Spalling: Patch repair was done using polymer modified cementitious and epoxy mortar. Major spalling was repaired using pressure grouting (prepacked aggregate or guniting, sprayed concrete).
b) Cracks: Epoxy injection was carried out for small (hair line) cracks. Larger cracks (width $>2\,mm$) were repaired using Portland cement grouting. Strengthening and reconstruction was done where extensive disintegration/loss had taken place.

3 CONCRETE DETERIORATION AT THE KNA BASEMENT WALL (AL-QESEER)

The KNA basement wall which is about 4 m below ground level, facing the Gulf sea, was affected by leakage through the expansion joints and other discrete weak areas. The concrete had already deteriorated and cracked, some reinforcement bars were exposed and corroded severely, joint sealants had been eroded and removed leaving no protection to the substrate beneath. Leakage from columns and suspended beams had also taken place in the basement area. (see Figures 8, 9, 10, 11). Basement leakage problem is considered to be an old problem since 1994.

KNA engineers met with Ministry of Public Works (MPW) specialists in order to solve this problem. MPW took two water specimens in order to determine the chemical composition of the leaked water, Table 1. According to the ionic concentration, it was concluded that the specimens were from low salinity water (ground water or brackish water). This test was performed by Government Center for Testing and Research in October-1994. At the same time the work team had met and decided to use piezometers to measure the increase and drop of the basement ground water level for a period of three months.

Figure 10. Reinforcement corrosion (2007).

Figure 11. Lower Basement Leakage (2007).

Three piezometers were fixed in the agricultural area surrounding the main building for a depth of 10 m below ground level, as shown in Figure 12 and readings were taken from Jan-1995 till Oct-1995.

In Dec-1995, MPW provided KNA engineers with a technical report with a problem analysis and some recommendations. The report referred to the data gathered from the piezometers, old borehole tests (1986) and monthly water consumption measurements compared with irrigation water consumption.

Table 1. Chemical analysis of the leaked water.

Parameter	Specimen 1	Specimen 2
Sulphates (ml.g/ℓ)	2150	3050
Soluble Sulphides (ml.g/ℓ)	0	0
Required chemical oxygen (ml.g/ℓ)	20	80
Suspended solids (ml.g/ℓ)	7	953
Colonic Bacteria (Colony/100 ml)	0	0

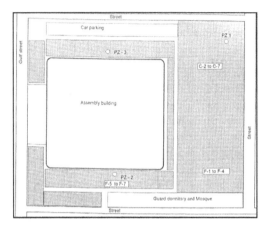

Figure 12. Piezometer locations (1995).

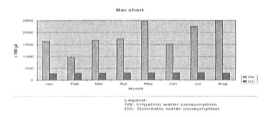

Figure 13. Monthly water consumption comparison between IW and DC (1995).

The two causes of the problems found were:

a) Tidal movements because of the building proximity to the Gulf sea. This cause was excluded, because the Gulf sea water level varies daily, but this was not the case with the basement water level.
b) Surface water due to: b1) Rain: this cause was excluded; because the basement problem remained even if there was no rain. b2) Domestic water consumed: The water consumed in WCs compared to the irrigation water consumed was 1: 10 (see Figure 13).
c) From Figure 14, it is noticed that, first, the ground water table level was lower than the basement level till the end of April-1995, and then it increased relatively. Secondly, the irrigation water level was higher than the basement level till the end of April-1995 and then suddenly increased resulting in an increase of the basement water level.

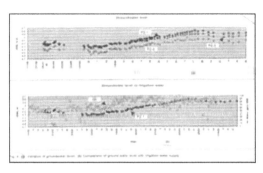

Figure 14. Groundwater level – ground water level vs. irrigation water (1995).

Figure 15. Concrete slab joint insulation (2006).

After the previous analysis, which MPW had stated in the technical report mentioned earlier, two recommendations were made:

a) Attempt to decrease irrigation water consumed and maintain a stable consumption level throughout the year.
b) Dig 4 deep ground holes or more to fix dewatering systems around the building.

At the time when the piezometric tests were being carried out, contact with specialized basement leakage companies was taking place in order to prepare curing systems and recommendations to solve this problem from their point of view.

It is important to say that the problem is still alive and active. The pictures that follow were taken in 2006–2007. Meetings between KNA and MPW engineers are still being held to study and decide the best solution for this problem which really affects concrete durability in the basement wall.

There are some recent preventive steps taken to stop the leakage such as:

a) Repairing and insulating the expansion joints of the concrete slabs in the ground floor along the VIP main entrance, up to 10 m away from the basement wall. 25 mm diameter sponge rods were used in the joints and sealed by mastic. (see Figure 15). However, it has been noticed that some of the mastic sealants have already started to fail.
b) Protecting the walkway on the VIP main entrance by waterproofing cementitious materials and concrete epoxy coatings (see Figure 16). However, it

Figure 16.   VIP entrance concrete slab insulation (2006).

Figure 17.   The camera beside the main hole (2006).

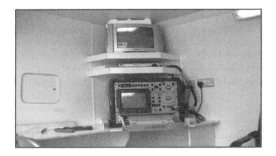

Figure 18.   The monitoring system (2006).

has been noticed that the epoxy coating cracked after less than a year removing layers underneath the protective waterproofing cementitious materials.

c) One of the possibilities of the leaked water source was that water could seep from a defect or a crack in the main rain pipe, along the width of the VIP main entrance at the ground floor, to the basement wall. The rain pipe was checked by MPW under the supervision of KNA engineers, with a developed and accurate video camera which was running inside the pipe and connected to a monitoring system , in turn connected to a TV monitor (see Figures 17 and 18). It was found that the pipe was free of defects and cracks.

d) Manholes, both water and electrical, in the VIP main entrance at the ground floor above the basement wall are being maintained and waterproofed

Figure 19.   Main holes maintenance (2007).

Figure 20.   Efflorescence on the basement floor (2007).

to ensure no seepage proceeds from them. Electricity manholes were raised 100 mm above ground level to prevent cleaning water entrance at this area (see Figure 19).

e) According to the MPW report in July-2006, it was concluded that the leakage water is due to cleaning the outer surface of the concrete slabs adjacent to the wall at the main VIP entrance.

f) MPW provided KNA engineers with specifications and suggestions from four different companies. All of them suggest injection curing with different materials. No action as yet has been taken

4   EFFLORESCENCE ATTACK

Efflorescence is a white crystalline or powdery, often fluffy/fuzzy deposit on the surface of materials like concrete, brick, clay tile, etc. It is caused by water seeping through the wall, floor, etc. The water dissolves salts inside the object while moving through it, then evaporates leaving the salt on the surface (Al-Jassim 2007).

It was noticed that efflorescence problems were attacking several locations in the KNA such as: basement floors, basement interior walls, basement columns, car park beams and columns, VIP main entrance concrete floor slabs etc. Pictures gathered from different affected locations are shown in Figures 20, 21, 22.

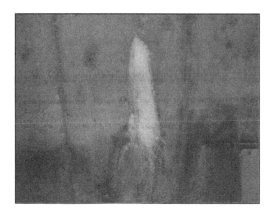

Figure 21. Basement concrete wall efflorescence (2007).

Figure 22. Leakage and efflorescence on some car park suspended beams (2007).

Figure 23. Basement column corroded steel bars (2007).

4.1 *Effects of efflorescence attack on reinforced concrete*

Efflorescence is adversely affecting the structural concrete since water contaminated with salts seeps through the permeable layer and stains the surface allowing ingress of aggressive materials inside the concrete body. Corrosion can occur aggressively if this is not stopped or repaired correctly and at the right time because of the buildup of corrosive materials. Figure 23 shows corroded steel bars in the KNA basement.

Owing to the ground water level investigated by the piezometers in (1995) mentioned previously it was noticed that, through July and August, ground water levels increased to about the basement level which is 4 m below ground level. It was noticed in this period that there was leakage/seepage in the basement floor concrete slab. The seeped water was ground water with soluble salts and sulphates and may also contain some other corrosive agents, due to the underlying soil layers indicated in the borehole test in (2006). It seeps even to the column in this area and crystallized beneath the concrete cover causing reinforcement corrosion, concrete cracks and spalls caused by the expansive steel volume. Walls surrounding this area are also affected by this whitish aggressive powder to about 1m high from basement floor. This problem is still being studied by KNA engineers.

5 KNA CAR PARK BUILDING CONCRETE DURABILITY PROBLEMS

Several problems have been manifested in the parking building from the basement to the roof. Insufficient concrete cover has exposed reinforcement bars in different suspended beams and walls. It is well known that concrete cover is the first defence line against ingress of aggressive agents. Insufficient cover adversely affects the durability and serviceability of the building by reinforcement corrosion, if the bars are not protected. It has been noticed also that water had seeped into some columns, even after rain had stopped for about a week, which may indicate that concrete permeability is low and ingress of contaminated water inside the concrete body had taken place. Concrete slabs in different locations of the parking garage are subjected to cracks patterned along the embedded reinforcement, both longitudinal and transverse, which are indicative of reinforcement corrosion under the concrete cover, but not yet spalled. Efflorescence, salt and chloride build-up, joints sealant deterioration, scattered cracks and other problems still attack the KNA car park and need to be repaired.

Water leakage and the micro climate of the effected concrete elements could be the main reasons behind such problems. Water proofing mechanisms used for insulating the parking roof since the re-construction phase were reported to be useless and inefficient. After the Iraqi invasion, the American military applied the following system directly on the exposed concrete surface to insulate the parking roof depending on the idea of water filtration:

a) Bitumen membrane 2 mm;
b) Fabric filter layer;
c) Gravels;
d) Tiled walk-ways.

Since the application of the filtration system, two main problems have occurred:

a) A sandy thick layer has precipitated on the filter preventing water drainage.

Figure 24. Affected weak positions

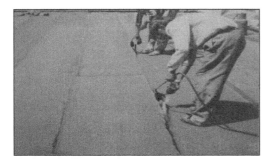

Figure 25. Burning the bituminous membrane for adhesion.

b) The tiled walk-ways were not well-leveled, therefore they were moving and destroying the layers beneath.

Figure (24) indicates the weak position of stones.

After the apparent failure of the above method, KNA decided to apply a new system. In year 2000, the old water proofing system was replaced by a new system. This operation cost about 80,000 KD. And in this system bituminous membranes were laid directly upon the concrete surface. The following were used:

a) Ordinary bituminous membrane
b) Gravelled bituminous membrane

Each layer was installed in a reverse direction to the other. Figure 25 shows the installation process of this system.

This system remained till 2006, and the regular maintenance applied to this system does not exceed patching the visible expected weak locations with new membranes (see Figure 26). In early 2006, a partial maintenance operation took place by rolling two layers of capcoat on the sealed membranes joints and installing a capsilver solar reflection protection.

Nevertheless, it has been noticed that the above water proofing system faces several problems: Leakage, concrete cracking, precipitation of water over and under the membranes adjacent to the concrete surface and deterioration of expansion joint water proofing materials and sealants (see Figure 27). It can be said that concrete damp surfaces should be exposed to ensure drying and not kept humid, which may accelerate the transport of aggressive soluble agents into the concrete pores.

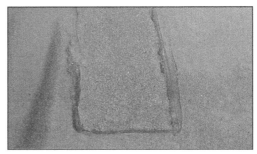

Figure 26. Patching the old bituminous membrane.

Figure 27. Deterioration of expansion joints sealants and damp concrete surface (2006–07).

Figure 28. Car park leakage (2007).

In general, the car park building is suffering from a combination of aggressive attack which affects concrete durability:

a) Water leakage, Figure (28)
b) Efflorescence, Contamination and build-up of corrosive salts and chlorides etc. Figure (29)
c) Differential cracks, Figure (30)
d) Expansion joint deteriorations and cracks at the ramps, Figure (31)

5.1 KNA Car Park Cracked Beam

This is an old problem, which was maintained by MPW in 1995. Initially, there was a water tank on the first floor directly above the affected beam. Owing to its loading on this beam, a shear crack had formed. MPW solved this problem by removing the water tank and supported the concrete beam by H-beams. The

Figure 29.   Aggressive agent buildup (2007).

Figure 32.   The cracked beam (2007).

windows opening directly to the Gulf sea. Relative humidity in such a coastal area is 95%. Moisture carrying aggressive species such as chlorides and salts can easily ingress into this wide crack. Efflorescence has also occurred, as well as the dusty contaminated winds which blow in and out. All of these agents collectively have worsened the state of this cracked beam, of great concern, since it is a vital structural element in the ramp case between the ground and the first floor of the car building. This problem is still being studied by KNA engineers.

6   CONCLUDING REMARKS

It can be summarized that, there are several problems in the KNA building regarding durability and serviceability of the structure. Therefore, special measures should be undertaken to remedy and maintain the affected members. Studies of the suggested remedial materials and methods are still going on, in order to decide the best economical and effective solutions. In the meantime, common sense routine maintenance regime is in place.

Figure 30.   Wall cracking (2007).

REFERENCES

Kuwait Facts and Figures (7th Edition) State of Kuwait – Ministry of Information
Haque, M.N., Al-Khaiat, H. and John, B., Proposal for a Draft Code for Designing Durable Concrete Structures in the Arabian Gulf. The Arabian Journal for Science and Engrg., Vol. 31, No. 1C, pp. 205—214, June 2006.
KNA Public Relations Office.
Whai, F.F., Structural Appraisal & Repair of Fire-Damaged buildings, Building Service Unit, KNA.
Al-Qeseer, N.K., KNA old Portfolios, Kuwait National Assembly – Technical Department Supervisor.
Al-Jassim, J.M., Private Communication, Kuwait National Assembly – Technical Department, Civil Maintenance Supervisor. 2007.
http://www.answers.com/topic/kuwait-oil-fires-persian-gulf-war (Accessed 5 April 2009)
http://factsfacts.com/MyHomeRepair/efflorescence.htm (Accessed 5 April 2009)

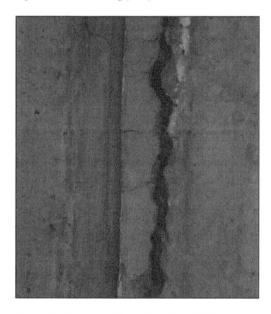

Figure 31.   Expansion joint wall cracking (2007).

problem has, however, appeared again. Shear cracks are present as shown in Figure 32.

Cracking is not the only problem in this case, but what is worse is that this beam is positioned near

Concrete Solutions – Grantham, Majorana & Salomoni (Eds)
© 2009 Taylor & Francis Group, London, ISBN 978-0-415-55082-6

Mass loss evaluation of a roller-compacted concrete dam: Understanding the problem for planning repair intervention

Alex Joukoski & Kleber Franke Portella
LACTEC, Curitiba, PR, Brazil

José Carlos Alves Galvão
UFPR/PIPE, Curitiba, PR, Brazil

Emerson Luiz Alberti
ELEJOR, Curitiba, PR, Brazil

Marcos Alberto Soares
COPEL-GER, Curitiba, PR, Brazil

Paola Tümmler
UFPR/DQ, Curitiba, PR, Brazil

ABSTRACT: Roller-compacted concrete (RCC) has been widely employed on recent hydroelectric dam constructions in Brazil. Its higher permeability, when compared to conventional concrete, it is well known by the technical community. Water infiltration in RCC structures is usually associated with leaching of solid material from cement and/or aggregates used in the concrete, which can lead to a significant mass loss over the years. Depending on the extent and distribution of mass loss, repair interventions need to be planned, in order to ensure structure stability and safety. A methodology for the evaluation of concrete mass loss, consisting of the product between flow and the difference of total ionic concentrations of percolated water samples collected in the dam inspection gallery and reservoir water samples, is proposed in this paper. After one year of periodical monitoring of a RCC gravity dam in Iguassu river basin, Southern Brazil, global results indicate about 170 kg of monthly mass loss.

1   INTRODUCTION

Roller-compacted concrete (RCC) is mainly characterized by its lower cement and water contents and lower consistence, if compared to conventional concrete. This properties lead to mixtures with reduced workability, requiring the use of special compacting methods and improved *in situ* quality control (Andriolo 1998).

Nowadays it is well known that RCC has been successfully used as a civil construction material in a variety of works, from road paving to hydroelectric power plant dams.

The earliest studies regarding RCC begun in the United Kingdom in the 1940's, initially applied to pavement development. In the following decades, evaluation of use in hydraulic structures was also studied, resulting in the first dams constructed with this technique. Shimajigawa and Willow Creek dams, respectively in Japan and United States, were the world's foremost barrages entirely built with RCC (Andriolo 1998)

In Brazil, the preliminary employment of roller-compacted concrete took place in the construction of Sao Simao dam (1977) and Saco de Nova Olinda dam

(1986), the latter being the first national dam totally made with this material (Andriolo 1998).

According to the International Water Power & Dam Construction yearbook (International Water Power & Dam Construction. Year Book 2007), by the end of 2007 there were more than 330 RCC dams in the world, of which almost 60 were located in Brazilian territory. One of these dams is the object of the study presented in this paper.

The assessed structure is part of a 246 MW hydroelectric complex in Iguassu river basin, in Southern Brazil. It is a 588 m long gravity dam, with 67 m maximum height. Its construction has employed about 480,000 m³ of concrete, and instrumentation sensors and drainage system (consisting of 286 drains) were systematically implemented in the dam bulk. The power plant operation started in August 2005.

2   EXPERIMENTAL PROCEDURE

The experimental procedure consisted of bi-monthly collection of water samples from previously defined collection points inside the inspection gallery and in

Table 1. Points of water sample collection.

Collection point type	Quantity	Remarks
Flow meters	2	Gather all seepage water from dam body
Joint drains	6	Two on the right bank side and four on the left
Curtain drains	9	Four on the right bank side and five on the left
Foundation drains	6	Three on each bank side of the dam
Joint cracks	2	Dripping points out of joint drains
Reservoir	3	Distinct depths: surface, middle, bottom

Figure 2. Typical triangular-shape flow meter, located inside dam inspection gallery.

Figure 1. Types of drains inside dam inspection gallery: foundation drain (top left); curtain drain (top right); joint drain (bottom).

the reservoir. These points included foundation, curtain and joint drains, as well as flow meters. A total of 28 places were selected for water sample collection, as specified in Table 1.

The drain selection was done in terms of individual flow, the ones with higher values being chosen. The different types of drains are shown in the photographs of Figure 1.

In the moment of collecting, temperature of each sample and flow of every point were measured.

After all samples had been collected in each campaign, they were brought to the laboratory for physicochemical analysis, according to APHA methodology

(APHA 1998), in order to determine the following parameters:

- pH,
- conductivity,
- total alkalinity,
- alkalinity to phenolphthalein,
- suspended solids,
- dissolved solids,
- total solids,
- turbidity,
- hardness,
- and ionic concentration (Ca^{2+}, Mg^{2+}, Na^+, K^+, Fe^{3+}, Al^{3+}, Si^{4+}, F^-, Cl^-, NO_3^-, NO_2^-, SO_4^{2-}, Br^-, and PO_4^{3-}).

The equipment used consisted of a Perkin-Elmer 4110 atomic absorption spectrophotometer and a Dionex 2010 I high performance ion chromatograph.

The results of ionic concentrations for each collection point were subtracted from the corresponding mean value of the reservoir samples. To determine the amount of ions leached out from concrete by every drain, the water flow was considered. In order to know the total ionic mass loss of the barrage body, only flow meter results were enough to allow such a determination, since both points are located in the lowest part of inspection gallery, damming up all seepage water from the structure. A typical triangular-shape flow meter can be seen in Figure 2.

Seepage water samples have been collected every 2 months since October 2007, and the study is going to continue for a further 2 years.

3 RESULTS AND DISCUSSION

3.1 Physicochemical analysis of water samples

The results of physicochemical parameters of water samples collected during each campaign in the pair of flow meters (named FM1 and FM2), as well as the respective measured flow values, are shown in the

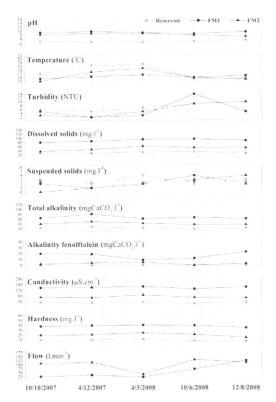

Figure 3. Variation along time of physicochemical parameters of water samples from FM1, FM2 and reservoir.

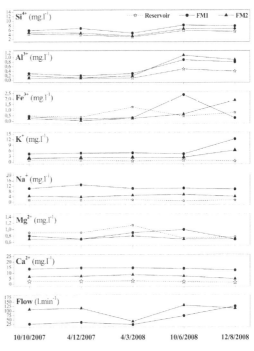

Figure 4. Variation along time of cations concentration of water samples from FM1, FM2 and reservoir.

graphs of Figure 3. FM1 gathers all seepage water from left bank side of the barrage, while FM2 dams up the water from right bank side. The correspondent mean values from reservoir samples are also shown.

The graphs showing the results of cation and anion concentration of water samples collected in the FM1 and FM2 flow meters and in the reservoir during each campaign are presented in Figures 4 and 5, respectively. Corresponding measured flow values are also shown.

3.2 Estimation of ion mass loss

The partial results of ionic mass loss, estimated for the whole dam, due to concrete leaching by water seepage through the dam bulk are presented in Figure 6. Data is available for each sample collecting campaign, which has been performed every 2 or 3 months since October 2007. In the main graph, total and individual ionic mass loss results of both FM1 and FM2 flow meters are shown. Information about flow and reservoir water level can also be seen in the bottom graph.

It is perceptible that ion mass loss and flow values have a direct relationship to reservoir level, which is a fairly normal behaviour, due to the consequent higher hydrostatic pressure.

Considering the average ionic mass loss value from the five campaigns so far done, which is around

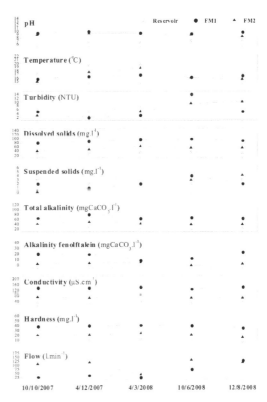

Figure 5. Variation along time of anions concentration of water samples from FM1, FM2 and reservoir.

285

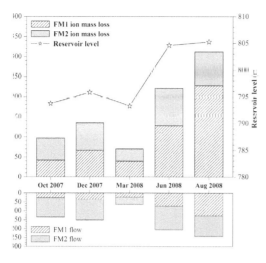

Figure 6. Estimated monthly ion mass loss for each campaign period.

$167 \, \text{kg} \cdot \text{month}^{-1}$, the estimated annual amount of ion mass loss may reach 2 tons.

The obtained value is rather normal if we bear in mind that it represents merely 0.00016% of the dam total concrete mass, which is about 1,272,000 tons. Besides, the estimated ion mass loss is a global parameter, since it includes all infiltration water that seeps through the structure and converges to flow meters.

Therefore, it is important to evaluate individual ionic mass loss for each drain – or, at least, for the ones presenting higher flow values – in order to detect expressive mass losses that may become structural problems at specific regions of the dam. Putting this procedure in practice is very important in terms of giving maintenance engineers the accurate information needed for immediate or future repair intervention.

4 CONCLUSIONS

Despite the study being at its beginning – only five campaigns have been done so far – the evaluation procedure has proved to be an appropriate method for estimating dam ion mass loss caused by water infiltration in the concrete dam bulk. Important information regarding global seepage levels could be obtained, besides the individual data obtained for the other 23 drains and joint cracks, not shown in this paper. Considering the average bi-monthly ionic mass loss results presented by both flow meters, an estimated annual value of about 2000 kg has been appraised. Moreover, a direct relationship between flow and ionic mass loss has also been noticed.

Periodical and constant monitoring of ion mass loss of drains, joints and cracks can point to local structural damage caused by concrete, mortar and/or cement leaching, so continuity of this study is quite essential. Not only for providing maintenance staff with precise data, but also for guaranteeing the dam stability and safety.

ACKNOWLEDGMENTS

The authors would like to thank LACTEC, ELEJOR, COPEL, ANEEL and CNPq (PIBITI) for providing technical and financial support for accomplishment of this work.

REFERENCES

American Public Health Association. *Standard Methods for the Examination of Water and Wastewater.* APHA, Washington, 1998.
Andriolo F. R. *The Use of Roller Compacted Concrete.* Oficina de Textos, São Paulo, 1998.
International Water Power & Dam Construction. *Year Book 2007.* IWPDC, United Kingdom, 2007.

Concrete Solutions – Grantham, Majorana & Salomoni (Eds)
© 2009 Taylor & Francis Group, London, ISBN 978-0-415-55082-6

Existence of deterioration in concrete balconies

J. Lahdensivu, I. Weijo & J. Mattila
Department of Structural Engineering, Tampere University of Technology, Tampere, Finland

ABSTRACT: Concrete structures exposed to outdoor climate are deteriorated by several different deterioration mechanisms, whose existence and progress depend on many structural, exposure and material factors. Under Finnish outdoor climate the freeze-thaw damage of concrete together with corrosion of reinforcing steel is the major deterioration mechanism that causes the need to repair of concrete facades. During ongoing project *Repair strategies of concrete facades and balconies*, it has been gathered the condition investigation data from 946 buildings to a database. 89% of protective pore ratios in Finnish concrete balconies are less than 0.20 and 78% are less than 0.15. Despite this, 56–70% of buildings, depending on the buildings location, have any visible freeze-thaw damage. This means, it is possible to have good results with thorough patch repair and protective coatings with balcony glazing in most of the cases. Protective measures slow corrosion rates by 30–80% and freeze-thaw damage stops completely.

1 INTRODUCTION

1.1 Background

Since the 1960's a total of about 44 million square metres of concrete-panel facades have been built in Finland as well as almost a million concrete balconies, Vainio et al (2005). In fact, more than 60% of Finnish building stock has been built in the 1960's or later. Compared with the rest of Europe the Finnish building stock is quite young.

Despite being of quite young age, the Finnish building stock has encountered several problems in its maintenance and repair. The structures have deteriorated due to several different deterioration mechanisms whose progress depends on many structural, exposure and material factors. Thus, the service lives of structures vary widely. In some cases the structures have required, often unexpected, technically significant and costly remarkable repairs less than 10 years after their completion. For that reason, Finland has, during the last 20 years, developed many new methods for maintaining and repairing these structures. The methods include a condition investigation practice and its extensive utilization, rational repair methods and their selection as well as first-rate repair products and appropriate instructions for managing repair projects.

The value of Finnish buildings and infrastructure is about 300 billion euros, Vainio et al (2002). The construction and real estate business accounts for more than 30% of the Finnish gross domestic product. Concrete structures have been repaired extensively in Finland since the early 1990s. During that almost 20 year period, about 10 percent of the stock built in 1960–1980 has been repaired once. It is estimated that the total annual value of building repair work in Finland

is about 5500 million of euros, of which about 30% involves external structures (walls, balconies, roofs, windows, etc.). The total annual volume of facade renovation is about 15 million m². In addition, 40,000 balconies are repaired annually and 4500 new balconies are added to old buildings. It is estimated that the volume of facade renovation will grow 2% annually, Vainio et al (2002 & 2005). Because of the great amount of those existing concrete structures, it is very important to solve their repair needs economically and in a technically durable way. This means, we have to use the most suitable repair methods for each case and it is also important to be able to determine the optimal time of those repairs.

1.2 Structures in concrete balconies

The most common type of balcony from the late 1960's in Finland consists of frame, slab and parapet elements which are prefabricated from concrete. Those balcony elements are used in so called "balcony-towers," which have their own foundations and the whole tower has been connected to building frames and designed only for horizontal loads. Typical thickness of frame elements is normally 150–160 mm and in parapets 70–85 mm. Parapets usually have quite heavy reinforcement near both surfaces.

1.3 Objectives

This paper is based on the authors' experiences from about 150 condition investigations of concrete structures, long-continued development of condition investigation techniques, and an ongoing project called *Repair strategies of concrete facades and balconies*. The general objective of this research is to study the factors that have actually had an impact on the service life, existence and progress of deterioration in concrete facades and balconies. The three sub goals of the research are:

1 To find out the factors that have actually had an impact on the existence of defects and progress of different deterioration mechanisms in concrete facades and balconies.
2 To find out the relative importance of said factors.
3 To provide new reliable data on the service-lives of concrete facades.

The research project started in March 2006 and it will continue until the end of March 2009.

2 CONDITION INVESTIGATION

Damage to structures, its degree and extent, due to various deterioration phenomena can be determined by a comprehensive systematic condition investigation. A condition investigation involves systematic determination of the condition and performance of a structural element or an aggregate of structural elements (e.g. a facade or balcony) and their repair needs with respect to different deterioration mechanisms by various research methods such as examining design documents, various field measurements and investigations and sampling and laboratory tests.

The wide variation in the states of deterioration of buildings, and the fact that the most significant deterioration is not visible until it has progressed very far, necessitate thorough condition investigation at most concrete-structure repair sites. Evaluation of reinforcement corrosion and the degree of freeze-thaw damage suffered by concrete are examples of such investigations.

3 THE DATABASE

Condition investigation strategies for concrete facades and balconies have been developed in Finland since the mid-1980s. A large body of data on implemented repair projects has been accumulated in the form of documents prepared in connection with condition investigations. About a thousand precast concrete apartment blocks have been subjected to a condition investigation, and painstakingly documented material on each one exists, including the buildings' structures and accurate reports on observed damage and need for repairs based on accurate field surveys and laboratory analyses.

During the ongoing project *Repair strategies of concrete facades and balconies*, the condition investigation data from 946 buildings has been gathered in a database. Those condition investigation reports have been collected from companies which have conducted such investigations as well as from property companies owned by cities.

4 DETERIORATION

4.1 General

The most common degradation mechanisms causing the need to repair concrete facades and balconies in Finland, and concrete structures in general, are corrosion of reinforcement due to carbonation or chlorides as well as insufficient freeze-thaw resistance of concrete which leads to, for instance, freeze-thaw damage, Pentti et al (1998).

In the following text there are the first results concerning mostly concrete balconies and small parts and also concrete facades. These results are based on analyzing our database.

4.2 Corrosion of steel

Reinforcing bars in concrete are normally well protected from corrosion due to the high alkalinity of the concrete pore water. Corrosion may start when the passivity is destroyed, either by chloride penetration or due to the lowering of the pH in the carbonated concrete. Steel corrosion in carbonated concrete or in chloride migrated concrete has been widely studied, Tuutti (1982), Parrott (1987), Schiessl (1994), Richardson (1988), Broomfield (1997), Mattila (2003) etc.

In condition investigation reports there is a lot of measured data concerning corrosion of steel in concrete. When we are studying carbonation induced corrosion we have to find out the distribution of cover depths of steel bars and carbonation of concrete. Average carbonation factor k [mm/$\sqrt{a}$] in balcony frames is 2.60 in outer surfaces and 2.91 in inner surface and similarly in parapets the factors are 2.10 and 2.15, respectively. The standard deviation in the carbonation factor is relatively high: between 1.31–1.51 as mentioned cases. Carbonation factors are in general remarkably smaller in buildings made in 1990 and after. This is mostly a consequence of raising the concrete strength up to C30–C45.

The distribution of cover depths of steel bars in balcony frame elements is shown in figure 1. If we think to use so called light repair methods, like patch repairs, we have to look the amount of cover depths which are ≤10 mm. Carbonation of concrete is in many cases advanced more than 10 mm and steel bars deeper than that very seldom cause visible corrosion damage, because the diameter in ordinary concrete elements varies usually between 6–10 mm and is mostly 12 mm.

Chlorides are not a general problem causing corrosion in Finnish concrete buildings, because only in four buildings out of 946 were there chlorides ≥0.05 weight-% of concrete. Concrete bridges and their edge beams is another story!

4.3 Active corrosion

Once the passivity is destroyed either by carbonation or by chloride contamination, active corrosion may start in the presence of moisture and oxygen, Parrott (1987). Corrosion may run for a long time before it

Figure 1. Typical precast concrete balcony from 1970s.

**Cover depths of steelbars, Balcony frame
Single measurements 32 676 piece from 653
buildings**

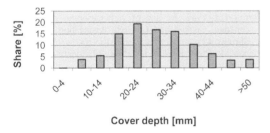

Figure 2. Distribution of cover depths of steel bars in balcony frames in the database.

can be noticed on the surface of the structure. Because corrosion products are not water soluble, they accumulate on the surface of steel near the anodic area Mattila (1995). This generates an internal pressure, because the volume of the corrosion products induced by carbonation is four to six times bigger than the original steel bars, Tuutti (1982).

Internal pressure caused by corrosion products leads to cracking or spalling of the concrete cover. Visible damage appears first on the spots where the concrete cover is smallest.

The amount and state of visible damage depends firstly on the age of the building when condition investigation was completed. In every case there is either smaller local damage or far advanced and wide spread damage, when the balcony was completed before

1990's. In 1990 and later built balconies there was local visible damage in 27% of buildings. Visible damage has no correlation on the situation of buildings; visible damage appears as well in coastal area as in inland.

4.4 Disintegration of concrete

Concrete is a very brittle material. It can stand only extremely limited tensile strains without cracking. Internal tensile stresses due to expansion processes inside concrete may result in internal cracking and, therefore, disintegration of concrete. Disintegration of concrete accelerates carbonation and in this way also steel corrosion. Concrete may disintegrate as a result of several phenomena causing internal expansion, such as freeze-thaw weathering, delayed ettringite formation (DEF) or alkali-aggregate reaction (AAR). In Finland freeze-thaw weathering of concrete is the most important disintegration mechanism, the others are very rare.

4.5 Freeze-thaw resistance of concrete

Concrete is a porous material whose pore system may, depending on the conditions, hold varying amounts of water. As the water in the pore system freezes, it expands about 9% by volume which creates hydraulic pressure in the system, Pigeon & Pleau (1995). If the level of water saturation of the system is high, the overpressure cannot escape into air-filled pores and thus damages the internal structure of the concrete, resulting in its degradation. Advanced freeze-thaw damage leads to total loss of concrete strength.

The freeze-thaw resistance of concrete can be ensured by air-entraining which creates a sufficient amount of permanently air-filled so-called protective pores where the pressure from the freezing dilation of water can escape. Finnish guidelines for the air-entraining of facade concrete mixes were issued in 1976, Anon (2002).

In those guidelines mentioned above, the protective pore ratio [p_r] should be ≥ 0.20 in the normal Finnish outdoor climate. According to condition investigations in Finnish balconies p_r is <0.10 in 59% of samples, between 0.10–0.14 in 19% of samples and between 0.15–0.19 in 11% of samples; the total number of samples is 1907. The distribution of the protective pore ratio is slowly getting better year after year since 1976 because of air-entraining in concrete mixes.

Moisture behaviour and environmental stress conditions have an impact on freeze-thaw stress. For instance, the stress on balcony structures depends on the existence of proper waterproofing. Despite the non freeze-thaw resistant concrete in Finnish balconies, the share of visible freeze-thaw damage is quite small, see fig. 3.

Protective pore ratio tests and thin-section analysis gives congruent information on the freeze-thaw resistance of concrete. Of course thin-section analysis gives more accurate information about pore size distribution and the degree of freeze-thaw damage e.g. beginning freeze-thaw damage.

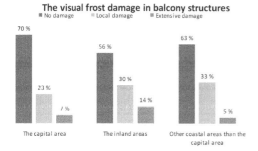

The visual frost damage in balcony structures
■ No damage ■ Local damage ■ Extensive damage

Figure 3. Distribution of visible freeze-thaw damage in Finnish balconies in the database , Weijo (2008).

5 REPAIR ALTERNATIVES

In Finland concrete balcony renovation methods are divided into three categories (repair principles):

- protective repair methods
- cladding
- demolition of balcony and rebuilding.

Protective repair methods are suitable mainly for structures where deterioration has just begun and the damage is not widespread. Possible protective repair methods suitable for concrete balconies are divided into:

- painting over the old paint
- protective painting after removal of old paint
- thorough patch repair and protective painting.

If the existing structures are more severely damaged, protective repair methods are no longer effective. In Finland different cladding methods are widely used with concrete facades but not with balconies. In balconies, in that case, whole balcony structure is usually demolished and replaced with new concrete balconies.

6 CONCLUSIONS

The most common deterioration mechanisms causing the need to repair concrete facades and balconies in Finland, and concrete structures in general, are corrosion of reinforcement due to carbonation or chlorides as well as insufficient freeze-thaw resistance of concrete which leads to freeze-thaw damage, Pentti et al (1998).

Concrete used in Finnish facades and balconies is not freeze-thaw resistant, 89% of protective pore ratios in concrete balconies are lower than 0.20 and 78% are lower than 0.15. Despite this 56–70% of buildings, depending on the buildings location, do not have any visible freeze-thaw damage. This means, it is possible to have good results with thorough patch repair and protective coatings with balcony glazing in most of the cases. According to the previous research project *Performance of protective measures in Finnish concrete facades and balconies*, Mattila & Pentti (2004), protective measures decrease effectively the moisture content in concrete and this slows the corrosion rate

by 30–80% and freeze-thaw damage stops completely. Again, it has been estimated it is possible to get 20–25 years more service life for concrete balconies with the protective measures mentioned before compared with doing nothing.

This is a remarkable result for property owners and in the perspective of global climate change, It has been estimated that heavy rains and freeze-thaw actions will increase remarkably during the following decades. If we use those protective measures early enough in our concrete structures, perhaps we do not have to demolish and rebuild all Finnish concrete balconies at all.

REFERENCES

Anon. 2002: Condition investigation manual for concrete facade panels. Helsinki. Concrete Association of Finland BY 42 178 p. (In Finnish)

Broomfield J. 1997: Corrosion of steel in concrete – understanding, investigation and repair. London. E & FN Spon. 240 p.

Mattila J. 1995: Realkalisation of concrete by cement-based coatings. Tampere, Tampere University of Technology, Structural Engineering. Licentiate's Thesis. 161 p. (In Finnish)

Mattila J. 2003: On the durability of cement-based patch repairs on Finnish concrete facades and balconies. Tampere, Tampere University of Technology, Structural Engineering. Publication 123. 69 p. (In Finnish)

Mattila J., Pentti M. 2004: Performance of protective measures in Finnish concrete facades and balconies. Tampere, Tampere University of Technology, Publication 450. 111 p.

Parrott L.J. 1987: Review of carbonation in reinforced concrete. Cement and Concrete Association. Wexham Springs. 42 p.

Pigeon M., Pleau R.1995: Durability of Concrete in Cold Climates. Suffolk. E & FN Spon. 244 p.

Pentti Matti, Mattila Jussi, Wahlman Jyrki 1998: Repair of concrete facades and balconies, part I: structures, degradation and condition investigation. Tampere, Tampere University of Technology, Structural Engineering. Publication 87. 157 p. (In Finnish)

Richardson M. 1988: Carbonation of reinforced concrete. Dublin, CITIS. 203 p.

Schiessl P, Breit W, Raupach M. 1994: Durability of local repair measures on concrete structures damaged by reinforcement corrosion. Mathora, V.M- (editor). Durability of concrete. Detroit. American Concrete Institute. Pp. 1195–1215

Tuutti K. 1982: *Corrosion of Steel in Concrete*. Stockholm. Swedish Cement and Concrete Research Institute. CBI Research 4:82. 304 p.

Vainio et al. 2002: Korjausrakentaminen 2000-2010 (Repair, maintenance and improvement work in Finland). Espoo 2002. VTT Research Notes 2154. 60 p. + app. 25 p. (In Finnish)

Vainio et al. 2005: Julkisivujen uudis- ja korjausrakentaminen. (Building and renovation of facades) Tampere, 2005. VTT 26 p. + app. 13 (In Finnish)

Weijo I. 2008: Freeze-thaw Resistance and Deterioration in Existing Concrete Balconies. Tampere. Tampere University of Technology, Structural Engineering. Masters Thesis. 172 p. (In Finnish)

Concrete Solutions – Grantham, Majorana & Salomoni (Eds)
© 2009 Taylor & Francis Group, London, ISBN 978-0-415-55082-6

Estimating the economic life span of a reinforced concrete building

Eddie S.S. Lam

Department of Civil and Structural Engineering, The Hong Kong Polytechnic University, Hong Kong

ABSTRACT: Due to the limited supply of land in Hong Kong, there is a sharp increase in redevelopment of the existing buildings. Developers may commence by an application to the Court for selling some or all of the premises in a building, pursuant to the Land (Compulsory Sale for Redevelopment) Ordinance. The main criterion justifying an order for collective sale under the Ordinance is related to the age or state of repair of the building. This is commonly assessed by considering, *inter alia*, the economic lifespan of the building. The economic lifespan is related to the state of repair and can be measured with the assistance of structural assessment. It is the purpose of this paper to explore some of the main points on structural assessment in relation to civil litigation. In general, information/data anticipated by the Court may differ substantially from what is expected by the Structural Engineer. For instance, the Court is interested in the accuracy and reliability of the test data as these are the territories very much challenged in the Court. Using a few case studies related to reinforced concrete buildings, it is demonstrated that a carefully planned structural assessment may alleviate most, if not all, of the uncertainties and/or assumptions that are in existence in the investigation.

1 INTRODUCTION

In an effect to tackle and control problems with building deterioration, the Land (Compulsory Sale for Redevelopment) Ordinance ("**the Ordinance**") was enacted on 7th April 1998 and came into operation in 1999. It facilitates, *inter alia*, redevelopment to be carried out in the private sector. Unfortunately, Hong Kong faced deep recession in 1998. As a result, the property market collapsed and any interest in collective sale of existing buildings for redevelopment ("**collective sale**") was diluted.

In 2004, the property market started to recover and there was sign of increasing activity on collective sale. For instance, 20 sites were launched for collective sale by tender in 2006 as compared with 14 sites in 2005. In 2007, there was a sharp increase in redeveloping the existing buildings due to the success of a large number of applications on collective sale.

The interest in collective sale was driven by three main factors. Firstly, the government has compelled owners to carry out maintenance on their buildings since 2000. Currently there are over 7,500 private buildings (about 20% of the total number of private buildings) which are aged 40 years or above in Hong Kong. In 2007, 690 statutory building repair orders were issued and about half were directed to old buildings. Secondly, the outbreak of Severe Acute Respiratory Syndrome or SARS in early 2003 has raised public concern on the possible dire consequences of building neglect. There is a genuine need to redevelop the existing buildings, especially those failing to provide the most basic hygienic amenity. Thirdly, old buildings are less attractive in the market as they fall short of the latest standard of modern buildings in providing a comfortable and convenient living environment.

Developers were and still are keen on redeveloping the existing buildings, thereby increasing the value of the properties. On the other hand, in a buoyant property market, owners that originally agreed to sell may suddenly demand higher prices. It is not uncommon for owners to oppose an application for collective sale, and the Court will have to decide whether or not to grant an order for collective sale.

2 THE ORDINANCE

The Ordinance provides a person who owns not less than 90% of undivided shares in a lot to apply to the Lands Tribunal ("**the Tribunal**") for a compulsory sale of the whole lot for the purpose of redevelopment. To strike a balance between facilitating private redevelopment and protecting individual property rights, the Tribunal will only make an order for collective sale if it is satisfied, *inter alia*, that the application for "redevelopment of the lot is justified ... due to the age or state of repair of the existing development on the lot". Section 4(2)(a)(i) of the Ordinance refers. The Tribunal has accepted that the economic lifespan test is an applicable test as propounded in Good Trader Ltd v Hinking Investments [2007] 3 HKC 219.

In a recent decision (Re Intelligent House Limited LDCS 11000/2006, unreported case, at pages 103–104), the Tribunal outlined the factors (though not intended to be exhaustive) to be considered in deciding whether a redevelopment is justified. As quoted from the judgment:-

"On the ground of age, the Tribunal is entitled to look at:

(a) Whether the old building has reached the end of its physical life.
(b) Whether the old building has reached the end of its economic lifespan... when the cleared site value of the lot significantly exceeds the existing use value of the building...

On the ground of state of repair, the Tribunal is entitled to look at:

(a) The state of repair of the old building is such that it has rendered the building a danger to the residents or the public at large.
(b) The state of repair of the old building is such that it has rendered the building coming to the end of its economic lifespan, in that it has become economically unworthy to repair. This includes situations where (a) the costs of repair exceed the existing use value of the building, or (b) the costs of repair significantly exceeds the enhancement value arising from or attributable to the repairs.
(c) ...

On the grounds of both the "age" and "state of repair" of the old building, the Tribunal is entitled to look at all of the above factors or tests collectively to see if that justifies redevelopment, even though when each of them is considered alone, it is insufficient to do so."

It follows that if a building is in a state of disrepair; reaching the end of its physical and/or economic life span; and/or becoming economically unworthy to repair, an order for collective sale shall be granted. It is not necessary to present expert reports with extensive and complicated coverage on structural matters. Such technical information is of little assistance to the Tribunal. Rather, the real questions to be answered fully may include, *inter alia*, the following:-

- What is the physical life span of the building?
- Further and/or in the alternative, has the deterioration reached a propagation phase?
- Is it cost effective to keep the building in a safe and functional state?
- Further and/or in the alternative, how much is it to keep the building in a safe and functional state?
- How much is it to maintain the economic life span of the building?

3 EVIDENCE/EXPERT REPORTS

3.1 *Expert evidence*

The Tribunal is always guided by expert evidence. It hears evidence from the expert witnesses representing the applicant(s) who applied for collective sale and the expert witnesses representing the respondent(s) who opposed the application. Consequently, expert witnesses plays a crucial role in the hearing.

An expert witness is distinguished from a factual witness in that he is allowed to express "his opinion on any relevant matter on which he is qualified to give expert evidence" and such opinion "shall be admissible in evidence" (Section 58 of Evidence Ordinance). The Court may reject evidence from a person purporting to be an expert witness if it has decided that the person does not have sufficient knowledge and/or experience to be qualified as an expert witness.

In carrying out an investigation, reporting the finding and expressing expert opinion, an expert witness is expected to make good use of his knowledge and expertise. He is obliged to act independently and to assist the Court (Lam 2007). In particular, he is expected to provide impartial assessment and must not "confirm" an opinion already formed. For instance, an expert witness has to be properly instructed "to carry out structural assessment on the condition of the building", and shall not be instructed to "investigate the appalling condition of the building".

Useful guidelines on how to conduct an investigation, prepare an expert report, etc., are referred to Gumpertz (2003), IStructE (2003) and Lam (2007).

3.2 *Structural assessment*

The Tribunal relies on, *inter alia*, a condition survey Report prepared by a Building Surveyor with advice from, *inter alia*, a Structural Engineer and a property manager. The Condition Survey Report will include visual inspection, on-site testing, maintenance records, improvement/repair records, building orders, unauthorized building works, structural assessment report, etc. It provides a basic understanding of the defects in the building, how the building was used, any unauthorized alteration (e.g. structural alteration, modifications on fire services) that could affect health and safety, state of repair and condition, etc.

A condition survey may comprise, *inter alia*, the following building parts and installation:-

1. Structural frame
2. Building envelope
3. Staircase and lobbies
4. Internal condition of individual flats
5. Drainage/plumbing system
6. Electrical installation
7. Fire services installation

To estimate the physical and/or economic life span of a building, a Structural Engineer (acting in the capacity as an expert witness) will be engaged to carry out a structural assessment (on items 1–4 above) to report the defects in the building; to investigate causes of the defects and to provide reasonable estimates on the maintenance cycle and costs of repair. He will inspect the structural frame and work together with the laboratories and other specialists, if necessary.

The structural assessment will start with a site visit to provide an overall impression of the condition of all the structural members. It will form an initial opinion as to the necessity for tests to be carried on-site

Figure 1. Pitting corrosion observed on the soffit of a floor slab immediate below a toilet area.

Figure 2. Vertical cracks/spalling at the corners caused by corrosion of main reinforcements.

and/or in the laboratories, and provide an estimate of the resources needed for the assessment.

The structural assessment must be thoroughly conducted with reasonable skill and care so that the resultant expert report can be authoritative. For instance, all the equipment must be properly pre-calibrated before the tests and all observations must be documented photographically and/or videoed for reference.

It is absolutely necessary to prepare a programme of work in advance, detailing scope of the structural assessment, e.g. dates of inspection, number and locations of tests, etc. Sometimes, details of the structural assessment will have to be first submitted to all parties for consent to commence the investigation.

Good understanding of the causes of common building defects may effectively limit the scope of the structural assessment and optimize the cost of investigation. For completeness, a short description on the causes of building defects that are frequently found in Hong Kong is given as follows.

In Hong Kong, deterioration of a reinforced concrete building is mainly due to the corrosion of reinforcement embedded inside the concrete (and in rare cases due to alkali-aggregate reaction in concrete). Corrosion of reinforcement is caused by a reduction of the alkalinity of concrete as a result of carbonation and/or chloride salts in the concrete. In particular, chloride contamination is fatal to the reinforcement. It causes pitting corrosion due to the catalytic action of the chloride ions inside concrete. As seawater is used locally for toilet flushing, chloride from the seawater may escape and penetrate into the floor slabs. Figure 1 shows the appalling condition of a floor slab immediately below a toilet area. Signs of pitting corrosion can be observed, suggesting that the main cause of spalling of the concrete is due to chloride attack. Chloride can also be airborne in areas close to the sea, like in many places in the southern part of Hong Kong.

Aiming for possible deterioration due to carbonation and/or chloride contamination, it is recommended to include at least the following inspection and on-site/laboratory tests in the structural assessment.

3.3 Inspection

Inspection is carried out or witnessed by an experienced Structural Engineer with due care and diligence to ascertain accuracy and reliability of all the findings. It is usually limited to the external appearance of the finishes/concrete and includes visual inspection, a crack survey and soundness test.

(a) Visual inspection: Visual inspection is by far the most cost effective method to identify signs of distress. It identifies spalling of concrete, delamination, signs of water seepage, overstressing, unauthorized building works, etc., and helps to decide any need for further testing. In a number of cases, it is possible to identify the defects from inspection alone and be able to suggest the probable cause(s) of defect while ruling out others.

(b) Crack survey: This provides a reasonable estimate of the extent and causes of defects. In general, crack widths less than 0.3 mm are considered to be non-structural and immaterial. When a crack is greater than 1.0 mm in width, it becomes significant and has structural implications. By examining characteristics of the cracks, it is possible to identify whether the reinforcement is corroded. Figure 2 shows an example of spalling of concrete. It is probably caused by severe corrosion of the main reinforcement as evidenced by patches of brown staining and the presence of vertical cracks at the corners in the direction of the main reinforcement.

(c) Soundness test: This identifies early signs of spalling of concrete. It is often used as a means to assess the general condition of concrete so as to select the locations for subsequent on-site testing.

3.4 Testing

Reasonable estimation of strength and durability of structural members can be obtained by carrying out tests on-site and/or in the laboratories. Methods of testing are well documented in the international standards, for instance BSEN, ISO, ASTM, and Bungey *et al* (2006).

Strength can be assessed by carrying out rebound hammer tests on-site together with compression tests on concrete cores collected from site. Defects that could possibly affect durability can be assessed by carrying out opening up inspection, carbonation depth tests, chloride profile tests, half-cell potential mapping and resistivity measurements. Microscopic properties of concrete can be examined through petrography or scanning electron microscopy ("**SEM**"). The following are short descriptions of the above-mentioned tests:-

(a) Rebound hammer test (BS1881: Part 202): This is sensitive to local variations in concrete, e.g. the presence of large aggregate near the surface, and is determined by averaging the rebound values obtained from a number of tests carried out in a 300 mm by 300 mm area. It is not intended to be a replacement for the compression test. Also, see Section 5.1 on proper application of the rebound hammer test.

(b) Compression tests (CS1:1990, Section 15): Compression tests are carried out on concrete cores collected from structural members. They provides a reasonable estimate of the in-situ strength of concrete. In addition, the size of voids, number of voids, presence of honeycombing, evidence of segregation, if any, can be identified from a core surface to assess the condition/quality of concrete.

(c) Opening up inspection: This refers to the inspection of the main reinforcement, for instance the bottom reinforcement at the mid-span of a slab. It provides accurate assessment on the degree of corrosion of reinforcement by measuring the diameter of some of the reinforcement embedded inside the concrete to calculate the loss in sectional area. In general, severe loss of cross-sectional area is assumed if the loss in sectional area is greater than 30%. The drawback is that it is expensive and very disturbing.

(d) Carbonation depth tests (BSEN14630:2006): These are carried out on concrete cores. A concrete core is split along its longitudinal axis and as near as possible across the diameter of the core. The freshly broken surface is then sprayed with phenolphthalein indicator solution to estimate the carbonation depth.

(e) Chloride profile test: This assesses the possibility of pitting corrosion. Samples of drilling powders are taken from concrete cores at 25, 50 and 75 mm depths, or at any depth as appropriate. By plotting the variation of chloride concentration against depth, chloride concentration at the surface of the reinforcement can be estimated. A 0.35–0.40% chloride ion by weight of cement is generally accepted as the threshold limit for a low risk of corrosion (for conventionally reinforced concrete. There are more stringent limits for pre-stressed concrete).

Figure 3. SEM image of microstructure of cement paste/binder showing incomplete hydration.

(f) Half-cell potential mapping (ASTM C876-91): This estimates the probability of occurrence of corrosion activity. The presence of corrosion activity can be detected by measuring the electrochemical potentials on the reinforcement from the surface of concrete with respect to a reference half-cell. Values of the potential and/or potential difference indicate the likelihood of corrosion activity.

(g) Resistivity (Bungey *et al*, 2006): Corrosion activity is predicted based on the resistivity of concrete. It is usually applied together with the half-cell potential mapping.

(h) Petrography (ASTM C856-04) and SEM (Stutzman, 2001): Using petrography, concrete in a two-dimensional thin-section is examined using a polarizing microscope. It provides information about the constituents of concrete, features of deterioration, mechanisms producing the deterioration, etc. Morphology of a three-dimensional surface of concrete can be observed through a SEM. It is principally applied to the examination of cement paste/binder microstructure and aggregate texture. An example is given in Figure 3 showing a concrete specimen with possible incomplete hydration of binder.

3.5 *Selecting the appropriate method of testing*

Tests will be conducted if they are well justified, e.g. to predict the physical and/or economic life span of a building. One of the main objectives in carrying out the tests is to group the structural members according to the rank of repair and frequency of repair so that the cost of repair can be systematically estimated based on the standard rates of repair.

Condition of concrete evaluated by visual inspection may provide an indication of possible defects or deterioration mechanisms which may be present in the structural members. This is of considerable value to the selection of appropriate methods of testing. Figure 4

Method of inspection or testing	General condition	
	Poor	Good
Visual inspection	Y	Y
Crack survey	Y	Y
Soundness test	N	Y
Rebound hammer test	N	Y
Compression test	Y	Y
Opening up inspection	Y	Y
Carbonation depth test	Y	Y
Chloride profile test	Y	Y
Half-cell potential mapping	N	Y
Resistivity	N	Y
Petrography and SEM	Y	N

Figure 4. Method of inspection/testing *v* general condition of a surface.

maps the methods of inspection/testing to the general condition of a surface.

In the figure, a "poor" condition is distinguished from a "good" condition by the presence of signs of distress, e.g. spalling of concrete, severe rust stains, cracks with crack width greater than 1 mm.

The number of samples may vary and depend on the type of testing. Two examples are given below.

- If the general condition of concrete, as judged based on skill and experience, varies substantially throughout the building, it is advisable to increase the number of samples.
- It is not uncommon that a "poor" condition with severe spalling or cracking is observed at a few sampling points, but the observation does not correlate well with the state of corrosion of the embedded reinforcement. It is probably due to insufficient sampling, leading to high probability of sampling error. This may invite adverse comments from the opposite parties on the unreliability of part, if not all, of the structural assessment and may jeopardize the whole investigation. It is worthwhile to consider carrying out additional tests at those sampling points and/or to increase the number of sampling points.

4 COST OF REPAIR

To provide a reasonable estimate of the economic lifespan of a building, the structural elements are categorized according to the rank of repair and frequency of repair.

4.1 Rank of repair

Repair works are ranked according to the severity of defects, health and safety issues, statutory requirements, etc. They are divided into "replacement", "specialty" and "traditional".

- "Replacement" is generally recommended if a structural member reaches the end of its economic life span.

- A structural member will be replaced unless it is cost effective to carry out a "specialty" type of repair. For instance, electro-chemical treatment (e.g. using sacrificial anodes) may be recommended to apply to areas with moderate chloride concentration, and ferrocement may be applied to strengthen overstressed members instead of complete replacement.
- "Traditional" methods of repair will be suggested if they can efficiently and economically restore a structural member. For instance, in areas where minor spalling of concrete is observed, in the absence of chloride concentration and the reinforcement is in reasonably good condition, it may be sufficient to improve durability of the concrete cover by "traditional" methods of repair (e.g. patch repair using non-shrink grout).

4.2 Frequency of repair

Frequency of repair is categorized into "urgent", "regular" and "routine" according to the condition of the structural members.

- "Urgent" repairs may include areas with extensive spalling of concrete, areas with high chloride content, members being substantially overstressed, etc.
- "Regular" repairs are recommended in areas with a high probability of occurrence of defects. As an example, half-cell measurements have identified a high probability of corrosion of reinforcement due to high negative potentials. It is recommended to carry out repair at "regular" intervals.
- "Routine" repair is recommended if no obvious defect is found in a structural member. Periodic inspection and maintenance is required to minimize possible wear and tear.

4.3 Estimating the cost of repair

Figure 5 relates rank of repair with frequency of repair. In general, "urgent" repair can be the most expensive whereas "routine" repair using commonly available materials and techniques is the most cost effective.

Based on data obtained from the structural assessment, structural members are categorized according to the rank of repair and in respect of the respective floor area, A_i, where,

$$\text{Total floor area of the building} = \sum A_i \qquad (1)$$

If necessary, structural members may further be separated into beams, slabs, columns, walls, etc., for ease of reference to the unit rate of repair.

The total cost of repair is subdivided into initial repair cost and future repair cost in the form of,

$$\text{Initial repair cost} = \sum A_i * R_i \qquad (2)$$

$$\text{Future repair cost} = \sum A_i * P_{ij} * Y / n_j \qquad (3)$$

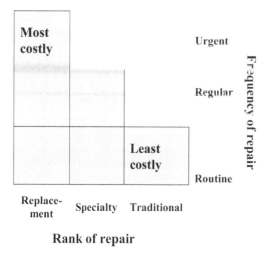

Rank of repair

Figure 5. Rank of repair *v* cost of repair.

$$\text{Total repair cost} = \sum A_i * (R_i + P_{ij} * Y / n_j) \qquad (4)$$

where R_i, P_{ij}, Y and n_j are the respective unit rate of repair for rank i; unit rate of regular/routine repair for rank i and frequency j; number of years of maintenance and period of "regular"/"routine" repair.

For example, 8,000 m² of floor area requires patch repair, the unit rate of patch repair is HK$750/m² and frequency of repair is 7 years. The total repair cost for a period of 20 years = 8,000*750*(20/7) or HK$17,200,000.

5 CASE STUDIES

5.1 *Rebound hammer test revisited*

Rebound hammer tests can only reflect properties of the surface zone of concrete. The depth of this surface zone is around 30 mm (BS1881: Part 202). However, carbonation may penetrate typically 10–20 mm from the surface of concrete (Neville, 1995) in old buildings and increase the hardness (i.e. in-situ strength) of concrete. If rebound values are taken directly from the surface of concrete, in-situ strength will be overestimated.

Figure 6 plots the estimated cube strength of concrete (at the age of around 45 years) based on two different sets of rebound values against cube strength obtained from concrete cores collected at each and every sampling location.

In the first set of data, rebound hammer tests were conducted at 20 different locations directly on the surfaces of concrete. As shown, strengths estimated by the rebound values can be more than 80% greater than those obtained from concrete cores and are unsatisfactory.

The second set of data was obtained from 10 different locations with the rebound hammer tests conducted 20 mm from the surface of concrete. The test data are

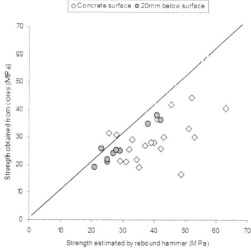

Figure 6. Rebound values v coring data.

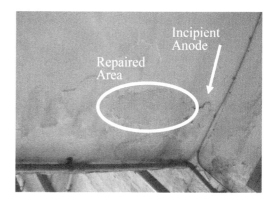

Figure 7. Incipient anode (courtesy of FOSROC).

in very good agreement with those values obtained from concrete cores and are within 10% difference.

In practice, the first 20 mm of concrete is removed to expose a fresh concrete surface at a depth less affected by carbonation for rebound hammer tests. In addition, the surface must be thoroughly smoothened and no test shall be carried out on a textured/rough surface.

5.2 *Maintenance record*

It is absolutely essential to inspect the maintenance records. Firstly, the maintenance records may disclose structural weakness that can adversely affect the physical and/or economic life span of a building. Secondly, it is not desirable to have many tests carried out on areas that were newly repaired. This can be avoided by inspecting the record of repair. Thirdly, it is useful to inspect the condition of recent repairs to draw fair opinion on the quality and method of previous repair.

An example of effective use of maintenance records is displayed in Figure 7 showing spalling of concrete in the neighborhood of a recently repaired area.

Figure 9. Presence of cracks on walls.

Figure 8. Delamination/presence of cracks on tiles.

It suggests possible formation of an incipient anode due to the presence of moderate to high chloride concentration. Hence, a chloride profile test was applied to the affected area.

5.3 *Assumptions made in the investigation*

Any assumption made in the investigation can be critically cross-examined by the opposite parties at trial. However, sometimes it is still possible to infer the causes of defects from other evidence.

When half-cell potential mapping and resistivity are used to estimate the probability of occurrence of corrosion activities, it is necessary to measure the carbonation depth and chloride profile in the vicinity of the testing area. This is because the presence of moderate to high chloride concentrations and/or carbonation in the testing area may affect the data and the opposite parties could challenge the reliability of the test data.

The prediction of possible defects in concrete based on crack patterns observed on the finishes is in itself an assumption. For instance, alkali-aggregate reaction may be identified by the presence of randomly orientated crack pattern over an extensive area. It is still desirable to carry out tests on some of the areas to verify the presence of alkali-aggregate reaction. Otherwise, the opposite parties may raise the question as to why no test was carried out to verify the opinion. It is a good practice to carry out petrographic examination on some of the defective areas to confirm the presence of alkali-aggregate reaction and to correlate the test data with the observations.

Figure 8 shows delamination of tiles from structural walls and the presence of cracks on the tiles. It was necessary to distinguish whether the defect was only cosmetic, due to normal tear and wear of the tiles or structural, due to spalling of concrete caused by corrosion of reinforcement. In that particular case, further investigation (e.g. opening up inspection) was not allowed, making it difficult to identify causes of the defect. As an alternative, tests were carried out on other locations with similar defects to infer causes of the defect.

Figure 9 shows the appalling condition of a flat at the ground floor of a 4-storey 50-year old building.

Figure 10. External walls with vertical cracks.

Lengthy diagonal cracks with up to 5 mm crack width were found on un-reinforced brick walls. Floor slabs were severely deformed with tilting to 1:40. In addition, large vertical cracks were observed on external walls at the upper floors, see Figure 10. All the evidence had consistently indicated excessive settlement of foundation at the material times. Furthermore, extensive internal redecoration was carried out in the flat about one year before commencement of soil nail installation works ("**the works**") near the foundation. In the circumstance, it was concluded on the balance of probability that the foundation was dragged down unevenly by the works. "Urgent" repair was recommended to preserve the physical and/or economic life span of the building. However, it was recommended to conduct expensive geotechnical investigation, and to include the cost of such investigation in the initial repair cost.

6 CONCLUSIONS

An order for collective sale under the Ordinance is commonly assessed by considering, *inter alia*, the economic lifespan of the building. The physical/economic lifespan of a building can be measured, *inter alia*, through structural assessment, including inspection and testing carried on-site and/or in the laboratories. Based on the data obtained from the structural assessment, repair costs can be estimated according to the rank of repair and frequency of repair.

Using a few case studies related to reinforced concrete buildings, it is well demonstrated that structural assessment has to be thoroughly conducted with reasonable skill and care so that the resultant expert report can be authoritative. It is necessary to ascertain the accuracy and reliability of the test data as these are the territories very much challenged by the opposite parties.

ACKNOWLEDGEMENT

The author would like to thank The Hong Kong Polytechnic University for providing part of the financial assistance (YE6-86) and technical support.

REFERENCES

ASTM C856-04, *Standard Practice for Petrographic Examination of Hardened Concrete*, ASTM International.

ASTM C876-91, *Standard Test Methods for Half-cell Potentials of Uncoated Reinforcing Steel in Concrete*, ASTM International.

BS1881: Part 202, *Concrete Test Hammer Guide*, British Standards.

BSEN14630:2006, *Determination of carbonation depth in hardened concrete by the phenolphthalein method*, British Standards.

Bungey J.H., Millard S.G. and Grantham M.G. (2006), *Testing of Concrete in Structures*, 4th Edition, Taylor & Francis, London and NY.

CS1:1990, Construction Standard, *Testing Concrete*, Volume 2, January 2002, HKSAR Government.

Gumpertz W.H. (2003), The Engineer As Expert Witness, *Proceedings of the 3rd Congress on Forensic Engineering*, Oct 19–21, 2003, San Diego, California, ASCE, pp. 462–476.

IStructE (2003), *Expert Evidence – A guide for expert witnesses and their clients*, 2nd edition, The Institution of Structural Engineers, October 2003.

Lam E.S.S. (2007), "My Recent Professional Experience as an Expert Witness", *Proc. of a Conf. on Latest Updates on Dispute Resolution in the Construction Industry in HK*, 3/3/2007, HK.

Neville A.M. (1995), *Properties of Concrete*, Fourth and Final Edition, Longman.

Stutzman P.E. (2001), "Scanning Electron Microscope in Concrete Petrography", *Proceedings of a Workshop on the Role of Calcium Hydroxide in Concrete*, The American Ceramic Society, Florida, November 1–3, 2000.

Concrete Solutions – Grantham, Majorana & Salomoni (Eds)
© 2009 Taylor & Francis Group, London, ISBN 978-0-415-55082-6

Durability model for concrete and concrete repairs under corrosion

G. Nossoni & R.S. Harichandran
Michigan State University, East Lansing, MI, USA

ABSTRACT: Prediction of the long term behavior of reinforced concrete elements requires the assessment and modeling of the steel reinforcement corrosion. In this paper, the corrosion rate of steel is modeled based on the cathodic reaction. The reduction reaction is modeled using the oxygen consumption at the cathode and oxygen diffusion through the barrier layers. The effect of different repair methods on the corrosion rate was also evaluated using the same model. The results show that a barrier such as a fibre reinforced polymer (FRP) overlay can decrease the diffusion of oxygen and reduce the corrosion rate considerably. The effect of different repair methods such as a traditional patch, and a patch with an FRP overlay, on the time for cracking of the concrete cover was evaluated using finite element analysis. The analysis shows that use of an FRP overlay on top of a traditional patch also delays cracking of the concrete cover.

1 INTRODUCTION

Corrosion of reinforcing bars in concrete is one of the main causes of structural deterioration and is one of the greatest maintenance challenges (Du et al. 2006, Beaudette 2001). Corrosion reduces the strength, durability, and service life of reinforced concrete structures. As the reinforcement corrodes, it expands causing cracking of concrete and spalling. Chloride concentration, carbonation, temperature, relative humidity, cover depth, and concrete quality are the major factors affecting the rate of corrosion. The transformation of metallic iron to rust can result in an increase in volume of up to 600%, depending on the final rust form (Mehta & Monteiro 1993). For corrosion initiated by chloride ions, the presence of both oxygen and chloride is required for the corrosion activity to continue, and corrosion may slow down considerably if a barrier could reduce the diffusion of moisture and harmful ions like chloride through the concrete.

In many cases a patching approach to concrete re-pair is adopted in which the damaged concrete is removed, the reinforcing steel is cleaned, and the area is patched with a patching compound. How-ever, even for the most promising patching products, cracking and full or partial delamination of the patching material from the concrete substrate due to corrosion is generally unavoidable (Baluch et al. 2002).

Patch materials have higher porosity than concrete and restrained shrinkage introduces micro-cracking in the patch. Both of these factors increase the availability of oxygen and chloride ions to the corrosion cell, thereby increasing corrosion in the repaired part and consequently causing premature damage to the patch material.

A model for chloride initiated corrosion is presented in this paper. The corrosion current is modeled as a function of the available oxygen and chloride. The model aims to predict the time for concrete cover or patch material cracking based on the many factors. The model first predicts the corrosion current based on the consumed and diffused oxygen to the bar surface, then calculates the build up of rust products at the bar surface, and finally the pressure applied on the concrete cover or patch, up to the time of the first crack. Experimental investigation of the durability and performance of concrete with and without a patch was performed using an accelerated corrosion test to confirm the model results. The effect of using a fibre reinforced polymer (FRP) overlay on top of a traditional patch as a barrier to diffusion of chloride and oxygen was also investigated experimentally.

2 CORROSION DAMAGE

2.1 Modeling the corrosion rate

A clear understanding of the electrochemical process involved in chloride initiated corrosion is needed to model the corrosion rate. Corrosion begins after the chloride concentration reaches a threshold value. The anodic reaction is

$$Fe \rightarrow Fe^{2+} + 2e^- \tag{1}$$

and the cathodic reaction is

$$O_2 + 2H_2O + 4e^- \rightarrow 4OH^- \tag{2}$$

The initial corrosion rate is highly dependent on the chloride concentration and is controlled by the anodic reaction (Nossoni 2009). However, corrosion cannot continue at this rate for a long time. The corrosion rate will slow down due to the lack of available oxygen and

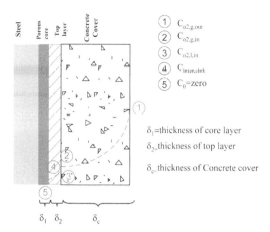

δ₁=thickness of core layer

δ₂_thickness of top layer

δ_c_thickness of Concrete cover

Figure 1.   Sketch of oxygen transport path through concrete and rust.

the cathodic reaction will then begin to control it. In this paper the corrosion rate is modeled as a function of the oxygen concentration. The limiting current at any time can be calculated through

$$i_{corr} = n_{O_2}.F.j_{o_2,consumed} \tag{3}$$

where $j_{O2,consumed}$ is the flux of the oxygen consumed at the cathode, $n_{O2} = 4$ is the "valency" of oxygen in Eq. 2 (i.e., number of moles of electrons consumed for each mole of oxygen), and F is Faraday's constant.

At the current limiting stage the oxygen flux into the cathode is a dominating factor. Figure 1 shows a simplified model of the different stages of oxygen diffusion through different porous layers that consist of the concrete cover (δ_1) and rust scale (δ_2 and δ_3). The corrosion scale is assumed to consist of two main layers, the inside part or core layer, and the outer part or top layer.

The core layer is a porous mass made up of small particles of different component phases. Fe(II) is expected to dominate the core part of the scale due to lack of hydroxide ions since the cathodic reaction will take place outside the core layer. The pores in the core layer act as corrosion pits and will have high chloride ion concentrations. The porous core is a conductive layer that can transfer electrons for other reactions that occur outside. The wet outside region of the scale is exposed to air and the ferrous hydroxide is rapidly oxidized to the ferric components Fe(OH)₃ and Fe(OH)₃+3H₂O. This layer is expected to be saturated with concrete pore water.

The flux of oxygen to the steel bar can be divided into three stages. In the first stage, oxygen diffuses through the concrete cover and will follow Fick's Second Law. At any time, the concentration of oxygen gas in the concrete adjacent to the bar is denoted by $C_{O2,g,in}$. Since only the dissolved oxygen can be consumed in the cathodic reaction, and the dissolved oxygen is in thermodynamic equilibrium with the oxygen partial

pressure in the gaseous phase, $C_{O2,g,in}$, Henry's Law yields the dissolved oxygen concentration as

$$C_{O_2,l,in} = C_{O_2,g,in} \frac{RT}{K_H} \tag{4}$$

$C_{O2,l,in}$ diffuses across the top layer of the rust scale where the cathodic reaction takes place. However, the top layer consumes some of the oxygen in the secondary non-electrochemical reaction, in which Fe(II) is oxidized to Fe(III):

$$4Fe(OH)_2 + 2H_2O + O_2 \rightarrow 4Fe(OH)_3 \tag{5-1}$$

$$4Fe(OH)_2 + O_2 \rightarrow 2\gamma - 4FeOOH + 2H_2O \tag{5-2}$$

Assuming that the top layer is a homogeneous porous medium, Fick's First Law can be used to predict the flux of the diffused oxygen through the rust layer:

$$j_{O_2,consumed} = (D_{o,r} \frac{\partial C_{O_2}}{\partial x}) \tag{6}$$

where $D_{o,r}$ is the diffusion coefficient of oxygen in the rust layer. $D_{o,r}$ can be calculated from the general equation for the diffusion coefficient in porous media (Ababneh et al. 2003):

$$D_m = \frac{2\left|1 - (V_P - V_P^c)\right|}{s^2}(V_P - V_P^c)^f \tag{7}$$

where V_P is the porosity of the cement paste, S is the specific surface area (surface area/bulk volume), $V_P^c = 3\%$ and $f = 4.2$. The final limiting current will be

$$i_{limit} = n_{O_2}.f.D_{o,r}.(\frac{C_{O_2,l,in} - C_{O_2,consumed}}{\delta_2}) \tag{8}$$

where $C_{O2,consumed}$ is the loss in oxygen concentration due to consumption of oxygen in the top layer (Nossoni 2009). Since the oxygen diffusion coefficient is higher than the chloride diffusion coefficient by a factor of almost 2 to 2.5 (Castellote 2001), when the chloride reaches the threshold value for the initiation of corrosion ($[Cl^-]/[OH^-] \geq 0.6$) (Mindess et al. 2003) the oxygen concentration profile in the concrete cover almost reaches a steady state condition as shown in Figure 2. However, after corrosion begins, the available oxygen adjacent to the steel bar is consumed very quickly and the corrosion is controlled by the rate of oxygen diffusion. Figure 3 shows the corrosion rate (corrosion current) as a function of time for concretes with water-cement ratios of 0.4, 0.5 and 0.6 that have progressively higher values of the oxygen diffusion coefficient.

The initial very short plateau region at the left end in Figure 3 is the anodic reaction controlled phase, but as the plots indicate for all three water-cement

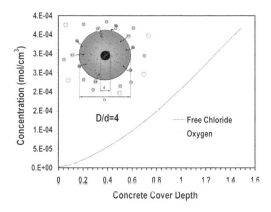

Figure 2. Oxygen and chloride concentration profile along the concrete cover at the initiation of corrosion.

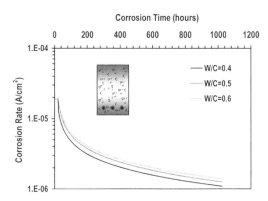

Figure 3. Corrosion rate for different water-cement ratios.

ratios, the corrosion current slows down depending on the oxygen diffusion coefficient and decreases with time gradually as the rust thickness builds up and slows down oxygen diffusion to the cathodic site. The corrosion current is higher for concrete with higher water-cement ratios since the diffusion coefficient of oxygen increases with the water-cement ratio.

2.2 Cracking time

The concrete surrounding the reinforcing bar will be subjected to an internal radial pressure due to the expansion of corrosion products. The pressure adjacent to the bar increases due to accumulation of the rust products, and the concrete will eventually crack and spall. The development of cracks and spalling of the concrete is highly dependent on the ratio of the concrete cover and distance between bars to the bar diameter (k).

Concrete cracks when the pressure due to the accumulation of rust products reaches a critical value. The critical pressure was calibrated using 3-D finite element analyses conducted with the general purpose ABAQUS software. Concrete was modeled using ABAQUS's "Concrete Plasticity Model." It was

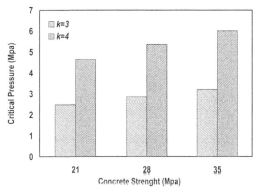

Figure 4. Critical pressure for concrete cracking.

assumed that the corrosion occurs uniformly around the steel bar, although it is known that this is not the case in reality (Mehta & Monteiro 1993).The pressure applied at the location of the bar that caused the concrete to crack was determined. Figure 4 shows the critical pressure as a function of concrete strength and bar diameter to concrete cover ratio.

As Figure 4 shows, the critical pressure for concrete cover cracking has a significant effect on the cracking time of the concrete cover. Increasing the concrete cover by almost one centimeter increases the critical pressure by a factor of 1.9, and corrosion-induced cracking of the cover can be delayed by years. Experimental work to establish the relationship between the corrosion rate and the critical pressure is ongoing.

3 EFFECT OF REPAIR

The effect of different repairs on the corrosion rate and critical cracking pressure was studied. In concrete repaired with a patch, unavoidable restrained shrinkage of the patch material leads to cracking of the patch. The main factor influencing the corrosion rate according to Equation 6 is the presence of oxygen, and after cracking of the patch the problem becomes worse because cracks facilitate the intrusion of oxygen and chloride ions to the steel bar. A barrier that can reduce the diffusion of oxygen should be able to decrease the corrosion rate considerably. The effect of using an FRP overlay as a secondary reinforcement on top of a traditional patch on the corrosion rate was studied. The FRP overlay also serves as a barrier for the diffusion of harmful species such as chloride ions and oxygen into the concrete and consequently should reduce the corrosion rate considerably.

3.1 Corrosion rate

The corrosion rate for (1) concrete, (2) concrete repaired with patch material, and (3) concrete repaired with both a patch and FRP overlay, was compared using this model assuming that the chloride concentration was the same for all three cases, and that the

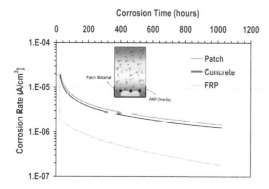

Figure 5. Corrosion rate for three cases.

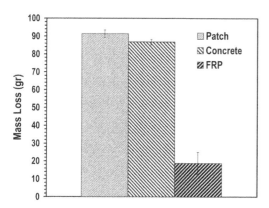

Figure 6. Corrosion mass loss for different cases.

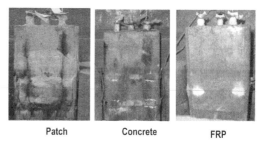

Patch   Concrete   FRP

Figure 7. Experimental test result for three cases.

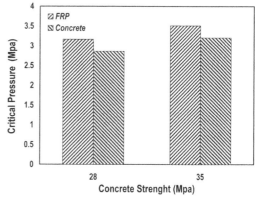

Figure 8. Critical pressure for cracking of concrete repaired with an FRP overlay.

only factor governing the rate of corrosion was the rate of diffusion of the oxygen into the concrete and rust layer.

It was assumed that the oxygen diffusion coefficient for concrete repaired with patch material was 1.5 times that of concrete, since the porosity of the patching mortar is slightly higher than that of concrete and micro-cracking due to restrained shrinkage further increases the diffusion coefficient. When an FRP overlay is used, the oxygen diffusion coefficient is almost zero at the FRP covered surface and the oxygen needs to travel a longer distance from the top surface or the sides of the concrete depending on the geometry of the repaired part. Figure 5 shows a comparison of the corrosion rates as a function of time for the three different cases for concrete with a water-cement ratio of 0.5.

The results show that the corrosion rate increases slightly by 14.2% compared to concrete when the patch material is used. However, the corrosion rate decreases dramatically by 86% when an FRP overlay is used on top of the patch material. Similar trends are displayed in Figures 6 and 7 which show experimental results from an accelerated corrosion test for all three cases (Nossoni & Harichandran 2007). In the accelerated corrosion test the average corrosion rate (proportional to the mass loss) increased by 7% when the patch was used and decreased by 78% when an

FRP overlay was used with the patch. The experimental results show very good agreement with the model. The model can be further calibrated/refined with additional experimental results.

3.2 Cracking time

Finite element analysis was performed for the case with the FRP overlay to investigate the effect of the overlay on the critical pressure and cracking time of the concrete cover. Figure 8 presents the FE results of the critical pressure, and indicates that using the FRP overlay can also increase the time to cracking of the concrete cover by increasing the critical pressure by 10% for different cases. When an overlay is used, the overall stress distribution changes and the crack propagates away from the region influenced by the overlay (Nossoni & Harichandran 2007). In other words, the FRP overlay strengthens the weakest region that is the concrete cover and increases the service life of the patch repair.

4 CONCLUSIONS

A new model is presented for predicting the corrosion rate of reinforcement in concrete structures. The model accounts for the diffusion of chloride and oxygen through the concrete and rust layers and is based on the underlying electrochemical reactions.

The formation of expansive rust products creates pressure that induces cracking of the concrete cover. Finite element analyses were performed to obtain the critical pressures required to crack the cover concrete for various bar diameter to cover thickness ratios.

Experimental investigation of the relationship between the corrosion rate and the pressure generated by the rust products is currently being performed. When this work is complete, the time for cracking of the concrete cover can be predicted.

The corrosion rate and the critical pressure required to crack the concrete cover were determined for plain concrete, concrete repaired with a polymer concrete patching compound, and concrete repaired with a patch and an FRP overlay. The results indicate that use of an FRP overlay on top of a traditional patch repair significantly reduces the corrosion rate and increases the critical pressure required to crack the concrete cover. The combination of both effects will therefore dramatically increase the durability of the patch repair enhanced with the FRP overlay. The results agree well with those obtained from accelerated corrosion tests on reinforced concrete specimens with and without repairs.

REFERENCES

Ababneh, A., Benboudjema, F. & Xi, Y. 2003. Chloride penetration in nonsaturated concrete. *Journal of Materials in Civil Engineering*. 15(2): 183–191.

Baluch, M.H., Rahman, M.K. & Al-Gadhib, A.H. 2002. Risks of cracking and delamination in patch repair." *Journal of Materials in Civil Engineering* 14(4): 294–302.

Beaudette, M.R. 2001. Investigation of patch accelerated corrosion with Galvashield Xp. *Interim Report*, January.

Castellote, M., Alonso, C., Andrade, C., Chadbourn, G.A. & Page, C.L. 2001. Oxygen and chloride diffusion in cement pastes as a validation of chloride diffusion coefficients obtained by steady-state migration test. *Cement and Concrete Research* 31: 621–625.

Du, Y.G., Chan, A.H.C., & Clark, L.A. 2006. Finite element analysis of the effects of radial expansion of corroded reinforcement. *Computers and Structures* 84(13–14): 917–929.

Mehta, P. & Monteiro, J. 1993. *Concrete, structure, properties, and materials* (Second edition). Englewood Cliffs: Prentice-Hall, 160–164.

Mindess, S., Young, J.F. & Darwin, D. 2003. *Concrete* (Second edition). New Jersey: Prentice Hall.

Nossoni, G. 2009, Holistic electrochemical and mechanistic modeling of corrosion-induced cracking in concrete structures. *Ph.D. Dissertation*, Michigan State University, East Lansing, Michigan.

Nossoni, G. & Harichandran, R.S. 2007. Improved durability of patched concrete bridges against corrosion by using an FRP overlay. *Proceedings (CD-ROM), 86th Annual Meeting of the Transportation Research Board*, Washington, D.C., Paper No. 07-2594.

Concrete Solutions – Grantham, Majorana & Salomoni (Eds)
© 2009 Taylor & Francis Group, London, ISBN 978-0-415-55082-6

Numerical modeling of cathodic protection in concrete structures

Rob B. Polder

TNO Built Environment and Geosciences, Delft, The Netherlands
Faculty Civil Engineering and Geosciences, Technical University Delft, The Netherlands

Willy H.A. Peelen

TNO Built Environment and Geosciences, Delft, The Netherlands

ABSTRACT: Concrete structures under de-icing and marine salt load may suffer chloride induced reinforcement corrosion, in particular with increasing age. Due to high monetary and societal cost, replacement is often undesirable. Durable repair is necessary, e.g. by Cathodic Protection (CP). CP involves applying a small direct electrical current through the concrete to the reinforcement from an external anode. The current causes steel polarisation, electrochemical reactions and ion transport. Thousands of CP systems have been installed worldwide and experience is positive. Corrosion is stopped for long periods of time, provided that systems are checked regularly. CP systems are designed from experience, which results in conservative designs and their performance is a matter of wait-and-see. Using numerical models for current and polarisation distribution, CP systems can be designed for critical aspects and made more economical. Previously principles and results of numerical calculations for design of CP systems were developed, neglecting beneficial effects of CP current flow. This paper addresses time dependent polarisation modelling applied to protection of local damage in bridges (e.g. below leaking joints). Preliminary results of validation based on a field trial system are discussed.

1 INTRODUCTION

Cathodic protection (CP) has been applied successfully to stop corrosion of reinforcing steel in concrete structures across the world for about 25 years. However, the number of cases where CP has been applied is low compared to the huge potential market of corroding structures. This is at least partially due to CP being an "unknown" technology for most users and to economical aspects. Generally, CP is a significant investment with a relatively long working life, improving safety and serviceability and saving large amounts of money in the mid to long term. CP may be "too big a step" for many owners. Consequently, CP is still a "promising" technique, despite its long track record. One way to improve the situation would be to make more economical CP system designs. This calls for innovations, e.g. using numerical modelling in the design stage, which allows making a "smart" design that is as slender as possible in terms of materials used, with reliable spatial distribution of protection. European research project ARCHES work package 3 aims at innovating CP design, to demonstrate the potential benefits of and to lower existing barriers for applying CP, in particular to bridges. It involves numerical modelling, installing smart CP systems to typical bridges in EU New Member States, validating the modelling based on field measurements and finally publishing a Guideline. A trial CP system was realised in Slovenia in July 2008, whose preliminary results are incorporated in this paper. Previous numerical modelling has proven its principle, but unrealistic results were obtained. In particular time-dependent (beneficial) CP effects need to be considered. This paper presents a possible approach to that issue and compares preliminary results from the ARCHES trial with model calculations.

2 NUMERICAL DESIGN OF CP FOR CONCRETE

Based on recent developments in numerical modelling of electrochemical corrosion processes in concrete (Redaelli et al, 2006), numerical design of CP systems seems feasible. A finite element model is made of a typical part of the structure to be protected, which ideally includes polarisation processes at the anode and at the steel and ohmic drop in the concrete. Using such a model, current densities and potentials can be calculated at every point in the modelled structure. By equating the calculated polarisation with depolarisation, the usual criterion for quality of protection, the model can be used to verify the spread of protection for various anode configurations and to predict local polarisations that can be tested experimentally.

The modelling is based on solving Laplace's equation in the (ohmic) concrete electrolyte space and boundary conditions that represent steel polarisation (based on Butler-Volmer type equations) with the

FEM package COMSOL multiphysics. Recent work has shown that the general modeling concept works (Polderf et al, 2008). However, calculations using active steel with a typical corrosion current density ($10\,\text{mA m}^{-1}$), produced much too high CP current densities compared to practical systems. One explanation is that the beneficial effects of CP on corrosion current density (Pedeferri, 1996) are neglected, one of which is pH increase at the steel due to generation of hydroxyl ions. A tentative model was set up to account for this effect.

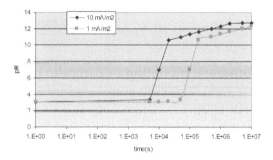

Figure 1. Development of pH inside pit with time at 1 and $10\,\text{mA m}^{-2}$.

3 TENTATIVE MODEL FOR CP INDUCED PH INCREASE

Cathodic polarisation involves hydroxyl ion production at the steel. The question is: at what time scale does the pH in a pit approach 13? The approach proposed here is simple electrochemistry of pH increase due to current flow, based on Faraday's law and ionic migration in concrete. The physical model comprises:

- flat geometry, 1 D
- 1 m^2 of steel surface per m^2 of concrete surface
- the complete steel surface is actively corroding
- pit depth is 1 mm (filled with acidic electrolyte)
- inside the pit pH is 3, or a solution of 0.001 M HCl
- outside the pit the electrolyte is 0.1 M NaOH, with ion transport numbers tCl = tCa = 0, tOH = 0.8 or higher (see below), e.g. 0.9 or 0.95; tNa = 0.2 or lower e.g. 0.1 or 0.05
- current entering the pit causes the reaction $2H_2O + O_2 + 4e \rightarrow 4OH^-$
- subsequently, the liquid is gradually neutralised by $OH^- + H^+ \rightarrow H_2O$.

3.1 Model calculations

The volume of acid in $1\,\text{m}^2 \times 1\,\text{mm}$ is 1 litre; its acid content at pH 3 is 0.001 mole (HCl). Assuming that all hydroxyl formed remains inside the pit, a current of $10\,\text{mA m}^{-2}$ (steel surface) produces 0.01 mole of hydroxyl at the cathode (1 mole is 96,500 Coulomb (symbol C), with 1 C is 1 As) in $96.500 / 0.01 = 10^4$ s. So in 2.8 hours the pH in the pit would reach 7. To get from pH 7 to pH 13, 0.1 mole of hydroxyl is needed, which takes 10^6 s. Thus, if all OH remains inside the pit the total time to pH 13 is $1.01 * 10^6$ s. However, ion migration will carry away a substantial fraction of the hydroxyl ions, at least for pH > 7. In NaOH solution, transport numbers are 0.8 for OH and 0.2 for Na. In concrete this may be different, due to build up of Na and OH concentration gradients and their subsequent back diffusion. Transport numbers for OH may change to 0.9 or 0.95 (Polder et al, 2002), effectively reducing OH build up. In the longer term, diffusion causes a steady state, which reduces net OH production to zero (Peelen et al, 2008). This appears to occur at a time scale of about 90 days at $10\,\text{mA m}^{-2}$. In any case, a steady state does not occur before the pH at the steel is above that of the surrounding electrolyte.

For $t_{OH} = 0.8$, 80% of all OH ions produced are carried to the anode by migration. So, 1/5 is available for neutralisation and pH increase. Consequently, a pH of about 13 would be reached at $10\,\text{mA m}^{-2}$ in 5×10^6 s or c. 60 days. Realising that steady state may be reached in 90 days, this is probably optimistic with regard to such a t_{OH} in the later parts of this period, which would probably change to 0.9 or 0.95 at higher pH. In any case, a pH of 12 would be reached in 6 days, at which time it is unlikely that a steady state is approached. Further, pH 12 would probably be sufficient to stop active corrosion. A semi-quantitative picture of pH with time for two current densities is presented in Figure 1. Stabilization at pH 12.7 occurs due to an increased OH transport number used in our modelling ($0.8 \rightarrow 0.95$).

3.2 Discussion of modelling results

At $10\,\text{mA m}^{-2}$ our model predicts a pH of 7 after 3 hours; a high pH (c. 12) and consequently a reduced corrosion rate as early as at 6 days; an even higher pH (12.7) is reached in 60 days. The amount of charge at 60 days is about $50\,\text{kC m}^{-2}$, which is nearly the same as the 60 to $80\,\text{kC m}^{-2}$ proposed by Glass and co-workers (Glass et al, 2007) to stop corrosion,. Moreover, in less than 3 hours, the pH would reach 7, which may already reduce the corrosion rate significantly.

It should be noted that the present calculations are based on a fully developed and undisturbed corrosion pit. In applications of CP in practice, spalled areas will be removed and the steel will be cleaned (although not to perfection). This probably takes away a large part of the acidic material from the corroding steel surface. Subsequently, such areas are repaired with alkaline and chloride free mortar. Consequently, the pH will be more likely to be above 7 than near to 3. Thus, the calculations represent a worst case situation. Such a situation may occur, however, in cases where concrete spalling has not occurred yet and no concrete is removed before applying CP.

Based on simple considerations, the (re)passivation of actively corroding steel in concrete may be expected to take place on the time scale of one week at typical current densities for CP operation.

4 COMPARISON TO FIELD DATA

4.1 *CP trial system*

Three CP test areas (TA1 through TA3) and a control area (TA4) were created on two cross beams of a bridge in Slovenia near Ljubljana. Leaking of joints had caused heavy corrosion, as indicated by potential mapping and by spalling of concrete, exposing the reinforcing steel. The beams measured about 1.5 m × 1.5 m. Each TA comprised about 5 m length of beam, totalling about 30 m² of concrete surface, containing slightly over 30 m² of steel surface area. In July 2000, cracked and spalled concrete was removed, the steel was cleaned superficially (as appropriate for CP) and open spots were repaired using a cementitious mortar. Reference cells were installed (activated titanium in TA1-3 and Mn/MnO₂ in TA4). CP anodes were installed: 5 activated titanium strips (De Nora Lida grid) in cementitious mortar of 5 m length horizontally on each vertical side of TA1; a conductive coating (AHEAD) all over the surface of TA2; and 5 titanium strips of 1.5 m length vertically on the head of TA3. On August 27, native potentials were measured and a voltage of 1.88 V was imposed on all three trial CP systems at 8:30 AM. After 4 hours, the current was switched off and potential shifts were measured. TA2 showed 400 mV shift into the negative direction and its voltage was reduced to 1.60 V. TA1 showed an average shift of about 100 mV, but 2 out of 4 cells shifted 50 mV or less, upon which the voltage was increased to 1.88 V. TA3 showed very small shifts (3 to 20 mV), upon which the voltage was increased to 2.1 V. Since then, these voltages were kept constant. Most recent available data are from October 10, 2008. On September 10, 2008, depolarisation was tested.

4.2 *Test results*

Currents were measured 18 times during the first 24 hours, then 30 times in the next ten days and from then almost daily for another ten days. Air temperature and relative humidity were measured simultaneously. In the first week, daily temperature variations were between 15 C (8 AM) and 25 C (4 PM). In the later period, temperatures dropped to about 10 C in the morning (when measurements were taken).

Currents for TA2 have been plotted in Figure 2. It shows that start up current (at 1.88 V) was about 100 mA, which after 10 minutes had decreased to 50 mA, dropping to 30 mA after 4 hours. 30 mA represents about 1 mA/m² of steel surface area. With a voltage of 1.60 V from 4 hours on, the current was initially about 20 mA, then steadily decreased to c. 15 mA. Figure 2 also shows the air temperature variation. Further analysis showed that currents from 24 hours after start up showed a nearly linear positive temperature dependence, see Figure 3. Currents during the first 4 hours (at 1.88 V), however, decreased with time while the temperature increased. Current between 4.5 and 14 hours fitted well in with currents measured between 24 and 119 hours (all at 1.60 V).

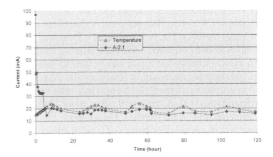

Figure 2. Development of CP current in TA2 and air temperature with time.

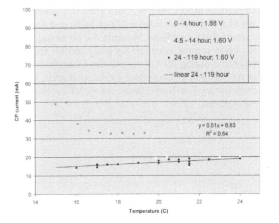

Figure 3. Development of CP current in TA2 with temperature.

The average 24 hour depolarisation on September 10 in TA2 was 280 mV.

5 DISCUSSION OF TEST AND MODEL RESULTS

The CP current in TA2 dropped during the first 4 hours, despite a temperature rise.

The current decrease from t = 0 to 10 minutes may be due to anode polarisation (which was not specifically tested). Taking that into account, the current decrease suggests that pH increase and repassivation may have taken place in a few hours. This agrees well with our tentative model's predictions for a quick rise of pH to about 7. Depolarisation testing after 14 days confirmed that the steel had become passive.

6 CONCLUSIONS

Numerical modelling of CP in concrete provides a promising tool to assist system design, allowing "slender" CP systems to be critically evaluated before they are installed. Previously, unrealistically high CP current densities were found by modeling. The work presented here suggests that pH increase in corrosion pits due to CP current flow occurs on a short time scale (of hours to days). The first results of a field trial that involves heavily repaired concrete elements, confirm this view. Probably the repair has helped to increase pH and suppress corrosion, at least partially. Applying CP to corroding elements without repair may behave differently and pH increase will occur more likely on a different time scale (weeks). Future work includes modeling of spatial distribution of current and polarisation and comparison to data from the ARCHES trial.

ACKNOWLEDGEMENTS

Our partners in ARCHES WP3 are gratefully acknowledged for their valuable help: Aljosa Sajna and Andrej Kranc of ZAG, Ljubljana, Slovenia, for obtaining permission to use the structure for trials and for testing before and during the CP period; Jan Leggedoor and Gerard Schuten of Leggedoor Concrete Repair, Gasselternijveen, The Netherlands, for installing the CP trial systems.

REFERENCES

Glass, G.K., Davison, N., Roberts, A.C., 2007, Pit realkalisation and its Role in the Electrochemical Repair of Reinforced concrete, *J. Corr. Sci. Eng.*, Vol. 9, paper 10

Pedeferri, P., Cathodic Protection and Cathodic Prevention, *Construction and Building Materials*, Vol. 10, No. 5, 1996, pp. 391–402.

Peelen W.H.A., Polder R.B., Redaelli E., Bertolini L., 2008, Qualitative model of concrete acidification due to cathodic protection, *Materials and Corrosion*, Vol. 59, 81–89.

Polder, R.B., Peelen, W.H.A., Nijland, T. Bertolini, L., 2002, Acid formation in the anode/concrete interface of activated titanium Cathodic Protection systems for reinforced concrete and the implications for service life, ICC 15th International Corrosion Congress, Granada, September 22–27 (CD-rom)

Polder, R.B., Peelen, W.H.A., Lollini, F., Redaelli, E., Bertolini, L., 2008, Numerical design for Cathodic Protection systems for concrete, *Materials and Corrosion,* Vol. 59, in press.

Redaelli, E., Bertolini, L., Peelen, W., Polder, R., 2006, FEM-models for the propagation of chloride induced reinforcement corrosion, *Materials and Corrosion,* Vol. 57, (8), 628–635.

Concrete Solutions – Grantham, Majorana & Salomoni (Eds)
© 2009 Taylor & Francis Group, London, ISBN 978-0-415-55082-6

Contextual repairs – A case study

Paul Sandeford
Principal Engineer, Materials Technology, GHD Brisbane

ABSTRACT: As repair practitioners, the desire is to present the optimum technical/financial solution to a deteriorated structure. This may be replacement or a full blown repair programme. For various reasons e.g. present budget constraints, disruption during replacement, planned changes in usage, this may not be practicable. Through understanding of both the client and community constraints, and the deterioration mechanisms it may be possible to undertake cost effective measures that will retard further deterioration and meet stakeholder needs. This is demonstrated through the case study of a 45 year old 2 lane road bridge in a North Queensland tidal environment. Various elements of the bridge were suffering from ASR, concrete spalling and corrosion of exposed steel. Barriers were not compliant with current codes and no records of apparent strengthening in the past existed. The bridge was the only access to a small beachside suburb, development in the area was restricted due to park lands and the council was under budgetary constraints. The load rating was reassessed, new barriers designed and remedial works and protective measures specified to extend the life of the bridge by 10-15 years with minimal local disruption at 30% of the cost of replacement.

1 INTRODUCTION

A Northern Queensland city council commissioned GHD Pty Ltd to perform a detailed condition assessment of the a small bridge serving a coastal residential area. The bridge consists of two lanes over two spans with an additional pedestrian lane.

The substructure comprises 1 central pier located in the creek, which consists of 7 square section piles driven into the bed supporting a concrete headstock beneath steel beams supporting the deck. Abutments at either end support the bank and end spans. A general view of the bridge is shown in Figure 1

The bridge provides access over a creek to a small beach side suburb. The location of the bridge is shown in Figure 2. There is no other access to the suburb apart from the road. At the time of the investigation the client

advised that further development of the suburb was under consideration but that environmental restrictions could preclude such development. The outcome of the process to approve the development was unknown and could take a significant period.

A previous visual inspection of the bridge had noted significant cracking and spalling, possibly due to corrosion of the reinforcing steel, in the tidal zones, abutments and barrier posts. It also noted significant corrosion loss of the bolts and nuts holding down the steel beams at the abutments. The scope of the investigation included:

- Review previous reports.
- Conduct on site visual, non-destructive testing and core sampling for laboratory testing.

Figure 1. General view of the bridge.

Figure 2. Location of bridge.

- Assess whether the bridge can continue to be used and with what load capacity.
- Recommend appropriate remedial actions if required.
- Prepare a preliminary cost estimate for any repairs recommended.

Through access to personnel previously working for the City Council it was established that a repair programme had been undertaken on the late 1980's including:

- Break out of spalled concrete on the central pier piles, repair and jacket using steel fibre reinforced concrete with a thickness of approximately 100 mm.
- Grit blast and repaint structural steel elements.
- Excavate and expose one abutment pile to establish corrosion state, none noted.

In addition to this repair programme the East corner of the South abutment had been repaired following impact damage.

2 RESULTS OF THE INVESTIGATION

2.1 Visual inspection and delamination survey

Visual observations throughout the site inspection were noted and relevant points are highlighted below

- Abutments – significant cracking was observed in all original cast abutments.
- Guard Rail – The posts showed significant spalling particularly at the base on the seaward side.
- Parapet – spalling of the cover concrete was observed in areas of low cover (<20 mm) on the seaward side in the area of the central expansion joint.
- Centre pier – minor visible cracking in the jacket repairs to the piles and one small area of spalling on the soffit of the beam.
- Steel beams – Pitting corrosion was noted over two areas, one each adjacent to the connections at the North abutment and centre pier. Minor surface corrosion was visible at interfaces of concrete and steel and at isolated spots in general areas.
- Bolt connections – significant corrosion was evident in bolted connections at abutments with the

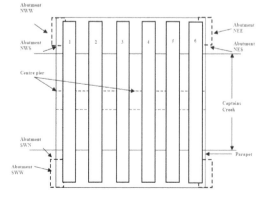

Figure 3. Schematic showing bridge components.

nuts completely corroded at some locations. The connections on the centre pier had been protected at some stage by wrapping using some form of fibre, possibly hemp, and application of a heavy layer of grease. This had been effective and minimal corrosion was observed.

- A Delamination survey indicated no significant delamination.

2.2 Non destructive testing

General observations from the covermeter survey included:

- Cover in the centre pier region varied largely; between 35–60 mm in the areas measured, low cover regions were observed on the bottom corners of the concrete headstock with covers as low as 18 mm.
- Cover at the abutments varied largely between 65 and 85 mm with minima as low as 40 mm in the areas measured.
- Cover at the parapet varied from 40–60 mm in the area measured.

Results from the Ultrasonic steel thickness tests included:

- Flange thickness in un-corroded areas was measured at 22–23 mm.
- Overall flange thickness varied between 17–23 mm in the areas measured, the lowest readings being observed in areas had been subject to heavy corrosion in the past but were still covered with intact coating at the time of inspection, presumably from the late '80's refurbishment.

Results from coating thickness included:

- Coating thickness varied between 870–1300 μm on the web region, 710–960 μm in the bottom flange region and 760–1270 μm in the top flange region.

Half cell potential readings indicated:

- Potential readings at all locations except the North East Abutment varied between −100 and −190 mV (Cu/CuSO$_4$).
- At the north east abutment average potential readings were more negative than −350 mV.

Results from the concrete resistivity survey included:

- Concrete resistivity varied between 116 and 330 kΩcm.
- The lowest resistivity values observed were at the east facing parapet (seaward side).

2.3 Sampling and laboratory testing

Where a bar was exposed by coring, its corrosion state was inspected where possible. The following observations were made:

- Exposed bars at the north east abutment appeared to have minor corrosion.
- Significant corrosion was evident on the exposed vertical bar sat the southwest abutment (west face). The north face showed minor corrosion with minimal loss of cross section.
- Exposed bars at the north west abutment showed some discolouration and corrosion.

The depth of carbonation was observed at four locations and varied from 5–20 mm.

Chloride levels in excess of 0.06% by mass of concrete (commonly accepted as the threshold for corrosion initiation) were observed at the reinforcement depth at the centre pier and on the surface of the South West Abutment. Raised chlorides (=>0.03%) were noted at the parapet and NE abutment.

Results from the ASR test using the uranyl acetate fluorescence method indicated a high probability that some ASR was taking place. As ASR in older structures in Northern Queensland is a known problem, this triggered the need for more detailed testing and further cores were taken for petrographic examination. This indicated:

- Low AAR present – extent of concrete matrix weakening due to AAR low.
- Aggregates from both samples indicated minimal degradation within the concrete matrix had occurred.

The lab carrying out the assessment rate samples they examine from 1 to 10, 1 being "normal homogeneous concrete" with few microcracks through to 10 "All cementitious value, coherence lost". The two samples examined rated as follows:

- Grade of deterioration 2. Slight deterioration, possibly through slightly excess voidage, excess micro-cracking, uneven paste composition, low levels of alkali-aggregate reaction, drying shrinkage, low temperature curing, possible slightly lean mixture.
- Grade of deterioration 3. Moderately low deterioration, possibly with enhanced voidage, microcracking frequency fairly high, excessive paste porosity, evidence of leaching or other forms of secondary alteration, possible lean mixture.

In order to asses the significance of the cracking, the size and location of the cracks were mapped in detail, see typical details in Figure 4. An analysis of the cracking in terms of millimetres of crack width per metre length required for assessment of AAR was carried out. Key findings included the following:

- Significant cracking was evident in both abutments with the upstream and beach sides being areas of maximum crack widths.
- Cumulative crack widths were generally 2–3 mm/m with maximum values as high as 4 mm/m.

3 DETERIORATION DISCUSSION

3.1 Corrosion of reinforcement

Significant corrosion damage was evident on many reinforced concrete guard rail posts and had compromised the effectiveness of the barrier.

Corrosion was observed in isolated low cover areas of the parapet concrete on the seaward side of the bridge. This appeared to be the result of a small area of low cover. Comparison of depth of chloride penetration with measured covers indicated that remaining areas were unlikely to be similarly affected in the short term.

Previous repairs to reinforced concrete piles remained effective at the time of the investigation. Surface cracking was observed, however, and the remaining life of the repairs was unclear.

The chloride contents located past cover depth in the concrete headstock were likely to have already initiated corrosion in the areas tested. This was expected to worsen over time with spalling becoming apparent, although the measured resistivity was high and corrosion rates therefore likely to be low.

3.2 Abutment cracking

Considerable cracking was evident in all original cast in situ abutments. The crack patterns observed could have been indicative of Alkali Silica Reaction (ASR) or shrinkage. No ASR gel product was observed on the concrete surface. Preliminary testing indicated the presence of ASR. However, the results of petrographic testing indicated that the level of ASR was too low to have caused the observed cracking in the abutments.

In any case, "Structural Effects of Alkali-Silica Reaction – Technical Guidance on the Assessment of Existing Structures" publication by the Institution of Structural Engineers indicates that the effect of ASR on the flexural strength of beams and slabs is likely to be minimal due to restraint of expansion by the reinforcement cage, providing surface expansion is less than 6 mm/m (i.e. total crack width in 1 m span is less than 6 mm). The observed maximum surface expansion was 4 mm/m.

The widespread "map" or three legged "Isle of Man" crack patterns are also consistent with shrinkage. The petrographic analysis suggested voidage was "moderately high" in both samples with an inferred water/cement ratio of >0.5. This would have lead to quite significant shrinkage. The observed cover of

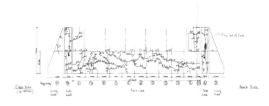

Figure 4. Typical crack width mapping.

65–85 mm and widely spaced reinforcement would have been insufficient to control shrinkage at the surface. Inadequate curing may have also contributed to the process. A typical abutment core sample revealed cracking initiated at the surface and ceasing toward the reinforcing bar. This was considered indicative of shrinkage cracking.

Cracking was therefore considered likely due to shrinkage, which would have ceased given the age of the structure. Owing to the significant amount of cracking present and exposure location, the risk of further reinforcement corrosion at crack locations was considered high. Chlorides at bar depth on the northeast abutment at a crack location were 0.04% w/w concrete, i.e. slightly lower than the threshold at which corrosion would be initiated. However, chloride concentration either side of the crack may have been higher, leading to localised corrosion. The presence of active corrosion was supported by the more negative potentials in this area. Corrosion in the NE abutment had apparently initiated and would lead to spalling in the medium term.

3.3 Steel beams

Two small areas of significant corrosion were observed at the North abutment and centre pier. Areas of minor corrosion were also observed at isolated positions along each beam. The loss of flange cross section was negligible. The previous corrosion repair still remained with minimal losses in coating thickness. At the time of the investigation, the effect of deterioration on the serviceability of the beams was minimal, however areas of concentrated corrosion were likely to worsen if not addressed by spot maintenance coating.

3.4 Bolt connections

Significant corrosion was observed at the abutment bolt connections to the extent of complete loss of cross-sectional area of the nuts at several locations. The center pier connections appeared to be sound due to measures used to alleviate corrosion during previous remediation. In terms of serviceability of the steel beams the loss of the nuts was not considered an issue as the bolts restrained horizontal movement. Further corrosion of the bolts at the abutments was considered likely in the short term as the nuts possibly protected them galvanically.

4 STRUCTURAL ASSESSMENT

4.1 Deck beams

It was noted that the bottom flange of all the RSJ deck beams had been plated with an approximate 20 mm thick doubling plate. This plate was not shown on the Department of Works drawings from construction time and appeared to have been installed in-situ to increase the bridge capacity. A more detailed assessment of the load capacity of the bridge was therefore carried out taking into account this modification.

Original drawings of the bridge indicated that it had been designed to satisfy the H20-S16-44 loading (32 tonne truck with a max axle load of 14 tonne). The details of the loads were obtained from the National Association of Australian State Road Authorities (NAASRA) Highway Bridge Specification, 1965.

The bridge was assessed against the load requirements of the current AS 5100.2 – 2004 Bridge design Part 2: Design Loads Code. The current bridge code also allows for existing bridges to be reviewed under previous codes. Based on the assessment, the bridge did not have sufficient capacity for M1600 or S1600 loads in accordance with AS 5100 Part 2 or a T44 truckload in accordance with the previous Australian Bridge Design Code (AUSTROADS).

To determine the capacity of the bridge, the configuration for the T44 truckload was used with reduced axle loads. Based on the assessment, the bridge had sufficient strength to satisfy the reduction of the T44 load to a 40.5 tonne Truck Load with Tandem axle loads of 9 tonne per axle 1.2 m apart.

4.2 Bearing bolts

The 'in-situ' bearing bolt arrangement at the central pier headstock also differed from the original drawings which depicted an expansion bearing connection where the steel beams rested on the central headstock. Such a connection would have permitted no transfer of longitudinal thermally-induced and/or traffic braking forces into the central pier headstock. Instead, the respective abutments would have resisted these loads – as is typically the case in bridges of similar construction.

The observed four-bolt connection allowed longitudinal load transfer to the central pier headstock. This effectively reduced the longitudinal load in the abutment bearing connections by approximately 50%. Therefore, the corroded two-bolt abutment bearing connections were still likely transferring longitudinal forces to the abutment headstocks. As the bolts appeared to have significantly more than 50% of their original capacity remaining, it was considered likely that they were able to resist their proportion of the original longitudinal design forces (ie: approximately 50%). The loss of the nuts was immaterial as no vertical restraint was required.

Due to access constraints replacement or repair of the bolts could not be undertaken without significant expense and bridge-down time. It was therefore recommended that the connections be cleaned and appropriate corrosion prevention measures taken to prevent any further section loss.

4.3 Guard railing

It was considered that, in their observed condition, numerous guard posts were unlikely to satisfy the requirements of even the lowest performance level of AS 5100. The code specifies geometric requirements for post and rail type barriers. The existing guard railing configuration did not comply with the code's post setback requirement which states that the traffic face of guard rail posts must be set back from the traffic face of the guard rail by a minimum of 100 mm. Corrosion had weakened the posts in a number of areas.

A risk assessment was undertaken in accordance with AS 5100.1 Appendix B to determine the performance level of the new traffic barrier. The following information provided by council allowed the performance level to be determined:

• 5000 vehicles/day;
• 3% commercial vehicles; and
• Threshold Limits, 60 km/h.

The performance level required for the new traffic barriers was determined to be low level. Off the shelf barriers or designs for barriers were not available to the new code so a design was developed.

5 RECOMMENDATIONS

Despite apparently significant damage, the investigation proved that the damage was either in isolated areas or not as significant as assessed from a purely visual inspection. Two repair options were therefore considered in addition to replacement. (all costs are in Australian dollars).

5.1 *Short term solution*

The bridge in its strengthened condition was assessed as having a capacity of. 40.5 tonne Truck Load with Tandem axle loads of 9 tonne per axle 1.2 m apart. The observed damage was unlikely to have significantly reduced this capacity.

The guard rail was assessed as non-compliant with AS5100 and would need to be replaced at a cost of $70,000 excluding any specific environmental control measures.

Aside from the guard railing works, the bridge could therefore continue in operation at this capacity with the following provisos:

• The abutment bolts were exposed to further corrosion and should be cleaned and protected with grease, as they were on the central pier (Approx $8,000)
• Additional deterioration of the abutments, parapet and central pier were anticipated, the rate of which was unclear. An annual inspection should therefore be carried out by a qualified Engineer and any deterioration assessed for its affect on capacity (Approx. $6,000 per annum).
• The bridge could continue on this basis for 2-5 years with a more detailed evaluation after 5 years or when significant worsening was observed.

5.2 *Medium term solution*

A longer life repair could be achieved through:

• Surface sealing cracks on the concrete abutment.
• Spot preparation and coating of corroding areas on the steel beams.
• Surface preparation and coating of the abutments to minimise further moisture penetration and hence corrosion and ASR activity.
• Patch repair of palling on the parapet.
• Application of silane water repellent to exposed concrete areas.
• Renewal of expansion joint sealant in the decks

The cost of these works would be of the order of $85,000 in addition to the guardrail replacement.

The bridge should still be inspected by a suitably qualified Engineer every 3 years and the frequency of the inspections increased if further unexpected deterioration is observed.

Following application of these measures the bridge was considered likely to remain serviceable for 5–10 years, possibly longer depending on the monitoring.

5.3 *Replacement of bridge*

Based on similar recent projects the cost to replace the bridge like with like would be $500,000–$600,000 excluding all diversions, temporary traffic arrangements or cost of additional road if the bridge were relocated slightly.

6 CONCLUSIONS

The client elected to implement the medium term solution and tenders were being called at the time of writing.

In addition to the desire to provide the optimum technical solution, careful consideration must be given to the needs of the local community. In this case the client and the residents were unsure of the future of the area and the bridge provided the only vehicular access to the suburb. The optimum overall solution was to keep the bridge open with an acceptable level of repair and minimum disruption to traffic until the future for the area became clearer.

REFERENCES

Structural effects of alkali-silica reaction: technical guidance on the appraisal of existing structures: (Institution of Structural Engineers, UK, 1992,);
NAASRA bridge design specification (National Association of Australian State Road Authorities, 1965)
AS5100.2 Bridge Design Part 2 – Design Loads.

Concrete Solutions – Grantham, Majorana & Salomoni (Eds)
© 2009 Taylor & Francis Group, London, ISBN 978-0-415-55082-6

Code for management and Rehabilitation of Coastal Concrete Structures (RCCS): Some case studies

Muhieddin Saleh M. Tughar
Department of Civil Engineering, Margeb University, Garaboli Branch, Garaboli, Libya

Ali F. Aweda
Department of Architecture Engineering, Alfateh University, Tripoli, Libya

ABSTRACT: After presenting a brief state of the art management system for design & protection, preservation and conservation of sustainable concrete structures in some coastal states in MENA and Mediterranean Countries, this paper shows how concrete codes need to be evolved. It reviews the basic significant differences and special features of the coastal states that require special attention when it comes to rehabilitation of coastal concrete structures (RCCS) in this region of the world. It briefly describes the fundamental aspects related to issues like: lack of much needed defined assessment and performance requirements; lack of specific health and safety regulations; lack of suitable materials for concrete making, workmanship quality control and finally, the prevailing widely varying and unique environmental conditions in the vast coastal states in MENA & Mediterranean Countries.

The paper demonstrates that in order to put the concrete industry in this region on the right track, it is essential to develop the coastal states code similar to the American, European and Asian codes. It is felt this would not only facilitate possible trade of concrete products and services and increased engineering construction activities within the coastal states, but also enhance much needed state of the art education, research and concrete practice in the region. In addition, the new ideal code should reflect the regional and national particularities on several levels namely; educational, technological, social, environmental, trade and economical levels. It also highlights lessons learnt and best practices from some of the interesting case studies in coastal states. An effective management system to safeguard the CCS against the impact of hot and humid environment requiring developing a concrete code is proposed. A UNIDO sponsored APCI concept cum approach that indicates a way forward to protect against progression of vulnerability and RCCS for coastal states is the main feature of the paper.

1 INTRODUCTION

Many types of concrete for recreational and coastal protection are subject to maritime conditions that are unique to countries bordering the sea. For many years, large quantities of resources have being spent on maintenance and rehabilitation of coastal concrete structures due to a lack of durability. World over, at present, there are several concrete codes of practice to guide design, construction and maintenance engineers in coastal states. For example, ACI-318 code, British standards, CEB-Fib Model code on the international level, and the South African Code, Algerian code, the Egyptian code and many others at regional level. Many coastal states in MENA & Mediterranean region which do not have their own codes have been using these international codes. In fact, some of these countries are using several codes with varying assessment and performance requirements at the same time, sometimes in the same project. Moreover, most of the local codes are in reality adopted and adapted from the well known international standards. For example, the South African code is based on European standards,

the Egyptian code is based on American standards and the Algerian code is based on French standards. The need for a unified regional code for concrete practice in coastal states region stems from the logic that because there are so many codes in one region with similar circumstances it would be more sensible to have one uniform standard. It makes some sense that coastal states with similar conditions and levels of development would require a standard that would closely reflect their particularities. Do coastal states possess the inertia, critical mass, scientific background, experience, research base and experts to carry on this task? The following would attempt to answer some important aspects of these fundamental questions.

2 AN OVERVIEW OF THE COASTAL STATES

The vast and varied coastal states with the highest population growth, wherein lives 70% of humanity with diversified natural resources, looks forward at the beginning of the twentieth century for real development on all fronts. Coastal states need to apply technology wisely, need to comprehend the lessons of their predecessor's examples in Europe, America and more recently Asian development. As an estimate, by 2050 more than 75% of the world population will live in coastal areas. To meet and serve these frequent and far reaching changes, large quantities of materials are required for construction of shelters and infrastructure. Concrete, for obvious reasons, lends itself as the only feasible material capable of meeting these needs. In general, the developing countries in coastal states are consuming about one third of the concrete produced annually, although these countries make up 85% of the coastal states population. If the consumption trends and over consumptive life style being followed in the industrialized coastal countries are to be imitated in the developing coastal state countries, the world will become unsustainable. The amount of materials needed for decent living is unimaginable.

There is a clear need to address this matter and to find alternatives to current consumption practices so that the world can be sustainable and material consumption and waste minimized. Concrete being the most important material of construction worldwide, plays a predominant role in coastal state welfare. In fact, concrete consumption is a real indicator of social progress and development in any coastal state country. If coastal states are to develop their infrastructure to support their future economic growth, an increase in concrete consumption will follow. There is real need to establish and develop one coastal state regional standard in the concrete industry to facilitate and accelerate economic development. The deterioration of concrete in coastal structures is usually a result of lack of durability rather than overloading, so there is a need to direct code requirements towards the production of durable structures. Coastal States Concrete Codes and other standards and codes from around the world need to discuss, along with the planning of construction works under coastal conditions, the maintenance and durability of coastal structures.

3 AN APPRAISAL OF CONCRETE CONSUMPTION IN COASTAL STATES

Portland cement remains the main indicator of concrete consumption in coastal states. From an economical viewpoint, it seems that there is no real alternative to Portland cement as a binder for construction and rehabilitation of concrete structures. With the projected growth of the concrete consumption, a corresponding increase in cement production is expected. In some of the European countries, Portland cement has been replaced completely by blended neat cement. In the United States, however, only 20 percent of cement used is blended cement. However, Portland cement, as we know it will be with us for a long time to come especially in coastal states. Cement can be considered as an indicator of concrete development and progress

in coastal state wide regions where lives two third of humanity. Cement being the most important ingredient for making concrete is now produced in many coastal states too. Statistics show the cement production for industrialized and developing countries alike. China by far is the leading cement producer, the United States of America ranks third in cement production after Japan. Presently, India ranks as the fourth world cement producing country.

Cement production is estimated to reach 2 billion tons annually. Production and consumption of cement in coastal states in developing countries is an indicator of the concrete industry activities. Under developed countries in the sub-Sahara region such as Uganda, for example, suffers from a shortage of cement that is imported from Kenya. Uganda's annual consumption, however, is only about half a million-tons. China's cement production is 380 million tons per annum and India is producing 170 million tons annually. This constitutes one third of the world population and will provide the largest construction arena, the world ever knew in recent history. Labour is relatively cheap as against materials which constitute 75 percent of construction cost against only 40 percent in the developed world. Capital and foreign exchange are limited in many developing coastal states countries, thus restricting imported goods such as reinforcing steel. In the next 20 years, the combined consumption of cement of India and China will exceed one billion tons per year. The rest of Asia will also consume another one billion tons per annum. Africa will consume about half a billion ton. Cement consumption clearly indicates and reflects the country's development (e.g. Nagab et al, 2001, UNIDO, 2002/2003/2005, UN, 2007, Samarai, 2008).

4 CONCRETE PROFESSION KNOWLEDGE IN COASTAL STATES AND BUILDING CODES

The practice of engineering in coastal states as a profession is, by its very nature based on knowledge. This knowledge comes from research and experience gained in practice. Codes are written based on knowledge and experience of local origin. Concrete codes are based on experience and experiments reflecting the educational, social; economical, technological and social welfare concerns of a society. Therefore, codes can be a powerful means of collecting and disseminating knowledge. This knowledge is translated into minimum requirements to provide the public with safety and enhance public economy. The American code for example, the most widely known, is based on the needs and requirements of the American construction industry at large. The European codes are local in nature, but the unification of Europe has made it possible to draft the new European code for Europe. The EC-02 code is based on the European experience. More recently the Asians have drafted a new Model code for Asia to reflect the Asian experience and to help facilitate their emergence on the economical level.

Codes are developed with key principles in mind, such as providing:

- Adequate structural safety and ductility
- Adequate durability and structural integrity
- Acceptable performance at service load levels.

5 CONCEPT OF STRUCTURAL SAFETY RELATED TO STRENGTH IN COASTAL STATE COUNTRIES

The main objective of structural design in coastal states is to produce a safe and serviceable structure, which can be constructed conveniently. It is expected to provide a high probability of survival under anticipated loads and aggressive hot and humid environment and which can be constructed at acceptable cost. The building code should be able to guide a structural engineer to produce such a structure. Safety, therefore, is the most important objective of any code. The code should also provide adequate safety not only against collapse but also against being unserviceable. Most codes follow a more deterministic approach with a rational probabilistic basis.

The partial safety factors in the design codes should result in acceptable probabilities of failure at a reasonable cost. Also, the structure must have a high probability of remaining serviceable under anticipated service loads. This is now accomplished by using the "limit state" approach". The structure is checked at the ultimate limit and serviceability limit states. In practice the structure is designed for the ultimate limit state and checked for the service limit state. Load and material factors are taken as unity. Even this approach has been challenged. Leading experts like Swamy (1997) suggest that design must be based on strength through durability rather than durability through strength. Variation in concrete strength is quite different from that observed in developed countries. Quality control changes drastically from one place to another. Applying a single factor would disadvantage the good contractor or would result in an unsafe design structure. Partial safety factors would be recommended, for good quality concrete or acceptable level concrete if the quality of concrete is questionable. In addition, in many coastal state countries, steel and cement are imported from abroad. Labour costs in these countries are much cheaper than developed countries. Therefore, design and construction practices adapted in developed countries which aim to minimize cost of labor at the expense of using more materials may not yield economical designs in coastal states countries.

6 CONCEPT OF STRUCTURAL INTEGRITY RELATED TO DURABILITY IN CODES

The compressive strength of concrete is not a good indicator of concrete durability. This being a test result,

it cannot account for construction practice, such as placing, compaction and curing. Therefore, a major concern of using the existing international codes in coastal states directly is the way each code deals with the durability problem in concrete to ensure structural integrity. ACI for example deals with this matter in a general way, whereas the Euro code which is based on the CEB-Fib model code is more specific. In ACI 318-05, only two classes of exposure are identified as exterior and interior with limitations on crack width for these conditions. Durability, according to this code, is achieved by limiting the water to cement ratio and specifying the concrete compressive strength. Also the ACI code gives limitations on chloride content of concrete. The Euro code on the other hand, elaborates on the issue of durability by specifying different exposure conditions leading to risk of structural integrity.

In brief, coastal states regions have long relied on foreign codes which were developed for other conditions and for different climatic conditions and construction practices. In the MENA & Mediterranean region, the combination of high temperature, dry winds, low precipitation leads to plastic shrinkage, drying shrinkage and thermal cracking (Tughar 2006, 2007). Also an increase in temperature leads to an increase in water demand, leading to lower strength and lower durability. Coastal climates are hot and humid, arid or tropical in general.. In addition to these environmental effects, concrete constituent's materials contain poorly graded aggregate and sand that in many cases are not suitable for concrete making. The lack of suitable material for concrete making, workmanship quality control and the prevailing environmental conditions in the coastal states require developing a special standard that could account for these specific problems.

7 AN OVERVIEW OF INTERNATIONAL COASTAL STATES REFERENCE CODE

There is a need to base the new code for coastal states on the existing codes of practice. These codes are well established and have been in use for some time. The choice between these codes however, is quite difficult. This is so because the most widely used codes are based on different approaches and basically local experiences and experiments[4-7].

The most important and widely used two international code documents nowadays are: European Committee for standardization for design of concrete structures (EC2) and the American Concrete Institute Building Code Requirements for structural Concrete ACI-318-05 and Commentary 318-05. These documents, when compared, show significant differences in many situations (Bindra et al, 2005). The ACI code, for example, which is based on the ultimate strength design method, until recently still considered the working stress design method as another alternative method. Only in the last edition of the code ACI-318-05, was the controversial decision to drop this method taken. On the other hand, the European code is based on the Limit state design method for an interim period. The margin of safety as related to imposed loads and reduction capacity material factors are quite different between these codes. The load factors for several codes show these differences. In addition, the ACI code has been changing at a rapid pace lately every three years. One of the recent significant changes in the ACI318-05 code is the inclusion of method and approach to the design of flexural members. Instead of using the concept of balanced reinforcement ratio and maximum reinforcement ratio, the code uses the concept of limit strain approach. But the code is still unclear in dealing with durability and serviceability requirements for deflection of slabs that meet the design requirement of the code, in addition to ever changing methods of design of slender columns. These changes have a sweeping effect on the coastal states construction industry level. Indeed it is somewhat difficult for many countries, especially in this part of the coastal states, where change is slow, to follow the ACI code's unwarranted and fast changes. The safety factors inherent in these codes are selected on the basis of overseas practice of concrete quality control (ACI 318-05).

7.1 Concrete compressive strength

The ACI 318-5 has no upper limit on compressive strength of concrete to be used in reinforced concrete structures, while the EC-2 code limits compressive strength to 90 MPa. ACI limits compressive strength only in some situations in the design equations, for example in development length and shear. The EC2 code starts changing expressions when the concrete compressive strength reaches 50 MPa. These codes also limit the use of concrete strength in certain design expressions. The ACI code starts changing some design expressions at concrete compressive strengths of 70 MPa.

7.2 Load and resistance factors

The ACI 318 Code uses strength reduction factors applied to the nominal strength capacity, which varies with the type of action (flexure, shear, axial load) force .The EC code uses partial safety approach with reduction factors applied separately for concrete and steel. Factors are also introduced to account for brittle behavior of high strength concrete in the EU-2 code.

7.3 Modulus of elasticity

Both ACI 318 and EU-02 codes use a secant modulus concept for the modulus of elasticity of concrete. The definition of the secant modulus, however, is different in the two codes. ACI318 bases the modulus on the specified compressive strength, whereas, the EU-02 code bases the secant modulus on an average compressive strength. ACI 318 gives an expression for the secant modulus of elasticity defined as the slope of

the line drawn from a stress of zero to the compressive stress equal to 0.45 of the compressive strength of concrete. The EC02 code gives a different expression: the secant modulus here is defined as the slope of the line drawn from zero stress to the compressive stress equal to 0.40 of the compressive strength of concrete.

7.4 Concrete tensile strength

The tensile strength of concrete determines the cracking load. This is needed to determine the minimum required steel reinforcement in the design of concrete structures in flexure. ACI 318 uses the modulus of rupture 'fr' to determine the flexural cracking moment while, the EC02 code uses the direct tensile strength as a basic parameter to calculate the tensile strength of concrete. In the absence of direct tensile tests, EC-02 provides an expression for the mean tensile strength in terms of compressive strength of concrete.

7.5 Minimum reinforcement for flexure

The primary role of minimum reinforcement is to provide adequate reserve of strength after cracking to prevent a sudden failure. ACI 318 uses an expression for this minimum value related to both 'fy' and 'fc'. It is noted that the EC-2 requires less minimum reinforcement for beams than does ACI-318. Unlike the requirement for EC-2, the minimum reinforcement requirements for slabs given by ACI 318 expressions are not a function of concrete strength.

7.6 Flexure and axial load

EU-2 permits the use of several equivalent stress distributions. The ACI-318 code for example at strengths above 50 MPa predicts significantly higher column capacities than the EU-2 code for the compression controlled regions.

7.7 Minimum shear reinforcement

The ACI-318 expressions give smaller amounts of steel reinforcement in shear than EU-02 for all concrete strengths.

7.8 Punching (Two way shear)

The ACI-318 expressions for punching shear depend on column aspect ratios. The critical shear section is taken at d/2 from the face of the column. EU-02 determines the resistance of concrete for punching shear at a critical section located at a distance 2d from the face of the column. EU-2 expressions for low reinforcement ratios give considerably lower punching shear resistance than ACI-318 expressions. At higher ratios of reinforcements, the two codes give similar results. EU-2 code provisions account for size effect and longitudinal reinforcement ratios whereas; the ACI-318 code does not.

Generally the EU-02 expressions give shorter development lengths than the ACI-318 expressions. The ACI-318 limits the value of square root of the compressive strength to a certain value, whereas, the EU-02 code cautions the designer of the increasing brittleness of high strength concrete. Cracking width limits chemical ingress in concrete.

Design codes are continually evolving documents using a consistent philosophy and the latest research results often reflecting the prior state of art and tradition of the country of origin. There are, however, significant differences in the code philosophies and approaches to design of concrete structures.

One may conclude that it is not possible and is structurally inaccurate to use a code written for a specific region and culture , used blindly under different conditions. Application of such codes may be not bring about the promised safe, durable and economical structures. In fact, this promise has not been achieved in its homeland. The durability crisis is heating the industrialized world more than ever, partly because they have been using these codes for so many years.

8 INTENDED OBJECTIVES OF THE NEW COASTAL STATES CONCRETE CODE

Coastal States Concrete Codes need to be developed with the following key principles in mind, such as providing:

- Adequate safety
- Adequate ductility
- Acceptable performance at service load levels

Codes from different countries have similar objectives; they have significant differences in their requirements for the design of concrete structures, which make it difficult to apply these codes without considering these differences. These significant differences in code provisions dictate the need for further experiments and analytical studies and harmonization of these codes.

The goal of the Coastal State Concrete Code is the development of a standard that will ensure the optimized design objectives for concrete structures in the coastal states. These design objectives include safety, serviceability, and durability, constructability and service life. Ideally, these objectives have to be fulfilled in accordance with local environment and local practice leading to sustainable concrete practice in the coastal states. The code must reflect the following aspects to be desirable. The code should also reflect the regional and national particularities on several levels, namely, educational, technological, social, economical and environmental levels. The code need to be based on current knowledge and experience of the concrete practice in the coastal region. Internationally, a code must accommodate the special physical and environmental features of the coastal states.

What is needed is a code that will help link these countries in close cooperation in the concrete industry. Ideally a code bridges the gap between knowledge and practice. A code needs to link the coastal states as a whole with other international codes such as ACI, Euro and the Asian codes. It must include precautions necessary to protect concrete from aggressive environments to enable it to attain its full potential for durability and strength.

In summary, the emphasis of the code for hot and humid coastal states must be based on durability design that should satisfy the following requirements:

- Minimum cement content
- Additional cementing, materials (fly ash slag, silica fume etc.)
- Mandatory curing practice
- Minimum compressive strength
- High durability
- Minimum concrete cover
- Minimum water cement ratio (maximum 0.50)
- Limit on chloride and sulfates contents, and
- Mandatory quality control

The anticipated coastal states concrete code needs to document in a comprehensive manner the aspects pertaining to cover the design and construction of structural concrete used in structures of all types. The quality and testing of materials used in construction have to be related to an appropriate standard. Use of the code will be by reference in general building codes of the appropriate coastal authority.

9 NEED FOR THE COASTAL STATES RECREATIONAL STRUCTURE CODE TO SERVE AS A LEGAL DOCUMENT

Recreational Concrete structures management in coastal states predominantly deals not only with sustainable high performance of the structures but also the reduction of coastal erosion and flooding. The techniques of coastal concrete structures management fall into two main categories, "hard" and "soft" engineering (Tughar, 2006). Hard engineering is the more traditional engineering response to erosion and involves the construction of structures which stop wave energy reaching the shore, or absorb and reflect the energy. These have often caused problems themselves, such as increasing erosion elsewhere, and soft engineering techniques have become more popular because of this. These techniques involve promoting natural systems such as beaches and salt marshes which protect the coast, and are usually cheaper to construct and maintain than hard engineering techniques, and may be self-sustaining.

Sustainable recreational concrete structures vital for tourism development require continuing commitment and action by all stakeholders at all levels of government, industry and the community. While there are regulations governing coastal structure development, the tourism industry is being encouraged to adopt voluntary management procedures, such as using environmental guidelines and codes of practice, rather than be strictly regulated. It is believed that self-regulatory techniques are likely to be more effective than statutory regulation in addressing specific environmental issues because they are flexible enough to adapt to changing circumstances. Also, the tourism industry is more likely to take responsibility and ownership for any self-regulatory approach. Since the days of Hammurabi, 4000 years ago, the public needed protection and wanted design to be based and governed by rule of law. Structural engineering practice, therefore, must be regulated by the rules of law, because it is directly related to public safety and economy. Coastal States building codes, when adapted by a certain authority, become law. When a building code becomes a law, it will be ultimately interpreted by judges. Judges interpretation of the code prevails. Structural engineering is not an exact science. Therefore, any structural design problem will have different solutions and problems that can be solved in different ways, making the practice of structural engineering design a law, implies that each engineering problem has only one unique acceptable solution. In codes, therefore, the authorities involved must be clearly stated. The 2000 International Building Code for example, states clearly the duties and powers of the building officials in enforcing the code and defines the responsibility of each partner. The code wording has to be clear so that even judges who are not engineers will not misinterpret the intent of the committee writing the code.

10 RCCS INITIATIVE AT MARGEB UNIVERSITY

RCCSI is an intellectual product of Libya's Margeb University's years of hard work and dedication by successive members of academia in cooperation with our partners around the country who volunteered their time and efforts to contribute for the cause of advancing the knowledge in concrete engineering for coastal recreational structures. Review of coastal amenities shows that concrete is by far the most widely used construction material in Libya (Tughar, 2006, 2007, 2008). Presently most consultants design and construct concrete structures based upon guides, standards and national concrete codes of practice that are derived from ACI, Euro Code or Asian Codes. Most national codes, norms, measures, guides, regulations and standards are neither normative documents and not developed according to consensus procedure of National Committee members of the responsible committee. These coastal structures face major problems of deterioration caused by unanticipated premature degradation of concrete materials produced by adopting and adapting either end product specification or method specification that are commonly employed to select concrete materials. In addition, prefabricated products produced from concrete material

neither have adequate competitive productive capacity nor the export potential market because they do not adequately conform to WTO agreement on technical barriers to trade (TBT) and application of sanitary and phyto-sanitary measures (SPS). What is needed is to assist these countries to produce concrete products that meet the quality requirement of the present markets and to upgrade them in order to tap future markets. The objective of this Margeb University initiative is to present an overview of the role of Concrete codes in the development process of Libya. We have examined the challenges being faced by Libyan, MENA and other Mediterranean countries construction industries in general and the concrete product sector in particular, due to protectionism, and intra-trade issues are also given. Criteria for selection of concrete materials appropriate for coastal region logically depend upon the position of the material within the structure, and on the changes which take place in the material properties during construction and after construction. Formal decision analysis and the scientific justification are guided by concrete codes. The anticipated Libyan concrete code is intended to serve as a guide to design and construct durable, serviceable, economical and safe concrete structures in the African region. Like other Southern and Eastern Mediterranean countries, Libya too faces a major challenge as we have little input into the international codes that serve as a basis for technical regulations and health and safety measures. We lack both the financial and human resources to play an active role in the deliberations of the relevant international bodies. We also typically lack the infrastructure needed to demonstrate acceptable conformity to the mandatory technical requirements. Only a few countries are able to participate in work on developing standards at international or regional level. The participation of this limited number of countries is also in most cases not effective, as it is not supported by the background research and analysis that are required to ensure that technical specifications of the concrete we produce and processes used are adequately taken into account in developing standards. Our needs are enormous and diverse. To address them comprehensively will require strong cooperation across the whole range of development partners. In addition we need technical assistance in relation to the WTO agreement on technical barriers to trade (TBT) and application of sanitary and phyto-sanitary measures (SPS).

11 AN APPRAISAL OF PROCEDURES FOR STANDARDS SETTING IN COASTAL STATES

An overview of the state of the art on concrete code standard setting shows that all coastal states require concrete to conform to the technical regulations that they apply to domestically produce ready mix concrete for performance, health safety and consumer protection.

The scope of the standard setting work includes preparing:

- Specifications for concrete in various geographical locations including hot and humid hostile and marine environments.
- A handbook of quality control for concrete materials.
- A specification for materials both for normal and hostile hot and arid desert regions and hot and humid marine environment, and
- Recommended practice for various applications.

The specifications using modified cement as binders would involve investigations on increased adhesion, durability, fatigue resistance & resistance to deterioration and degradation. In certain situations it requires field performance studies of pilot trial sections using recommended materials in different desert regions and hot and humid marine environment in Africa on a long term basis

Appraisal shows that that most national coastal state's concrete guides, standards and codes of practice lack input from reliable databases. Most of them are developed without adequately adhering to the conventional six step process for standard setting (preliminary stage, proposal stage, preparatory stage, committee stage, enquiry stage, approval stage, publication stage). Thus they are incoherent, piecemeal and "cook-booky" and desperately require periodic revision. Several factors combine to render a standard out of date: technological evolution, new methods and materials and new quality and safety requirements. Of course some of the coastal countries do follow ISO established general rules: like all standards need to be reviewed at intervals of not more than five years. On occasions it is necessary to revise a standard earlier.

Appraisal shows that most coastal states need technical assistance to:

- Overcome the problems they encounter in participating effectively in international standardization activities
- Meet effectively the technical requirements specially in their concrete prefabricated products export markets, and
- Build capacities for deriving full benefits from WTO agreements on TBT and SPS

The coastal states concrete code (CCC) needs to be drafted by an international non-profit non-governmental organization. This organization must be made up of volunteers from all over the world who are interested in the progress and development of coastal states. We have intention to involve all individuals and associations which have the experience and motivation to help in this task. We plan to cooperate to start drafting the initial code within five years. The scientific community in Libya will act as a focal point for such action. Several symposiums and workshops will be organized in different coastal states in the years ahead to establish a solid base for this work.

12 LIBYAN INITIATIVE BASED ON A COMPREHENSIVE QUESTIONNAIRE SURVEY

In order to identify the technical assistance needs of coastal states countries under a Libyan initiative framework, a comprehensive study is underway as an extension of an on-going questionnaire survey, the preliminary responses of the results of which have been already presented elsewhere in some of the international and regional conferences (Tughar, 2005–2008). Respondents from both coastal states have relatively more developed national institutions engaged in standardization conformity activities and also those at widely different stages of development whose experiences at national levels in the area of standardization and conformity are at a nascent stage. It should be noted that the basis of selection is for analytical purposes only to assess broadly the technical assistance needs at widely different stages of development. The intention is not to be interpreted as involving a value judgment on the actual level of development in standardization activities in each of these countries.

Survey responses demonstrated an immediate need to resolve difficulties of most coastal states hindering the export of concrete products due to shortcomings in their standardization and conformity assessment structure. They lack both the financial and human resources to play an active role in their deliberations of the relevant international bodies. They also lack the infrastructure needed to demonstrate acceptable conformity to the voluntary and mandatory technical requirements in their export market. Most of the respondents suggest needs for technical assistance to deal with technical regulations and sanitary and phytosanitary (SPS) measures required by export markets in relation to WTO Agreements on Technical Barriers to Trade (TBT) and the Application of SPS.

Almost all the respondents strongly urge the need to address these comprehensively and this requires strong cooperation across the whole range of development partners. Nearly all of them feel that the way forward to coastal states challenges is the use of AU/NEPAD supported African Productive Capacity Initiative framework that defines the need to establish an African Productive Capacity Facility (APCF). APCF is a set of resources dedicated to the support of initiatives like ways to draft the anticipated African Concrete Code. It would consist of Loans, Loan guarantees, grants, technical assistance, fiscal measures and contributions in kind. It is envisioned that many types of technical assistance like engaging retired experts to offer training and advice. They suggest that we need to learn from good practices evolved from several of the ongoing UNIDO international programs for adapting a new competitive and global production networks. A value chain approach to establish African concrete codes in line with the UNIDO approach is suggested by some experts as a way forward to establish a code. They all propose a need to implement fourteen steps from vision to action as proposed by

Conference of African Ministers of Industries (CAMI) (UNIDO, 2003). Finally it is felt that the suggested APCI framework would help development of appropriate CCC codes for putting the coastal states on a sustainable path. Review shows that the needed model coastal concrete code must incorporate a performance-based methodology – one that clearly describes the required performance of the structure being designed, constructed and maintained during its service. Keeping this in view, the Libyan initiative divides it into 3 levels – Common Level 1, Level 2 and Level 3.

The Common Level 1 document provides the framework for and the basic principles underlying the code across the 3 Parts: Part 1-Design, Part 2-Materials and Construction and Part 3-Maintenance. The Level 2 documents specify the required performances and the ways and means for the structure to achieve such performances. It contains provisions that are common to all coastal states. The working Level 3 documents are to be prepared by each country that adopts the code by incorporating their own national coastal concrete engineering practices.

13 OBSERVATIONS TO CCC SETTING

Paper has demonstrated that development of a Coastal Concrete Code (CCC) in line with an AU/NEPAD supported APCI has the potential to enhance productive capacity of coastal states. The present state of the art shows that till this day there is no reliable concrete code for the unique prevailing situation covering special areas like coastal states.

The author of this paper has no authority on making decisions but only offers suggestions on this matter (Tughar 2006–2008). He may however submit some obvious thoughts – but rudimentary ones, where ambitious projects need wealthy means. The first component of this wealth is the devotion and enthusiasm of the protagonists and participants in the project which have to be secured in the case of CCC. The second component is an appropriate *organizational* scheme: A Central Committee, Working Groups, Harmonization and approvals, etc. The third component is of an *economic* nature. Apart from symposia of scientific character, large workshops may also be needed among official representatives of interested coastal states, so that a gradual transnational acceptance of the CCC to be enhanced.

An ideal code needs to be based on experience and experiments largely of local conditions relevant to their original country. It requires undertaking a series of innovative bio-science and self organizing concrete material based case study approach to develop environment friendly green concrete codes. It is worth mentioning that existing national codes are very helpful as Nationally Determined Values, and will further accumulate experience useful for further actions (Amazo, 2005, Tassios, 2007).

The goal of an ideal Coastal Concrete Code (CCC) should remain relevant to coastal regional needs to ensure the optimized objective for the concrete structures in the vast coastal states. These design objectives of the intended code need to include safety, serviceability, durability, constructability, and service life.

The Libyan initiative based upon a series of innovative concepts being initiated to document successful practices for evolving coastal states concrete code based upon the unified field of all the laws of nature is the way forward. It is being achieved by establishing a network of Public Private Partnership Committees focusing on competitiveness and employment with the objective to build a regional observatory on competitiveness in the construction sector through a quality concrete export market in line with WTO initiatives. It would also require preparing a road map in line with the AU/NEPAD strategic plan supported by:

1. National concrete code action plans,
2. Specific public private partnership programs and projects and
3. Peer review committees focusing on competitiveness and employment through quality concrete products.

It would help assure sustainable concrete production using global production networks similar to the problem free and prevention oriented administration of our ever-expanding galactic universe. It will require a regular consultation between coastal states and non-state actors/institutions to mobilize the resources and promote advocacy.

It is recommended that a value chain approach to establish a Coastal Concrete Code be adopted in line with the UNIDO approach and with regard to the CSD (United Nation Commission on Sustainable Development). Fourteen steps from vision to action as proposed by Conference of African Ministers of Industries (CAMI) (UNIDO 2003) may also be taken as a common framework for implementing the APCI.

Finally, it should be said that joint efforts towards a common coastal concrete code seems to be the only realistic solution or option for harmonization of production and commerce as well as development of these coastal states!

REFERENCES

ACI 318-05, American Concrete Institute, Building Code Requirement for Structural. Concrete. *and Commentary (ACI 318R-05).*

Amazo E., Development of Productive Capacities. *Sub-Regional UNIDO/NEPAD Conference Presentation,* Tunis, Tunisia, 2005, 26–28 Sept.

Bindra S. P., & Tughar S. M., Civil Engineering Education for Information Society-A Case Study. *World Forum On Information Society, WFIS,* Tunis, Tunisia, 2005, 14th.–16th. Nov.

Mehta P. K. & Burrows, R. W., Building Durable Structures in the 21st Century. *Concrete International,* 2001, March.

Nagab, A. & Bindra S. P., Towards Sustainable Concrete Technology in Africa. *Proc. 1st. Int. Conf. on Structural Engineering, Mechanics and Computation, SEMC1,* Elsevier Service Ltd., South Africa, 2001.

National Economic Strategy, An Assessment of the Competitiveness of the Libyan Arab Jamahiriya. *The General Planning Council of Libya-Cambridge Energy Research Associates, CERA, Monitor Group Report*, U.K., 2006.

Samarai, M. A. and Qudah, L. M., Durability And Quality of Construction: Challenges Facing UAE. *Proc. 7th. Int. Concrete Congress – Construction Sustainable Option: Role of Concrete In Global Development*, HIS BRE Press, CTU, University of Dundee, Dundee, U.K, 2008, 7th.–10th July.

Siess C. P., Research, Building Codes and Engineering Practice. *Republished in Concrete International,* 2004, V. 26, No. 2, Feb. Swamy, R. N., Design for Durability-an-Integrated Material / Structural Strategy. *Paper presented in the ordinary general meeting of the Institution of Structural Engineers*, Yorkshire Branch, held in University of Bradford, 1997, 12th. Feb

Tassios T. P., CEB Model Code as a Sound Basis for Codes in Developing Countries. *Proc. 2nd. Int. Conf. on Concrete Technology in Developing Countries*, Tripoli, Libya, 1986, 27th.–30th. Oct

Tassios T. P., Towards African Concrete Code. *Proc. 8th. Int. conf. on Concrete Technology in Developing Countries.* Hammamat, Tunisia, 2007, 8th.–9th. Nov.

Tughar, S. M. & Bindra, S. P., Management of Coastal Recreational Concrete Structures: Some Case Studies. *Proc. 2nd. Int. Conf. on the Management of Coastal Recreational Resources, Foundation of International Studies- ICoD, University of Malta*, Malta, 2006, 25th–27th. Oct.

Tughar, S. M., et al., Concrete of Maritime Structures-A Case Study of Libyan Harbors. *Proc. 3rd. Nat. Conf. on Building Materials & Structural Engineering,* Misurata-Libya, 2006, 21st–23rd. Nov., (In Arabic).

Tughar, S. M., et al, Coastal Concrete Structures: Some Case Studies. *Proc. 3rd. Int. Conf. on Recent Development in Structural Engineering, Mechanics and Computation, SEMC3,* Millpress Science Publishers, University of Cape Town, Cape Town, South Africa, 2007, 7th–10th. Sept.

Tughar, S. M., & et al, Management of Recreational Concrete Structures: Some Case Studies In Coastal States. *Proc. 6th. Alexandria Int. conf. on Structural and Geotechnical Engineering, AICSGE6,* Alexandria University, Alexandria, Egypt, 2007, 15th.–17th. April.

Tughar, S. M, & et al, Salt-Induced Reinforcing Steel Corrosion in Concrete Structures. *Proc. 8th. Int. Conf. on Concrete Technology in Developing Countries,* Hammamat, Tunisia, 2007, 8th–9th. Nov.

Tughar, S. M, and Mustafa H. Z., Effects of Using Water and Super Plasticiser as Retempering Agents on Workability and Compressive Strength For Plain and Super Plasticised Concretes. *Proc. of 7th. Int. Concrete Congress- Construction Sustainable Option: Precast Concrete – Towards Lean Construction,* HIS BRE Press, CTU, University of Dundee, Dundee, U.K, 2008, 7th.–10th. July.

UN Department of Economics and Social Affairs. *CSD Indicators of Sustainable Development,* 3rd Edition, August, 2007, www.un.org/esa/sustdev/natlinfo/indi

UNIDO, Competing Through Innovation and Learning. *Industrial Development Reports 2002/2003,* Vienna, 2003.

UNIDO, Influencing and Meeting International Standards. *UNCTAD/WTO Commonwealth Secretariat,* Annual Report, Vienna, 2005.

UNIDO, African Productive Capacity Initiative-From Vision to Action, Main Report. *African Union, NEPAD, 16 CAMI Meeting*, Vienna, 2003, Nov.

www.accreditation-libya.net

www.lncam.org

Concrete Solutions – Grantham, Majorana & Salomoni (Eds)
© 2009 Taylor & Francis Group, London, ISBN 978-0-415-55082-6

COIN – and durability of structures

Ø. Vennesland & U. Angst
Department of Structural Engineering, Norwegian University of Science and Engineering, Trondheim, Norway

T.F. Hammer
Sintef Building and Infrastructure, Trondheim, Norway

ABSTRACT: COIN stands for Concrete Innovation Centre and is one of 14 Centres for research based innovation (CRI) – which were set up after an initiative by the Research Council of Norway. The purpose of the centres for research based innovation is to build up and strengthen Norwegian research groups that work in close collaboration with partners from innovative industry and innovative public enterprises. It is further to support long-term research that promotes innovation and the competitiveness of Norwegian industry. The duration of these centres is 8 years with an evaluation after 5 years. The vision of COIN is creation of more attractive concrete buildings and constructions. Attractiveness implies aesthetics, functionality, sustainability, energy efficiency, indoor climate, industrialised construction, improved work environment, and cost efficiency during the whole service life. This may be fulfilled by developing advanced materials, efficient construction techniques and new design concepts combined with more environmentally friendly material production. COIN shall be staffed from Sintef, NTNU and industrial partners with altogether about 25 man-labour years. In addition there shall be 8-10 PhD-students continuously working within it as well as many MSc-students. It is further hoped to attract many international guest researchers for COIN.

1 INTRODUCTION TO COIN

COIN stands for Concrete Innovation Centre and is one of 14 Centres for research based innovation (CRI) – which were set up after initiative of the Research Council of Norway (www.sintef.no/coin). The purpose of the centres for research based innovation is to build up and strengthen Norwegian research groups that work in close collaboration with partners from innovative industry and innovative public enterprises. It is further to support long-term research that promotes innovation and the competitiveness of Norwegian industry. The duration of these centres is 8 years with an evaluation after 5 years.

Both SINTEF and NTNU have a history for close contact with the concrete industry in Norway and the building industry in Norway has earlier documented a great potential for added value through innovation. It should also be mentioned that concrete work is an important part of construction works and the whole industry has many challenges related to environmental issues. We also see that recruiting is a greater challenge for the building industry than for many other industries.

The objective of COIN is to bring the development a major leap forward by developing advanced materials combined with efficient and sustainable construction techniques and design concepts and to increase the level of competence and strengthen the degree of innovation in the concrete business.

The vision of COIN is creation of more attractive concrete buildings and constructions. Attractiveness implies aesthetics, functionality, sustainability, energy efficiency, indoor climate, industrialised construction, improved work environment, and cost efficiency during the whole service life. This may be fulfilled by developing advanced materials, efficient construction techniques and new design concepts combined with more environmentally friendly material production.

The corporate partners are leading multinational companies in the cement and building industry and the aim of COIN is to increase their value creation and strengthen their research activities in Norway. The industrial partners are: Norcem AS, Rescon Mapei AS, Borregaard LignoTech, maxit Group AB, Unicon AS, Aker Solutions ASA, Spenncon AS, Veidekke Entreprenør AS, Skanska Norge AS and Norwegian Public Roads Administration.

The objectives of COIN are to increase the reputation of concrete to make it the natural and environmentally correct choice, appear as the leader in research and application, to attract leading international companies to increase its research activities in Norway and to contribute to increased education of researchers and masters.

COIN shall be staffed from Sintef, NTNU and industrial partners with all together about 25 man-labour years. In addition there shall be 8-10 PhD-students continuously working within it as well as many MSc-students. It is further hoped to attract many international guest researchers for COIN.

2 PROJECTS IN COIN

There are five projects in COIN /1/. The projects and the part projects are shown.

- **Advanced cementing materials and admixtures**
  - Cements with lower CO_2 emission during production
  - Admixtures to control hydration development
  - Cements and admixtures to prevent cracking
  - Alternative pozzolanas
  - Cements with lower porosity

- **Improved Construction Technique**
  - Concretes with high fibre content to be used in load carrying structural parts
  - Concretes and production techniques to give good looking surfaces
  - Technology for production of optimal crushed aggregate

- **Innovative Construction Concepts**
  - Design and verification basis for utilization of fibres in load carrying structures
  - Development of superlight high performance aggregate and concrete
  - Hybrid structures – development of new material combinations, e.g. to be used in arctic environment

- **Service Life Design**
  - Reliable tools for service life design included design of load capacity in structures under reinforcement corrosion or ASR
  - Preventive measures to increase service life (surface treatments, inhibitors, low corrosive reinforcement, etc)

- **Energy efficiency and comfort**
  - Utilization of the thermal mass to reduce the need for cooling/heating of buildings

3 PHD-STUDENTS IN COIN

At all times it is planned to have 8-10 PhD students. At the moment (August 2008) there are eight and the PhD students are well distributed among the projects (All projects have two students – except project five that was started recently. On durability there are two at the moment, Jan Lindgård working with ASR and Ueli Angst on critical chloride content. At the moment one PhD student is working in resistivity of concrete. Recently an advertisement for 6 new students has been made public. The advertisement was published in the home pages of NAV (The Norwegian Labour and Welfare Administration), Jobbnorge (Work in Norway), Teknisk Ukeblad (a weekly technical paper) and the NTNU home page. The English version is in the EU portal of the NTNU page. A reference announcement is in the written issue of Teknisk Ukeblad August 8.

4 PART PROJECTS ON CONCRETE DURABILITY

4.1 Service life modelling and prediction

- Develop operational and reliable tools for service life prediction of concrete structures exposed to chloride induced corrosion

Service life modelling and prediction is mainly to identify models for chloride ingress and corrosion and to collect data for relevant model parameters. Other part projects will deliver input parameters to the identified Service life models. It is the intention to have statistical quantification of model parameters and reliability based service life prediction for chloride induced corrosion. As is seen, corrosion of steel in concrete is the main cause (in COIN) for changes of the service life of reinforced concrete structures.

4.2 Critical chloride content and corrosion process

- Study both in laboratory and in the field new and used methods for obtaining critical chloride content
- Develop a method for critical chloride content
- Investigate stainless steels as concrete reinforcement
- Investigate the relation between concrete resistivity and the corrosion process

A PhD student, Ueli Angst from Switzerland is engaged. He has written most of a state of the art report on Critical Chloride Content and contributed much to a workshop that was held June 5 and 6 this year.

4.3 Electrical resistivity

- Establish relations between resistivity and reinforcement corrosion, and thereby develop service life models that include resistivity. Standardised methods for measuring resistivity will be suggested

This part project is now seeking a PhD student. The main goals of the project are to model the time dependent behaviour of resistivity, to identify any threshold values of resistivity and establish relations between resistivity and reinforcement corrosion, and thereby develop service life models that include resistivity. It is the intention to do laboratory tests versus field tests and to see whether there is a need for any compensation. It is further the intention to contribute to standardisation of test methods and recommendations.

4.4 ASR – mechanisms and performance based concepts

- To provide improved understanding of the accelerated AAR effects when concrete is subjected to elevated temperature and moisture regimes during various laboratory (performance) testing, compared to field conditions

A PhD student, Jan Lindgård from Norway is engaged. The main goals of the PhD study are to evaluate accelerated ASR effects when concrete is subjected to elevated temperature and moisture regimes during various laboratory "performance" testing, compared to field conditions, to produce a state-of-the art paper covering the key parameters that may influence the speed and extent of alkali aggregate reactivity of a concrete and address the research needs on important parameters. It is further the intention to summarize and evaluate the performance testing performed in Norway during the last ten years and to document and evaluate the influence of the selected issues/parameters for study on the results and outcome from the performance testing with the various test methods selected. Together with the supervisor, Terje Rønning of Norcem, the PhD project shall give input to the RILEM TC ACS-P sub-group as basis for the development of a future international agreed performance testing concept for evaluation of the alkali reactivity of binders/aggregate/mix design combinations.

4.5 Preventive measures

– The objective is to study the effect of different preventive measures, i.e. surface protection, low-corrosive steel reinforcement and corrosion inhibitors, on the service life of concrete structures exposed to corrosion inducing substances.

The intention with this project is firstly the use of laboratory study to map factors that are contributing to ageing of hydrophobic impregnations (factors that may cause ageing are temperature and temperature fluctuations, moisture and detrimental chemical substances such as alkalis. The mapping will be performed using accelerated test regimes). In this phase chemical and mechanical methods of characterization to will be used to study the test samples. Finally the intention is to study the effect of corrosion inhibitors on the critical chloride content.

4.6 Residual service life & load bearing capacity

– The objective is to model and predict residual service life and load bearing capacity of concrete structures with on-going reinforcement corrosion for different limits states

The intention of this project is to come up with models for the propagation period (corrosion process) and modelling of corroding structures. The intention is to test the instrumentation and monitoring methods that are used in the assessment of corroding structures. The intention is further to collect data on corroded concrete elements under field exposure and thereby predict the residual service life and capacity of the structure. Finally numerical simulation (FE analysis) will be made of corroded concrete structures

5 LABORATORY EXPERIMENTS FOR DETECTING CRITICAL CHLORIDE CONTENT IN REINFORCED CONCRETE

The knowledge of chloride threshold values for initiation of reinforcement corrosion in concrete is important for service life predictions and service life design. A lot of research has been devoted to finding such values. The reported results for critical chloride contents scatter over a wide range, e.g. from 0.02 to 3.08% when expressed in the form of total chloride by weight of binder (Angst et al, 2008). This is mainly related to the variety of possible measurement techniques, both for field studies and laboratory setups. At present, no standardised or accepted testing method for critical chloride content exists.

A literature review over nearly 40 references reporting chloride threshold values has concluded that many studies were not practice-related (Angst et al, 2008). Main pitfalls in laboratory work on critical chloride content have been identified to be the procedure to introduce chlorides and the quality of the steel-concrete interface. The latter is affected by both the rebar characteristics and the properties of the matrix. In many works, smooth rebars have been used instead of ribbed bars and the rebars have been prepared by sandblasting, polishing, cleaning, etc.

Experiments have been conducted in alkaline solutions, or in cement paste, mortar or concrete. It is evident that the properties of the steel-concrete interface in the case of, for instance, smooth and polished rebars embedded in cement paste differ from reality. This difference is considered to be important since the steel-concrete interface has been identified to be one of the major influencing factors with regard to critical chloride content (Cigna et al, 2002). Also, compaction is usually better for laboratory concrete in comparison with real concrete. With regard to the chloride introduction, several techniques are available to accelerate chloride penetration in order to avoid time-consuming experiments as in the case of pure diffusion. However, the situation is different from reality, e.g. in the case of mixed-in chlorides where the steel might not be able to initially passivate.

The literature evaluation (Angst et al, 2008) also revealed that most critical chloride contents have been measured in terms of total chloride content by weight of binder; free chloride contents have primarily been reported in the case of experiments dealing with solutions of porous cement paste/mortar/concrete. This is mainly due to the limitations of the pore solution expression technique, which is usually the one used for obtaining samples for analysis of free chloride concentrations in concrete pore solution. Up to now, there is a lack of information on the critical chloride content in dense concrete (low w/c ratio, alternative cement types) on the basis of free chloride.

5.1 Current PhD project on critical chloride content

The current project aims at measuring the critical chloride content based on both free and total chloride contents in laboratory concrete samples with a dense matrix. Emphasis is put on realistic and practice related conditions with regard to the factors mentioned in the previous section.

For the measurement of the free chloride content, embedded "chloride sensors" are used. These sensors consist of Ag/AgCl electrodes and have been used earlier by other researchers (Elsener et al, 2003).

5.2 *Chloride sensors – direct potentiometry in concrete*

The use of ion selective electrodes (ISE) in direct potentiometry is well established and has long been used in many fields such as analytical chemistry to determine the ionic activity of a certain species in aqueous solutions by a potential measurement. However, the situation in a cement based material such as concrete is more complicated. With regard to this, preliminary investigations have been undertaken by the present author. Some of the considerations and results are presented here; a more detailed description will follow in the full paper.

5.2.1 *Sensitivity and error sources*

Potentiometric measurements are very sensitive to variations in potential. The sensitivity directly follows from the slope of measured calibration curves or from the theoretical slope of 59.2 mV/decade. If only small deviations in potential are considered (in the range of only a few mV), it can be calculated that an error in potential of 1 mV results in a relative error in chloride concentration of ca. 4 percent (Koryta, 1972)

Several phenomena have been identified to disturb accurate measurements of sensor potentials when measuring in cement based materials (Angst, 2008). When an external reference electrode is used by establishing contact to the concrete surface with a wetted sponge the situation can be schematically depicted as in Fig. 1. Several components contribute to the measured potential, namely the *liquid junction potential* at the interface between inner solution of the reference electrode and the wetting agent in the sponge E_{RE-S}, the *liquid junction potential* at the boundary between the wetting agent in the sponge and the concrete pore solution E_{S-C}, as well as *membrane potentials E_C* across the concrete, and eventually *iR drops*. Both liquid junction and membrane potentials are all so-called *diffusion potentials* and arise due to concentration gradients in combination with differences in mobility between the diffusing ions. The significance of diffusion potentials for electrochemical measurements in concrete has been recently discussed by the author in ref. (Angst, 2008)

5.2.2 *Consequences for experimental setups*

In order to measure reliable free chloride contents in concrete with potentiometric sensors one has to be very careful with regard to the setup. Liquid junction potentials at the interface between an external reference electrode and the sample surface can be minimised (but never completely avoided) if the reference electrode is contacted to the concrete with an appropriate solution (Angst, 2008). With regard to membrane potentials the position of the reference

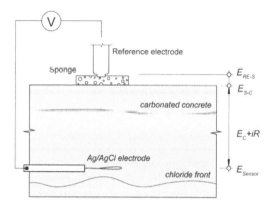

Figure 1. Disturbing phenomena when measuring potentials in concrete.

electrode appears to be decisive: To minimise contributions from membrane potentials arising from internal concentration gradients the reference electrode should ideally be placed at the same depth as the chloride sensors (e.g. by use of embedded reference electrodes).

5.3 *Conclusions on the PhD project*

The following sums up the current project plans and the results obtained so far:

1 The current project aims at using a practice-related testing setup for determining the critical chloride content in dense concrete, based on both free chloride and total chloride contents.

2 Ag/AgCl electrodes can be used as ion selective electrodes embedded in concrete to measure the free chloride content non-destructively. However, the technique is highly sensitive to errors in potential measurements arising from phenomena such as diffusion potentials. This is not a problem of the sensor itself, but of potential measurements through concrete as such.

REFERENCES

Angst U. and Vennesland Ø.. Critical chloride content in concrete – state of the art. In *2nd Int. Conf. on Concrete Repair, Rehabilitation and Retrofitting*. Cape Town, South Africa. 2008, (accepted for publication).

Angst U., Vennesland Ø. and Myrdal R.. Diffusion potentials as source of error in electrochemical measurements in concrete. *Materials and Structures* (2008), (in press).

Cigna R., Andrade C., Nürnberger U.,. Polder R, Weydert R. and. Seitz E, ed. *COST 521: Final report "Corrosion of steel in reinforced concrete structures"*. Luxembourg, 2002.

Elsener B., Zimmermann L. and Böhni H.. Non destructive determination of the free chloride content in cement based materials. *Materials and Corrosion* 54 (2003), 440–446.

Koryta J.. Theory and applications of ion-selective electrodes. *Analytica Chimica Acta* 61 (1972), 329–411.

sintef.no/coin (2008-08-15)

Concrete Solutions – Grantham, Majorana & Salomoni (Eds)
© 2009 Taylor & Francis Group, London, ISBN 978-0-415-55082-6

Structural repair of defects and deterioration to extend service life

Jonathan G.M. Wood

Structural Studies & Design Ltd, Chiddingfold UK

ABSTRACT: Without structurally effective repairs we cannot extend the service life of our deteriorating infrastructure. Structurally effective repairs require a fundamentally different approach to cosmetic and corrosion control repairs. Structural risk can arise from design and construction errors and/or developing deterioration. The nature of concrete conceals most design and construction defects and the deterioration of the reinforcement, so a rigorous review of design, detailing and construction records is needed to guide the site investigation and testing programme. This must be focussed on the identification of the often very localised "at risk" details and on providing the data for structurally assessing their current and future reserves of strength.

For structural repair the load sharing of dead and live load between the original concrete and reinforcement and the repair materials can govern the ultimate load behaviour. Dimensional strain incompatibility between original concrete and repair materials is often the cause of premature repair failure. There are reservations about the compatibility of EN 1504 CE materials with substrate concrete. A more structure specific performance based approach to specification is recommended. The paper is based on a wide range of case studies of investigation and structural remedial works including some heritage structures.

1 INTRODUCTION

As our infrastructure ages we are increasingly faced with problems of deciding when and how to repair or replace structures as corrosion, frost, alkali aggregate reaction (AAR/ASR), etc degrade their reserves of strength. Design codes provide no explicit margin for the loss of strength from the inevitable deterioration with time.

Much of our concrete infrastructure from the 1950s to 1970s is now deteriorating at an accelerating rate. The early stages of the deterioration were regarded as superficial, merely requiring cosmetic and surface corrosion control measures. Now deeper and more extensive deterioration, combined with other deficiencies, makes it necessary to carry out repairs which are structurally effective.

If we cannot achieve this, we will be faced with the costs and disruption of demolition and reconstruction.

In most cases structural hazards arise in details where severe deterioration is concentrated in locations where the reinforcement detailing is sensitive. Henderson (2002) in a review of deterioration in 200 car parks gives many examples. Because the damage is localised, the expense of high quality structural repairs can be justified.

Many structures were designed to codes which overestimated strength, particularly in relation to shear. Risks are the greater because of the variability in quality of as-built concrete. The variation in concrete strength and the cover to reinforcement is far greater than that assumed in old and current codes.

There are the additional risks of gross errors, beyond the statistics of variability, from the uncertain skill of designers and contractors.

Investigation and remedial works to control risks in deteriorating concrete structures must be focussed on areas of structural and deterioration sensitivity identified by quantifying the variability. Bad details create the environment for localised, often hidden, accelerated corrosion.

Cutting out and repairs seldom restore original strength and often weaken structures. Remedial procedures must be based on structural considerations. The variability of performance and durability found in repairs is far worse than in most original construction.

The best laboratory is the real world and applying rigorous scientific analysis to it. The short term oversimplified unrealistic conditions in laboratories do not enable the fundamental variability and the interactions between different deterioration processes found with repairs to real structures to be understood.

When planning and executing remedial works all these factors need to be considered in the evaluation of structural risk from concrete deterioration and from the repair, Wood (2006). This paper focuses on the risks that arise from investigation, cutting out and the repair process.

2 RISK FACTORS IN REMEDIAL WORKS

Detailed evaluations of the performance of repairs to structures have demonstrated a range of often

Figure 1. De la Concorde overpass after collapse. The deteriorated repaired half joint and lower part of the cantilever can be seen hanging down.

Figure 2. Tuckton bridge. Built 1904–05 repaired in 1950s, 1960s, 1970s, 1990s, investigated for further remedials 2006.

Figure 3. Tuckton bridge. Cracking of failing 1996 repair in 2006.

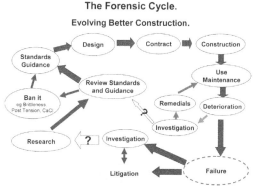

Figure 4. The forensic cycle.

interacting risk factors. These risks were dramatically demonstrated by the collapse (5 dead), of the de la Concorde overpass, Quebec (2007), Wood (2008a).

This report should be read by all concerned with the management, assessment and inspection of deteriorating infrastructure and Appendix 1 provides a check list.

Two areas of risk arise from the investigation. The greater risks arise from investigations which fail to identify design and construction deficiencies requiring remedial work in the as-built structure, and/or from deterioration developing in structurally sensitive locations. The second risk is from damage from coring or cutting out during the investigation.

The investigation of the condition of the 1904 Hennebique Tuckton Bridge (Wood and Grantham and Wait, 2007), is an example of rigorous investigation. Small diameter coring and testing with petrography were used on this historic structure where successive remedial works to localised chloride induced corrosion from poor drainage of deicing salts.

Once remedial work is initiated, the cutting out of material prior to repair inevitably weakens a structure and alters the stress distribution. It can also physically damage the substrate to which the repair must bond.

Finally the repair material needs to have an adhesion, stiffness and a strain history from casting, shrinkage, thermal effects and creep which enable it to carry a sufficient share of the loads in the structure under dead and live load conditions. Sometimes ductile redistribution prior to the ultimate limit state can be considered. The risks of strength loss and spalling hazards from repairs delaminating need explicit consideration.

To quantify the sensitivity of the critical parts of the structure to these risks they must be explicitly appraised, before remedial works start, by an engineer for the full range of risk scenarios:

'as built',
'as deteriorated',
'as cut out for repair',
'as repaired' and
'with repair delaminated'

The proper utilisation of data from forensic investigations (Fig 4) of deteriorating structures, Wood (2008b), and of the performance of repairs provides a basis for more accurately evaluating risks and how we can mitigate them. The proper analysis of success and failures of repairs in day to day maintenance is as important as data from major failures.

The CONREPNET results reported at the Concrete Solutions Conference in St Malo in 2006 (Mathews et al, 2006), provide a good picture of the magnitude of repair and remedial works. They also provide (Tilly, 2006), an overall picture of the frequency of premature failure of repairs, with 60% failing after 10 years.

If we are to understand the strengths and limitations of EN1504 Concrete Repair (Davies et al, 2006), we need to look in more detail at the specific reasons for repair failures on site. Data from the detailed site investigations of long term performance provide a necessary counterbalance to theoretical treatments and laboratory research.

3 IS THE STATE OF THE ART IMPROVING?

The papers to the 2006 St Malo Concrete Solutions conference provide a good indication of the state of the art of concrete repair. Many of them highlight limitations in our current knowledge and shortcomings in our procedures for long term management of deteriorating and substandard concrete structures.

It is interesting to compare the 'St Malo' 2006 state of the art with the 1986 CIRIA guidance, (Pullar-Strecker, 1986, updated 2002). The 'General principles for the use of products and systems' in EN 1504-9 are very similar to practice in the 1980s. However many parts of the other sections are less coherent and practical.

Some of the 60 test procedures, of variable value, in EN 1504 are little changed from the 1980s, others are new. The arbitrary limits now associated with these tests and CE designation are widely regarded as inadequate for some applications and excessive and expensive for others.

There is a fundamental conflict between the CE criteria and the engineering requirement to tailor the performance of repair materials and coatings to the specific characteristic of the structure and its deterioration state. This is made worse by the clumsy Construction Products Directive which can require unsuitable CE materials to be accepted while preventing the use of materials tailored to the required performance.

A review of concrete repair products on the market shows that they have maintained the tradition of elaboration of their expensive repackaged cocktail of ingredients to produce a limited range of 'off the peg' products tailored to meet CE requirements. Selecting from these to achieve the specific performance requirements of a repair contract is, at best, difficult.

We need to replace current CE designation based on arbitrary limits with a range of basic mortars and coatings with properties defined by appropriate EN 1504 tests. The CE would report the range of results from quality control testing of each product. It should be a certificate of testing not an indication of suitability. Repairs can then be specified on performance requirement related to the test data.

Repair mortar mixes can then be adapted to suit each structure's requirements, by combining with aggregates and, when appropriate, specific admixtures, to suit the application technique. Some usual combinations could be tested and included in CE data. We need not complicate mortar composition by trying to achieve higher resistance to carbonation or chloride ingress than adjacent un-repaired areas. Better resistance overall can be more easily achieved by coating.

Because repair performance depends on the overall compatibility of materials, application method and structural form, pre-contract site trials and testing of materials combinations, application methods and operatives are advisable for large contracts. Pull off tests, with petrographic examination of both sides of the fracture by removing the base of the core, can be particularly valuable.

4 STRAIN COMPATIBILITY OF SUBSTRATE WITH REPAIR

The fundamental need for strain compatibility of repair was spelt out thrice, nearly 2000 years ago, in the Bible, in Matthew Ch 9 v16, Mark Ch 2 v21 and Luke Ch 5 v36.

"No man putteth a piece of new cloth unto an old garment, for that which is put in to fill up taketh from the garment, and the rent is made worse".

Emberson and Mays (1990, 1996) have validated the importance of this for concrete repair.

If we are to prevent repairs from cracking and delaminating, we must first determine the initial stress and strain state of the concrete and how it will vary with temperature, moisture changes and stress cycles. Neville (1995), provides the classic source, bringing together research data on stiffness and thermal, shrinkage and creep strains of concrete and particularly on their sensitivity to water/cement ratio (w/c) and aggregate/cement ratio (a/c) of the concrete.

These strains need to be considered with the stress and strain range determined from the structural assessment for dead and live loading.

An example of detailed evaluation of strain characteristics including creep, E, and wetting and drying strains of concrete relative to repairs by RMCS Shrivenham is summarised in Section 8 of the full report to HSE on the Pipers Row collapse, Figure 5, (Wood, 2002). This also reports on the BRE petrographic and analytical diagnosis of the concrete deterioration and the repair failures. However the punching shear failure initiated at an adjacent column where the severe deterioration had not been identified.

5 MATCHING REPAIRS TO CONCRETE

The first requirement of the repair material must be matching Young's Modulus ($\sim$30 GPa) and Coefficient of Thermal Expansion ($\sim$10 $\times$ 10^{-12} per C°)

Figure 5. Delamination of repairs at pipers row.

Table 1. Shrinkage (microstrain) of prisms to 50% relative humidity. (Table 9.3, Neville, 1995).

At w/c → a/c ↓	0.4	0.5	0.6
3	800	1200	–
4	550	850	1050
5	400	600	750
6	300	400	550

with the concrete. For this cementitious materials with a reasonably high aggregate content have a clear advantage.

The shrinkage of cementitious materials, as they dry from as cast condition to equilibrium with the environment, is the major cause of repair cracking and delamination. The concrete being repaired will usually have stabilised.

Minimising the repair shrinkage by reducing water cement ratio (w/c) and maximising aggregate cement ratio (a/c) content is essential. Typical values of shrinkage are given in Table 1.

These shrinkages need to be compared to the tensile strain to cracking for concrete and mortars of typically 150 to 200 microstrain.

In many instances concrete repairs will remain in a wetting and drying range of 90 to 75% RH, so shrinkages will be reduced. However the surface layers of repairs directly exposed to the sun will seasonally dry to below 50% RH.

There are limits to the extent to which water/cement ratio can be reduced and aggregate content increased without problems arising with placing and compacting the repair. However, for example, using SBR latex to improve bond and other properties also enables water/cement ratios of below 0.4 and aggregate/cement ratios of 5 to be achieved with good compaction.

The shape, grading and rock type of aggregate is also important in matching strain. Depending on the size of repair and the reinforcement configuration and clearances, aggregate grading up to 10 mm is preferred, but 5 mm maximum is sometimes necessary.

Many manufacturers still advertise 'Shrinkage compensated' concretes. These produce an early age expansion over the first day or so of hydration by the formation of ettringite.

To achieve this expansion (typically 1000 microstrain) careful control of mixing and placing is required and the risk of damage from late ettringite formation must be avoided. Short term tests will demonstrate the free expansion phase, but the behaviour in most repairs is very different.

As the expansion develops and is restrained in the plane of the repair, the expansion occurs in the free dimension outwards to the surface. In the longer term, as drying out occurs, shrinkage develops, as with any cementitious mix, creating tensile strains and cracking.

A long term natural drying shrinkage test to quantify this was included in CIRIA TN 141 CIRIA (1993). Data on long term drying shrinkage at 50% RH as a function of water/cement ratio and aggregate content should be provided for all repair materials applying for CE compliance. Data for early age expansions are also needed when 'Shrinkage Compensation' is claimed.

The use of fibres in the repair mix is seldom advantageous, except in complete overlay screeds. In most cases, with a typical repair depth of 50 to 75 mm, shrinkage produces fine surface cracking distributed over the repair and at the edges. If a coating is provided these cracks are sealed.

Introducing fibres enhances the tensile strength and suppresses the fine cracking so that a larger crack develops at the perimeter. This can initiate a peeling delamination if shrinkage is high. Fibres can also inhibit compaction.

The cementitious material in the repair needs to be consistent with the cement of the concrete. A finer grind of OPC to speed hydration can be helpful, also PFA to assist workability. There is no point in using silica fume or high cost refined cementitious materials to produce a durability for the repair which is an order of magnitude greater than the adjacent un-repaired concrete.

The most important factor in achieving good cementitious performance is the water cement ratio when placed and the proper wet curing to fully develop hydration and strength before drying and shrinkage starts. PFA and ggbfs in mixes require longer term wet curing to develop good hydration and carbonation resistance.

6 CUTTING OUT TO THE REPAIR TO CONCRETE INTERFACE

The interface of the repair with the underlying substrate tends to be the weakest zone of a repair.

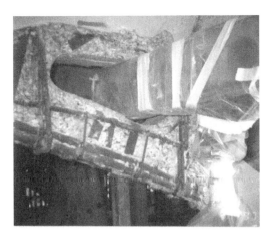

Figure 7. Localised pitting corrosion of links in a half joint.

Figure 6. It would have been better to remove all concrete and recast than the patch repairs which have failed.

Determining the depth and extent of cutting out required to ensure that carbonated or chloride contaminated concrete is removed is the first criteria. This depth is far more variable than simple spot tests suggest.

Getting a clear space behind the reinforcement so that repair mortar can be fully compacted is the second criteria. By this stage the loss of section may be weakening the structure. This may require substantial propping for safety and so that the repair can take its share of dead load.

Traditionally the jack hammer was the tool of choice and this often led to serious damage to reinforcement and the bruising and micro cracking of the substrate concrete, particularly with blunt tools. Petrographic examination of failed repairs and of the fracture surfaces of low strength pull off tests shows this clearly. This percussive procedure did not properly clean the reinforcement, particularly where chlorides had created pitting. Some old specifications referred to the 'use of emery cloth to clean the back of reinforcement'!

High pressure water jetting and diamond cutting, with selective use of percussive tools, enables a more surgical approach to preparations for repairs to be taken. However the profile of the surface to which the repair must bond must permit good compaction onto a bond coat without trapped voids.

7 PLACING AND CURING THE REPAIR

It is very difficult to place, compact and finish a repair to achieve the full potential of the materials used. The repair materials need to be optimised to suit the orientation and procedure for placing to be used. The long term performance of the repair can be prejudiced by unsuitable procedures like the use of flowable mixes.

Once well worked onto the substrate and built up to profile the repair needs to be retained in place and protected from drying out until it has fully developed is strength and bond.

The influence of the propping on the dead load stress state of a repair has been studied in detail by Canisius (2002). This clearly shows the importance of propping for structural effectiveness, as well as for safety.

When a section has seriously deteriorated the option of fully supporting the structure while removing the whole concrete section and recasting with concrete is a more reliable and structurally effective option then extensive patching.

In many structures there are deficiencies in reinforcement either from the original design, construction errors or from the effects of corrosion on steel section or bond. At de la Concorde, in Canada, all these factors contributed to the collapse.

Secondary reinforcement, stirrups in columns and the corners of shear steel in beams, Figure 7, are particularly vulnerable.

There are many ways of inserting or supplementing the reinforcement for a concrete structure, but this must be a topic for other papers.

8 CONCLUSIONS

- Durable structurally effective repairs are essential for extending the service life of our infrastructure.
- The specification of structural repair systems for concrete structures must be based on performance criteria for the stress and strain conditions in the repair material, at the interface and in adjacent unrepaired concrete elements
- To ensure that repairs carry dead load, as well as live load, propping during cutting out, casting and until strength is achieved is important. This is also necessary for safety.
- Repair materials need to be tested for their long term strain compatibility with the concrete of the structure.
- Long term shrinkage strains are a frequent cause of repair failures and test data on this is a prime requirement for a performance specification

- Increasing aggregate content and reducing water cement ratio of repair materials enables shrinkage to be reduced.
- Bruised substrates from percussive tools and poor bonding from bad application are the other principle causes of repair failures
- Many proprietary repair materials have overcomplicated formulations, some of which, like fibres, flowable mixes and expansive cements, can be detrimental as well as expensive.
- Specifications need to be performance based, so as to match the repair material to the concrete of the specific construction element.
- The concept of a standard CE mix required by a Construction Products Directive is fundamentally flawed and should be abandoned.
- Full propping with total recasting of concrete should be considered as an alternative to extensive patching when structures are seriously weakened.
- EN 1504 Part 9 is soundly based but the rest needs to be rigorously reviewed and simplified. It should provide tests which can be used in performance specifications of mixes and methodologies tailored to the wide variety of structural requirements.
- Repair systems can only be validated by tests on long term performance in the field. Forensic investigation of failures is also required to achieve better methodologies and materials.

9 APPENDIX 1

Guidelines for structural investigation and repairs, developed following Pipers Row collapse, Wood. (2002). It is interesting that the report on the de la Concorde collapse, Quebec (2007) identified that procedural failures under almost every heading in this list contributed to that fatal collapse

A check list for those involved in assessment, inspection and repair of deteriorating structures.

1. Check as built drawings and maintenance records.
2. Carry out a structural review to identify the key areas of structural weakness and/or structural sensitivity to deterioration as a basis for inspection procedures. This should cover both strength and risks from spalling.
3. Check for any features:
   - for which factors of safety may be inadequate for actual construction method and quality,
   - where the structural form is not explicitly covered by Codes,
   - which may be vulnerable to progressive collapse.
4. Establish by inspection and testing:
   - any departures from as built drawings,
   - any indications of defective or substandard construction,
   - indications of severe local environments from ponding, waterproofing breakdown, seepage etc.
   - the current trends of deterioration and likely long term trends.
5. Identify where and when protection, strengthening and/or repair may become appropriate and cost effective as part of the long term maintenance programme.
6. Ensure that before repairs are carried out that there is:
   - a full specification and procedure for repair, propping and testing
   - a Structural Engineer's check of the structure: 'as built', 'as deteriorated', 'as cut out for repair, with propping if required', 'as repaired, with propping if required' and 'with repair delaminated'.
7. Insist on a full recorded survey of condition before problems are hidden below patch repairs, coatings or waterproofing.

REFERENCES

Davies, H. and Robery, P.C. 2006. European standards for repair and protection of concrete, M Grantham, et al, Ed. Proc. Concrete Solutions Conf. St Malo, BRE Press, 2006.

Canisius, T.D.G. et al, 2002. Concrete repair patches under propped and unpropped conditions. FBE Report 3, BRE, March 2002.

CEN. European Standard on Concrete Repair' ENV 1504. BSI.

CIRIA "Standard tests for repair materials and coatings for concrete", Part 1 'Pull-off tests' TN 139, Part 2 'Permeability tests' TN 140, Part 3 'Stability, substrate compatibility and shrinkage tests'. TN 141 1993.

Emberson, N.K. & Mays. G.C. 1990 1996. Significance of property mismatch in the patch repair of structural concrete. Parts 1 and 2 Mag. of Concrete Research V42, No 152, pp 147–170, Sept. 1990, and Part 3 MCR V48, No 174, pp 45–57, Mar. 1996.

Henderson, N.A. et al. 2002. 'Enhancement of whole life cycle performance of existing and future car parks', Mott MacDonald PII Report. www.planningportal.gov.uk/uploads/odpm/4000000009277.pdf

Mathews, S.L. & Morlidge, J.R. 2006. What is wrong: Concrete repair – Solution or problem,' pp 3–10 and associated CONREPNET papers, M Grantham, et al, Ed. Proc. Concrete Solutions Conf. St Malo, BRE Press, 2006.

Neville, A.M. 1995. Properties of Concrete. 4th Edition. Longmans.

Pullar-Strecker, P. 1987. Corrosion damaged concrete assessment and repair, CIRIA Butterworth.

Pullar-Strecker, P. 2002 Concrete Reinforcement Corrosion: from assessment to repair decisions. ICE Design and Practice Guide, Thomas Telford

Quebec. 2007. Report of the Commission of inquiry into the collapse of the de la Concorde overpass. Gouvernement du Québec, 15/10/07. www.cevc.gouv.qc.ca/UserFiles/File/Rapport/report_eng.pdf

Tilly, G P. 2006. Past Performance of Concrete Repairs , pp. 11–15 M Grantham, C. Lanos and R. Jauberthie, Editors, Proc. Concrete Solutions Conference, St Malo, BRE Press, 2006.

Wood, J.G.M. 2002. Pipers Row Car Park, Wolverhampton: Quantitative Study of the Causes of the Partial Collapse on 20th March 1997, with 20 page summary.

SS&D Contract Report to HSE, on www.hse.gov.uk/research/misc/pipersrow.htm.

Wood, J.G.M. 2003. Pipers Row Car Park Collapse: Identifying risk. Proc. Structural Faults + Repair, 2003. précis in Concrete, Oct. 2003.

Wood, J.G.M. 2006. Evaluation of structural risk from concrete deterioration and repair. pp 757–766, M Grantham, et al, Ed. Proc. Concrete Solutions. St Malo, BRE Press, 2006.

Wood, J.G.M, Grantham M.G. & Wait S. 2007. Tuckton Bridge, Bournemouth. An Investigation of Condition after 102 years. pp 457 – 468 Proc. Concrete Platform 2007 ed. Russell M. and Basheer P., Queens University, Belfast and Structural Faults 2008.

Wood, J.G.M. 2008a. Implications of collapse of de la Concorde Overpass. The Structural Engineer 86/1, pp16–18, 8th January 2008.

Wood, J.G.M. 2008b. Application of Lessons from Investigations of Deteriorating Structures to the Management of Infrastructure. 4th Int. Conf. on Forensic Engineering, ICE, London.

Repair with composites

Concrete Solutions – Grantham, Majorana & Salomoni (Eds)
© 2009 Taylor & Francis Group, London, ISBN 978-0-415-55082-6

Effect of ECC thickness at soffit of concrete beams before strengthening with CFRP

A.M. Anwar

Department of Environmental Engineering, United Graduate School of Agriculture Sciences, Tottori University, Tottori, Japan

K. Hattori & H. Ogata

Department of Environmental Engineering, Faculty of Agriculture Sciences, Tottori University, Tottori, Japan

ABSTRACT: The invention of ultra ductile engineered cementitious composites (ECC) with a strain hardening behaviour makes ECC attractive for a broad range of applications. In the current research, sound and damaged plain concrete beams were prepared; the beams were replaced from their tension side by layers of ECC with different thicknesses. The specimens were examined destructively directly and after strengthening with additional sheets of carbon fibre reinforced polymers (CFRP). The results show enhancement in both the member load carrying capacity and its ductility with the increase in ECC thickness. In contrary, the increase in ECC thickness in systems with additional external strengthening by CFRP did not show significant differences in their overall performance; however, the flexural capacity of these beams was better than their corresponding cases without CFRP. Moreover, the application of thin layers of ECC before strengthening with CFRP was good in preventing or delaying the occurrence of the undesirable interfacial debonding mode of failure.

1 INTRODUCTION

Carbon fibre reinforced polymers (CFRP) have been widely used to repair and strengthen concrete structures. The interfacial debonding of CFRP is the critical failure mode, and probably governs the overall strength of the strengthened members (Gao et al. 2006). Attempts to minimize the occurrence of the debonding mode of failure were conducted; Xiong et al (2007) used U-shape fibre strips near the ends of the main CFRP sheet. Another group of researchers, Malek et al. (1998), Smith & Teng (2001) calculated the ultimate shear and normal stresses along the adhesive aiming to avoid reaching their failure limits.

The development of high performance fibre reinforced cementitious composites and particularly the engineered cementitious composites (ECC) by Li (2003), nearly two decades ago, attracted many researchers to utilize ECC for repairing and strengthening purposes. ECC has a strain hardening behavior when exposed to tensile forces. It also exhibits a strain capacity up to 200 times that of concrete (Maleej & Leong 2005). The many cracks with very small crack width are among its superior characteristics. ECC was recommended by many researchers as a repair material, Kanada et al (2002) used ECC in a spray form for rehabilitation of walls. Fischer & Li (2003) used ECC for strengthening of columns under cyclic loading.

In the current research, CFRP sheets were used to strengthen and repair concrete beams with different thicknesses of ECC applied at their soffit. A simple experiment was first done to examine the bonding efficiency of CFRP applied on the ECC substrate. The investigation showed that the bonding of CFRP on ECC was better than in the case of concrete substrates. The study expanded to evaluate the effect of ECC thickness, applied at the soffit of concrete beams, with and without strengthening by external CFRP sheets. The experiment was again repeated to repair initially cracked beams. Contrast in terms of the maximum load carrying capacity and the mode of failure for each combination was highlighted. It was noted that as the thickness of ECC increased, both the ductility and the section loading capacity increased as well. It was shown that the change in the ECC thickness before further strengthening with CFRP did not show a significant difference in the section loading capacity, it was however, capable of preventing the occurrence of the undesirable interfacial mode of failure.

2 EXPERIMENTAL WORK

2.1 *Materials*

The mix proportions for both concrete and ECC are shown in Tables 1 and 2, respectively. The concrete mix was designed as per the American Concrete Institute, ACI 211.11 (1982). The ECC used was in the form of ready mixed powder. The ECC incorporated high modulus polyethylene fibres. The fibre length was 12.7 mm and diameter 38 microns, tensile strength was 1690 MPa and elastic modulus, 40,600 MPa.

Table 1. Mix proportions of concrete.

Aggregate size (mm)	Slump (mm)	Air content (%)	W/C (%)	S/a (%)	Unit content (kg/m³)			
					Water	Cement	Sand	Gravel
20	100	2	56	38	207	370	635	1050

Table 2. Mix proportions of ECC per unit volume.

Ready mixed powder (Kg)	Water (Kg)	Fibre volume fraction (%)	Super plasticizer (Kg)	Anti-shrinkage agent (Kg)	Surface active agent* (Kg)	Slump flow (mm)	Air content (%)
50	11.2	2	0.54	0.49	0.1	500	10 ± 4

*Diluted with 25 times of water mass.

A polycarboxylic-acid-based super plasticizer was used. A bio-saccharide type viscous agent was also added to provide compatibility between fluidity and fibre dispersibility. In addition, an alcohol-type shrinkage reducing agent was also added. The ECC mixture was brought from Kajima Technical Research Institute in Japan. Full details of the ECC design have been reported by Kanda et al. (2006). The measured compressive and flexural strengths of concrete at 28 days were 35 MPa, and 4 MPa, respectively. The corresponding values for the ECC were 27 MPa, and 9 MPa, respectively. At 56 days, the compressive and flexural strengths of concrete showed little enhancement and were 36 MPa, and 4.5 MPa, respectively. As supplied by the manufacturer, the unidirectional CFRP sheets used were of 0.111 mm thickness, tensile strength of 3.4 GPa and modulus of elasticity of 3.4 GPa. The bonding epoxy had a flexural strength of 40 MPa, tensile strength of 30 MPa and shear strength of 10 MPa.

2.2 CFRP on ECC and concrete substrates

Before conducting the study, it was necessary to check the adequacy of pasting the CFRP on the ECC substrate. Thus, a preliminary experiment with a configuration similar to that shown in Figure 1a was conducted. The assembly consisted of two blocks, each of dimensions ($100 \times 100 \times 200$ mm) and joined together using one layer of CFRP sheet (280×90 mm) attached to the base of the blocks, leaving approximately a 2 mm gap between both blocks. This assembly was repeated twice, one for ECC, the other for concrete. The assemblies were subjected to a four point loading bending test of a bending span equal to 300 mm. The modes of failure for both the concrete and ECC assemblies are shown in Figures 1b, c, respectively. The interfacial debonding failure mode was observed in the case of the concrete while shear failure with no interfacial debonding was observed in the case of ECC. This might be attributed to the ductile behavior of ECC, where the absorbed energy during loading was released through a ductile shear failure; i.e. the ductility of ECC worked on redistributing the flow of forces shifting them from the high stressed bottom fibre to less stressed portions. In this study, a diagonal tension failure occurred which revealed that the bonding strength between the CFRP and the ECC substrate was even higher than the tensile strength of the ECC.

The brittleness of concrete did not allow compatibility in the deformation between the CFRP sheet and the concrete substrate; this consequently led the lower corner edge of one of the concrete prisms to be broken followed by interfacial debonding failure. The strains at the mid-span and at cut-off were measured and shown in Figure 2. It was also noted that the ECC assembly failed at a load value higher than that of concrete (38% more than the concrete). To check the flow of stresses along the CFRP sheet in both assemblies, the difference in the strain values was plotted for each assembly and shown in Figure 3. Both ECC and concrete showed nearly the same shear flow until a load level around 20 KN: a shear failure was then started in the case of ECC with a sudden release to the strains along the CFRP sheet. For concrete, a constant increase in the shear flow in the adhesive was observed until a sudden interfacial mode of failure occurred.

2.2.1 Reasons for the good bonding between ECC and CFRP

The reasons for the strong bonding between ECC and CFRP were due to 1) the micromechanical behavior of ECC contaminated with the strain hardening trend, where the ECC can develop two orders of magnitude strain higher than that of concrete. This behavior led the ECC and CFRP to create compatible strains, consequently the interfacial shear stresses in the adhesive was decreased and finally, decelerated the debonding failure mode. 2) ECC creates many but micro-size cracks in the most stressed portions which decreases the stress gradient between the CFRP sheet and the ECC substrate at the vicinity of cracks, and 3) the remaining small amount of fibre on the polished surface of the ECC, along with the high amount of fibres just underneath the prepared surface, created a strong reinforced adhesive compound and helped in decelerating the debonding failure. The previous explanation

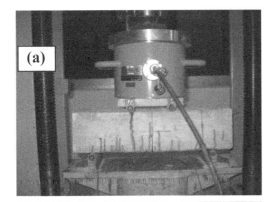

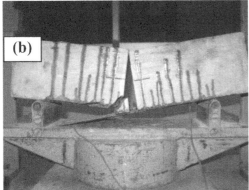

Figure 1. a) Test configuration, b) interfacial debonding in case of concrete substrate, and c) shear failure in case of ECC substrate.

might not be applicable in the case of concrete where one or more localized cracks appeared followed by the concentration of stresses in the vicinity of the cracks and subsequently induced sudden interfacial debonding failure.

2.3 Layered concrete/ECC beams

The overall flexural performance of concrete beams with different ECC thickness with/without the application of CFRP sheets at the soffit of beams was evaluated. Twelve specimens of dimensions (400 × 100 × 100 mm) were prepared. Two specimens

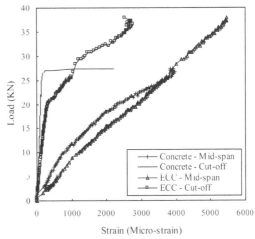

Figure 2. Measured Strains for both concrete and ECC.

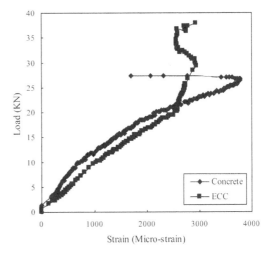

Figure 3. Difference in stain values between mid-span and cut-off for concrete and ECC.

were kept as control and cast to the full beam depth with concrete and ECC and named as C100, and E100, respectively. Three specimens were prepared with a partial replacement of concrete by ECC from the bottom; the replacement depth was 10, 30, and 50% of the total specimen depth and named C90E10, C70E30, and C50E50, respectively. The previous five specimens were again replicated with the application of CFRP sheets and denoted as follow, C100-CFRP, E100-CFRP, C90E10-CFRP, C70E30-CFRP, and C50E50-CFRP, respectively. To prevent the expected shear failure in the ECC specimens, two extra specimens for ECC and concrete were prepared with CFRP strips wrapping the shear zones and named as E100-CFRP/Sh, and C100-CFRP/Sh, respectively.

2.4 Application of ECC to pre-cracked beams

In addition, six initially cracked beams of the same size were prepared. Each beam contained a 50 mm

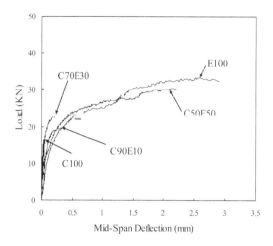

Figure 4. Load-deflection curves for layered specimens without CFRP.

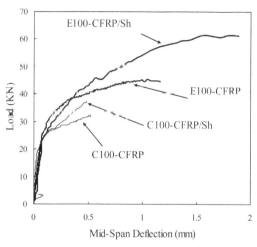

Figure 6. Load-deflection curves for specimens with additional shear reinforcement.

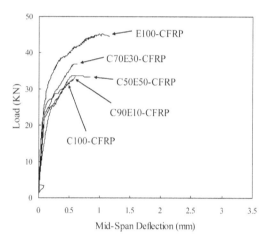

Figure 5. Load-deflection for layered specimens with CFRP.

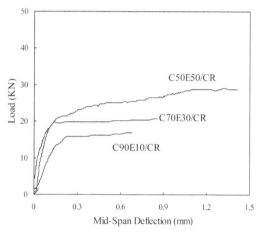

Figure 7. Load-deflection curves for repair of the cracked beams with ECC only.

depth crack and with a thickness less than 2 mm. Three specimens were repaired by placing ECC layers with thicknesses 10, 30, and 50 mm. The specimens were denoted as C90E10/CR, C70E30/CR, and C50E50/CR, respectively. The repair length was intentionally selected as 280 mm, i.e. 20 mm lesser than the bending span. It was thought that this combination might check the adequacy of the bonding between the repaired material (ECC) with different rigidity and the concrete substrate, as well as checking the capability of ECC to bridge the internal forces between the crack edges. The same combinations were again repeated after further strengthening by external CFRP sheets and were denoted as C90E10/CR-CFRP, C70E30/CR-CFRP, and C50E50/CR-CFRP, respectively. The specimens of this attempt were first cast and cured for 28 days. After which, the repair with ECC was done. The repaired specimens in their final form were again kept for an additional 28 days to cure.

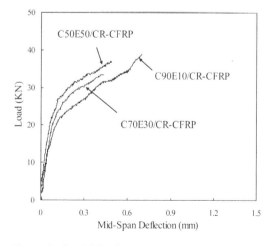

Figure 8. Load-deflection curves for repair of cracked beams with both ECC and CFRP.

342

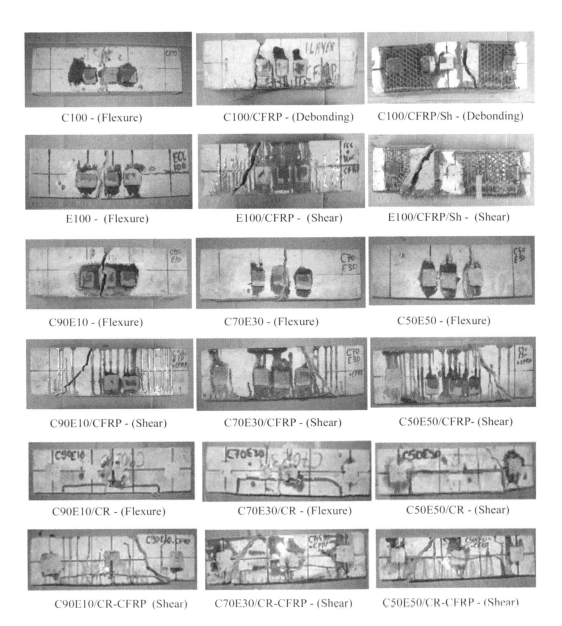

C100 - (Flexure)   C100/CFRP - (Debonding)   C100/CFRP/Sh - (Debonding)

E100 - (Flexure)   E100/CFRP - (Shear)   E100/CFRP/Sh - (Shear)

C90E10 - (Flexure)   C70E30 - (Flexure)   C50E50 - (Flexure)

C90E10/CFRP - (Shear)   C70E30/CFRP - (Shear)   C50E50/CFRP- (Shear)

C90E10/CR - (Flexure)   C70E30/CR - (Flexure)   C50E50/CR - (Shear)

C90E10/CR-CFRP (Shear)   C70E30/CR-CFRP - (Shear)   C50E50/CR-CFRP - (Shear)

Figure 9.   Modes of failure for the examined specimens.

All the specimens in this paper were destructively examined using a four point bending regime. The deflection at mid-span was measured and the load-deflection curves were plotted and compared.

3 RESULTS AND DISCUSSION

3.1 *Overall performance of the examined beams*

Figures 4 and 5 show the load deflection curves for the layered specimens with and without CFRP strengthening, respectively. It was shown that with the increase in the ECC thickness, there was a corresponding increase in the beam load carrying capacity. The enhancement in the section capacity for C90E10, C70E30, and C50E50 was 20, 41, and 88% more than the control beam, C100, respectively. It was clear that the ductility of the beams increased with the increase in the ECC thickness; this could be seen from the maximum deflection achieved by each specimen. Good bonding, with no spalling, was also observed for specimens C90E10, C70E30, and C50E50 which revealed the strong contact between the ECC and the concrete substrate without the use of additional adhesives. Pure flexure cracks were common to all the previous examined beams.

For layered specimens with external strengthening with CFRP sheets, the load deflection curves, for all the specimens except E100-CFRP, do not show significant variation in the peak load value. It was also observed that the application of the ECC layer before pasting the CFRP sheet was significant in delaying the interfacial debonding or even preventing this failure mode. The existence of the ECC layer acted as a transition zone between the brittle concrete and the high tensile strength CFRP sheets, with a high probability to form local cracks. The very good contact between the ductile ECC and the CFRP sheet led to dissipation of the absorbed energy during loading through weaker planes in the specimens. In this study, the shear failure was dominant and replaced the common interfacial debonding failure. For C100-CFRP, the undesirable interfacial mode of failure was the governing mode, the crack started at the location of one of the applied point loads followed by complete debonding of the CFRP sheet.

Thus, it was suggested to re-examine the specimens again after confinement the shear zones. The confinement of the shear zone was found to be critical for the case of ECC beams where the section capacity was increased significantly as shown for E100-CFRP/Sh. The same shear mode of failure occurred, without any debonding of the CFRP sheet. The examination of C100-CFRP/Sh did not show significant change from C100-CFRP in both loading capacity and mode of failure. Figure 6 shows the load- deflection curves for beams with additional shear reinforcement.

Similar observations were noted for beams repaired with ECC. The specimens with replacement depth of 10, and 30 mm exhibited a flexural failure mode, while that with a 50 mm ECC layer exhibited a shear mode of failure. This might be attributed to the decrease in the concrete shear resisting section combined with the presence of a high flexural rigidity repairing layer. It was also noted that ECC was able to bridge the internal forces between the crack edges. The combinations between ECC and CFRP did not show a significant change in the overall performance of the examined specimens with changing the thickness of ECC. It was however, able to increase the section loading capacity significantly as well as to prevent the occurrence of the undesirable debonding mode of failure. Figures 7 and 8 show the load deflection curves for the initially cracked beams after repair. The modes of failure for the examined beams are shown in Figure 9.

4 CONCLUSIONS

In the current study, the effect of ECC thickness in strengthening and repair of concrete beams before applying external CFRP sheets was conducted. The study started with a comparison between applying CFRP on ECC and concrete substrates. The following were among the main findings:

1. For systems with no external CFRP, as the depth of ECC increased, the section loading capacity and ductility was enhanced proportionally.

2. For systems with external CFRP, the difference in the ECC thickness did not change the overall flexure capacity significantly. It was however, better than systems without CFRP.
3. The shear confinement by means of wrapping the beams at their ends did not affect the mode of failure for both concrete and ECC beams, but it helped the ECC beam to achieve higher load carrying capacity.
4. The use of ECC as a transition zone between concrete and CFRP was mainly effective in preventing the undesirable debonding mode of failure without deficiency in the member loading capacity.
5. The change in the curing age between concrete and ECC for the repaired beams did not show any deficiency in their bonding.

REFERENCES

ACI Committee 211, *Standard practice for selecting proportions for Normal, Heavyweight, Mass concrete* (ACI 211.1-32), ACI manual of concrete practice, 1982.

Fischer, G. & Li. V.C. 2003. Deformation Behavior of Fibre-Reinforced polymer Reinforced Engineered Cementitious Composites (ECC) Flexural Members under Reversed Cyclic Loading Conditions, *ACI Structural Journal* 100 (1): 25–35.

Gao, B., Kim, J. & Leung, C. K.Y. 2006. Strengthening Efficiency of Taper Ended FRP Strips Bonded to RC Beams. *Composites Science and Technology* 66: 2257–2264.

Kanda, T., Saito, T., Sakata, N. & Hiraishi, M. 2002. Fundamental Properties of Direct Sprayed ECC. *Proceedings of the JCI International Workshop on Ductile Fibre Reinforced Cementitious Composites (DFRCC) – Application and Evaluation, October 2002*. 133–141 Takayama, Japan.

Kanda, T., Tomoe, S., Nagai, S., Maruta, M., & Kanakubo, T. 2006. Full Scale Processing Investigation for ECC Precast Structural Element, *Journal of Asian Architecture and Building Engineering, JAABE* 5 (2): 333–340.

Li, V.C. 2003. On Engineered Cementitious Composites (ECC) – A Review of the Material and Its Applications. *Journal of advanced Concrete Technology* 1 (3): 215–230.

Maleej, M. & Leong, K.S. 2005. Engineered Cementitious Composites for Effective FRP-Strengthening of RC beams, *Composites Science and Technology* 65: 1120–1128.

Malek, A.M., Saadatmanesh, H. & Ehsani, M.R. 1998. Prediction of Failure Load of R/C Beams Strengthened with FRP Plate Due to Stress Concentration at the Plate End. *ACI Structural Journal* 95 (1): 142–152.

Smith, S.T. & Teng, J.G. 2001. Interfacial Stresses in Plated Beams. *Engineering Structures* 23: 857–871.

Xiong, G.J., Jiang X., Liu, J.W. & Chen, L. 2007. A Way for Preventing Tension Delamination of Concrete Cover in Midspan of FRP Strengthened Beams, *Construction and Building Materials* 21: 402–408.

Concrete Solutions – Grantham, Majorana & Salomoni (Eds)
© 2009 Taylor & Francis Group, London, ISBN 978-0-415-55082-6

Non-linear finite element modeling of epoxy bonded joints between steel plates and concrete using joint elements

L. Hariche

Department of Civil Engineering, Djelfa University, Algeria

M. Bouhicha

Civil Engineering Research Laboratory, Laghouat University, Algeria

ABSTRACT: The issue of maintenance and repair of existing structures has become a major issue, particularly extending the service lifespan of reinforced concrete structures. Amongst various methods developed for strengthening and rehabilitation of reinforced concrete (RC) structures, external bonding of steel or fibre reinforced plastic (FRP) strips to the beam has been widely accepted as an effective and convenient method. However, the development of a design method that can properly describe and predict the behaviour of a strengthened RC beam is an extremely difficult task because of the relatively large number of likely failure modes that have been reported by researchers. One of the most critical failure modes encountered in plated beams is the premature and brittle rupture of the plate to concrete bond. This debonding of the plates tends to occur when the bond strength is reached locally. Consequently, the integrity of the strengthening system does not depend solely on the plate material but also on the properties of the interfaces involved in the joint, namely, the plate-adhesive and the adhesive-concrete interfaces.

In this paper, a non-linear finite element model is developed for studying the behavior of epoxy adhesive bonded concrete-steel joints. The predictions of the model have been validated with experimental test data found in the literature. The study was conducted for varying plate and adhesive thicknesses, and the behaviour of the joint was assessed in terms of load carrying capacity and the shear stress distribution at the interfaces and in the adhesive. The results show that a non-linear finite element model may reliably predict the behaviour of epoxy adhesive joints.

1 INTRODUCTION

One of the greatest challenges facing the construction industry is the rehabilitation of concrete structures. The maintenance and repair of concrete structures raises serious concerns for most countries, as it is associated with heavy expenditure. Hence, it is of paramount importance to develop inexpensive and reliable repair and strengthening techniques.

A number of repair techniques exist today. It should be highlighted, however, that no quantitative design rules have yet been developed which would allow a designer to choose the most cost effective and durable concrete repair materials and system in a given situation (Swamy 1987) This is due to the complex nature of the repair problem where critical decisions have to be made in the selection of repair materials and systems to be used which will hopefully avoid the interminable sequences of repairing the repairs (Charif 1983). Furthermore, conventional repair systems that use cement based materials seem to suffer from problems related to the compatibility between new repair material and the existing structure (Hariche 2002). Several studies have been carried out to alleviate this problem and to propose a set of performance criteria for selecting

the repair materials (Swamy 1987, Aprile 2001 and Muravljov 1994).

The technique of epoxy bonding of steel or FRP plates to the surfaces of RC beams has received great attention in recent years because it offers various desirable attributes such as efficient form of mechanical retrofitting, is inexpensive, and causes minimal disruption to moving traffic. The extensive experimental investigations, carried out by researchers so far, to assess the structural behaviour of plated beams indicate that significant improvements in the structural performance can be obtained in both service and ultimate situations (Teng 2002 and Siera-Ruiz 2002). Moreover steel, as well as FRP, plates have been bonded to the surface of RC beams in several ways. For instance, they have been attached to the beam soffit or tension zone in order to increase the flexural strength and stiffness. To enhance flexural and shear capacities, plates have been applied either to the sides of the beams or to both the sides and soffit to form angle or channel plated sections (Charif 1983).

However, the design of plated beams is a complicated issue as researchers have reported as many as 30 possible failure modes (Charif 1983). The most critical of these are related to the debonding or peeling

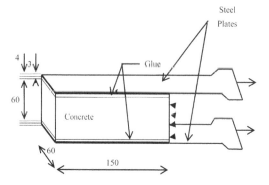

Figure 1. Experimental set-up for the joint test (Charif 1983).

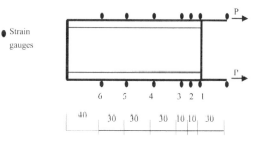

Figure 2. Strain gauge positions (Charif 1983).

Table 1. Pull out test results (Charif 1983).

Effect of concrete strength Glue thickness = 1.0 mm		Effect of glue thickness Concrete strength = 45 N/mm^2	
Concrete strength (N/mm^2)	Mean shear stress (N/mm^2)	Glue thickness (mm)	Mean shear stress (N/mm^2)
25.1	2.17	0.5	2.42
44.9	2.52	1.0	2.52
60.2	2.77	1.6	2.38
71.3	3.33	3.0	2.83

Mean shear stress = Failure load/joint bonded area

of the plate, leading to a brittle premature failure of the strengthened beam. Practical ways of preventing this type of failure as well as approximate numerical and analytical methods, based mainly on the assumption of linear elastic behaviour of the materials, that could be used in the analysis of the strengthened beams have been proposed in the literature (Muravljov 1994, Teng 2002 and Siera-Ruiz 2002). Nevertheless, it should be highlighted that the above failure modes are strongly influenced by the integrity of the bond between the plate and the concrete (Muravljov 1994). Thus, in order to satisfactorily study this failure mode, it is important to use models that describe accurately the stresses that develop at the interfaces of the epoxy joint. Exact closed form solutions of the shear distributions that take place at the interfaces are unfortunately rather complicated (Teng 2002), and numerical approximation is therefore necessary to gain an insight into the behaviour of these structural joints.

A non-linear finite element model is developed herein to study the behaviour of epoxy joints used in bonding steel plates to concrete. This model has been validated against data from test results for a pull out joint specimen found in the literature (Charif 1983), and a parametric study was carried out to assess the different parameters that may affect the behaviour of the joint. Although the main emphasis of this research was on steel plates, the model can be easily adapted to FRP bonded plates (Aprile 2001).

2 EXPERIMENTAL SETUP

2.1 Details of test specimens

The geometry of the test specimens (Charif 1983) is shown in figure 1. The test variables were the adhesive thickness, which was varied from 0.5 to 3 mm, and the concrete cube strength, which varied from 25.1 to 71.3 N/mm2.

The specimens were instrumented with strain gauges positioned on the central line of the plate surface as shown in figure 2.

For each investigated variable, six specimens were tested and the average values of the tests were taken as shown in table 1.

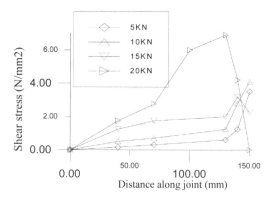

Figure 3. Experimental shear stress distributions along the joint (Charif 1983).

2.2 Experimental results

The local shear stress along the bonded plate length is given by:

$$\tau = \frac{\Delta F}{b.\Delta L} \qquad (1)$$

Where ΔF: is the variation of longitudinal force between two consecutive strain gauges positions that are distant by ΔL.

The shear stress distribution, shown in figure 3, along the plate is exponential at low load levels. At higher loads (exceeding 60% of the ultimate load) the joint starts failing at the most stressed end and the shear in this region becomes zero.

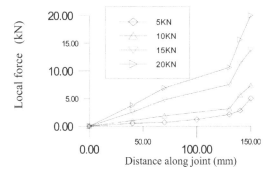

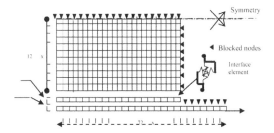

Figure 4. Experimental local force distribution along the joint (Charif 1983).

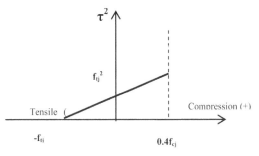

Figure 5. Finite element mesh used (Hariche 2002).

The distribution of local forces in the steel plates, presented in figure 4, shows a similar trend as the one observed in local shear stress again being exponential for low applied loads.

3 FINITE ELEMENT MODELING OF THE JOINT

The formulation of a finite element model to simulate the behaviour of the joint requires the choice of an element that enables easy preparation of the data needed and adequate interpretation of the results to be made after the analyses. Simple examination of the experimental set-up indicates that a state of plane stress can be considered. The above conditions are perfectly satisfied by the use of the four node isoparametric element. Moreover, in order to represent the gradual failure of the joint, interface elements (Goodman 1969) as shown in figure 5 were used to model the steel-adhesive and concrete-adhesive interfaces.

Several analyses were carried out using increasingly finer meshes of the type shown in figure 5 to ensure the convergence of the results. The adopted mesh in this study consisted of 1061 nodes and 994 elements that include 122 interface elements. The material properties that have been used for the validation model were those determined during the experimental investigation, and are presented in table 2, corresponding to a concrete cube compressive strength of 44.9 N/mm2.

The stiffness of the springs forming interface elements were computed as follows (Barbes 1982):

$$K_n = K_t = 10^5 \rho\, R_{max} \qquad (2)$$

Table 2. Materials properties for the validation model.

Material	E (N/mm2)	ν	Thickness (mm)
Steel	200 000	0.3	3
Adhesive	1500	0.499	1
Concrete	36300	0.16	60

Figure 6. Failure criterion for the concrete surface.

Where:

$\rho = E_{glue}/E_{concrete}$ or E_{glue}/E_{steel} depending on the interface

$R_{max} = $ Maximum coefficient in the global stiffness matrix

E_{glue}, $E_{concrete}$, and E_{steel} are respectively the elasticity modulus of the adhesive, concrete, and the steel.

A failure criterion, shown in figure 6, was adopted for the surface of the concrete. It is similar to the criterion used in checking the shear capacity of prestressed concrete at the serviceability limit state (Barbes 1982). The formulation of the criterion is as follows:

$$\tau^2 = f_{tj}(f_{tj} + \sigma) \qquad (3)$$

Where: $\tau = $ Shear stress acting on the interface

$\sigma = $ Normal stress acting on the interface (compression positive, tension negative) with $\sigma \leq 0.4\, f_{cj}$

$f_{tj} = $ Tensile strength of concrete at day j

$f_{cj} = $ Cylinder compressive strength of concrete at day j

The tensile strength of concrete and the modulus of elasticity were estimated using the equations (BPEL 1991):

$$f_{tj} = 0.6 + 0.06\, f_{cj}$$

$$E_{ij} = 11000 \sqrt[3]{f_{cj}} \qquad (4)$$

3.1 Model validation:

A finite element program was developed that performed non-linear analyses of epoxy joints. The load was applied gradually to the model in increments of 0.5 kN. For each increment, the stress field acting on the adhesive-concrete interface was checked. The gradual failure of the joint was captured through the modification of the interface element stiffness (K_n and K_t) as described below (Hariche 2002):

If $\tau \geq \tau_{lim} = \sqrt{f_{tj}(f_{tj} + \sigma)}$ Sliding occurs and $K_t = 0$
If $\sigma \geq -f_{tj}$ Peeling occurs and $K_n = K_t = 0$

The increase in the applied load was maintained until the complete failure of the interface. The computed failure load obtained for the validation model was equal to 22.5 kN. This is in excellent agreement with the corresponding experimental value of 22.68 kN (error of 0.8%) and thus confirms the validation of the model.

The computed distributions of shear stresses in concrete, adhesive, and steel elements adjacent to the interface are shown in figure 7 where it is found that the numerical model reproduces the experimental stress distributions. It should be noted here that the shear stresses shown were evaluated at the centre of the elements forming the mesh (Aprile 2001).

While the concrete and the adhesive transfer the load basically through shear stresses, the plate seems to be subjected mainly to direct tensile stresses as is clearly indicated in figure 8.

The shear force acting at the adhesive-concrete interface is presented in figure 9 for different loading levels. It can be noted that for lower load levels, an exponential distribution of the stresses starting from the peak value toward the non-stressed end is recorded.

3.2 Effects of concrete strength

The investigation into the effects of concrete strength on the load carrying capacity of the joint was carried out using the model used during the validation analysis. The results obtained are summarized in table 3.

3.3 Effects of adhesive thickness

The properties of materials and the mesh used during this analysis were identical to that of the model used for validation. The only changing parameter was the adhesive thickness, which was varied from 0.5 to 3.0 mm. Table 4 shows that.

3.4 Effect of plate thickness

Figure 10 shows that the computed joint capacity varies almost linearly with the plate thickness. It should be noted here, as before, that only the thickness of the plate varied, the other parameters were identical to those of the validation model. Further, the plate thickness seems to be the most important parameter affecting the joint capacity after the tensile concrete strength. The increase in plate thickness seems also to lead to a more uniform distribution of stresses along the adhesive-concrete interface by reducing the peak value. The initiation of the separation is, as a result, delayed leading to a higher failure load.

3.5 Anchorage length

The critical anchorage length is defined as the bonded joint length beyond which any increase in length does

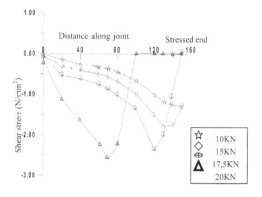

(a) Shear stresses distribution in the plate steel concrete along the joint (Hariche 2002)

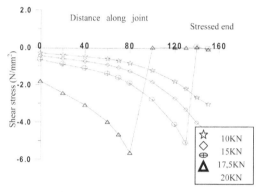

(b) Shear stresses in the adhesive

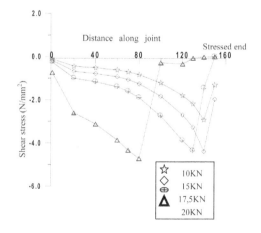

(c) Shear stresses in the concrete

Figure 7. Shear stresses distribution in the concrete, the adhesive, and the steel along the joint.

not lead to an increase in the load bearing capacity of the joint (Siera-Ruiz 2002). This has many practical and economic applications in patch repair of concrete elements. To investigate this, the validation model was used again but with varying length of the bonded joint.

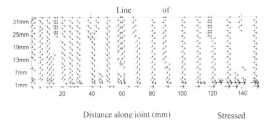

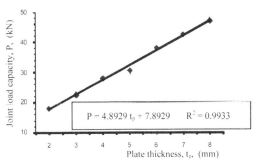

Figure 8. Directions of the major principal stress in the joint model.

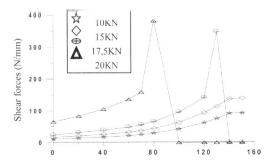

Figure 9. Shear force distributions along the concrete-adhesive interface.

Table 3. Effect of concrete strength on the load capacity of the joint.

Concrete properties (N/mm^2)			Failure Load (kN)		
f_{cube}	f_t	E	Exprimental	Numerical	Error (%)
25.1	1.8	30251.5	19.53	19.0	2.5
44.9	2.7	36300.0	22.68	22.5	0.8
60.2	3.5	40021.0	24.93	26.0	4.3
71.3	4.0	42343.4	29.97	31.0	3.5

Table 4. Effect of adhesive thickness on the load capacity of the joint.

Adhesive thickness (mm)	Experimental load (kN)	Numerical load (kN)	Error (%)
0.5	21,78	21	3.0
1.6	21.42	23	0.8
2.0	–	25	–
3.0	25.47	28	8.0

Starting from the stressed end of the joint, the adhesive elements were gradually removed to vary the bonded length of the joint from 150 to 90 mm. It can be seen from the results given in table 4 that the computed

Figure 10. Variation of the joint load capacity with the plate thickness.

value of the joint load bearing capacity does not significantly change for bonded joint lengths higher than 120 mm. Thus the joint anchorage length seems to amount to 80% (40 × plate thickness) of the initial joint length.

4 CONCLUSIONS

This study has shown that nonlinear finite element analysis may reliably predict the behaviour of epoxy bonded steel plate to concrete joints. Detailed information on the behaviour of such joints can be obtained from this analysis. The following results were observed:

- Failure of the joint was always initiated in the concrete surface. Therefore, the load capacity of the joint is strongly correlated to the shear strength of concrete. The failure of the joint is gradual and propagates along the joint with the increase in loading.
- The distribution of shear stresses along the bonded joint follows an exponential trend. The peak value of this stress tends to be concentrated in a very small region of the joint.
- The increase of the glue thickness does not seem to have a very significant effect on the load carrying capacity of the joint (Muravljov 1994 and Teng 2002).
- The existence of an anchorage length of the joint is confirmed during this study. It was found to be equal to 40 times the plate thickness.
- The load capacity of the joint studied herein may be estimated by:

P = 4.851tp + 2.283ta + 5.386ft − 8.936

Where:
tp = plate thickness
ta = adhesive thickness
ft = tensile strength of concrete

The above results confirm and complement the available experimental observations and show the importance of using numerical models to analyse the behaviour of epoxy joints.

REFERENCES

Aprile, A., Spacone, E., Limkatanyu, S., "Role of bond in RC beams strengthened with steel and FRP plates", ASCE Struc. Eng. Jrnl., Vol.127, No.12, 2001, pp1445–1452

Barbes, A., Franck, R., "Utilisation de la MEF en mécanique des sols dans le domaine elasto – plastique", Rapport de recherche LCPC, No.116, Sept. 1982.

Charif A , "Structural behaviour of reinforced concrete beams strengthened by epoxy bonded steel plates". Ph.D. Thesis, Sheffield University, 1983.

DTU , "Règles de calcul des ouvrages en béton précontraint aux états limites", BPEL91, Eyrolles, Paris, 1991.

Goodman, R.E., Taylor, R.L., Breeke, T.L., "A model for the mechanics of jointed rocks", ASCE, Soil. Mech. Div, Vol. 94, No.3, 1969, pp637–659.

Hariche L, "Simulation non linéaire du comportement des joints en résine époxydique utilisés dans les structures en BA renforcées par collage des plaques d'acier, Mémoire de Magister, Juillet 2002, Université de Tiaret.

Muravljov, M., Krasulja, M., "Modèles physiques et numériques pour evaluer la capacité de charge de cisaillement de joints d'adhérence acier – béton", Materials and Structures, Vol27, No.165, 1994, pp40–53.

Siera-Ruiz, V., Destrebecq, J.F., Crediac, M., "The transfer length in concrete structures repaired with composite materials: a survey of some analytical models and simplified approaches", Composite Structures, Vol.55, 2002, pp445–454

Swamy, R. N., Jones, R., Bloxham, J. W., "Structural behaviour of reinforced concrete beams strengthened by epoxy steel plates", The Structural Engineer, Vol.65 A(2), 1987, pp59–68

Teng, J.G., Zhang, J.W., "Interfacial stresses in RC beams bonded with a soffit plate: a FEM study", Construction and Building Materials, Vol.16, 2002, pp1–14

Concrete Solutions – Grantham, Majorana & Salomoni (Eds)
© 2009 Taylor & Francis Group, London, ISBN 978-0-415-55082-6

Experimental and analytical investigation of concrete confined by external pre-stressed strips

S. Mohebbi
Department of Civil Engineering,University of Tehran, Tehran, Iran

H. Moghaddam & M. Samadi
Department of Civil Engineering, Sharif University of Technology, Tehran, Iran

K. Pilakoutas
Department of Civil and Structural Engineering, University of Sheffield, Sheffield, UK

ABSTRACT: Strengthening is often a necessary measure to overcome an unsatisfactory deficient situation or where a new code requires the structure or a member of it to be modified to achieve new requirements. In the engineering practice such restraint to lateral dilation, indicated by the name of confinement, has been traditionally provided to compression members through steel transverse reinforcement in the form of spirals, circular hoops or rectangular ties. Steel and concrete jackets are other techniques for providing additional confinement for compression members, too. This paper presents the results of an experimental and analytical study on the application of strapping techniques for retrofitting of concrete columns. The experimental program included axial compressive tests on cylindrical and prismatic small-scale columns which were actively confined by pre-stressed metal strips. Test results showed a significant increase in strength and ductility of columns due to active confinement with metal strips. The effect of various parameters such as pretensioning force in the strip, number of strip layers wrapped around the specimens and spacing of confining strips, on the strength of the concrete is studied. Nonlinear Finite element models of the tested specimens were also made and analyzed. The observed stress-strain behaviors of columns with different levels of confinement are compared to those obtained from the finite element method.

1 INTRODUCTION

Recent earthquakes in urban areas have repeatedly demonstrated the vulnerability of older structures to seismic deformation, not only the traditional ones but also those made with reinforced concrete materials, with deficient shear strength, low flexural ductility, and insufficient lap splice length of the longitudinal bars. Very often, these structures also have inadequate seismic detailing, and, in many cases, very bad original design, with insufficient flexural capacity. The most critical mode of failure in RC structures is column shear failure. To prevent this brittle failure, the columns need to have guaranteed shear capacity both at their ends, in potential plastic hinge regions, where concrete shear capacity can degrade with increasing ductility demands and in the column center portion, between flexural plastic and/or existing built-in column hinges.

Many researchers have shown that confinement of concrete columns at the potential plastic hinge regions increases significantly the failure strain of concrete and therefore the overall ductility and strength (Alcocer et al. 1993) (Ghobarah et al. 1997) (Xiao et al. 2003) (Khoury et al. 1991). Steel plate jackets and reinforced concrete jackets have been widely used to strengthen the RC columns.

In this paper, an easy retrofit technique for concrete is presented. The main aim of this research was quantification of the enhancement of concrete strength and ductility by the application of the technique. The results of experimental and analytical studies on the performance of the technique are discussed. This study focused on application of this technique for high strength concrete.

The technique used for strengthening concrete columns in this study, involves post-tensioning high-strength packaging straps around the column (by using standard strapping machines used in the packaging industry) and subsequently locking their ends in a metal clip.

Commercially available strapping tensioners and sealers make it easy to pretension the strip and fix the strip ends in the clamps. The available straps have widths of 10 to 50 mm and thicknesses of 0.5 to 1.12 mm. In terms of strength, high strength strips in excess of 10000 kg/cm^2, are available in the market. The strips are tensioned to 25 percent of their yield stress. Hence, an effective lateral stress is applied on the column prior to loading. This has many benefits

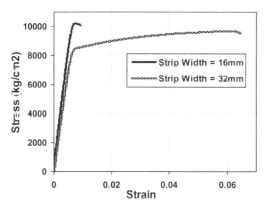

Figure 1. Stress-strain relationship of confining strips used.

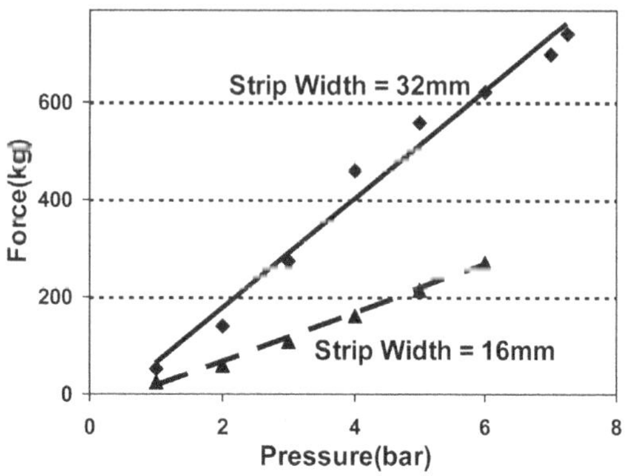

Figure 2. Results of calibration tests, the relation between tensioning force and air pressure.

such as full utilization of the strip capacity and prevention from premature crushing of the confined concrete, as would be the case with improperly tightened strips.

The low cost of strips and speed and ease of application of the strapping technique make this method efficient for use as a repair and strengthening technique for RC structural members. An RC column would normally require six man days' work to be jacketed whilst a maximum of two days' work is required for external strapping, which clearly demonstrates the cost saving when using the proposed technique. (Frangou et al. 1995)

The principal objective of the work presented in this paper is the verification of the effectiveness of the repair and strengthening technique for RC members by utilizing the strapping method. This objective has been achieved through the following two tasks:

1) Confirmation in practice that the technique can be applied for the repair and strengthening of columns under axial load.
2) Investigation of the suitability of Nonlinear Finite Element Methods for confinement of concrete for the purpose of strengthening by the strapping technique.

2   EXPERIMENTAL PROGRAM AND OBSERVATION

The concrete specimens were fabricated in the structure and concrete Laboratory at the building and housing research center (BHRC). The material used for the concrete specimens included type I portland cement, local sand and gravel. The maximum size of the gravel was 12 mm. No additive was used in any of the mixes.

Experiments included 25 cylindrical and 15 prismatic concrete specimens. The column models were made of a relatively high-strength concrete with no air-entrainment. The concrete reached an average uniaxial compressive strength of about 50 MPa.

The specimens were removed from the forms after 2 days and put into water to be moist cured. The cylindrical and prismatic specimens were tested 428 days after casting. Two sizes of metal strips were used for strengthening of the specimens. The specimens were strengthened by using 16 × 0.5 mm and 32 × 0.8 mm strips. In addition to the difference in width and thickness, the material behavior of the strips was also dissimilar. In Figure 1, the stress-strain behavior of the strips used, which were obtained from standard tensile tests, are shown. As can be seen in these figures, although both strips had similar strengths, the elongation of the strips, that is an important characteristic of the confining elements, were quite different. The 32 mm wide strip had a larger ductility making it more suitable for application as a confining element for concrete.

One of the important parameters in this study was to compare the active and passive external lateral confinements by this technique. In order to do so, some of the cylindrical specimens were tensioned only to 40 kg (which will be called passively confined specimens hereafter) while a tensioning force of 250 kg was applied in pretensioning the other specimens (which will be called actively confined specimens hereafter). In fact, the metal strips of the latter specimens were tensioned to 0.31 of their yield strain. Two pneumatic tensioners were used to strap the two strips.

The tensioning force in the strips was calibrated by means of special setups. Once the air pressure before the tensioner was set to a certain value, the tensioning force in the strip was monitored by means of a tensile load cell and a data logger. Figure 2 shows the relationship between air pressure and the tensioning force in the 16 mm and 32 mm strips and the fitted regression lines. It can be seen that a linear relation exists between these parameters. After tensioning the strip, the two ends of the strips were fixed together by sealing the clamp.

This results in some loss in pretensioning force. The loss in tensioning force due to the sealing process was also monitored for some strips. The average percentage

Figure 3. Setup of axial testing of a prismatic specimen.

Figure 4. Prismatic and cylindrical specimens after testing.

of losses was 19% and 31% for 16 mm and 32 mm strips, respectively.

After the concrete column models had been cured, the metal strips were strapped around the specimens. Axial compression tests were conducted using a testing machine with a capacity of 1,780 KN in the Concrete Laboratory of BHRC. The load was increased until significant strength decay was recorded, which indicated failure of the specimens. Figure 3 illustrates the test setup for a typical loaded column specimen.

A total of six displacement transducers were made to obtain the longitudinal and transverse strains. Three 50 mm CDP displacement transducers were mounted on steel rods between the top and bottom plates to measure the axial displacement of the column specimen's top surface. In addition a special setup with two 25 mm CDP displacement transducers was made to measure the relative displacement over the middle 3/4 height of the specimens. Also a DP tape measure type displacement transducer was used to measure the circumferential strain of the specimens.

FLA-5-11 Strain gages from the TML Company were attached to the strips to obtain the strain of strips during the test. A 200 ton load cell was located at the end of the specimen to measure the load at desired intervals together with other data.

Compressive tests were conducted on the specimens. The axial load, with a load rate of 178 KN/min, was increased monotonically until the specimens failed. This loading rate is equivalent to 0.25 MPa/sec. The ASTM standard loading rate for compressive strength of cylindrical concrete specimens is within the range of 0.14–0.34 MPa/sec. Therefore, the selected loading rate falls within the range of the ASTM standard. Figure 4 illustrates Prismatic and cylindrical specimens after test.

The results obtained from strain gauges as well as displacement transducers were analyzed. It was observed that the axial stress and the confining pressure kept increasing until the value of lateral strain reached the yield strain of the strips in a circumferential direction. The specimens reached their maximum strengths when one or more of the strips yielded. After the peak stress, the strips ruptured one by one resulting in the loss of axial stress. Column specimens with

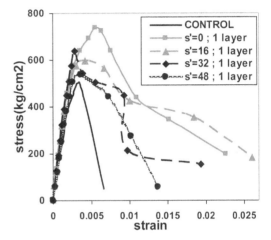

Figure 5a. Stress-strain behavior of passively confined cylindrical specimens with 1 layer strips.

two layers of the metal strips gained larger strengths as well as a larger ultimate axial strain as compared with column specimens with one layer of the metal strip.

The observed stress-strain behavior of some of the test specimens are drawn in the following figures. In these figures, the results of specimens that were actively confined with only one layer of 16 mm strip confined with strips with different clear spacing (s') are shown. Figure 5 shows the results of cylindrical specimens while in figure 6, results of the prismatic specimens are shown.

It can be concluded from the figures that:

1) This technique has been able to increase the strength of concrete up to 2.3.
2) An increase in the spacing between the strips, has always led to increase in strength of confined concrete.
3) The concrete confined with double layer metal strips has generally shown better enhancement in concrete strength than confinement with single layers.
4) Active confinement resulted in more increase in concrete strength than passive confinement. This is primaril because, whilst the ordinary passive confinement is mainly utilized after the core

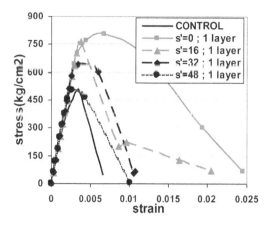

Figure 5b. Stress-strain behavior of actively confined cylindrical specimens with 1 layer strips.

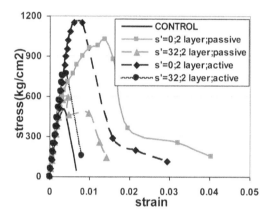

Figure 5c. Stress-strain behavior of passively and actively confined cylindrical specimens with 1 and 2 layer strips.

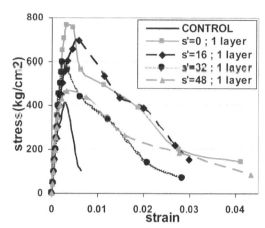

Figure 6a. Stress-strain behavior of actively confined prismatic specimens with 1 layer strips.

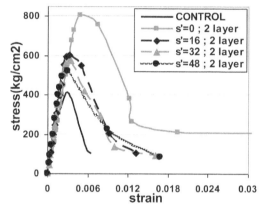

Figure 6b. Stress-strain behavior of actively confined prismatic specimens with 2 layer strips.

concrete has dilated (which means that some cracks have occurred in it); the active confinement influences the core concrete even before load application.

3 NONLINEAR FINITE ELEMENT MODELING

Finite element models of the tested specimens were made by using eight node solid elements in the ABAQUS program. The concrete damaged plasticity model of the program was used for modeling the nonlinear behavior of concrete. This model uses concepts of isotropic damaged elasticity in combination with isotropic tensile and compressive plasticity to represent the inelastic behavior of concrete. It consists of the combination of non-associated multi-hardening plasticity and scalar (isotropic) damaged elasticity to describe the irreversible damage that occurs during the fracturing process. The concrete damaged plasticity model requires that the elastic behavior of the material be isotropic and linear.

The model is a continuum, plasticity-based, damage model for concrete. It assumes that the main two failure mechanisms are tensile cracking and compressive crushing of the concrete material. Under uniaxial tension, the stress-strain response follows a linear elastic relationship until the value of the failure stress is reached. The failure stress corresponds to the onset of micro-cracking in the concrete material. Beyond the failure stress, the formation of micro-cracks is represented macroscopically with a softening stress-strain response, which induces strain localization in the concrete structure. Under uniaxial compression the response is linear until the value of initial yield. In the plastic regime, the response is typically characterized by stress hardening followed by strain softening beyond the ultimate stress.

The degradation of the elastic stiffness is characterized by two damage variables, which are assumed to be functions of the plastic strains, temperature, and field variables. The damage variables can take values from zero, representing the undamaged material, to one, which represents total loss of strength. The

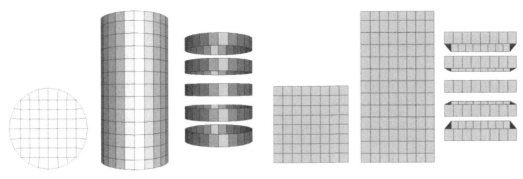

Figure 7a. Elements of a cylindrical model.

Figure 8a. Elements of a prismatic model.

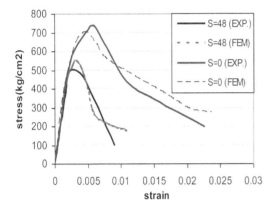

Figure 7b. Comparison of NFEM and experimental results in cylindrical specimens.

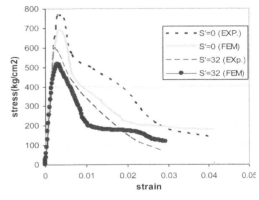

Figure 8b. Comparison of NFEM and experimental results in prismatic specimens.

concrete damaged plasticity model assumes nonassociated potential plastic flow. The flow potential G used for this model is the Drucker-Prager hyperbolic function. The FE models of tested cylindrical as well as prismatic specimens were made in ABAQUS.

The model of cylindrical and prismatic specimens consisted of solid elements for concrete with the above mentioned plasticity model and shell elements for modeling metal strips with stress-strain relationship as observed in tensile tests of strips. The bottom surfaces of models are restrained and the load was applied by incrementally increasing the displacement of nodes of the top surface. Figure 7 shows the mesh details and results of nonlinear finite element models of some of tested cylindrical specimens compared with observed stress-strain behavior.

In Figure 7b analytical and experimental results for two columns with a strip spacing of 48 and 0 mm are compared. As can be seen in this figure, there is relatively good agreement between analytical and experimental results. However, the NFEM has underestimated the experimental results. It should be mentioned that the pretensioning force in the strips in actively confined specimens has been applied in the model before application of the incremental displacement. Similarly in figure 8, mesh and details for steel and concrete elements and the results of nonlinear

finite element models of some of the prismatic specimens are shown. A comparison between analytical and experimental results showed that the nonlinear finite element method had underestimated the post-peak part of stress-strain behavior of confined specimens.

4 CONCLUSIONS

The applied technique for strengthening of concrete columns could increase strength and ductility of concrete considerably. The technique was able to increase the peak strength and its corresponding strain of concrete up to 230 percent. The gain in ductility of the confined concrete was very sensitive to the ductility of the metal strip used. Active confinement resulted in better enhancement of strength and ductility of confined concrete than passive confinement. The efficiency of confinement in cylindrical specimens, i.e. the gain in strength and ductility, was greater than that of prismatic ones. The damaged plasticity model was capable of estimating the behavior of confined concrete with reasonable accuracy. However it underestimated the results of both actively and passively confined concrete.

ACKNOWLEDGMENT

The authors would like to thank BHRC for their kind support.

REFERENCES

Alcocer, S.M. & Jersa, J.O. 1993. Strength of reinforced concrete frame connections rehabilitated by jacketings, *ACI Structural Journal, American Concrete Institute*, Detroit, MI, USA 90(3): 249–261.

Frangou, M. & Pilakoutas, K. 1995. Structural repair/strengthening of RC columns, *Construction and building materials* 9(5): 259–266.

Ghobara, A. & Aziz, S.T. & Biddah, A. 1997. Rehabilitation of reinforced concrete frame connections using corrugated steel jacketing *ACI Structural Journal* 4(3): 283–291.

Khoury, S.S. & Sheikh, S.A. 1991. Behavior of Normal and High Strength Confined Concrete Columns with and without Stubs. *Research report, UHCEE*, Department of Civil and Environmental Engineering, University of Houston, Tex., 345 91(04)

Xiao, Y. & Wu, H. 2003. Retrofit of reinforced concrete columns using partially stiffened steel jackets. *Journal of Structural Engineering, ASCE* 129(6): 725–732.

Concrete Solutions – Grantham, Majorana & Salomoni (Eds)
© 2009 Taylor & Francis Group, London, ISBN 978-0-415-55082-6

Effect of elimination of concrete surface preparation on the debonding of FRP laminates

D. Mostofinejad & E. Mahmoudabadi

Department of Civil Engineering, Isfahan University of Technology (IUT), Isfahan, Iran

ABSTRACT: The use of fibre-reinforced polymer (FRP) composites to strengthen existing concrete structures is rapidly expanding. One of the important reasons that lead to initial failure is non-preparation of the concrete surface before attaching the FRP laminate. In some cases, surface preparation causes a lot of problems such as the high cost for preparation and lack of access to the member surface. In this paper, an attempt is made to present the experimental results of the influence of concrete surface treatment. The specimens were investigated under tensile stress due to four-point loading. Concrete surface preparation was used for about half of the specimens and then they were strengthened with FRP laminates. Other specimens had no surface preparation and were also strengthened with FRP. Comparison between surface-prepared and non-surface-prepared specimens showed that the surface preparation provides about 5 to 15 percent higher capacity at the point of failure due to debonding of the FRP laminates.

1 INTRODUCTION

1.1 General

Strengthening and repair of structures have been vastly used in structural activities. Strengthening of structures is used to increase the load carrying capacity of the structures and improving the operative condition (Fukuyama, 2000).

One of the materials which is used for rehabilitation and strengthening of structures in recent years has been FRP composite (Fibre Reinforced Polymer). Employment of fibre reinforced polymer (FRP) composites to repair and retrofit concrete elements has been steadily increasing. In general, FRP is externally bonded to concrete by means of adhesives while transfer of the interfacial stresses to concrete is governed through bond. The ultimate load carrying capacity of the retrofitted members is directly influenced by bond, and for this reason the subject has received much attention. Research has been performed for determination of local bond behavior in terms of the characteristic properties, factors affecting these properties, and the mechanism of bond failure (Teng et al 2002).

Various advantages of these kinds of materials such as high strength capacity, low weight, easy installation, etc, lead to using FRP laminates in structural strengthening and rehabilitation.

The purpose of surface preparation is to remove contamination and weak surface layers, to change the substrate surface topography and/or introduce new surface chemical groups to promote bond formation. An appreciation of the effects of surface preparation may be obtained from surface analytical or mechanical test techniques. Surface preparation generally has a much greater influence on long term bond durability than it does on initial bond strength, so that a high standard of surface preparation is essential for promoting long term bond performance (Kamada et al, 2000).

1.2 Surface preparation of concrete

The plane of the surface(s) to be treated (horizontal, vertical, overhead, etc.) has a large bearing on the selection of an appropriate method. The choice of the method, or combinations of methods, depends upon the costs, the scale and location of the operation, access to equipment and materials, and health and safety condition (Hutchinson, 1993).

In essence, the purpose of surface preparation is to remove the outer, weak and potentially contaminated skin together with poorly bound material, in order to expose small- to medium-sized pieces of aggregate (Fig. 1). This must be achieved without causing micro cracks or other damage in the layer behind which would lead to a plane of weakness and hence a reduction in strength of the adhesive connection. Any large depressions, blow holes and cracks must be filled with suitable mortars prior to the application of a structural adhesive. This will ensure a relatively uniform bond line thickness in order to maximize the efficiency of shear stress transfer[4].

The basic sequence of steps in the process of surface preparation should be:

- To remove any damaged or substandard concrete and reinstate with good quality material.
- To remove laitance, preferably by grit blasting.
- To remove dust and debris by brushing, air blast or vacuum.

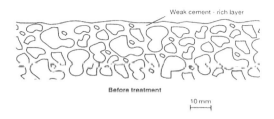

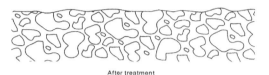

Figure 1. Schematic idealization of concrete following surface preparation (Hutchinson, 1987).

Figure 2. Prepared surfaces.

Additional steps could include:

- Further cleaning, with a suitable solvent, to remove any remaining contaminant.
- Drying the surface to be bonded, if necessary.
- Application of an adhesive-compatible (epoxy) primer, if necessary.

Other techniques such as wire brushing, grinding, bush hammering, acid etching, water jetting or flame blasting are not recommended for plate bonding applications. In particular, the use of pneumatic tools can cause significant damage to the underlying concrete (Edwards, 1987)

The recommended method of laitance removal, by blasting, is fast, plant intensive and operator dependent. There exist a multiplicity of types of blast media, media sizes, blast pressures and types of equipment. With dry systems, oil and water traps should be used to prevent contamination from air compressors; dry systems may be open, or closed with vacuum recovery and recycling of the blast media. Generally a fairly micro rough but macro smooth surface is generated, together with a lot of dust; particles of blast media may also be left embedded in the surface. Clearly this dust and debris must be removed prior to bonding. Wet blasting is another option which overcomes some of the problems associated with the dust, but of course a wet concrete surface is left which must be dried prior to bonding (Hutchinson, 1993).

1.3 Surface preparation of fibre reinforced polymer composites

For concrete strengthening applications, the material may be treated offsite, enabling a variety of potential techniques to be used. However, relatively large areas of strip material will need to be treated in a reliable and consistent way. The surface of composite material may be contaminated with mould release agents, lubricants and fingerprints as a result of the production process. Further, the matrix resin may include waxes, flow agents and 'internal' mould release agents which can be left on the surface of a cured composite. The main methods of surface preparation for composites are solvent degreasing, mechanical abrasion and use of the peel-ply technique, and these methods are often used in combination (Hutchinson, 1993).

2 EXPERIMENTAL PROCEDURE

The experimental program involved evaluation of the effect of concrete surface treatment. The average nominal compressive strength of the concrete used for the manufacture of the specimens was 35 MPa and was determined by testing four 150*150*150 mm concrete cubes in compression (Fig. 3). The concrete mix design is given in Table 1. The experimental specimens were thirty 100*100*500 mm concrete beams. 1.1 mm thick unidirectional CFRP fabrics were cut into 70 mm wide and 360 mm long strips and then adhered to concrete test beams. The mechanical properties of the fabric provided by manufacturer were 231 GPa modulus of elasticity and 1.7 percent ultimate strain. The specimens were prepared according to the manufacturers' specifications. The concrete surface was first ground with a specific grinding machine and then cleaned by air blasting. Then the surface pores were filled by resilient epoxy. Sikadur-C31 was used for filling the pores as shown in Fig. 2. The method used for bonding of the CFRP sheets to the concrete blocks consisted of in-place adhesion and curing of the composite to the concrete surface. Sikadur 300 is in fact a two-component epoxy, which was used as the adhesive layer. The adhesive layer was placed by controlling its thickness. The specimens were allowed to be cured for at least 3 days prior to testing.

The loading setup shown in Fig. 4 was used for tensile stress under four-point loading. Two LVDTs were used for measuring the midspan displacement of the concrete beams (Fig. 4). The displacement measurements were used to draw load-displacement diagrams.

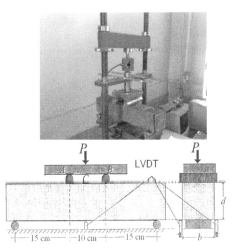

Figure 4. Test setup.

Figure 3. Cube specimens under compressive test.

Table 1. Concrete mix design (kg).

Cement	Water	Fine aggregate (0–5 mm)	Coarse aggregate (5–10 mm)
440	207	974	736

Figure 5. Specimen without surface preparation.

3 EXPERIMENTAL RESULTS

The specimens were about 30 samples with dimension of 100*100*500 mm loaded in the test machine up to failure. The complete response for each specimen was described by plotting the applied loads versus corresponding displacements. All specimens bonded with FRP laminates failed in the same manner. After a crack occurred on the bottom of the concrete beams around the midspan of the specimens, the FRP laminate essentially stressed until the debonding phenomenon occurred. The type of debonding in all specimens was end-debonding (Fig. 5 and Fig. 6).

Figs. 7 and 8 show the load versus the displacement at midspan for different specimens. As is illustrated in the mentioned figures, every figure has 4 curves as a symbol of all specimens which indicate the load

359

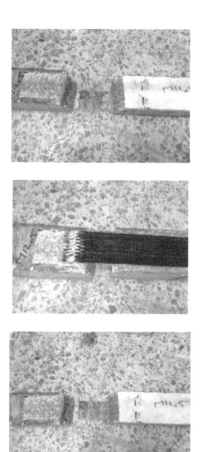

Figure 6. Specimens with surface preparation.

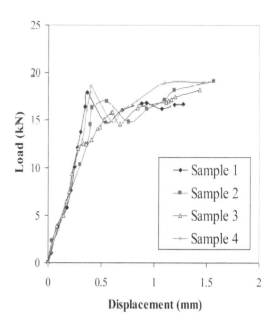

Figure 8. Load-displacement curve for surface prepared specimens.

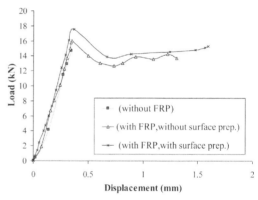

Figure 9. Load-displacement curves for different concrete beams.

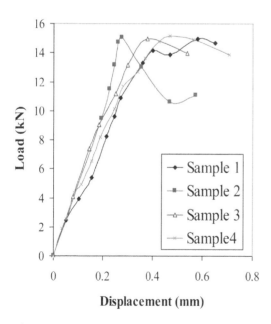

Figure 7. Load-displacement curve for unprepared specimens.

versus the displacement for specimens with specific properties (surface prepared, non-surface prepared and specimens with no FRP laminates). Fig. 9 contains 3 curves: one is for the specimens which have no FRP sheet, the others are about the effect of surface preparation (non-surface prepared and surface prepared specimens). The average ultimate loads for the prepared and non-prepared surface specimens were about 18.6 kN and 15.1 kN, respectively. Specimens having no FRP had an ultimate load of 14 kN. Therefore as a result of the experiments, it is concluded that the specimens with surface preparation had between 5 to 15 percent higher load capacities in ultimate strength compared to non-surface prepared specimens.

360

4 CONCLUSIONS

In this paper, an experimental study on the effect of the surface treatment on the strength of concrete specimens externally bonded to FRP sheets was performed. The following conclusions can be drawn from this study:

1. Externally bonding FRP laminates to the prismatic non-reinforced concrete specimens increased the ultimate fracture load.
2. Comparison between surface-prepared and non-surface-prepared specimens showed that the surface preparation provides about 5 to 15 percent higher capacity at the point of failure due to debonding of the FRP laminates.
3. The type of debonding in all specimens was the same and all specimens were debonded in an "end-debonding" manner.

REFERENCES

Edwards S. C. *Surface coatings in repair of concrete structures.* London, 1987, ch. 10.

Fukuyama H., Sugano S., Seismic Rehabilitation of RC Building. *Cement and Concrete Composites*, 2000, Vol. 22, No. 3, 59–79.

Hutchinson A. R. *Review of methods for the surface treatment of concrete.* MTS Project 4, Report No 2 on the Performance of Adhesive Joints, Dept. of Trade and Industry, London, 1993.

Hutchinson A. R. *Strengthening of reinforced concrete structures.* London, 1993, ch. 3, 70–74.

Kamada T., Wictor C. Li., The Effects of Surface Preparation The Fracture Behavior of ECC/Concrete Repair System. *Journal of Cement & Concrete*, 2000.

Teng J. G., Chen J. F., Smith S. T., lam L., FRP Strengthened RC Structures. *Journal of Composites for Construction*, 2002, Vol. 6, No. 3, 232–245.

Concrete Solutions – Grantham, Majorana & Salomoni (Eds)
© 2009 Taylor & Francis Group, London, ISBN 978-0-415-55082-6

Structural performance of reinforced concrete beams strengthened with PBO Fibre Reinforced Cementitious Mortars (FRCM)

L. Ombres

Department of Structures, University of Calabria, Italy

ABSTRACT: Innovative composites made of Fibre Reinforced Cementitious Matrix (FRCM) represent very effective solutions for strengthening reinforced concrete beams. Due to the large slip at the fibre/cementitious matrix interface, a more ductile behaviour and different failure modes were experimentally observed in FRCM strengthened beams with respect to those strengthened with FRP (Fibre Reinforced Polymers). These structural aspects are analysed in the paper by means of the results of an experimental investigation carried out on 9 reinforced concrete beams strengthened with a PBO (Polypara-phenylene-benzo-bisthiazole) FRCM. Midspan deflections, crack widths, crack spacing and failure modes were analysed varying the strengthening configurations, the internal steel reinforcement ratio and the concrete strength. A comparison of the experimental and predicted values was made in terms of load-deflection, load-crack width, load-crack spacing curves. The results obtained are presented and discussed in this paper.

1 INTRODUCTION

Due to their favourable features, such as ease and speed of application and good mechanical properties the Fibre Reinforced Polymers (FRP) system is widely used for strengthening of reinforced concrete beams. Generally, FRPs, in the form of sheets of unidirectional fibres, are bonded to concrete surfaces by organic resins (epoxy resin). The use of organic resins entails some disadvantages such as the lack of fire resistance, the degradation under UV light and the low compatibility with the substrate material.

To reduce or eliminate the afore-mentioned disadvantages, alternative bonding systems founded on the replacement of the organic resin with inorganic ones, like cement-based mortars, have been proposed and used as concrete strengthening systems.

Such bonding systems, made by fibre nets embedded into an inorganic stabilised cementitious matrix are named Fibre Reinforced Cementitious Mortars (FRCM). The use of fibre nets instead of unidirectional fibre sheets guarantees a good bond between the cementitious mortar and composite fibres.

Results of some studies and researches carried out both on flexurally strengthened reinforced concrete beams (Bruckner et al., 2006; Triantafillou, 2004; Triantafillou & Papanicolau, 2006)., and on confined concrete elements (Triantafillou et al. 2006; Ombres, 2007) have been evidence for the effectiveness of the FRCM system for strengthening of concrete structures. In comparison with FRP strengthened members, a remarkable increase of the durability and significant structural performance were observed in FRCM strengthened members.

The performance of the FRCM systems is dependent on the mechanical properties of the composite fibres; generally carbon or glass fibre nets are used. The use of ultra-high strength fibre nets, such as the PBO nets, can improve structural performance of the FRCM.

Mechanical properties of the PBO (Polypara-phenylenc-bcnzo-bisthiazole) fibres are, in fact, very significant: the elastic modulus and tensile strength are both higher than that of the high strength type of carbon fibres. PBO fibres have great impact-tolerance and energy absorption capacity superior to other kinds of fibres; in addition PBO fibres demonstrate high creep and fire resistance (Wu et al., 2001).

The use of the PBO FRCM system in strengthening of concrete structures is not completely defined even if the few available test results on PBO FRCM strengthened reinforced concrete beams (Di Tommaso et al., 2007; Di Tommaso et al., 2008) gave evidence of good performance both in terms of strength and ductility.

In the present paper, the structural behaviour of reinforced concrete beams externally strengthened with PBO-FRCM is analysed. An experimental investigation has been carried out on PBO FRCM strengthened beams varying some mechanical and geometrical parameters such as the percentage of PBO, the percentage of internal steel reinforcement and the compression strength of the concrete.

Mid-span deflections, crack widths, crack spacing and failure modes were analysed varying the strengthening configurations, the internal steel reinforcement ratio and the concrete strength. A comparison of the experimental and predicted values was made in terms of load-deflection, load-crack width, and load-crack

Table 1. Configuration of tested beams.

Beams	f'_c (MPa)	A_s (mm²)	A'_s (mm²)	Fibre type	Layer
T1-0	27.44	339.30	157.00	–	–
T1-1	27.44	339.30	157.00	PBO	1
T1-2	27.44	339.30	157.00	PBO	1
T1-3	27.44	339.30	157.00	PBO	1
T1-4	27.44	339.30	157.00	Carbon	1
T2-0	27.73	157.00	100.53	–	–
T2-1	27.73	157.00	100.53	PBO	1
T2-2	27.73	157.00	100.53	PBO	2
T2-3	27.73	157.00	100.53	PBO	3

Table 2. FRCM properties: manufacturer's values.

	Nominal thickness (mm)	Elastic Modulus (GPa)	Tensile strength (MPa)	Compression strength (MPa)
PBO	0.0450 (long.) 0.0225 (transv.)	270	5800	–
Carbon	0.0450 (long.) 0.0450 (transv.)	240	3500	–
Mortar	–	6.0	3.50	29.0

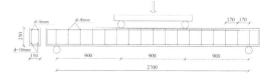

Figure 1. Test set-up.

Table 3. Ultimate loads and failure modes for tested beams.

Beams	Failure load (kN)	Failure modes
T1-0	75.78	Concrete crushing
T1-1	91.00	Concrete crushing
T1-2	87.60	Concrete crushing
T1-3	87.42	Concrete crushing
T1-4	81.30	Concrete crushing
T2-0	43.56	Concrete crushing
T2-1	54.24	Concrete crushing
T2-2	64.06	Intermediate crack debonding
T2-3	71.39	Intermediate crack debonding

spacing curves. The results obtained are presented and discussed below.

2 EXPERIMENTAL INVESTIGATION

The experimental investigation was carried out on simply supported reinforced concrete beams strengthened with the FRCM system. Two groups of beams, named T1 and T2, were tested; the percentage of internal steel reinforcement and the compression strength of the concrete were different for each group. The shear reinforcement of the beams was designed in order to prevent premature shear failure. The tested beams were loaded by two concentrated loads; the span was 3000 mm long and the concentrated loads were applied 900 mm from the supports. All beams had a rectangular section 150 mm wide and 250 mm high.

The tested specimens had different configuration of the FRCM reinforcement according to Table 1. Beams T1-1 and T2-1 were unstrengthened and used as control specimens. The T1-4 beam was strengthened with one Carbon fibre net; other specimens were reinforced with one, two or three PBO fibre nets embedded into a special cementitious mortar. The FRCM reinforcement was extended along the whole span length of the beams.

2.1 Materials properties

The average compressive strength of concrete determined on standard (150-mm diameter × 300-mm length) cylinder specimens, was 27.44 MPa and 27.73 MPa for the T1 and T2 beams, respectively.

The yield strength of the internal steel reinforcement of the T1 beams was 608.17 MPa and 620.73 MPa for 12-mm and 10 mm diameter rebars, respectively; while the yield strength of the internal steel reinforcement of the T2 beams was 525.90 MPa and 535.60 MPa for 8-mm and 10 mm diameter rebars, respectively.

Finally, the mechanical properties (manufacturer's values of tensile strength, elastic modulus are and ultimate tensile strain) of the FRCM system (Carbon, PBO net and inorganic mortar) used as externally bonded reinforcement are reported in the Table 2.

2.2 Test setup

The arrangement of the test setup used is shown in Figure 1. Beams were instrumented with linear variable differential transducers at mid-span and at the loading points to monitor deflections.

Strain measurements were made by strain gauges bonded on the concrete, steel rebars, carbon and PBO net surfaces. In particular at the mid-span, two strain gauges were bonded, after local sandblasting, to the concrete compression surface while two strain gauges were bonded, before concrete casting, to the internal tensile steel rebars. The measurement of the carbon and PBO strains was made by eight strain gauges distributed along the length of the net.

The load was gradually applied by means of a hydraulic jack and measured with a local cell. Crack formation and propagation were examined at each load step. Beam deflection, strains and load values were monitored by means of a data acquisition system.

2.3 Test results and discussion

The ultimate load values and the failure modes observed during the tests are reported in Table 3 for all beams tested.

As reported in Table 3, the flexural capacity of the FRCM strengthened beams increased with the

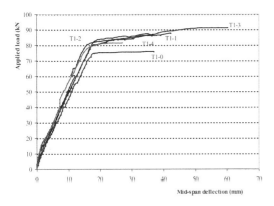

Figure 2. Load-deflection curves at the mid-span. T2 beams.

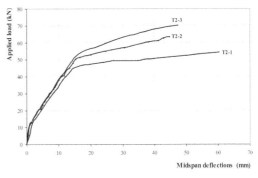

Figure 3. Load-deflection curves at the mid-span. T2 beams

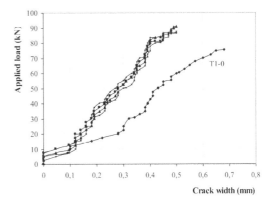

Figure 4. Load-crack width curves. T1 beams.

number of PBO net layers. The ultimate loads of beams strengthened with one layer of PBO net were 20% higher than that of the control beams T1-0 and T2-0. In spite of the occurrence of premature failures, in comparison with the control beams, an increase of 47% and 63% of the ultimate capacity was obtained for T2-2 and T2-3 strengthened beams, respectively. In addition, the ultimate load of the T1-4 beam strengthened with Carbon net, was less than those of the T1-1, T1-2 and T1-3 beams strengthened with PBO net.

As reported in Table 3, unstrengthened beams T1-0 and T2-0 collapsed by concrete crushing after yielding of internal steel rebars. The T1-4 beam collapsed by concrete crushing at mid-span.

The failure of beams strengthened with one layer of PBO net was similar to that of unstrengthened beams; concrete crushed at the loading point after yielding of internal steel rebars. At failure a perfect bond between the FRCM system and the concrete was observed both at the ends and at the intermediate widely cracked zones (T1-1 and T1-4 beams).

At failure the measured PBO net strains were in the range 0.007–0.01496 mm/mm; all values were lower than the ultimate strain, 0.0215 mm/mm.

The failure mode for T2-2 and T2-3 beams was intermediate flexural crack-induced interfacial debonding. In both beams debonding initiated in the high bending moment region as a result of the high width of the flexural cracks and then developed towards one of the end of the FRCM system. The mean value of the PBO strain at the beginning of the debonding was nearly 0.010 mm/mm. At failure the ends of the FRCM system remained perfectly bonded to the concrete.

In the T2-2 beams debonding started when the applied load was 50 kN and it propagated slowly until the failure load, 65 kN. The same behaviour was observed in the T2-3 beam; the debonding initiated under the applied load of 60 kN while the failure load was 70 kN.

Load–deflection curves in the mid-span of the tested beams, drawn in Figures 2 and 3, shows evidence of the effectiveness of the FRCM system both in terms of strength and ductility.

By the analysis of the curves drawn in Figure 2, it appears that beams strengthened with PBO net exhibit both ultimate load and displacement values higher than those of the beam strengthened with Carbon net; this result confirms the improvement of the structural performance of the PBO-FRCM system compared to that of the Carbon-FRCM system.

Figure 3 shows load-displacement diagrams of the T2 beams; curves of T2-2 and T2-3 beams are compared with the diagram relative to the T2-1 beam. As shown in the Figure, in spite of the premature failures, both beams showed good ductility.

The curves reported in Figure 4, show the effectiveness of the FRCM system on the crack width of beams; compared to the control beams, in all strengthened T1 beams, the mean crack widths were sensibly reduced.

Curves showing applied load versus crack numbers are drawn in Figure 5: it is evident that the number of cracks in FRCM strengthened beams is higher than that of the control beam.

The strain distribution along the depth of the midspan section of the T2-2 beam is reported in Figure 6. Diagrams refer to applied loads varying from 0.30 P_u and 0.75P_u P_u being the collapse load of the beams.

By the Analysis of diagrams it appears that the strain distribution can be considered almost linear along the depth of the section. As shown previously, at

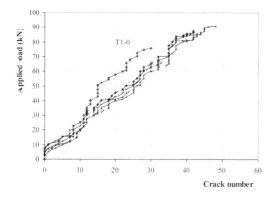

Figure 5.   Load-crack number curves. T1 beams.

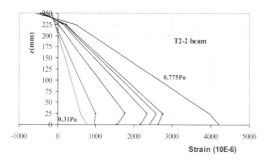

Figure 6.   T2-2 beam: strain distribution at the mid-span.

failure, a perfect bond of the FRCM layer to the concrete was observed; consequently, the high strains in the PBO net indicate a slippage between the net and the cementitious mortar that is not influential on the FRCM-to-concrete bond.

3   MODELING AND COMPARISONS

Experimental results, previously described, confirm that the structural modelling of FRCM strengthened beams can be shown to follow Bernoulli's Hypothesis (i.e. perfect bond FRCM-to-concrete). Test results, in fact, showed that slippages between the FRCM system and the concrete are significant only at the collapse of beams.

In addition, the premature failure of beams due to intermediate crack induced debonding, occurs similarly to that of FRP strengthened beam; consequently, the structural performance of FRCM strengthened beams can be determined by means of models usually adopted for the analysis of FRP strengthened beams.

To verify the validity of the above-mentioned models, a comparison between experimental results and predictions of the models was carried out: results are reported in the Table 4.

Theoretical predictions of the ultimate moment (M_{uth}) were determined assuming the mechanical properties of materials reported in Table 2. The ultimate moments of the T2-2 and T2-3 beams were determined assuming for the PBO strain equal to the

Table 4.   Theoretical – experimental comparison.

Beam	M_{uth} (kNm)	$M_{uexp.}$ (kNm)	$M_{uth}/M_{uexp.}$
T1-1	41.116	40.950	1.004
T1-2	41.116	39.420	1.043
T1-3	41.116	39.339	1.045
T2-1	24.400	24.300	1.004
T2-2	28.820	29.250	0.985
T2-2	32.120	31.500	1.020

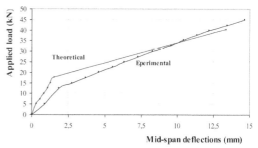

Figure 7.   T2-1 beam: experimental-theoretical comparison.

debonding strain value proposed by the Italian Code DT200/2004 (National Research Council, 2004)

$$\varepsilon_{fdd} = \frac{f_{fdd2}}{E_f} = \frac{k_{cr}}{E_f \gamma_{fd}} \sqrt{\frac{2 E_f \Gamma_{fd}}{n_f t_f}} \qquad \text{Eq 1.}$$

being

$$\Gamma_{fd} = \frac{0.03}{\gamma_c} k_b \sqrt{f_{cm} f_{ck}} \quad \text{the fracture energy,} \qquad \text{Eq. 2}$$

$$k_b = \sqrt{\frac{2 - \frac{b_f}{b}}{1 + \frac{b}{400}}} \geq 1 \; ; \qquad k_{cr} = 3.00 \qquad \text{Eq. 3}$$

n_f is the numbers of PBO layers having thickness t_f, E_f is the PBO elastic modulus;

f_{cm} and f_{ck} the mean value of the tensile strength and the characteristic value of the compression strength of the concrete, respectively;

γ_c and γ_f are partial safety coefficients;

b and b_f the width of the section and the PBO reinforcement, respectively.

The results obtained show the good agreement between theoretical predictions and experimental results; in all examined cases, the variation of theoretical predictions with respect to experimental values is less than 5%. In particular, from Table 4, it emerges that the prediction of the moment corresponding to the occurrence of the intermediate crack induced debonding is very accurate both for T2-2 and T2-3 beams (the variation with respect to the experimental values is less than 2%).

The comparison between theoretical and experimental mid-span load – deflections curves is shown in the Figure 7. Theoretical deflections were determined by integration along the beam axis of the moment-curvature diagram. In particular a simplified tri-linear moment curvature diagram was used to calculate beam deflections.

From the analysis of the results of the comparison, it emerges that the model found based on the perfect bond between the FRCM and the concrete is able to predict the serviceability behaviour of the beam well. The experimental and theoretical curves are, in fact, in good agreement for load values less than $0.70\ P_u$. The same trend was observed for all tested beams.

4 CONCLUSIONS

The structural performance of reinforced concrete beams strengthened with FRCM was analysed in the paper; the obtained results allow us to draw the following concluding remarks:

- the use of the FRCM system improves sensibly the flexural capacity of strengthened reinforced concrete beams. By tests described in the paper the ultimate capacity of strengthened beams increased from 20% to 63% with respect to the value relative to un-strengthened beams;
- the failure modes of FRCM strengthened beams are depending on the percentage of fibre net reinforcement. For beams strengthened with one layer of PBO or Carbon net (geometrical percentage of reinforcement $\rho_f = 0,0000188$) the failure was due to concrete crushing after internal steel yielding while a perfect bond FRCM-to-concrete was observed in spite of slippage between the PBO net and the cementitious mortar. In the presence of higher PBO reinforcement ($\rho_f = 0,0000376$ for the T2-2 beam and $\rho_f = 0,0000564$ for the T2-3 beam) a premature failure due to intermediate crack debonding occurred;
- predictions furnished by models usually adopted to analyse FRP strengthened reinforced concrete beams are in good agreement with experimental values.

Even if further analyses both theoretical and experimental are needed, from results described in the paper clearly emerges that the strengthening FRCM system is very effective both from strength and ductility point of view and it allows obtain significant structural performances of strengthened concrete elements.

REFERENCES

Bruckner, A., Ortlepp, R.& Curbach M. 2006. Textile reinforced concrete for strengthening in bending and shear. *Materials and Structures*, vol. 39, 741–748.

Di Tommaso, A., Focacci, F., Mantegazza, G. & Gatti A. 2007. FRCM versus FRP composites to strengthen rc beams: a comparative analysis. *Proceedings of the International Symposium on Fibre Reinforced Polymers Reinforced Concrete Structures , FRPRCS-8, Patras, Greece.*

Di Tommaso, A., Focacci, F., Mantegazza, G. 2008. PBO-FRCM composites to strengthen R.C. beams: mechanics of adhesion and efficiency, *Proceedings of the International Conference on "FRP Composites in Civil Engineering", CICE2008, Zurich.*

National Research Council, 2004. *Guide for the design and construction of externally bonded FRP systems for strengthening existing structures- CNR-DT200"*, Rome, Italy.

Ombres, L. 2007. Confinement effectiveness in concrete strengthened with fibre reinforced cement based composite jackets. *Proceedings of the International Symposium on Fibre Reinforced Polymers Reinforced Concrete Structures, FRPRCS-8, Patras, Greece.*

Wu, Z.S., Iwashita, K., Hayashi, K., Higuchi, T., Murakami, S, & Koseki, Y. 2001. Strengthening PC structures with externally prestressed PBO fibre sheets. In J. Teng (ed.) *FRP Composites in Civil Engineering, Proc. Int. Conf. CICE2001, Honk Hong*, Elsevier Science. 1085–1092.

Triantafillou, T. 2004. Recent developments in strengthening of concrete structures with advanced composites textile-reinforced mortar (TRM) jacketing. Proceedings of the International Conference on *Structural Composites for Infrastructure Applications,* Alexandria, Egypt

Triantafillou, T. & Papanicolau C.G. 2006. Shear strengthening of reinforced concrete members with textile reinforced mortar (TRM) jackets, *Materials and Structures.*

Triantafillou, T., Papanicolau C.G., Zissimopulos P & Laourdekis T. 2006. Concrete confinement with Textile-Reinforced Mortar Jackets. *ACI Materials Journal*, 103(1), 28–37.

Concrete Solutions – Grantham, Majorana & Salomoni (Eds)
© 2009 Taylor & Francis Group, London, ISBN 978-0-415-55082-6

An experimental study on the effect of steel fibre reinforced concrete on the behaviour of the exterior beam-column junctions subjected to cyclic loading

Anant Parghi & C.D. Modhera
Department of applied mechanics, S.V. National Institute of Technology, Gujarat, India

ABSTRACT: A total of ten full-scale RC exterior beam column joints were tested under cyclic loading and their performance was examined in terms of lateral load capacity, joint strength, ductility, residual strength, and established performance criteria. The first two specimens were made with normal concrete, two specimens with 1.5% dosage of fibre, the third two specimens with 3.0% fibre, the fourth two specimens with 4.5% fibre and the last two specimens with 6.0% of fibre by mass of concrete respectively. It was found that the energy absorption capacity increased by 54%, 86%, 204% and 133% for various mixes of SFRC 1.5%, 3.0%, 4.5% and 6.0% dosage of fibres respectively. The experimental results indicated that fibre reinforced concrete is an appealing alternative to conventional confining reinforcement.

1 INTRODUCTION

The recent Gujarat earthquake in 2001 revealed once more the importance of the design of reinforced concrete (RC) structures with ductile behavior. Ductility can be described as the ability of reinforced concrete cross sections, elements and structures to absorb the large energy released during earthquakes without losing their strength under large amplitude and reversible deformations .Generally, the beam-column joints of a RC frame structure subjected to cyclic loads such as earthquakes experience large internal forces. Consequently, the ductile behavior of RC structures dominantly depends on the reinforcement detailing of the beam-column joints. Numerous investigations have been reported about the behavior and reinforcement detailing of beam-column joints under reversed cyclic loading. Some of these include Pessiki (1990), Kurose *et al.* (1988), Kitayama *et al.* (1991), Aoyama (1985), Fuji and Morita (1991), Paulay *et al.* (1989), and Paulay (1989). In these papers, factors affecting the behavior of RC beam-column joints were studied. In brief, the results of these investigations showed that the shear strength and ductility of RC beam-column joints increased as the compressive strength of concrete and the amount of transverse reinforcement increased. Moreover, for adequate ductility of beam-column joints, the use of closely spaced hoops as transverse reinforcement was recommended in various earthquake codes for RC structures. Confining the concrete by closely spaced hoop reinforcement increased not only the ductility of the concrete sections at beam-column joints but also the strength of these sections. On the other hand, the cross sections of beams and columns close to the joints in RC structures under

the effect of strong earthquake motion were subjected to large bending moments and shear forces. Consequently, a large amount of longitudinal and transverse reinforcements of beams and columns should pass through these junctions. However, it is tedious to install the transverse reinforcement.

Because of placement difficulties, the beam-column joints of RC structures can-not be fully controlled by civil engineers and it is not easy to handle this situation with care according to the design drawings. Numerous researchers have attempted to reduce the workmanship difficulties by simplifying the reinforcement lay-out in the joints. In several experimental investigations (Recommendations ACI-ASCE Committee 352, 1985; Jindal and Hasan, 1984; Craig *et al.*, 1984; Katzensteiner *et al.*, 1992; Filiatrault *et al.*, 1994; Filiatrault *et al.*, 1995), the use of steel fibre reinforced concrete (SFRC) was proposed as additional reinforcement instead of squeezing stirrups in the beam-column joints. In many of these investigations, SFRC was used in certain parts of the joints together with normally spaced transverse reinforcement instead of squeezed stirrups. The effects of various parameters on the behavior of joints have been studied experimentally, such as the type of loading, the amount of steel fibre in the concrete mix, the method of loading, and the amount of transverse and longitudinal reinforcements. These experiments showed that beam-column joint specimens with normally spaced stirrups and SFRC at the joints, displayed higher capacity for shear forces and bending moments, dissipated more energy and showed more ductile behavior than conventional ductile beam-column joints of plain concrete. All the test specimens were made with steel fibre reinforced concrete due to its higher shear strength than that of conventional structural concrete. Even though it is rather well known as the first research project on shear properties of SFRC was started three decades ago (Batson *et al.*, 1972), SFRC is only occasionally used in structural elements of buildings. In fact, steel fibre in concrete can significantly increase the shear strength of structural concrete (Lim *et al.*, 1987, Casanova *et al.*, 1994). It will be shown here that adding even a small amount of steel fibre can significantly improve fracture properties of concrete, thus improving ductility, overall behaviour in tension as well as an element's performance in shear.

2 RESEARCH OBJECTIVES

This paper reports an experimental study carried out to investigate the behavior of joints made of SFRC. In previous experimental investigations, the amount of the steel fibre, spacing of transverse reinforcement, type and aspect ratio of fibre and loading, application points of the cyclic loads and- the scale of specimens have been separately taken into consideration as experimental parameters. In the present study, ten specimens representing an exterior beam column joint subjected to cyclic loading were tested under displacement controlled loading (Figure 1). Specimen no. 1 and no. 2 were completely composed of plain concrete while the joint and the confinement zones of the beam and the column of Specimens no. 3, 4, 5, 6, 7, 8, 9 and 4 were cast with Steel Fibre Reinforced Concrete. However, all of the seismic code requirements at these zones related to the spacing of transverse reinforcement were not considered. In all Specimens the requirements of the Indian Earthquake Code regarding the spacing of stirrups were followed. The results obtained from the tests of both SFRC and plain concrete specimens were compared in terms of the amounts of accumulated, dissipated and stored energy, as well as damage during the tests.

2.1 *Material properties and concrete mixes*

Concrete mixture designs were used as per Indian Standard 10262-1982 and they are given in Table 2. The laboratory test results of the concrete cylinders revealed that the average compressive strength of the plain concrete and SFRC varied between 22 MPa and 31 MPa and between 25 MPa and 37 MPa, respectively. In all test specimens, the longitudinal and transverse bars used were high-yield steel and mild steel, respectively.

The mechanical properties of the steel are given in Table 1. The yield strength of the transverse and longitudinal ribbed reinforcement was found to be 530 MPa from the tension tests performed in the laboratory. The crimped shape steel fibres having a length of 30 mm and a diameter of 0.5 mm and thus an aspect ratio of 60 with a yield strength of 552 MPa were added into the plain concrete mix at a 1.5% to 6.0% by mass of concrete. As is well known, the addition of

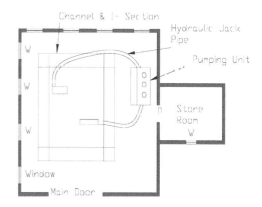

Figure 1. Plan of the heavy structure laboratory.

steel fibres of higher percentage into concrete mix makes the workability of concrete difficult. Therefore, superplasticizer was added to the concrete mix and the maximum size of coarse aggregate was limited to 20 mm for maintaining the strength and workability of the concrete.

2.2 *Experimental setup and testing procedure*

To meet the objectives of this experiment, ten beam-column joints were constructed to a full-scale. Test fixtures, test specimens, and seismic loading are described herewith.

2.3 *Test fixture*

The Heavy Structure Laboratory of the Applied Mechanics Department at S.V. National Institute of Technology has test fixtures, consisting of three 20-ft (6.1-m) long steel I-beams embedded in the concrete slab floor. The geometry of the specimens is given in Figure 2. Two of the I-beams were used to anchor the half-scale models of beam-column joints, as seen in the laboratory layout depicted in Figure 1.

The steel formwork was horizontally placed on the laboratory floor and the concrete was cast into this formwork while trying to avoid the mixing of plain concrete and steel fibre reinforced concrete. However, thorough compaction was made by tamping rod after casting so that the concrete was compacted properly and no segregation took place. In the experimental setup, the test assembly was placed at the loading frame with the column horizontal and the beam vertical. Both ends of the column were arranged to be simply supported to simulate inflection points of the columns at the mid-storey. Cyclic loading was applied to the tip of the beam by hydraulic jack displacement control.

2.4 *Plan*

To meet the objectives of this experiment, ten beam-column joints were constructed to a half-scale. Test fixtures, test specimens, and seismic loading are described herewith.

Table 1. Mechanical properties of steel.

Reinforcement type	Dia. of reinforcing bars	c/s area (mm ^{2})	Ultimate tensile stress (kg/mm^2)	% elongation
Stirrups/ties	6 mm	28	56.58	34
Beam longitudinal	8 mm	50	25.70	21
Column longitudinal	12 mm	113	53.03	22

Table 2. Characteristics of the concrete mixtures.

Material	Unit	Plain concrete	SFRC
Cement	Kg/m^3	344	344
Aggregate	Kg/m^3	1152	1152
Sand	Kg/m^3	739	739
Superplasticizer	% By mass of cement	0.10	0.10
Water	litre/meter^3	172	172
Steel Fibre	% By mass of concrete	–	1.5 to 6.0

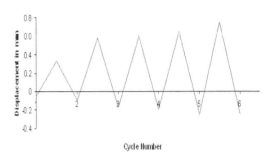

Figure 3. Simulated quasi-static earthquake loading.

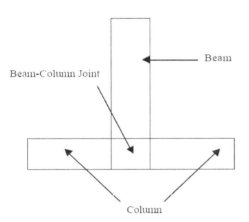

Figure 2. Definition of the elements comprising a test specimen for a beam-column joint

2.5 Test fixture

The Heavy Structure Laboratory of Applied Mechanics Department at S.V. National Institute of Technology has test fixtures, consisting of three 20-ft (6.1-m) long steel I-beams embedded in the concrete slab floor. Two of the I-beams were used to anchor the half-scale models of beam-column joints, as seen in laboratory layout depicted in Figure 1.

In this experimental work a small loading frame was fabricated. It is important the joint frame should not deflect so proper rigidity was provided by the help of welding. Steel channel bracing was provided and the bottom side on the hydraulic jack was fixed. Both the left hand side was horizontally fixed and right hand side hydraulic jack was also fixed.

3 EXPERIMENTAL PROGRAMMES

All specimens have been tested in the laboratory. beam-column junctions were cast for the same mix proportion of ingredients, obtained from the first stage of the experimental work. The behavior of beam-column junction was studied after a 28 day curing period.

A total 10 of beam-column junctions of size as shown in Figure 2 were cast for the purpose of testing. The size of specimen for columns was 300 mm by 300 mm and size of beams was 300 mm by 230 mm. The behavior of Beam-column junctions was studied under cyclic loading. Load versus deflection graph were plotted for study.

The general arrangements of the experimental setup are shown in Figure 2. An axial compressive load of 150 kN was applied to the column to represent normal force. Mechanical dial gauges were placed at 5 different points on each specimen and on the loading frame to measure the deformations and displacements of the beam-column joint under cyclic loading as shown in Figure 3.

At each displacement level, the first three-cyclic load was applied once at the tip of the beam until the occurrence of the first residual displacement. After that residual displacement level, at each displacement level, the load was cycled three times at each loading step up to the failure of the specimen. The loading steps for the test specimens and the number of loading cycles on the specimens are given in Figure 3.

3.1 Experimental results, general behaviour and failure mechanism

Experimental results are evaluated in relation to the behaviour of joints. Although numerous quantities

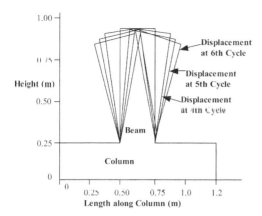

Figure 4. Beam displacements during simulated earthquake.

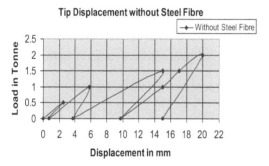

Figure 5. Hysteresis loop for beam-column joint #1 without SFRC.

were measured, only the main parameters of the results are given and discussed below.

The cracks were seen usually near the joints in all types of SFRC specimens during the application of load. As the loading increased, the additional crack formed propagated with widening. The specimen of the joint prepared without fibre addition showed a wider crack width near the joint during testing, while lower crack widths were observed for the SFRC specimens than that of the plain one. It can be seen that a brittle shear failure occurred in the joint region of specimens cast with plain concrete.

It was also noted that the core and cover of concrete were intact. The load displacement results were recorded for all specimens. The graphs obtained are presented in Figs. 5–8. The details are given below: Hysteresis loops of load versus beam deflection, for plain concrete joints #1, with 150-mm spacing are shown in Figure 5 Hysteresis loops of load versus beam deflection, for SFRC joints #2 with 1.5% steel fibre and 150-mm, spacing, are shown in Figures 6 Hysteresis loops for SFRC joints #3, with 3.0% steel fibre and 150-mm spacing, are shown in Figures 7 Hysteresis loops for SFRC joints #4, with 4.5% steel fibre and 150-mm spacing, are shown in Figures 8.

According to these evaluations, it is seen that SFRC used in the critical regions of beam-column joints

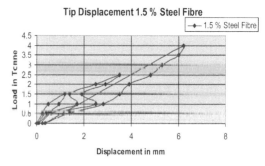

Figure 6. Hysteresis loop for beam-column joint #2 with SFRC 1.5%.

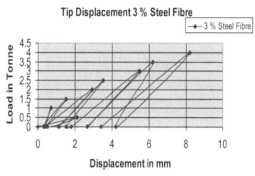

Figure 7. Hysteresis loop for beam-column joint #3 with SFRC 3.0%.

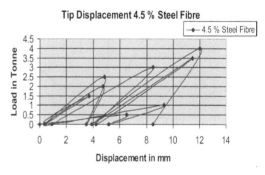

Figure 8. Hysteresis loop for beam-column joint #4 with SFRC 4.5%.

increased the strength capacity for bending moment and shear forces.

The behavior of reinforced concrete specimen of plain concrete (P_0) was shown to be more brittle than that of SFRC during the first two cycles. Load-displacement hysteresis showed an elastic nature up to a certain extent. Strain hardening was observed in the third cycle.

The trend of specimen of SFRC with 1.5% steel fibres (P1) behaviour is more ductile than that of no fibres. Addition of fibres increases ductility as well as load-displacement hysteresis showing elastic nature even in third cycle.

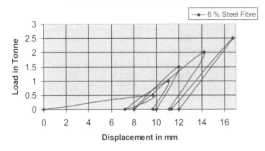

Figure 9. Hysteresis loop for beam-column joint #3 with SFRC 6.0%.

The behavior of the specimen of SFRC with 3.0% fibre (P₃) was similar to specimen of SFRC with 1.5% fibre content for first two cycles, but large strain hardening was observed in the third cycle.

The behavior of the specimen cast with 4.5% Steel fibre was more ductile as compared to specimens cast with 1.5% and 3.0% steel fibre. The residual deflection remained after both first and second cycles was less, within a strain hardening effect observed for the third, fourth, fifth and sixth cycle with maximum deflection.

The behaviour of the specimen cast with 6.0% Steel fibre was less ductile as compared to specimens cast with 1.5%, 3.0% and 4.5% steel fibre. The observed deflection of the first and second cycle was maximum and took less load compared to the others with 1.5%, 3.0% and 4.5% steel fibre and for the third, fourth, fifth and sixth cycle maximum deflection was observed.

The use of SFRC prevents the occurrence of shear cracks. Therefore, it can be proposed that SFRC be used together with normally spaced stirrups so that no shear cracks occur in beam-column joints under reversed cyclic loading. The widths and locations of cracks in the specimens at different displacement levels are given in Figure 10.

The beam-column joints for all test specimens showed minor cracks starting at the 2nd cycle and ending with the last cycle of the test.

3.2 Column cracking

Visible Cracking in the column was best resisted by the SFRC specimens with 3.0% steel fibre, although the 1.5% of SFRC joint exhibited superior behavior. For SFRC specimens, minor beam cracks began forming during the 2nd cycle and continued to grow until the end of testing. The SFRC specimen #3 with 3.0% developed minor beam cracks during the 3rd cycle, however some of those cracks began opening up during the 5th cycle and continued to open up during the 6th cycle. The plain concrete specimens did not fare as well as the SFRC specimens. Minor beam cracks started forming during the 3rd cycle. However, spalling of concrete began during the 4th cycle. Major spalling of beam concrete occurred during the 5th and 6th cycles.

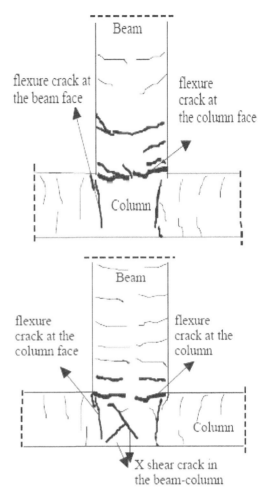

Figure 10. Crack propagation of the specimens in the confinement regions of the beam and column beam cracking.

3.3 Joint cracking

Joint cracking was best resisted by the SFRC specimens with 1.5% and 3.0% steel fibre. Minor cracking of the joint began during the 1st cycle. The joint cracks became more extensive during the 3rd cycle and increased in density during the 4th cycle. A crack across the top of the joint opened during the 5th cycle and continued to open up during the 6th cycle. Minor cracks began forming during the 1st cycle and became more extensive during the 3rd cycle. A crack across the top of the joint opened during the 4th cycle and continued to open up until the end of testing. The plain concrete specimens exhibited inferior performance. Minor cracks began forming during the 1st cycle. Some of these cracks opened during the 3rd cycle and led to spalling of joint concrete during the 4th cycle. The spalling became more extensive during the 5th and 6th cycles. Figure 9 illustrates the above-mentioned discussion. Spalling and cracking in the SFRC joints was confining by the steel fibres, as shown in Figure 9,

Table 3. Cumulative work done.

Spe. ID	Work done (N.m) in cycle						Work done (N.m)	% increase in work done
	1st	2nd	3rd	4th	5th	6th		
P0	1.25	5.8	22.5	15	–	–	44.55	–
P1.5	2.8	5.25	7.8	12	15.9	24.8	68.55	54
P3.0	1.05	2.25	8.75	16.3	21.7	32.8	82.9	86
P4.5	3.25	9.33	9.4	25.5	40.07	48	133.55	204
P6.0	4.82	11.05	18	28.4	41.75	–	104.02	133

which allowed for a better bond between steel and concrete. It preserved a good portion of its strength. This can increase the effectiveness of joint reinforcement.

3.4 Seismic strength

From the work done illustrated from in Table 3 it can be observed that the seismic strength increased by 54%, 86%, 204% and 133% for various mixes of SFRC 1.5%, 3.0%, 4.5% and 6.0% dosage of fibres respectively. It reflects that the energy absorption capacity is increased with the SFRC specimens.

The stored energy absorbed through the elastic behavior of the specimens during loading was given back to the system in the course of unloading of the beam-column joint specimens. This energy reserved by elastic behavior was defined as stored energy. The amounts of stored energy of the specimens were separately obtained by subtracting the amount of total dissipated energy from the amount of total energy for each specimen. The stored energy capacities of the specimens are shown in Table 3.

4 CONCLUSIONS

This paper describes the experimental study conducted on beam-column joints to determine the potential increase in joint hoop spacing in conventional concrete joints that might be achieved using steel fibre concrete in place of a conventional concrete joint region. During the study the following observations were made.

The results indicate that both the ductility and strength capacity is increased by adding fibres. This is allowed to reduce the stirrups requirement in the joint. Furthermore, the usage of SFRC can reduce the cost of steel reinforcement and its installation, and the difficulties in placing and consolidating the concrete in the regions of the beam-column joints. Moreover, the use of SFRC and transverse reinforcement in the critical regions can be recommended, in view of the total dissipated and stored energy. Thus, SFRC can be seen as an appealing alternative to conventional confining reinforcement. However, it is well known that ductile behavior and the strength capacity of beam-column connections depend on the percentage of fibre content, aspect ratio of the fibres, fibre type, and the regions of SFRC used in joints, the strength of the concrete,

and fibre orientation in the concrete mix. It is recommended that for exterior beam-column joints, in which ease of construction is desired, steel fibres can be used up to 3.0% by mass of concrete. Using fibres the joint gives better performance during earthquake loading; it can reduce the requirements for closely spaced ties. Hence the spacing of stirrups can be increased, which allows for good placement of concrete and reduces the chances of honeycombing in the joint. Capacity design guidelines are needed to make safe and efficient use of this technology in earthquake-resistant design.

ACKNOWLEDGMENTS

The authors acknowledge the Departmental Operating Cost Committee at S.V. National Institute of Technology, Surat for providing grants for this research. The support of SHAKTIMAN STEWOLS & CO., NAGPUR (INDIA) which supplied the steel fibres, is also appreciated. Thanks are also due Mr. Momahbhai Patel, Mr. Babubhai Gamit and Mr. Bharat More for his help during experimentations. The authors are also thankful to Dr. H S Patili and Dr. J A Desai (Prof. & Head) for valuable comments during experimental works.

REFERENCES

ACI Committee 408, Opportunities in bond research, American Concrete Institute Journal, Proceedings, Vol.67, No.11, Nov. 1970, pp. 857–867.

Balaguru, P.N., and Shah S.P.,-1992, "Fibre Reinforced cement composites", Mc-Graw Hill New York.

Batson Ball. C., Bailey, L., Landers, E., Hooks, J.-Flexural fatigue strength of steel fibre reinforced concrete beams, J. Amer. Conc. Inst., Proc., V. 69, No. 11, November 1972, pp. 673–677.

Bayasi, Z., R. Bhattacharya, and M. Posey. "Fibre Reinforced Concrete: Basics andAdvancements," Proceedings, Symposium on Advancements in Concrete Materials, Bradley University, 1989, pp. 1–1 to 1–27.

Bayasi, Z., and H. Kaiser. "Steel Fibres as Crack Arrestors in Concrete." The Indian Concrete Journal, April 2001.

Ehsani M. R. and Wight J.K., "Exterior Reinforced Concrete Beam-To-Column Connections Subjected To Earthquake Type Loading", ACI Journal, July-August 1985, pp. 492–499.

Ehsani R. Mohammad and Alameddine Fadel, "Design Recommendations for Type 2 High-Strength Reinforced

Concrete Connections", ACI Structural Journal, Vol.-88, May–June 1991, pp. 277–291.

Filiatrault Andre, Pineau Sylvain, and Houde Jules, "Seismic Performance of Code-Designed Fibre Reinforced Concrete Joints", ACI Structural Journal, Vol.-92, September-October 1985, pp. 543–551.

Fillatrault Andre, ladicani karim and Massicotte Bruno, " Seismic Performance of Code-Designed Fibre Reinforced Concrete Joints", ACI Structural Journal, Vol.-91, September-October 1994, pp. 564–571.

Ganesan.N and Indira P.V. "Latex Modified Steel Fibre Reinforced Concrete Beam- Column Joints Subjected To Cyclic Loading", ICJ Structural Journal, Vol.-74, July-2000, pp. 416–420.

Gregor G. G., Fischer and. Victor C .LI, "Deformation Behavior of Fibre Reinforced Polymer Reinforced Engineered Cementations Composite (ECC) Flexural Members Under Reversed Cyclic Loading Conditions", ACI Structural Journal, January–February 2003, pp. 520–532

Guimaraes Gilson, Kreger Michael E. and Jirsa James O., "Evaluation of Joint-Shear Provisions for Interior Beam-Column-Slab Connections Using High-Strength materials", ACI Structural Journal, Vol.-89, January–February 1992, pp. 89–98.

Hannant D.J., 1978 "Fibre cements and fibres concretes", John Wiley & Sons, New York 219 pp.

I.S:456-2000, Plain and Reinforced Concrete Code of Practice (Fourth Revision) Fourth Reprint October 2001, Bureau of Indian Standards, New Delhi.

I.S:4031-1988, "Method of Physical Test for hydraulic cement", Bureau of Indian Standards, New Delhi.

I.S: 1893 (Part 1): 2002, Criteria for Earthquake Resistant Design of Structures Part 1 General Provisions and Buildings. (Fifth Revision), Bureau of Indian Standards, New Delhi.

I.S: 13920: 1993, Ductile Detailing of Reinforced Concrete Structures Subjected to Seismic Forces – Code of Practice 2002–03, Bureau of Indian Standards, New Delhi.

I.S:10262-1982 Recommended Guidelines for Concrete Mix Design Fifth Reprint March-1998, Bureau of Indian Standards, New Delhi.

I.S: 2386-1963, Methods of tests for aggregate for concrete, Bureau of Indian Standards, New Delhi.

I.S: 383-1979, Specification for coarse and fine aggregate from natural sources for concrete, Bureau of Indian Standards, New Delhi.

I.S. 516-1959 Methods of tests for strength of concrete, Bureau of Indian Standards, New Delhi.

I.S 5816-1970 Methods of tests for split tensile strength of concrete cylinder, Bureau of Indian Standards, New Delhi.

Johnston, C. "Fibre Reinforced Concrete." Significance of Tests and Properties of Concrete and Concrete-Making Material, ASTM STP 169C, 1994, pp. 547–561.

Krishna Raju N., Basavarajaiah B. S., "Compressive Strength and Bearing Strength of Steel Fibre Reinforced Concrete", ICJ, June-1977, pp. 183–188.

Kukreja C. B., Kaushik S. K., kanchi M.B. and Jain O.P., ICJ, July-1980, pp. 184–189.

Leon, R.T., Shear Strength and Hysteretic Behaviour of Beam-Column Joints, ACI Structural Journal, V.87, No.1, Jan–Feb, 1990, pp. 3–11. (Chap1)

Mahajan M.A., 2004, "Study Of Beam-Column Junction Using High Performance Concrete Under Cyclic Loading" M.E. Dissertation, Applied Mechanics Department, S.V.N.I.T. Surat.

Macgregor, J.G., Reinforced Concrete Mechanics and Design, Prentice Hall Inc., 1988.

Mayfield, B., Kong, K.F. and Bennison, A. "Corner joint details in structural lightmass concrete," Journal of American Concrete Institute, May 1971, Vol. 65, No.5, pp. 366–372.

Michael G., 2001, "Application of Steel Fibre Reinforced Concrete in Seismic Beam-Column Joints", M.S. theses, Faculty of San Diego State University – California.

Modhera C.D., 2001, "Some Studies on Partially Set Fibre Reinforced Concrete Under Sustained Temperature Cycle Using Selfing Concept, Ph.D. Theses, Department of Civil Engineering, I.I.T. Bombay.

Neville A. M., Properties of Concrete, Pearson Education Asia Pvt. Ltd., England.

Nataraja M.C., Dhag N.and Gupta A.P., "ICJ, July 1998, pp. 355–356.

Nilson, I.H.E., Losberg, R. Reinforced concrete corners and joints subjected to bending moment, Journal of Structural Division, ASCE, June 1976, Vol.102, No.ST6, pp. 1229–1253.

Park, R., and Paulay, T., Reinforced Concrete Structures, John Wiley and Sons, 1975, 786p.

Paulay, T. and Priestley, M.J.N., Seismic Design of Reinforced Concrete and Masonry Buildings, John Wiley and Sons, 1992, 767p.

Parra-Montesions Gustavo J.,Peterfreund Sean W.and Ho Chao Shih, "Highly Damage-Tolerant Beam-Column Joints through Use of High-Performance Fibre-Reinforced Cement Composites", ACI Structural Journal, May-June 2005, pp. 487–496.

Parghi A M, 2006, "An Experimental Study on Behaviour of An Exterior Beam-Colum Junction Subjected to cyclic Loading, M.Tech, Theses, Department of Applied Mechanics, S.V.N.I.T.Surat.

Paulay Thomas, "Equilibrium Criteria for Reinforced Concrete Beam-Column Joints", ACI structural journal, November–December 1989, pp. 635–643.

Raffaelle Gregory S. and Wight James K, "Reinforced Concrete Eccentric Beam-Column Connection Subjected to Earthquake Type Loading", ACI Structural Journal, January-February 1985, pp. 45–55.

Recommendations for design of beam-column-joints in monolithic reinforced concrete structures, American Concrete Institute, ACI 352R-02, ACI ASCE, Committee 352, Detroit, 2002.

S.P., 22:1982, Explanatory Hand Book on Code for Earthquake Engineering, Bureau of Indian standards, New Delhi.

S.P. 23-1982, "Hand book on concrete mixes (based on Indian standards)", Bureau of Indian standards, New Delhi, 144 pp.

S.P., (S&T), 34-1987, "Hand book on Concrete Reinforcement and Detailing", Bureau of Indian standards, New Delhi, 204 pp.

Shah S.P. and Shah Ahmed, High Performance Concrete: Properties and Applications, McGraw-Hill, Inc., New York

Shetty M.S., Concrete Technology Theory and Practice, S. Chand & Company Ltd, New Delhi.

shiohara, H., New model for shear failure of RC interior beam-column connections, Journal of Structural Engineering Division, ASCE, V. 127, 2001, pp. 152–160.

Soao Wen Bin, "Reinforced Concrete Column Strength at Beam/Slab and Column Intersection", ACI Structural Journal, January–February 1994, pp. 3–10.

Soni D.G., 2004, An Experimental Study on Shear Strength test Method for SFRC, M.E. Dissertation, Applied Mechanics Department S.V.N.I.T. Surat.

375

Subramanian, N., and Prakash RAO, D.S. Seismic Design of Joints in RC Structures, The Indian Concrete Journal, February 2003, Vol. 77, No. 2, pp. 883–892.

Soroushian, P., Z. Bayasi, Narayanan and Darwish, "Strength and Ductility of Steel Fibre Reinforced Concrete under Bearing Pressure." Magazine of Concrete Research, Dec. 1991, pp. 243–248.

Stewols and Company, Steel fibre Reinforced Concrete Leaf left, Nagpur-India

Tan Kiang Hwee and Saha Mitun Kumar, "Ten Year study on Steel fibre-Reinforced Concrete Beams under Sustained Loads", ACI Structural Journal, May–June 2005, pp. 472–480.

Tsonsos A.G., Tegos I.A. and G. Gr. Penelis, "ACI Structural Journal, January February 1992, pp. 3–12.

Wahab-Abdul-Hasim M. S, "ACI Structural Journal, July–August 1992, pp. 367–374

Concrete Solutions – Grantham, Majorana & Salomoni (Eds)
© 2009 Taylor & Francis Group, London, ISBN 978-0-415-55082-6

Strengthening and repair of reinforced concrete structures using composite material

Anant Parghi & C.D. Modhera

Department of Applied Mechanics, S.V. National Institute of Technology, Gujarat, India

ABSTRACT: After recent major destructive earthquakes in China and India there is increased awareness for the need to evaluate and improve seismic performance of existing reinforced concrete buildings. This paper deals with the effect of externally bonded unidirectional GFRP wrap on RC members for ultimate load carrying capacity. The experimental study was organized in two phases. In phase one, the behaviour of 12 RC circular columns (150 mm × 680 mm) under uniaxial compressive stress was investigated to determine the ultimate load carrying capacity of the columns. The tests were organized so that the first three were wrapped with a single layer of externally bonded unidirectional GFRP, then three with a double layer and three with a triple layer. Three specimens were left unwrapped as control columns. The wrapping was done in the hoop direction of the columns. The ultimate load carrying capacity of the retrofitted columns, was estimated using the Richard et al. model for confinement of circular columns. In Phase two, a set of 12 RC beams with dimensions 150 mm × 200 mm × 1200 mm long were cast. GFRP was applied only in the flexure area of the beam. Nine RC beams were retrofitted using GFRP wraps with single, double and triple layers in a longitudinal direction as external reinforcement. The structural behaviour of the RC beams after application of GFRP in the flexure area was observed for ultimate load carrying capacity of the beams and mid span deflections were measured. The application of externally bonded GFRP on concrete surfaces using epoxy resins between the concrete surface and the GFRP wrap produced confinement on the concrete specimens and enhanced the load carrying capacity of the RC elements.

1 INTRODUCTION

Deterioration of reinforced concrete structures due to corrosion of the rebars or continual upgrading of service loads (increase of the traffic load on bridges for example) has resulted in a large number of structures requiring repairing or strengthening. Various methods are available to repair or strengthen those structures. External bonding of steel plates to damaged reinforced concrete structures is one of these methods and has been shown to be quite an efficient and a well-known repair or strengthening technique. It has been largely studied in France (L'Hermite, 1967), (Bresson, 1971) and intensive research performed in the beginning of the eighties (Theillout, 1983) resulted in French rules concerning the design of those structures (SOCOTEC, 1986). The use of composite materials represents an alternative to steel as it can avoid the corrosion of the plates. FRP materials are also very lightweight, have a high strength to weight ratio and are generally resistant to chemicals. The price of these materials, especially of Carbon Fibre Reinforced Plastics (CFRP), could represents a drawback but the ease in handling the material on construction sites, due to the light weight, helps to reduce labour costs. This technique has been largely investigated especially in Switzerland (Meier, 1995) where existing structures have been retrofitted using epoxy-bonded composite materials.

2 EXPERIMENTAL PROGRAM

2.1 *Test specimens*

In this study, tests were performed to evaluate the percentage increase in strength of FRP wrapped specimens. The experimental study was organized in two phase. All twelve columns 680 mm long and diameter 150 mm had internal reinforcement provided by six 8 mm diameter rebars (yield strength: 415 MPa). Six 6 mm diameter rebars with 126 mm spacing stirrups were also used. The twelve rectangular beams were tested in order to evaluate the effect of externally bonded composite-material reinforcement on the flexural capacity of RC beams. All twelve beams had a span length of 1000 mm (beams were 1200 mm long) and cross-sectional dimensions of 150 *times* 200 mm. These beams had internal reinforcement provided by four 8 mm diameter rebars (yield strength: 415 MPa). The average compressive strength of the concrete at the day of the test was found to be 29 MPa.

2.2 *Composite materials*

2.2.1 *Wrapping procedure on beams and columns*
Wrapping of the specimens is an important stage and requires skillful labour, and here also wrapping was done by the skillful labour of the company. The various steps for wrapping were as follows.

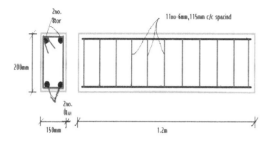

Figure 1.   Geometry and reinforcement details in of beam.

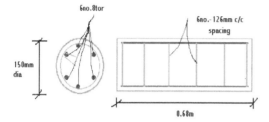

Figure 2.   Geometry and reinforcement details of column.

Figure 3.   Surface preparation of the specimen.

2.2.2   Surface preparation

Surface preparation is the most important step in the stage of wrapping; proper surface preparation ensures good bonding of wrapping with the substrate concrete. No surface unevenness more then 1 mm is allowed before wrapping: if proper surface preparation was not made then air pockets would have resulted. These would lead to bond failure of the FRP with the concrete. For surface preparation all the sharp edges of the beams were rounded off to get al smooth radius of 15 mm and this was done to avoid stress concentration at the sudden change in the cross-section. This step is shown in figures 3 and 4.

2.2.3   Putty leveling

Putty was applied on the surface on any concave area of the concrete specimens. Concavity of the surface leads to air pockets and lack of full contact of the FRP and concrete.

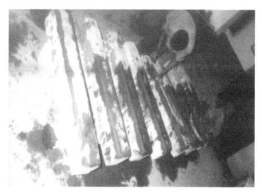

Figure 4.   Preparing of beam edge for a radius about 20 mm.

Figure 5.   Preparation of primer (Mixing of base & hardener).

In this study during application of GFRP in a U-shape on beams in the flexure area it was essential to radius the corner to avoid stress concentration. So sharp edges were removed by chisel and the corner radius around 20 mm was made with SBR modified mortar used for repair. The surface was made even and on that surface putty application was applied for proper bonding between fibre and concrete surface.

3   BONDING OF THE COMPOSITE MATERIALS

3.1   Primer coat application

When leveling of surface was completed and when no un-evenness more than 1 mm was left then a primer coat was applied. The primer was a two part compound and mixed in proportion of 1: 0.1: mixing was done in a small pan in the required quantities. A proper bond was made between the FRP and substrate concrete. The application primer is shown in figure 6.

The first coat of saturant was applied on the concrete after the primer had hardened. The saturant was also a two part compound and they were mixed in 1.5: 1 proportions base:hardener.

Figure 6. Application of primer (first coat) on concrete specimen. First coat of saturant.

Figure 7. Saturant mixing comp. A and comp. B.

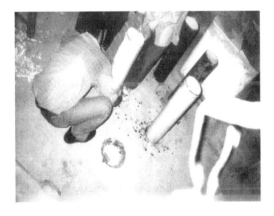

Figure 8. Application of GFRP on column.

4 FRP FABRIC APPLICATION

In these investigations we used a short column. Generally short columns fail in shear and a crack was generated parallel to the loading axis and tensile stress occurred perpendicular to the direction of crack. So

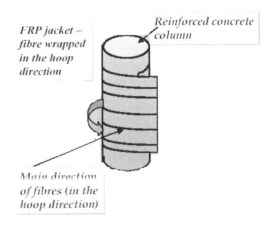

Figure 9. Direction of fibre wrapping.

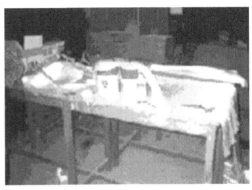

Figure 10. GFRP wrap materials.

GFRP was applied in the direction of tensile stress as shown in figure 9.

FRP cloths were pre-cut and applied on the concrete where the first coat of saturant was applied. The cutting of the FRP cloth also required some care. At the time of cutting of the FRP cloth it is advisable to wear hand gloves.

After the cloth was wrapped around the specimen by rubber roller, the cloth was pressed, so the saturant came out from inside to the top. For wrapping, the minimum over-lap was 150mm also required for good performance of FRP, otherwise the failure of the specimen would initiate from the joint. This kind of failure was observed in the beam.

4.1 Second coat of saturant

On the FRP cloth, the second coat of saturant was applied. That was also pressed with a rubber roller so that the saturant would be evenly distributed to every point of the FRP cloth and no portion of the cloth remained dry. After full application of the saturant, the specimens were kept for a period of 3 days for hardening, before testing. The cost of the primer was 80 Rs,/m^2; the saturant cost 300 Rs/m^2; the E glass fibre cost 230 Rs/m^2; the putty cost 80 Rs/m^2. The application cost was 130 Rs/m². Therefore the overall total cost of wrapping was 880° Rs/m².

5 TESTING PROCEDURE OF BEAM AND COLUMN (*Unwrapped and wrapped*)

All the specimens were tested on a hydraulically operated compression testing machine. Its capacity of applying load was 200 ton in compression. All the columns, cubes, cylinders, were tested in this machine. For testing columns, it was first ensured that the smooth face of the specimen was the loading face. After that, the upper plate was lowered till the distance between the face of the specimens and plate was 5 mm. The application of load was applied at a constant rate until the specimens failed.

In the case of columns, these were tested axially on the compression testing machine The column specimens were loaded up to failure only for axial loads, for both wrapped and unwrapped (control) specimens. Then the averages of failure load were compared to see what percentage increase in strength had been achieved. The test set up of the column is shown in Figure 16.

For testing of beams, all beams were tested up to failure under three point bending. Firstly the supports and the point where loading is to be applied were marked. At the center the deflection was required to be measured so for that a point was also marked. The support points were at a distance 15 cm from the edges at both sides. The application of load was at a constant rate (2000 kN) for testing of beams. The load was applied up to failure of the specimen and the failures of the specimen were observed. During the test, the vertical deflection at midspan (1) and the applied load were measured. Mechanical dial gauges were used to measure deflection. These instruments are represented on the test set-up shown in Figure 11.

A first series of beams was preloaded to a percentage of the ultimate load of the control beam. This percentage had been defined as a damaging degree and corresponded to an average width of the cracks of 0.5 mm. These three beams had been repaired using GFRP. Then they were tested up to failure. With this first series we wanted to study the effect of precracking and of the plate thickness on the flexural behaviour of the beams.

The second series consisted of strengthening the beams using three different composite materials as well as studying the effect of a single or double layer of carbon sheets.

6 TEST RESULTS

Table: 1 presents the ultimate load carrying capacity of the RC beams, maximum mid-span deflection, maximum stiffness and ultimate moment of resistance. Experimental results suggested that with FRP application on beam specimens we were able to surpass the original capacity of the control beam. The mid-span

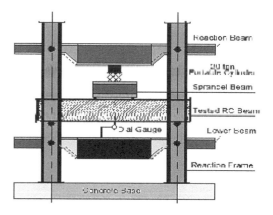

Figure 11. Experimental test set up of beam.

Table 1. Comparison of result of beams with various application of GFRP layer.

Description	Control	Unidirectional		
Type of FRP	–	GFRP	GFRP	GFRP
No. of Layers	–	1	2	3
Ultimate Load (KN)	60	88	100	120
Maximum Stiffness (KN/mm)	56.25	62.74	72.22	82.92
Max.M.O.R. (kN.m)	9.0	13.2	15.0	18.0
Percentage increase in ultimate load	100	146	166	200

Table 2. Comparison between experimental and theoretical capacities of RC columns.

Series	No. of Layers	(Ton) Load Measured	Theoretical Prediction Richard et. al. [1929] (Ton)	Measured Strength (%)
Control	–	29.86	–	100
Uni directional	1	77.5	79.47	253
	2	120.33	129.08	402
	3	166.67	178.69	558

load was applied gradually until the ultimate load carrying capacity was reached and the RC beam failed. The mid-span deflection of the beams was measured using 0.01 mm accuracy by mechanical dial gauge. The applied load and accompanied mid span deflection were recorded at equal intervals (2 kN). The ultimate load was recorded.

Results from the ten beams are summarized in Table 3 and plots of load-versus-deflection for typical beams are shown in Figure 14.

6.1 Beam behaviour

6.1.1 Glass fibre materials
When loaded, the control beam developed flexural tensile cracks in the constant moment region at a load of

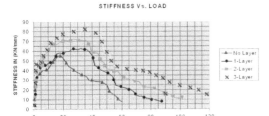

Figure 12.   Stiffness vs. load.

20 kN. At loads near 80 kN, the tensile reinforcing steel yielded. The beam failed in flexure due to crushing of extreme compression fibre at a load of 90 kN.

All composite repaired beams or externally reinforced beams showed significant increases in flexural stiffness and ultimate capacity as compared to the control beam. These precracked beams were seriously damaged as internal steel had yielded in all cases. Before being repaired they exhibited an average midspan deflection of 10 mm and many flexural cracks. Midspan deflections for beams after the second flexural test were respectively 25 and 21 mm. It suggests that the increase of the ultimate capacity depends on the plate thickness. At the ultimate stage the preloaded repaired beams showed better mechanical characteristics than the control beam.

Beams had increases in ultimate load of respectively 10% and 45% over the control beam, while midspan deflection, composite strains and crack widths were reduced. High stiffness was reached by using glass fibre rods, which behaved similarly to a plate. Another beam behaved in a linear fashion whereas the other beams exhibited two different behaviours. The repaired or externally reinforced beams showed linear behaviour between the first cracking and a point, which did not always correspond to the yielding of the tensile steel. After that point, those beams seemed to behave in a plastic fashion which really surprised us as we know that composite materials exhibit linear elastic behaviour up to failure.

GFRP Beams had increases in ultimate load of respectively 51% and 58% over control beam, while CFRP beams failed. Midspan deflection, tensile strains in the carbon sheets and cracks width was also reduced and an increase of the stiffness of the beams was noticed. The bonding of a second layer of carbon sheets can lead to an increase of the ultimate load and of the stiffness of the beam. Only 1.2 mm carbon sheets were available and we had to bond a double layer of them to take into account the thickness parameter. We can conclude from the tests that the influence of this parameter on the flexural behaviour of the beams is obvious.

The influence of the external reinforcement on the development of the cracks was really good. The cracking was more diffuse and the opening of the cracks was in the region of 0.1 mm (several millimeters for the control beam). Nevertheless the decrease of strains

and cracking showed some drawbacks as they were a tool for the designer to predict if the failure of the structure was close or not. Whereas glass fibre strengthened beams exhibited a kind of plastic behaviour, carbon fibre reinforced beams behave elastically up to failure. This lack of ductility is dangerous as it leads to a brittle failure of the beams.

6.1.2   Failure modes

The failure modes, which have been observed on the beams, are different from that of a classical reinforced concrete beam (concrete crushing or failure of the internal steel). Moreover the failure modes of the repaired beams are different from those of the externally reinforced beams.

6.1.3   Failure modes of the repaired beams

Before being repaired the beams exhibited open cracks, midspan deflection and the internal rebars had yielded. During the second flexural test, the opening of the existing cracks increased. This failure also occurred in a sudden manner.

6.1.4   Failure modes of externally reinforced beams

All externally reinforced beams using glass or carbon fibre materials failed in the same manner. We attended to the failure of a concrete layer along the internal reinforcement. The concrete was not initially precracked and the development of the cracks during the reinforcement test was highly influenced by the carbon sheets. The first cracking was delayed and more diffuse. Shear cracks occurred at the ends of the plates or of the sheets for values of the load between 70% and 80% of the ultimate load. Then cracks widened at the midspan by using existing flexural cracks. Finally, the sudden propagation of a horizontal crack in the concrete-steel bond region occured. This crack ran along the weakest surface, which is the concrete-steel interface. It lead to the failure of the beam as soon as the crack opened and separated the concrete cover from the rest of the beam. It is interesting to note that the weakest point of the assemblage concrete bond composite material was not the concrete composite interface but the concrete internal steel interface.

We observed the same failure mode for all externally reinforced beams. The load-versus-deflection diagram for these brittle failures was not obvious as the beam seemed to behave plastically. The failure of a brittle type was more obvious for the carbon sheet reinforced beams. Fig. 13 and 14 shows the sudden decrease of load, which suggests the failure of the beam. The full flexural capacity of the externally bonded beams was not reached and the high elastic strain of the carbon sheets was not really used.

6.1.5   Theoretical behaviour of RC circular columns

An estimation of the theoretical ultimate load carrying capacity of the retrofitted RC columns was carried out using the Richard et al. [1929] model, the strength confinement factor was set to $K_1 = 4.1$. The unconfined concrete compressive strength was taken based on the behaviour of the control column.

Table 3. Test result of beam specimen.

GFRP Layer	Ultimate Load (kN)	Max. Moment (kN.m)	Max. stiffness (kN/mm)	Deflection at Failure (mm)	Ultimate load Incr (%)
No Layer	60	9	56.25	8	100
Layer 1	88	13.20	62.74	11.90	146.67
Layer 2	100	15	72.22	9.20	166.67
Layer 3	120	18	82.92	8.40	200

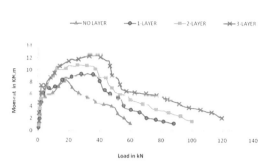

Figure 15. Moments vs. load.

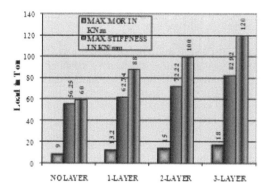

Figure 13. Various layers of GFRP wrapped beam w.r.t. control/unwrapped beam.

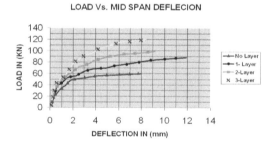

Figure 14. Load-versus-midspan deflection plot for reinforced beams using carbon fibre sheets.

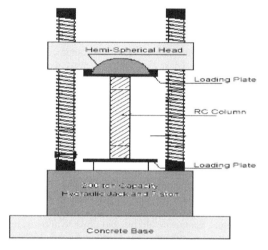

Figure 16. Experimental test set up of circular column.

6.2 Results of column tests

Table 4. Load carrying capacity of column for various layers of GFRP.

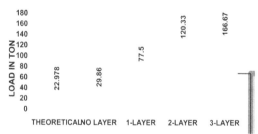

Figure 17. Load carrying capacity of various layers of GFRP application on column specimen.

Specimen	Theoretical column strength by limit state method in (Ton)	No layer (Ton)	layer 1 GFRP (Ton)	layer 2 of GFRP (Ton)	layer 3 of GFRP (Ton)
Specimen 1	22.978	29.8	75	125	165
Specimen 2	22.978	35.2	77	112	162
Specimen 3	22.978	24.6	80.5	124	170
Average	**22.978**	**29.86**	**77.5**	**120.33**	**166.67**

7 CONCLUSIONS

Glass fibre or carbon fibre bonded beams transformed the ductile flexural failure of the control beam into a brittle failure, which cannot agree with utilization in civil engineering works. Therefore we need to optimize this method of reinforcement and in particularly to develop ductile failures (this could be achieved by using an anchorage system). A global design strategy requires the determination of the interface law, which governs the behaviour of the beams. Indeed the knowledge of the stress-strain laws of the different materials

Table 5. Test result of circular RCC column.

GFPR	Theoretical Load carrying capacity of column (Richard et. al)	Experimental Load carrying capacity of column (Ton)	Ultimate stress (N/mm²)	%age increase in capacity
No Layer	22.978	29.86	16.89	100
Layer 1	79.47	77.5	43.85	259
Layer 2	129.08	120.33	68.08	402
Layer 3	178.69	165.67	93.74	554

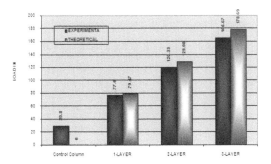

Figure 18. Theoretical and experimental comparison result of RCC circular column of various application of GFRP layer.

(concrete, steel, bond and composite) is not enough. This interface shear law will be soon obtained by using a numerical-experimental approach.

The results of tests performed in this study indicate that a significant increase in the flexural strength can be achieved by bonding GFRP plates to the tension face of reinforced concrete beams. The gain in the ultimate flexural strength was more significant in beams with lower steel reinforcement ratios. In addition, plating reduced crack size in the beams at all load levels. The successful application of this technique requires a careful preparation of concrete surface and the selection of a tough epoxy. Plating somewhat reduced the ductility of the beams. However, additional analytical and experimental studies must be undertaken to establish criteria for predicting the limiting load that causes the concrete layer between the longitudinal rebars and plate to fail. In addition, the effects of environmental factors, e.g., temperature and moisture on the epoxy joint, as well as the performance of upgraded beams under fatigue loading should be examined.

- The load carrying capacity of RC members increases as application of more GFRP layers are applied. The theoretical expression given by Richard et. al. for circular retrofitted columns matched with experimental work.
- Experimental results suggested that the ultimate load carrying capacity of retrofitted beams & columns were increased significantly over that of the control specimen.

Single Layer

Double Layer

Triple Layer

Figure 19. Failure pattern of column after application of various no. of GFRP layer.

- It was observed that there was an increase in ductility and failure was not catastrophic, which plays an important role in protection against earthquake forces.
- It was observed that there was an increased Moment of Resistance of RC beam, which means it can carry more load.
- From experimental programs the load carrying capacity of axial member increase w.r.t. the Control column was 253%, 402%, and 558% for single, double and three layers of GFRP respectively.

383

ACKNOWLEDGEMENTS

The experimental work was carried out with the financial support of Departmental Operation Cost and the Materials and Structural Laboratory in the Department of Applied Mechanics, S.V. National Institute of Technology, Surat whose support is greatly appreciated. Thanks are also due Mr. Momahbhai Patel, Mr. Babubhai Gamit and Mr. Bharat More for his help during experimentations. The Authors are also thankful to the Dr. J A Desai (Prof.& Head) for valuable comments and assistance during writing the papers.

REFERENCES

Abdel-Hady H. Hosny and Dr. Amr A. Abdelrahman "Research Progress In Egypt On Strengthening Concrete Structures With Frp" *Netcomposites Workshop*, Dec 2003

ACI Committee 318 2002. Building Code Requirements for Structural Concrete (ACI 318-02), *American Concrete Institute*, Farmington Hills, Mich., 391 pp.

AIJ, Architectural Institute of Japan 1997. Design Guidelines for Earthquake-Resistant R.C. Buildings Based on Inelastic Displacement Concept (draft, in Japanese).

Antonopoulos, C.P.& Triantafillou, T.C. *ACI Structural Journal*, Vol. 91, No.5, 552–563. 2002.

Antonopoulos, C.P.& Triantafillou Analysis of FRP-strengthened RC beam-column joints, *Journal of Composites for Construction*, Vol. 6, no.1, pp. 41–51., 2002

Antonopoulos, C.P.& Triantafillou, T.C. 2003. Experimental Investigation on FRP-Strengthened RC Beam-Column Joints, *J.l of Composites for Construction*, 7(1), pp. 39–43 Aycardi, L., E., Mander, J., B. & Reinhorn, A.M. 1994. Seismic Resistance of R.C. Frame Structures Designed Only for Gravity Loads: Experimental Performance of Subassemblages,

ACI Structural Journal, Vol. 99, N. 6. Calvi G.M., Magenes G. & Pampanin S. 2002a. Relevance of Beam-Column Damage and Collapse in RC Frame Assessment.

Beres, A., Pessiki, S., White, R., Gergely, P. 1996. Implications of Experimental on the Seismic Behaviour of Gravity Load Designed RC Beam-Column Connections, *Earthquake Spectra*, Vol. 12, No.2, May, pp. 185–198.

Bing. L, Yiming W. & Tso-Chien P. 2002. Seismic Behavior of Non-Seismically Detailed Interior Beam-Wide ColumnJoints. Part I: Experimental Results and Observed Behavior,

Bresson, J. (1971), L'application du béton plaqué, Annales de l'I. T. B. T. P., No. 278, 1971.

Calvi, G.M., Magenes, G. & Pampanin, S. *.J. of Earthq. Engng (JEE), Special Issue 1*, pp.75–100 2002b. ExperimentalTest on a Three Storey R.C. Frame Designed for GravityOnly, *12th ECEE*, London, p. 727.

Crawford, John E. and L. Javier Malvar, Kenneth B. Morrill "Reinforced Concrete Column Retrofit Methods For Seismic And Blast Protection"

Dolce, M., Cardone, D.& Marnetto, R. 2000. Implementationand testing of passive control devices based on shape memoryalloys. *Earthq. Engng. & Struct. Dyn.*, 29, 7, 945–968.

Dodd, L.L., & Restrepo, J.I. 1995. Model for Predicting CyclicBehavior of Reinforcing Steel. ASCE *Journal of StructuralEngineering*, Vol. 121, No. 3, pp. 433–445.

Eurocode 8, 2003 Design Provisions for Earthquake Resistanceof Structures, *European Committee for Standardization*, Brussels.

FIB (Federation International du Beton) 2001. Externally Bonded FRP Reinforcement for RC structures, *fib Bulletin n.14*, Lausanne.

Gergely, J., Pantelides, C. P., Nuismer, R.J. & Reaveley, L.D. 1988. Bridge Pier Retrofit Using Fibre Reinforced Plastic Composites, *Journal of Composite Constructions*, ASCE, 2(4), 165–174.

Gergely, J., Pantelides, C. P. & Reaveley, L.D. 2000. Shear Strengthening of RC T-Joints Using CFRP composites, *Journal of Composite Constructions*, ASCE, 4(2), 56–64.

Hakuto, S., Park, R. and Tanaka, H. 2000. Seismic Load Tests on Interior and Exterior Beam-Column Joints with Substandard Reinforcing Details, *ACI Structural Journal*, V. 97, N.1, 11–25.

Hollaway L.C. and Leeming, M.B.[1999] Woodhead Publishing, Cambridge, England "Strengthening of reinforced concrete structures"

Holzenkämpfer, P. 1994. Ingenieurmodelle des verbundes geklebter bewehrung für betonbauteile, *Dissertation*, TU Braunshweig (in German).

L'Hermite, R. (1967), L'application des colles et résines dans la construction. Le béton à coffrage portant, Annales de l'I. T. B. T. P., No. 239, Nov. 1967.

Mander, J.B., Priestley, M.J.N. and Park R., 1988. Theoretical Stress-Strain Model for Confined Concrete, *ASCE Journal of the Structural Division*, Vol. 114, No. 8, pp. 1804–1826.

Meier, U. (1995), Strengthening of structures using carbon fibre/epoxy composites, *Construction and Building Materials*, Vol. 9, No. 6, pp. 341–351, 1995.

Mukherjee Abhijit and Mangesh V. Joshi "Seismic retrofitting technique using fibre composites. *Indian Concrete Journal*, Dec 2001

SOCOTEC (1986), Recueil béton armé, Démolitions – Réparations – Renforcements, Tôles collées : justification par le calcul, juillet 1986.

Theillout, J. N. (1983), Renforcement et réparation des ouvrages d'art par la technique des tôles collées, PhD Thesis, Ecole Nationale des Ponts et Chaussées.

Concrete Solutions – Grantham, Majorana & Salomoni (Eds)
© 2009 Taylor & Francis Group, London, ISBN 978-0-415-55082-6

High strength ferrocement laminates for structural repair

M. Jamal Shannag
King Saud University, Riyadh, Saudi Arabia

ABSTRACT: This paper investigates the suitability of high strength ferrocement laminates containing several layers of welded wire mesh (WWM) for strengthening plain concrete cylinders. The overall response of the specimens in compression, in terms of load carrying capacity, axial displacement, axial stress and strain, lateral displacement, are described. Test results indicated that wrapping cylinders of 150 mm diameter and 300 mm height, with 2 layers of WWM ferrocement jackets showed about 16% increase in axial load capacity, 32% increase in axial strain and 30% increase in lateral displacement compared to control specimens without jackets; whereas the specimens wrapped with 4 layers of WWM showed about 30% increase in axial stress, 70% increase in axial strain and about 163% increase in lateral displacement. The findings of this investigation may encourage the construction industry to explore the potential applications of ferrocement as an alternative repair technique for concrete structures.

1 INTRODUCTION

The deterioration of existing concrete structures in many countries necessitates the need for developing cost-effective and long term repair and retrofit solutions that can be implemented in practice. A practical method of repair should take into consideration, the amount of damage, the shape of the member, materials of repair, construction cost, time and practicality. Several repair/retrofit techniques have been used for restoring the load carrying capacity of damaged concrete structural elements. These involved strengthening beams and columns by epoxy bonding of steel plates, external fixing of high performance fibre reinforced concrete jackets, or ferrocement laminates, and bonding of fibre reinforced polymer sheets to existing damaged concrete. Among these techniques, ferrocement laminates have received considerable attention from the research community in recent years due to their versatility, high mechanical performance and low cost compared to other techniques (Fahmy et al, 1990, Ong et al, 1992, Razvi et al, 1989 and Winokur et al, 1982).

Ferrocement has been successfully used in new structures, repair and rehabilitation of existing structures and marine environments. There has been increasing activity with ferrocement construction throughout the world including many countries such as USA, Canada, Australia, China, India, Thailand, Mexico, and Indonesia, (Naaman, 2000, Shannag, 2008, Shannag & Bin Ziyyad, 2007, Wang et al 2004). In China alone the tonnage of ferrocement vessels had reached about four million by 1989, (Shannag, 2008). The typical applications of ferrocement construction include water tanks, boats, roofs, silos, pipes, floating marine structures and low cost housing.

The main objective of this research is to investigate the suitability of locally available welded wire mesh (WWM) for increasing the load carrying capacity of plain concrete specimens. This objective can be accomplished through testing plain concrete cylinders of 150×300 mm under axial loading after wrapping them with thin ferrocement jackets containing 2 or 4 layers of medium spacing welded wire mesh (WWM) and a high strength cementitious matrix. It is expected that the results of this research should provide an efficient, low cost repair technique for preserving and renewing aging or damaged concrete structures and encourage the construction industry to discover the potential applications of ferrocement in numerous structures.

2 EXPERIMENTAL PROGRAM

The experimental program was planned to investigate the suitability of locally available welded wire mesh (WWM) for increasing the load carrying capacity of plain concrete cylinders.

2.1 Steel wire mesh

Woven galvanized steel square mesh with a wire diameter of 0.63 mm and a wire spacing of 12.5 mm was used. The mesh was tested in the laboratory following the guide for the design, construction, and repair of ferrocement reported by ACI Committee 549 (1988). The following properties were determined; equivalent yield strength = 374 MPa; ultimate strength = 530 MPa; elastic modulus = 106 GPa; density = 7.8 g/cc.

2.2 Plain concrete cylinders

The plain concrete cylinders were prepared using an ordinary concrete mix designed according to the ACI 211 (2005) method to achieve a 28-day compressive strength of 25 MPa and slump of 70–100 mm (S2 consistence in Europe). The concrete mix used consists of 300 kg/m^3 portland cement, 700 kg/m^3 crushed limestone, 600 kg/m^3 washed sand, 450 kg/m^3 silica sand, and 195 kg/m^3 free water.

After 28 days of moist curing, the cylinders were wrapped with 20 mm thick ferrocement jackets containing 2 or 4 layers of 12.5 mm square spacing welded wire mesh (WWM) and a high strength cementitious matrix. Finally the specimens were tested under uniaxial compression using a 2000 kN capacity universal testing machine.

2.3 Ferrocement jacket preparations

The mortar mix used for preparing the ferrocement jackets in this investigation had a 28 day compressive strength of about 63 MPa, and a flow of 132%. The mix proportions were 1: 2: 0.1: 0.1: 0.4: 0.035 by weight of cement, silica sand, silica fume, fly ash, water, and superplasticizer respectively.

All specimens wrapped with WWM were first placed in specially designed moulds that provide a 20 mm thick space around the specimens, and a 10 mm space from the top and bottom edge of the specimen to prevent any direct axial load on the external ferrocement jacket during test. Secondly the moulds along with the specimens were placed on the top of a vibrating table, followed by pouring the mortar matrix and vibrated at low speed to achieve full compaction and thus ensure that the mortar matrix encapsulated the WWM completely. The specimens were cured for 7 days using wet burlap, followed by another 7 days of drying at room temperature to be ready for testing.

After casting, the specimens were covered with wet burlap and stored in the laboratory at 23°C and 65% relative humidity for 24 hrs. and then demoulded and placed in water. Each specimen was labeled as to the date of casting, mix used and serial number. The specimens were then taken out of water a day before testing and dried in air. The specimens were tested under uniaxial compression using a 1800 kN capacity Forney testing machine available at King Saud University Materials Laboratory.

2.4 Preparation for testing and instrumentation

After at least 28 days from the day of specimen casting and at least 14 days after applying the mortar layer, the specimens were prepared for testing by capping their top surface with a thin gypsum layer to ensure parallel surfaces and to distribute the load uniformly in order to reduce any eccentricities. Axial displacements were measured by using two vertical linear variable differential transducers (LVDTs) of 100 mm range, installed at 180° apart around the specimen surface, in addition to another two vertical LVDTs installed at 180°

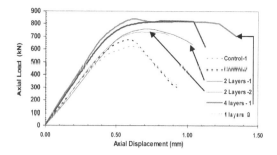

Figure 1. Axial Load-axial displacement curves for tested specimens.

attached to the head of the machine and the load cell to measure the total displacement and axial strain of the specimen.

The lateral displacements of the specimens were measured by using two horizontal LVDTs of the same range installed at 180° apart attached to the surface of the specimen, in addition to two external horizontal strain gauges to measure the strain at the jacket surface. The wires for the strain gauges, the load cell, and the LVDT's were attached to a data acquisition system and checked for readings. The load was applied at a loading rate of approximately 5 kN per second using a hydraulic testing machine of 2000 kN capacity and the specimens were tested under pure axial compression.

3 RESULTS AND DISCUSSION

The resulteing load-axial displacement curves of all tested specimens are shown in Fig. 1. It can be noticed that the behavior of cylindrical specimens with ferrocement jackets either with two or four layers of WWM are better than those of the control specimens. The jacketed specimens' curves have higher load capacity at higher axial strains as compared to the control specimens. For 4-layer ferrocement jackets the axial strain reached up to 0.0035. In addition, slightly higher axial stiffness can be observed in the case of specimens with 4-layers of WWM, because of the lateral confinement provided by the ferrocement jackets. The lateral displacement of all tested specimens at their mid height were plotted against the axial load: the resulting curves are shown in Fig. 2. It can be noted that the lateral displacement at maximum load is higher for jacketed specimens as compared to the controls. Such higher lateral displacements in jacketed specimens are attributed to the propagation and widening of the vertical cracks that are associated with yielding of the WWM, while sustaining its maximum axial load.

A summary of the average results of all the tested specimens is given in Table 1, in terms of the maximum axial stress, maximum axial strain and lateral displacement at maximum load. Results indicate about 16%

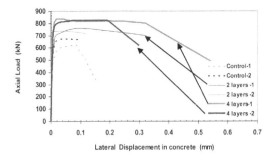

Figure 2. Axial load-lateral displacement curves for tested specimens.

Table 1. Summary of average results for the Specimens tested.

Specimen Type	Average peak axial stress (MPa)	Average maximum axial strain	Average maximum lateral displacement (mm)
Control Specimens	36.5	0.00209	0.1005
Specimens with 2 layers of WWM	42.32	0.00275	0.1305
Percentage increase for 2 layers of WWM	15.9%	31.6%	30.0%
Specimens with 4 layers of WWM	46.93	0.00356	0.264
Percentage increase for 4 layers of WWM	28.6%	70.3%	163.0%

increase in axial load capacity that is associated with about 32% increase in axial strain and 30% increase in lateral displacement for specimens wrapped with 2 layers of WWM ferrocement jackets. The specimens with 4 layers WWM showed about 30% increase in axial stress, 70% increase in axial strain and about 163% increase in lateral displacement. The increase in the specimen's lateral displacement was mainly attributed to the widening of the vertical cracks as the WWM was suffering from yielding. It is worth mentioning that increases in axial stress and axial strains were almost only due to the confinement offered by the ferrocement jacket, since the jackets did not reach the end of the specimens. The general behavior during test and the failure mode clearly demonstrates that the transverse wires were subjected to hoop tension and thereby, produced passive confinement pressure. As the load increased, much pronounced and wider cracks appeared which sometimes ended in separation and bulging of the mortar. Some of the meshes retrieved from the crushed specimens showed broken horizontal wire indicating yielding of these wires due to hoop tension.

4 CONCLUSIONS

Based on the test results of this investigation the following conclusions can be drawn:

1. Wrapping plain concrete specimens with an externally applied 20 mm thick high strength ferrocement laminate containing two or four layers of WWMs seems to provide an effective confinement in terms of axial load carrying capacity and lateral displacement, and ductile failure mode compared to control specimens.
2. Test results indicated that wrapping cylinders of 130 mm diameter and 300 mm height, with 2 layers of WWM ferrocement jackets showed about 16% increase in axial load capacity that is associated with about 32% increase in axial strain and 30% increase in lateral displacement for specimens wrapped with 2 layers of WWM ferrocement jackets compared to control specimens without jackets; whereas, such percentages increase for the specimens wrapped with 4 layers of WWM showed about 30% increase in axial stress, 70% increase in axial strain and about 163% increase in lateral displacement.
3. Based on the findings of the present investigation, high strength ferrocement laminates (jackets) containing a specific number of welded wire meshes can be considered as a promising material for maintenance and rehabilitation of concrete structures.

ACKNOWLEDGEMENTS

The author would like to acknowledge the support received from the research center of Engineering College at King Saud University.

REFERENCES

ACI COMMITTEE 318, 2005. Building Code Requirements for Structural Concrete, (ACI 318M-05). American Concrete Institute, Farmington Hills, Mich., USA.

ACI Committee 549, 1988. Guide for the design, construction, and repair of ferrocement, ACI Structural Journal 85(3): 325–351.

Fahmy, E., Shaheen, Y., & Korany, Y., 1997. Use of Ferrocement laminations for Repairing Reinforced Concrete Slabs, Journal of Ferrocement 27(3): 219–232.

Mansur, M., & Paramasivam, P., 1990. Ferrocement Short Column under Axial and Eccentric Compression, ACI Structural Journal, 87(5): 523–529.

Naaman, A.E., 2000. Ferrocement and laminated cementitious composites, TechnoPress 3000, Ann Arbor, Michigan, 372 pages.

Ong, G., Paramasivam, P., and Lim, E., 1992. Flexural Strengthening of reinforced Concrete Beams Using Ferrocement Laminate, Journal of Ferrocement, 22(4): 331–342.

Razvi, S., & Saatcioglu, M., 1989. Confinement of Reinforced Concrete Columns with Welded Wire Fabric, ACI Structural Journal, 86(5): 615–623.

Shannag, M. Jamal, 2008. Bending behavior of ferrocement plates in sodium and magnesium sulfates solution, Cement and concrete Composites, 30, 597–602.

Shannag, M.J., & Bin Ziyyad, T., 2007. Flexural response of ferrocement with fibrous cementitious matrices, Construction and Building Materials 21, 1198–1205.

Wang, S., Naaman, A.E., & Li, V.C., 2004. Bending Response of Hybrid Ferrocement Plates with Meshes and Fibres, Journal of Ferrocement, 34(1): 275–288.

Winokur, A., & Rosenthal, I., 1982. Ferrocement in Centrally loaded Compression Elements, Journal of Ferrocement 12(4): 357–364.

Bonding characteristics between FRP and concrete substrates

H.A. Toutanji, M. Han & J.A. Gilbert
Civil and Environmental Engineering Department, the University of Alabama in Huntsville, Huntsville, AL, USA

ABSTRACT: This study attempts to create a more accurate model for predicting the interfacial bond behavior between FRP and the concrete substrate. A simple maximum shear stress model was developed by integrating a shear-slip power law curve. The derivation is completed by combining the integrated results, which represent the interfacial surface energy, and the different surface energy expressions. The ratio of τ_{max} over s_0 is assumed as a function with respect to six material properties: the elastic modulus of the FRP plate E_f, the thickness of the FRP plate t_f, the shear modulus of the adhesive layer G_a, the thickness of the adhesive layer t_a, the shear modulus of the concrete substrate G_c, and the reference thickness of the concrete substrate t_{ref}. Few studies have reported that the shear modulus and the thickness of the adhesive layer are important properties in bond models. The current study takes into consideration the shear stiffness of the adhesive layer and develops an equation for calculating the ratio of τ_{max} over s_0 based on available data from previous studies.

1 INTRODUCTION

In previous studies, the failure modes were categorized into two groups: 1) where the full composite action of the beam is developed and 2) where premature debonding occurs (Toutanji et al. 2007). In the second group, the FRP cannot function as it should. The energy loss between the concrete and the composite during debonding prevents the strengthened beam from reaching its ultimate flexure capacity. Thus, the debonding failure mode continues to be the focus of most related studies. Generally, most previous research about debonding failure models can be categorized into three groups: 1) empirically derived models from a large amount of experimental data; 2) theoretical fracture analysis models; and 3) fracture mechanics based models with empirically derived parameters. In the first group, all models are based on simple tests such as single or double lapped shear tests, which are designed for obtaining certain parameters. Unfortunately, the real failure mechanism is much more complicated than the designed tests. In order to apply empirical models, many usage limitations must be set for each specific case. Otherwise, predicted results would seriously deviate from reality. The theoretical models were derived from fracture mechanics which reflect interfacial responses accurately and critically describe failure initiation and propagation. Obviously, these types of models, which involve many parameters, are not practical for actual engineering design. Fracture mechanics based models with empirically derived parameters will be the focus of this paper.

2 FRACTURE MECHANICS MODEL τ_{max}

According to Nakaba et al.'s study (2001), the shear-slip response relationship (Fig. 1) can be expressed by the following power law:

$$\frac{\tau}{s} = \frac{\tau_{max}}{s_0}\left[\frac{n}{(n-1)+(\frac{s}{s_0})^n}\right] \qquad (1)$$

Where $n = 3$ for concrete with a compressive strength range over 24–58 MPa. $s_0 =$ relative slip between the concrete and the FRP corresponding to maximum bond stress, $\tau_{max} =$ maximum shear stress. Substituting $n = 3$, the expression for the shear-slip response in Equation 1 becomes:

$$\tau = \tau_{max}\frac{s}{s_0}\left[\frac{3}{2+(\frac{s}{s_0})^3}\right] \qquad (2)$$

The area beneath the curve (Fig. 1) is equal to the value of the interfacial fracture energy G_f:

$$G_f = \int_0^\infty \tau_{max}\frac{s}{s_0}\left[\frac{3}{2+(\frac{s}{s_0})^3}\right]ds \qquad (3)$$

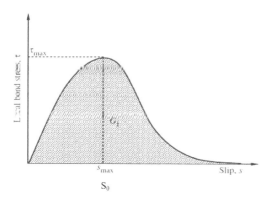

Figure 1. Popvic's expression based stress-slip relationship concrete and FRP (Nakaba et al. 2001).

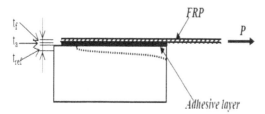

Figure 2. The interface deformation of FRP-concrete material system under shear test.

The ratio of the maximum shear, τ_{max}, over the corresponding slip, s_0, depends upon the properties of the concrete-adhesive-composite material system and is independent of the slip variation ds. Thus, this integration, Equation 3 becomes Equation 4:

$$G_f \approx \tau_{max} s_0 (2.214) \qquad (4)$$

Assuming that the maximum shear stress, τ_{max}, and the corresponding slip, s_0, have linearly increasing relationships and the ratio of τ_{max} over s_0 is related to the material system properties.

Figure 2 shows a simple description of the material system in a standard pure shear debonding test. The interface of the FRP-concrete material system consists of the FRP composite (thickness t_f), the adhesive layer (thickness t_a) and the concrete substrate (reference thickness t_{ref}), as shown in Figure 2. Each one of them affects the interfacial bond behavior and the ultimate transferable load. Thus, the ratio of τ_{max} over s_0 should be a function that includes variations of the FRP, the adhesive and the concrete substrate. In a pure shear test, the FRP sheet is under tension. This tension in the FRP plate is transferred to the concrete surface mainly through shear stresses in the adhesive layer. Thus, the dominant stress is tension in the FRP, and shear stress in both the adhesive layer and the concrete substrate within a very thin layer adjacent to the adhesive. Assuming the ratio of τ_{max} over s_0 is a function

of "R". "R" is a function of three variations: $E_f t_f$, K_a, and K_c.

$$s_0 = \frac{\tau_{max}}{R(E_f t_f, K_a, K_c)} = \frac{\tau_{max}}{R(E_f t_f, K')} = \frac{\tau_{max}}{(E_f t_f)^\alpha K'} \qquad (5)$$

Where α is the exponential parameter of the stiffness of the FRP plate and K' is the equivalent material system shear stiffness. Material shear stiffness is defined as $K = G/t$, where G is the material shear modulus and t is the material thickness. The material system, however, includes adhesive and concrete substrate. Thus, K' can be expressed as:

$$K' = \frac{1}{\dfrac{1}{K_c} + \dfrac{1}{K_a}} = \frac{K_c K_a}{K_c + K_a} \qquad (6)$$

Where G_a is the shear modulus of the adhesive; G_c is the shear modulus of the concrete; t_a is the adhesive thickness; t_{ref} is the reference distance in the concrete where it is influenced by the shear stress exerted by the FRP. The concrete shear modulus G_c can be expressed by the elastic modulus of concrete E_c and the Poisson's Ratio of concrete υ. Thus, the shear modulus can be expressed as $G_c = E_c/2/(1 + \upsilon)$.

To be able to define the parameter α in the function $R(E_f t_f, K')$, a database containing 24 test data was collected from two independent studies (Dai et al. 2005, Bizindavyi & Neale 1999). This database contains five different fibres and six epoxy systems including one type of primer. The shear stiffness, G_a/t_a, of the bond layer (adhesive layer, or adhesive + primer layer if the primer had been used) ranges from $0.45\,GPa$ to $1.14\,GPa$. A previous study shows that the bond strength of the FRP sheet-concrete interface can be efficiently enhanced by applying a soft adhesive layer which has shear stiffness between $0.14\,GPa$ and $1.0\,GPa$ (Lu et al. 2001). However, the properties of the adhesive layer were not always reported in the existing studies and none of these studies were particularly focused on soft adhesive. Therefore, based on the available data, the current study focuses on interfacial bond behavior, and in particular, maximum bond stress with a relatively soft adhesive layer.

The strain distribution was directly obtained from references (Dai et al. 2005, Bizindavyi & Neale 1999). The maximum shear and the corresponding slip within the FRP-concrete interface under any stage of load were calculated in the course of this study. It was found from available data that the maximum shear occurs within the area approximately $25\,mm$ away from the loaded end when the current external load is approximately 66% of the ultimate load. Experimental observation shows that the ratio of the maximum shear τ_{max} over the corresponding slip s_0 ranges from 79.6 to 109.4, and the median value is 94.1 as shown in Figure 3.

Following the assumption in Equation 5, it was found that when α is equal to 0.155 (with the standard deviation of 0.018 based on 24 data), the function

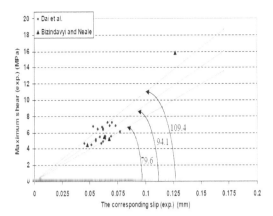

Figure 3. The ratio of the maximum shear (exp.) over the corresponding slip (exp.).

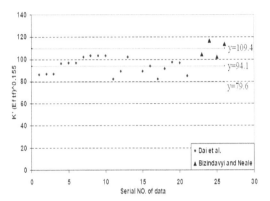

Figure 4. The assumed value for the ratio of maximum shear over the corresponding slip for each available data.

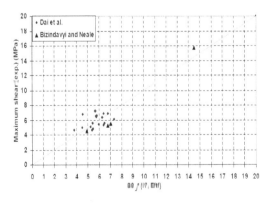

Figure 5. The comparison of $\tau_{max(exp.)}$ and $S_{0(exp.)}[(E_f t_f)^{0.155} K']$.

$R(E_f t_f, K')$ best fits the criteria (Fig. 4). Figure 5 shows the comparison of $\tau_{max(exp.)}$ and $S_{0(exp.)}[(E_f t_f)^{0.155} K']$. According to a statistical evaluation of Figure 5, the ratio of $S_{0(exp.)}[(E_f t_f)^{0.155} K']/\tau_{max(exp.)}$ has a mean

value, a median value and a standard deviation of *1.02, 1.00* and 0.17, respectively.

Therefore,

$$s_0 = \frac{\tau_{max}}{R(E_f t_f, K')} = \frac{\tau_{max}}{K'(E_f t_f)^{0.155}} \qquad (7)$$

After s_0 is defined, the fracture energy expression can be further described as:

$$G_f \approx \tau_{max} s_0 \cdot 2.214 = 2.214 \frac{(\tau_{max})^2}{R(E_f t_f, K')} = 2.214 \frac{(\tau_{max})^2}{K'(E_f t_f)^{0.155}} \qquad (8)$$

According to the work done by Toutanji et al. (2007), interfacial fracture energy can be described by the bilinear model:

$$G_f = 0.014 f_c' \qquad \text{when } 0 \leq f_c' \leq 46.2\,\text{MPa} \qquad (9)$$

$$G_f = 0.65 \qquad \text{when } f_c' \geq 46.2\,\text{MPa} \qquad (10)$$

Substituting these relations into Equation 8

$$\tau_{max} = 0.08\sqrt{f_c'} f(E_f t_f, K') \text{ when } 0 \leq f_c' \leq 46.2\,\text{MPa} \quad (11)$$

$$\tau_{max} = 0.55\sqrt{f(E_f t_f, K')} \text{ when } f_c' \geq 46.2\,\text{MPa} \quad (12)$$

3 CONCLUSIONS

For the purpose of engineering design, a simple maximum shear stress model has been developed by integrating the Nakaba shear-slip power law curve. The derivation is done by combining the integration result, which represents interfacial surface energy, and the same physical parameter which was derived in Toutanji et al.'s previous work (2007). The mechanical properties of the adhesive layer, the concrete strength, and the stiffness of the FRP have been taken into account in the new model.

The newly proposed model is in agreement with other independent studies (Yuan &Wu 1999, Lu et al. 2005, Dai et al. 2005) as follows:

1. The maximum bond stress τ_{max} increases almost linearly with $(f_c')^{0.5}$. (However, Equation 12 indicates that τ_{max} is independent of the concrete strength after f_c' is higher than 46.2 MPa.).
2. With decreasing shear stiffness of the adhesive layer K_a, both the maximum interfacial bond stress and the shear stiffness of the material system decrease, which leads to an improvement of the transferable load capacity.
3. The maximum bond stress slightly increases with increasing FRP stiffness. This is the main reason that most existing models have completely ignored the contribution of FRP. For the sake of accuracy, the new model takes into consideration the effect of the FRP plate even though it is small ($E_f t_f$ is only to the power of 0.155).

REFERENCES

Bizindavyi, L. & Neale, K. W. 1999. Transfer lengths and bond strengths for composites bonded to concrete. *Journal of Composite Construction* 3(4): 153–159.

Dai, J. et al. 2005. Development of the nonlinear bond stress-slip model of fibre reinforced plastics sheet-concrete interfaces with a simple method. *Journal of Composite Construction* 9(1): 52–62.

Han, M. et al. 2008. Bond behavior between FRP composites and RC beams based on fracture mechanics. *Forth International Conference on FRP Composites in Civil Engineering (CICE2008)*, 6 pages.

Lu, X. Z. et al. 2005. Bond-slip models for FRP sheets/plates bonded to concrete. *Engineering Structures* 27(6): 920–937.

Nakaba, K. et al. 2001. Bond behavior between fibre-reinforced polymer laminates and concrete. *ACI Structural Journal* 98(3): 359–367.

Toutanji, H. A. et al. 2007. Prediction of interfacial bond failure of FRP-concrete surface. *Journal of Composites for Construction* 11(4): 427–432.

Yuan, H. & Wu, Z. 1999. Interfacial fracture theory in structures strengthened with composite of continuous fibre. *Proceedings of Symposium of China and Japan: Science and Technology of the 21st Century*: 142–155.

Testing

Concrete Solutions – Grantham, Majorana & Salomoni (Eds)
© 2009 Taylor & Francis Group, London, ISBN 978-0-415-55082-6

The potential of terrestrial laser scanning for detecting the deterioration of building facades

F. Al-Neshawy, S. Peltola, J. Piironen, A. Erving, N. Heiska, P. Salo & M. Nuikka
Helsinki University of Technology, Finland

A. Kukko
Finnish Geodetic Institute, Masala, Finland

J. Puttonen
Helsinki University of Technology, Finland

ABSTRACT: Deterioration of building facades is a common phenomenon with multi-storey buildings. The objective of this paper is to explore the potential use of the terrestrial laser scanning technique to detect the deterioration on the surfaces of the building facades and to quantitatively measure the dimensions of the damaged areas. This paper is focussed on the detection of the bowing of marble cladding and the surface delamination of brick building facades. Field measurements were carried out using a terrestrial laser scanner and a tacheometer as reference. Measurements of the bowing of the marble panels were also carried out manually with a so-called "bow–meter". The results show that the terrestrial laser scanning technique gives an accurate and reasonable method for measuring the bowing of marble panels and the delamination of brick building facades. Terrestrial laser scanning is not a replacement for the existing condition survey techniques, but an alternative, which can be employed to complete many surveying tasks on large surfaces because of the spatial coverage of the point clouds and non-touching measurement principle.

1 INTRODUCTION

The reasons for deterioration of building structures are multiple and involve complex interactions between natural processes, the service environment, the material properties and the quality in design, detailing and construction. The deterioration of building structures can be classified generally into defect, damage and deterioration. [Huovinen et al. 1998] The deterioration of the building facades is typically assessed by visual inspection and destructive and non-destructive tests. The disadvantages of manual inspections lie in the fact that they are expensive, produce a huge amount of data and are time consuming. It is clear that a faster, more objective survey would offer a desirable alternative to these condition surveys.

The Terrestrial laser scanning technology has been well developed in recent years. The scan speed exceeds tens of thousands points per second. High performance scanners are used widely to record defect information and structural damage. The detail information on the defects recorded by the 3-D laser scanners can be used in a digital format. The digitized image can be further manipulated using colouring schemes to magnify the defects [Chang et al. 2008]. As a result, terrestrial laser scanning is becoming more feasible as a data collection method for applications in different industrial and construction fields.

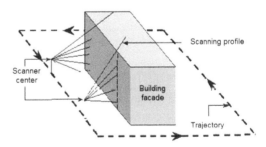

Figure 1. The principle of terrestrial laser scanning of building facades.

The data received from laser scanning has been widely used in survey applications, global positioning, maintenance of historical sites and structural monitoring. The last two decades have seen the emergence of various scanning technologies and various defect inspection methods and algorithms have been developed using terrestrial laser scanning. [Liya 2006] & [Van Gosliga et al. 2006]

Terrestrial laser scanning is a special technique in which the building is scanned as shown in Figure 1. The result is a high-resolution cloud of points. Terrestrial laser scanning systems use a directed laser beam for distance measurement to collect spatial information.

A laser scanner creates a model consisting of a large number of points with x y z coordinates. The point cloud is a regularly sampled spatial representation of the real world.

The x y z coordinates relate the points, measured on real world objects, to the origin of the scanner, or more often to a project coordinate system used to tie several scans together. The ability of a laser scanner to capture large amounts of data quickly and with a fine resolution means that the real world can be accurately modelled. [Bornaz et al 2004]

The results of this research show that the terrestrial laser scanning technique gives an accurate and a reasonable method for measuring the bowing of marble panels and the delamination of brick building facades, which could be also used for measuring the bowing and defects of the concrete elements of building facades.

2 METHODS AND MEASUREMENTS

2.1 Calculation methods

This paper is focusing on detecting the bowing of marble cladding and the surface delamination of brick building facades using a terrestrial laser scanning system. Deterioration of marble panels involves several parameters and properties. Shape deformation is the most obvious phenomenon, where the panels bow either convexly or concavely out of their original plane. Along with bowing follows also permanent volume changes i.e., the marble expands. [Grelk et al. 2007] The bowing of marble panels was calculated using equation 1.

$$B = \left(\frac{d}{L}\right) * 1000 \tag{1}$$

where B = the bowing expressed in (mm/m); d = the measured value of bowing in (mm); and L = the measuring distance in (mm).

The bowing of the concave and convex marble panels was calculated by fitting a second order curve to the laser scanning data in the vertical and horizontal direction as shown in Figure 2.

The value of bowing d, (mm) was calculated with equation 2.

$$d = y_{max} - \left(y1 + \frac{(y_n - y_1) * (x - x_1)}{x_n - x_1}\right) \tag{2}$$

The value of the measuring distance L in (mm) was calculated with equation 3.

$$L = \sqrt{(y_n - y_1)^2 + (x_n - x_1)^2} \tag{3}$$

where y_{max}; y_1; y_n; x; x_1; and x_n are shown in figure 2.

The most common signs of damage in masonry walls are deterioration of joint mortar and delamination and cracking of exterior brick units. Erosion and

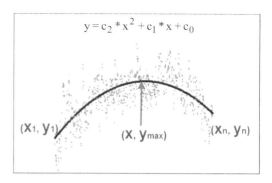

Figure 2. Quadratic polynomial curve fitting of laser scanning data.

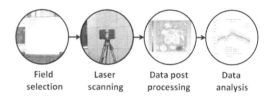

Field selection | Laser scanning | Data post processing | Data analysis

Figure 3. Flow chart of the laser scanning technique.

cracked mortar is the result of freeze-thaw cycles and the attack of harmful chemicals. The aging process accelerates if the mortar is not sufficiently compacted, or if the joint profile allows the water to stay there. When this water freezes, it damages both mortar and brick. Degradation of brick often involves delamination and cracking of the exterior brick surfaces as a result of weathering and freeze-thaw cycles. [Newman 2001]

2.2 Field measurements

The field measurements were carried out on two selected brick and marble panel building facades. A flow chart for the execution sequence of the terrestrial laser scanner technique is shown in Figure 3.

The field work was carried out with a FARO LS 880HE80 terrestrial laser scanner, Leica TCA2003 S.No 438743 robot tachymeter and manually. The distance measurement of the terrestrial laser scanner is based on a phase difference technique. A systematic distance error is +/− 3 mm when measuring at a distance of 25 meters. The achieved pixel resolution for the terrestrial laser scanner was approximately a 1 mm grid, as the scanner was three meters away from the object. Additionally, for the marble wall, few separate point clouds were merged together with the help of several target circles located around the wall.

The filtering of noise and deletion of additional and unnecessary points of the point clouds were performed with FARO Scene software. Geomagic Qualify and Realworks Survey software were used for further analysis of the data. For the purpose of a more in-depth analysis with other mathematical software, the coordinate system of the point cloud was transformed. The

Figure 4. FARO LS 880HE80 terrestrial laser scanning system.

Figure 5. Leica TCA2003 tachymeter and the location of measured marble panels.

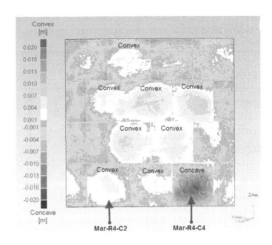

Figure 6. Terrestrial laser scanning data, colouring by the magnitude of the deformation from the planarity.

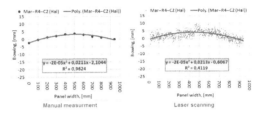

Figure 7. The horizontal convex bowing in marble panel Mar-R4-C2.

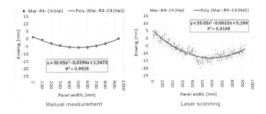

Figure 8. The horizontal convex bowing in marble panel Mar-R4-C4.

origin of the system was transferred to the lower left corner of the fitted plane of the facade. Likewise, the point cloud was thinned out in order to manage the measurement data better, resulting in 4, 5 and 7 mm grids to the data.

A tachymetric survey was performed to measure the three dimensional location of points on the marble cladding building facade as a reference for the laser scanning survey. Three locations on the wall were selected for the measurement of bowing, this selection being made according to the amount of observed bowing. Nine Leica reflective sticker targets (20×20 mm) in 3×3 regular patterns were fixed on each of the three marble panels. The measured value of bowing was computed using Leica Axyz software (version 1.3).

Measurements of the bowing of the marble panels were also carried out with a so-called "bow–meter". The "bow–meter" is a 1400 mm straight bar with a digital calliper that allows the distance from the edge of the bar to the panel surface to be measured accurately.

3 RESULTS AND DISCUSSION

3.1 Bowing of marble panels

As seen in Figure 6, eight of the examined panels demonstrated convex bowing and one panel demonstrated concave bowing. A reason for the panel bowing could be water penetrating behind the marble panels, which increases and speeds the deterioration of the structure behind the panels. Failure of the lateral fixing of the panels could also be a reason for the marble panel bowing.

Results of the onsite bowing measurements acquired with the laser scanning system and manually are shown by horizontal cross-sections for two of the marble panels in Figure 7 and Figure 8.

Table 1 presents the results of the onsite bowing measurements using terrestrial laser scanner,

Table 1. Results of measurements carried out on the marble facade using terrestrial laser scanning system tachymeter and bow-meter.

Method	Marble panel	Type of bowing	B (mm/m)
Laser scanning	Mar–R4–C2 (Hal)	Convex	6
	Mar–R4–C2 (Val)	Convex	9
	Mar–R4–C4 (Hal)	Concave	−6
	Mar–R4–C4 (Val)	Concave	−9
Tachymeter	Mar–R4–C2 (Hal)	Convex	8
	Mar–R4–C2 (Val)	Convex	8
	Mar–R4–C4 (Hal)	Concave	−9
	Mar–R4–C4 (Val)	Concave	−10
Bow-meter	Mar–R4–C2 (Hal)	Convex	5
	Mar–R4–C2 (Val)	Convex	7
	Mar–R4–C4 (Hal)	Concave	−6
	Mar–R4–C4 (Val)	Concave	−7

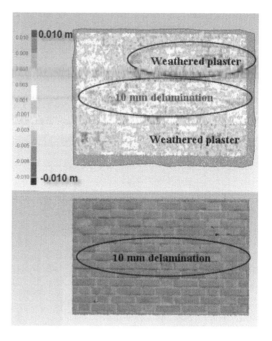

Figure 9. Terrestrial laser scanning data, colouring by the magnitude of the delamination of the bricks.

tachymeter and bow-meter. The comparison between the results from laser scanning, tachymeter survey and bow-meter showed a slight difference.

The magnitude of convex bowing (B = 6–9 mm/m) was measured on the panel (Mar–R4–C2) on the fourth row of panels from the top. On the bottom row of panels, the magnitude of concave bowing (B = −6– −9 mm/m) was measured on the panel (Mar–R4–C4). The high magnitude of the concave bowing could be due to the failure of the panel anchoring to the wall.

3.2 Surface delamination of bricks

Terrestrial laser scanning technique was used to determine the surface delamination of bricks. The development of an algorithm for exhibiting a wide range of features related to building facade deterioration including the delamination of bricks is under progress, but is out of the scope of this paper. However, Figure 9 shows an example of the delamination of bricks and the deterioration of joints detected from the laser scanning data. The maximum delamination of the bricks was about 10 mm out of the wall. The weathered plaster of the joint is about 10 mm depth. The surface delamination of the bricks could be the result of salt crystallization or weathering of the joints.

4 CONCLUSIONS

In this research, we proposed the terrestrial laser scanning technique as an automatic deterioration detecting system, which can be used on large surfaces because of the spatial coverage of the point clouds and non-touching measurement principle.

In this paper, we investigated mainly the bowing of marble cladding panels and the delamination of the brick building facades using laser scanning technology. The field work was carried out with a bow-meter, FARO LS 880HE80 terrestrial laser scanner and Leica TCA2003 S.No 438743.

The terrestrial laser scanning technique gives an accurate and a reasonable method for measuring the bowing of marble panels. The comparison between the results from laser scanning, tachymeter survey and bow-meter showed a slight difference. The surface delamination of bricks and the deterioration of joints is also detected using the laser scanning system.

Laser scanning is not a replacement for existing condition survey techniques, but an alternative which provides location based information on the building defects and deterioration.

ACKNOWLEDGMENTS

The authors would like to thank Hannu Heinonen from Nordic Geo Center Oy for co-operation and providing software for processing the laser scanning data.

REFERENCES

Huovinen, S., Bergman, J. and Hakkarainen, H. (1998). Deterioration defects and repair methods of facades, Helsinki university of technology, *Laboratory of structural engineering and building physics, Report 78*, 69 p.

Chang K-T., Wang E., Chang Y-M. and Cheng H-K. (2008). Post-Disaster Structural Evaluation Using a Terrestrial Laser Scanner. FIG Working Week 2008. 14–19 June 2008, Stockholm, Sweden. Available online at: *http://www.fig.net/pub/fig2008/papers/ts05c/ts05c_06_chang_etal_2905.pdf* [Accessed 14.10.2008].

Liya T. (2006). Automated Detection of Surface Defects on Barked Hardwood Logs and Stems Using 3-D Laser

Scanned Data. Doctoral Dissertation. Faculty of the Virginia Polytechnic Institute and State University. Available online at: *http://scholar.lib.vt.edu/theses/available/etd-09202006-145847/unrestricted/dissertation-lithomas-06.pdf* [Accessed 14.10.2008].

Van Gosliga, R., Lindenbergh, R., Pfeifer, N. (2006) Deformation analysis of a bored tunnel by means of terrestrial laser scanning, ISPRS Technical Commission Symposium, September 2006, Dresden, Germany. *International Archives of Photogrammetry, Remote Sensing and Spatial Information Systems,* Volume XXXVI, Part 5, pages 167–172.

Bornaz, L. and Rinaudo, F. (2004). Terrestrial Laser Scanner Data Processing. *International archives of photogrammetry remote sensing and spatial information sciences,* Vol 35, Part 5, pp. 514–519.

Grelk, B., Christiansen, C., Schouenborg, B. and Malaga K. (2007). Durability of Marble Cladding – A Comprehensive Literature Review. *Journal of ASTM International, Vol. 4,* No. 4. Available online at: *http://www.astm.org/JOURNALS/JAI/PAGES/289.htm* [Accessed 14.10.2008].

Hall D. (2008). Building defects, Inspections and Reports. Available online at: *http://www.buildingdefects.com.au/defect-salt-deterioration.html* [Accessed 14.10.2008].

Newman A. (2001). Structural Renovation of Buildings: Methods, Details, and Design Examples. *McGraw-Hill, New York, pp.* 505–539.

Concrete Solutions – Grantham, Majorana & Salomoni (Eds)
© 2009 Taylor & Francis Group, London, ISBN 978-0-415-55082-6

Monitoring the chloride concentration in the concrete pore solution by means of direct potentiometry

U. Angst, C.K. Larsen & Ø. Vennesland
NTNU Norwegian University of Science and Technology, Faculty of Engineering Science and Technology, Department of Structural Engineering, Trondheim, Norway

B. Elsener
ETH Zurich, Institute for Building Materials, Switzerland

ABSTRACT: Determination of the chloride concentration at the depth of the rebar is important when studying the mechanisms of chloride-induced corrosion. It is generally assumed that only the chloride ions dissolved in the concrete pore solution can initiate corrosion, while those bound by the constituents of the cement paste are considered harmless. Attempts to use ion selective electrodes in concrete, to non-destructively monitor the free chloride concentration, have thus been made by several researchers. In the present work, results from laboratory experiments with such chloride sensors are presented. It was found that the presence of concentration gradients such as pH differences markedly influences the measurement of the sensor potential and thereby the accuracy of this chloride measurement technique. This has to be taken into account carefully when designing an experimental setup involving the use of ion selective electrodes.

1 INTRODUCTION

There are many reasons for the presence of chloride in reinforced concrete. Most commonly, marine exposure conditions or use of road salt lead to ingress of chloride into a structure. The transport occurs through the pores of the concrete, where chloride ions dissolved in the pore liquid can move by several mechanisms such as diffusion and capillary suction. Once the reinforcement is reached, these so-called aggressive ions might attack the reinforcement and present a risk for chloride-induced corrosion.

Cement paste removes chloride ions from the pore solution by chemical and physical binding; this fraction of the total chloride content is considered to be rendered harmless with regard to corrosion initiation. The so-called free chlorides, however, are still mobile and available for interaction with the oxide layer or, e.g. in the case of carbonated concrete, directly with the steel surface. Knowledge of the concentration of these free chloride ions is thus valuable when studying chloride-induced corrosion in reinforced concrete.

Several techniques are known to determine the free chloride content in concrete: Pore water expression (Barneyback & Diamond 1981, Tritthart 1989) is often regarded as the most accurate method, but the application is limited to rather porous samples with fine aggregates. Moreover, a certain minimal water content is required in order to obtain enough pore liquid for further analysis. Pore water expression is thus usually not applicable to practical concrete samples as they often contain large aggregates and might be in a too dry moisture state. Another approach is to immerse crushed or powdered concrete samples in a solvent (e.g. distilled water) for a certain time and subsequently measure the chloride concentration in the solvent. The results of such leaching techniques have been found to be highly dependent on the parameters of the procedure such as type of solvent, temperature, extraction time, and binder type (Arya et al. 1987, Arya 1990). If water is used as solvent, the leached out amount of chloride is often called "water soluble chloride" and assumed to be equal to the free chloride concentration. Both pore water expression and leaching techniques require destructive sampling and can thus not be used for continuous monitoring of the free chloride content. In addition, these methods give average values of the sample volume under investigation, which leads to erroneous results in the case of concentration gradients.

Having these drawbacks in mind, attempts have been made to measure the free chloride concentration in the pore solution by means of embedded ion selective electrodes (Atkins et al. 1996, Elsener et al. 1997, Elsener et al. 2003). Silver/silver chloride electrodes are sensitive to (primarily) chloride ions and ideally exhibit a certain electrochemical potential that depends on the chloride ion activity in the solution and on temperature. Thanks to the small dimensions of such chloride sensors they allow highly localised measurements, e.g. accurately at the depth of the reinforcement, rather than average values over the comparatively large concrete volume under investigation when using other techniques. Moreover, the chloride

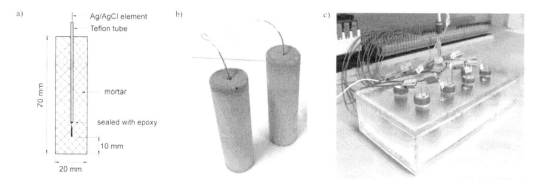

Figure 1. Sample geometry and experimental setup during immersion.

concentration can be monitored non-destructively and continuously.

In the present paper, results from laboratory experiments with Ag/AgCl electrodes embedded in mortar samples are reported. A major error source, namely diffusion potentials, for application of this technique has been identified.

2 EXPERIMENTAL

2.1 Mortar samples

Two series of mortar samples were cast with two different w/c ratios of 0.4 and 0.6. The maximum grain size of the sand was 2 mm and the mix proportions by mass were cement : sand = 1 : 3. Ordinary Portland cement (CEM I 52.5 N) and small amounts of superplasticizer (BASF Glenium 151) in the range 0.5–1.0% by cement mass were used. The samples were of cylindrical shape with diameter 20 mm and length 70 mm. A chloride sensor was mounted in a Teflon tube and placed centrally with a distance of 10 mm to the bottom face of the mortar sample (Figure 1a,b). The chloride sensors consisted of a silver wire covered with silver chloride on the tip; they were produced by a manufacturer of commercial reference electrodes.

One day after casting, the samples were demoulded, wrapped in plastic and kept at 50°C for one week (to accelerate hydration) and thereafter at room temperature. At ages of 28 days (w/c = 0.6) and 21 days (w/c = 0.4) the plastic was removed and the samples were dried at 50°C for 2 days and then stored at room temperature until immersion.

2.2 Chloride uptake

Synthetic pore solutions with different amounts of sodium chloride were prepared according to Table 1. The composition of the solutions was selected based on pore water expression on similar specimens in order to minimise differences in alkalinity between the mortar samples and the storage solutions.

Three mortar samples with w/c = 0.4 and three with w/c = 0.6 were immersed in each solution. In addition,

Table 1. Solutions for immersion tests.

Set	Artificial pore solution	Chloride conc
A	0.2 M KOH + 0.15 M NaOH + sat. Ca(OH)$_2$	0.1 M NaCl
B	0.2 M KOH + 0.15 M NaOH + sat. Ca(OH)$_2$	0.3 M NaCl
C	0.2 M KOH + 0.15 M NaOH + sat. Ca(OH)$_2$	0.5 M NaCl

chloride sensors covered with a 1–2 mm thin layer of cement paste (approximately w/c = 0.4) and blank sensors were also immersed (three sensors of each type per solution) (Figure 1c). The potential of the chloride sensors was logged vs. a saturated calomel electrode (SCE) that was placed in the same solution during the measurement. For every combination of solution and type of immersed sample, three parallel potentials were recorded by the data-logging equipment. The temperature in the laboratory was constantly at 20–22°C.

3 RESULTS AND DISCUSSION

3.1 Capillary suction

Figure 2 depicts the results from data-logging during the first hours directly after immersion in the solution with 0.1 M NaCl. The results for the other two solutions with 0.3 M and 0.5 M NaCl were similar. Both the blank chloride sensors and those covered with cement paste reached their equilibrium potential relatively fast (Figure 2a): After ca. 1 h these sensors exhibited a potential around 45–50 mV SCE, which corresponds to the chloride concentration of 0.1 M.

In the case of embedded sensors (Figure 2b), the initially recorded potentials do not reflect the chloride concentration at the electrodes since other effects resulting from capillary suction are dominant and make the potential readings unreliable. The electrical resistivity in the initially dry samples leads to a large *IR*

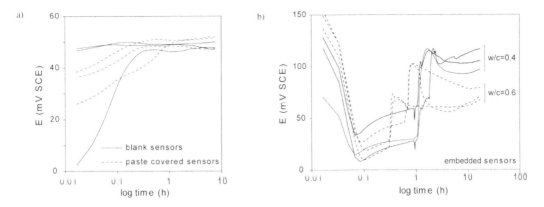

Figure 2.   Observed chloride sensor potentials immediately after immersion in synthetic pore solution + 0.1 M NaCl.

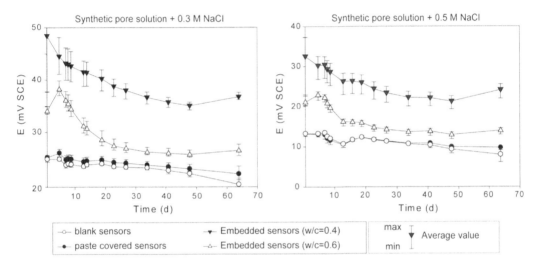

Figure 3.   Observed chloride sensor potentials (average of three values) vs. time.

drop and no stable potentials can be measured. After a while, the resistivity decreases, but then the movement of liquid through the pore system of the permselective cement paste leads to streaming potentials (Myrdal 1997) and dominates the measurement. It has also been reported by other authors that moisture gradients in concrete significantly affect potential measurements (Arup & Klinghoffer 1997, Schiegg et al. 2007).

The sharp potential increase shown in Figure 2b might be associated with chloride containing solution reaching the sensor. This happens earlier in the more porous sample with w/c = 0.6. Subsequently, the values appear to stabilise around 100 mV SCE in the case of samples with w/c = 0.4 and around 70 mV SCE in the case of samples with w/c = 0.6. These potentials are higher than those exhibited by the blank and paste covered sensors, and, since higher potentials correspond to lower chloride concentrations, indicate that the solution reaching the Ag/AgCl electrodes has a lower chloride content than the storage solution.

This is as expected since chloride is adsorbed by the constituents of the mortar when penetrating the samples.

3.2   Later stages

After the initial water suction phase, potentials were measured periodically. The results for the first 2 months are shown in Figure 3 for the samples immersed in 0.3 M and 0.5 M NaCl containing solutions (the results in the case of 0.1 M NaCl were similar). The potentials of all the embedded sensors gradually decrease over time, whereas the blank and cement paste covered sensors remain rather constantly at their equilibrium potentials. After ca. 30 days, a tendency towards lower values is observed for the latter; at about the same time, the embedded sensor potentials stop decreasing and later even start to increase. Both phenomena, the potential decrease in the case of blank and paste covered sensors as well as the potential increase of the embedded sensors, can be explained

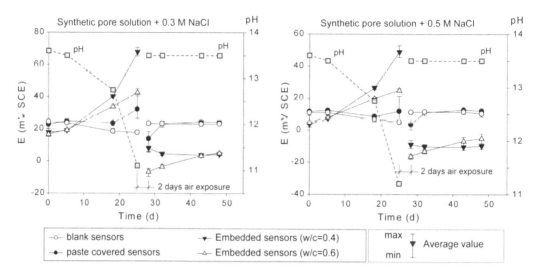

Figure 4. Observed chloride sensor potentials (average of three values) vs. time after renewal of the solutions and measured pH.

by CO_2 being slowly dissolved in the storage solutions and thereby lowering the pH.

A lower pH leads to an increase in chloride activity, which is reflected by the tendency towards slightly lower potentials of the sensors that are in direct contact with the solution (Figure 3). In the case of sensors covered with cement paste the effect is less pronounced, which might be due to local buffering of the pH by the layer of alkaline hydration products on the electrode surface.

Carbonation of the solutions also has an effect on the potential readings in the case of embedded sensors: a decrease in pH in the external solution results in a pH gradient towards the embedded chloride sensor, since the sample will still be highly alkaline in its interior. As a result of such a concentration gradient, diffusion potentials E_{dif} are established between the external solution and the chloride sensor. This value adds to the true potential of the chloride sensor and leads to an erroneous measurement. It can be calculated from theoretical considerations (Angst et al. 2009) that E_{dif} is positive in the case of a lower pH in the external solution, i.e. the recorded potential is higher than the true potential of the chloride sensor (as was observed in the current experiments).

These effects of pH differences were confirmed during two later stages, where 1) carbonation of the solution was provoked by leaving the box shown in Figure 1 open and thus guaranteeing access to air, and 2) carbonation was eliminated by keeping the complete setup in a closed environment with artificial CO_2-free air (79.2% nitrogen + 20.8% oxygen). For both stages, the solutions were initially renewed and the pH was subsequently monitored by means of indicator strips and use of a glass electrode. The results of the potential and pH measurements are depicted in Figure 4. In the first stage, the pH decreased from initially ca. 13.5 to about 11.0 within 25 days; in the second stage,

carbonation was successfully prevented and the pH was constantly high. As is visible from the graph, the measured potentials of the embedded sensors increase simultaneously with the drop in pH, in the case of w/c = 0.4 by ca. 50 mV and in the case of w/c = 0.6 by ca. 20 mV. Upon renewal of the solutions after 25 days, the potentials of the embedded sensors immediately jump to much lower values, even below the potentials of the blank sensors, which would imply a higher chloride concentration in the mortar samples than in the external solution. The sharp drop in potentials is interpreted as follows: by renewing the solutions, the pH gradient is inversed and thus the diffusion potentials have the opposite sign. This interpretation is supported by the fact that diffusion potentials are established immediately, whereas any effects at the sensor itself (in terms of activity) would require diffusion to take place through the nearly saturated mortar samples.

3.3 Implications for the use of chloride sensors

It is important to remember in this context that potentiometric measurements are very sensitive (Angst et al. 2009), Koryta 1972). This follows directly from the slope of Nernst's law, which, at room temperature, has a theoretical value of 59.2 mV/decade. A potential difference of ca. 60 mV corresponds thus to one order of magnitude in terms of chloride concentration; a small potential error of 1 mV equals ca. 4% in chloride concentration (Koryta 1972).

It was shown in the present experiments that pH gradients along the measurement path between the reference electrode and a chloride sensor affect the measurement considerably: diffusion potentials can easily reach a few tens of millivolts. The error in chloride concentration associated with this is unquestionably intolerable. In addition, it has to be kept in mind that any kind of concentration gradient leads to diffusion potentials. For instance, the chloride profile through the mortar samples present in the actual setup theoretically also causes diffusion potential errors. At high pH values, however, these can be considered to be small.

4 CONCLUSIONS

In the present article, experimental results of the application of ion selective electrodes as chloride sensors in concrete have been discussed. The presence of concentration gradients such as pH differences between the reference electrode and the chloride sensors give rise to diffusion potentials that drastically affect the accuracy of the chloride sensors. The sensor itself responds properly to the chloride activity in solution; it is rather the accurate measurement of the sensor potential that presents difficulties when applying ion selective electrodes in concrete. Experimental setups involving Ag/AgCl electrodes for determination of free chloride concentrations have to be designed carefully in order to avoid this error source.

ACKNOWLEDGEMENTS

The authors acknowledge the support of COIN (www.sintef.no/coin).

REFERENCES

Angst, U., Vennesland, Ø., & Myrdal, R. 2009. Diffusion potentials as source of error in electrochemical measurements in concrete. *Materials and Structures* 42: 365–375. (DOI:10.1617/s11527-008-9387-5).

Arup, H., & Klinghoffer, O. 1997. Junction potentials at a concrete/electrolyte interface. In *Proc. EUROCORR '97*. Trondheim, Norway. European Federation of Corrosion.

Arya, C., Buenfeld, N.R., & Newman, J.B. 1987. Assessment of simple methods of determining the free chloride ion content of cement paste. *Cement and Concrete Research* 17: 907–918.

Arya, C. 1990. An assessment of four methods of determining the free chloride content of concrete. *Materials and Structures* 23: 319–330.

Atkins, C.P., Scantlebury, J.D., Nedwell, P.J., & Blatch, S.P. 1996. Monitoring chloride concentrations in hardened cement pastes using ion selective electrodes. *Cement and Concrete Research* 26: 319–324.

Barneyback, R. S., & Diamond, S. 1981. Expression and analysis of pore fluids from hardened cement pastes and mortars. *Cement and Concrete Research*: 11: 279–285.

Elsener, B., Zimmermann, L., Flückiger, D., Bürchler, D., & Böhni, H. 1997. Chloride penetration – non destructive determination of the free chloride content in mortar and concrete. In *Proc. RILEM Int. Workshop "Chloride penetration into concrete"*. Paris.

Elsener, B., Zimmermann, L., & Böhni, H. 2003. Non destructive determination of the free chloride content in cement based materials. *Materials and Corrosion* 54: 440–446.

Koryta, J. 1972. Theory and applications of ion-selective electrodes. *Analytica Chimica Acta* 61: 329–411.

Myrdal, R. 1997. Potential gradients in concrete caused by charge separations in a complex electrolyte. *CORROSION/97*, Houston, NACE, Paper No. 278.

Schiegg, Y., Büchler, M., & Brem, M. 2007. Potential mapping technique for the detection of corrosion in reinforced concrete structures: Investigation of parameters influencing the measurement and determination of the reliability of the method. In *Proc. EUROCORR '07*. Freiburg im Breisgau, Germany. European Federation of Corrosion.

Tritthart, J. 1989. Chloride binding in cement. I. Investigations to determine the composition of porewater in hardened cement. *Cement and Concrete Research* 19: 586–594.

Concrete Solutions – Grantham, Majorana & Salomoni (Eds)
© 2009 Taylor & Francis Group, London, ISBN 978-0-415-55082-6

The detection of micro-cracks in concrete by the measurement of ultrasonic harmonic generation and inter-modulation

P.R. Armitage
Theta Technologies Ltd, The Innovation Centre, University of Exeter

L.V. Bekers
Sia Celtniecibas pakalpojumi, LV, Latvia

M.K. Wadee
School of Engineering, Computer Science and Mathematics, University of Exeter

ABSTRACT: The ultrasonic testing of concrete structures has posed many problems. Conventional methods such as pulse echo and pitch catch are of limited use due to its composition as aggregates will cause scattering and multiple reflections. Alternative ultrasonic methods have been recently investigated that examine the shape of the waveform as it traverses through a complex material, the idea being not to locate a single defect but to determine the overall mechanical properties within a certain region.

In damaged materials, particularly ones that have micro-cracking, the stress-strain relationship does not obey Hooke's Law of elasticity, stress is not proportional to strain. It is not linear, resulting in distortions to a pure ultrasonic sine wave traversing through it. The degree of this distortion is measured by examining the spectral content of the waveform, second, third and higher harmonics will be present and are related to the degree of micro-structure damage. Additional practical advantages in detecting non-linearity may be achieved by transmitting the sum of two ultrasonic sine waves into a material from one transducer and examining the spectra for inter-modulation products. This paper details experiments on small samples of concrete using both harmonic and inter-modulation spectral analysis.

1 INTRODUCTION

The testing of concrete structures has posed many problems. Conventional ultrasonic transmission and pulse echo methods have limitations due to the nature and composition of concrete since they cause multiple reflections and non-direct ray paths. Alternative ultrasonic methods have been recently investigated that examine the shape of the waveform as it traverses through or over a complex material, the idea being not to located a single defect but to determine the overall mechanical properties within a certain region. These methods are known as Nonlinear Elastic Wave Spectroscopy (NEWS). Lacouture (Lacouture et al 2003) details a NEWS method to monitor the curing process of concrete, by means of transmitting and receiving an 8 kHz sine wave signal through the setting concrete. Van Den Abeele (Van Den Abeele et al 2001) outlines various NEWS techniques to measure micro-scale damage in building materials, including concrete. In one of the NEWS methods two different frequencies are transmitted into the material via two separate transducers and a third transducer is used as a receiver.

Non-linear acoustic methods seek to determine how an ultrasonic waveform changes when it propagates through or over the surface region of a medium. These changes are directly related to the stress-strain relationship and the hysteretic properties of a material and are not unduly effected by the ray path. In damaged materials, particularly ones that have micro-cracking, the stress-strain relationship does not obey Hooke's Law of elasticity, stress is not proportional to strain: it is not linear. In addition, these materials often have a stress-strain relationship that is non-symmetric, that is the reaction to compression forces will have different properties to that of tensile forces: this is a result of the cracks opening and closing under tensile or compressional loads. Figure 1 shows two photographs of sectioned micro-cracked concrete samples, the cracks having been formed by chemical degradation and mechanical damage. These photographs, provided by Geomaterials Research Services Ltd, were taken using epifluorescence illumination with a Zeiss Axioskop polarizing photomicroscope.

The stress-strain curve for non-linear behavior is illustrated in Figure 2. The result of non-linearity is that any stress loading that is in the form of a pure sine wave will produce a strain that is distorted as it traverses the material. This is illustrated in Figure 3. The degree of this distortion is measured by examining the spectral content of this distorted waveform, second, third and

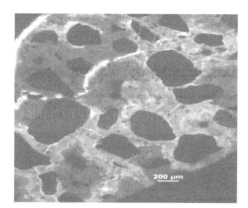

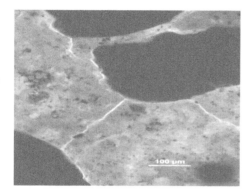

Figure 1. Photographs of micro-cracking in concrete.

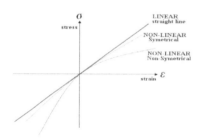

Figure 2. Non-linear stress v strain.

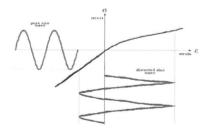

Figure 3. Waveform distortion.

higher harmonics will be present and are related to the amount of damage.

Greater sensitivity to non-linear effects can be achieved by transmitting complex waveforms into the material, for example a waveform that is composed

Through the material

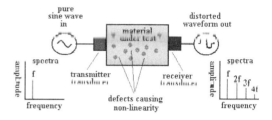

Over the surface

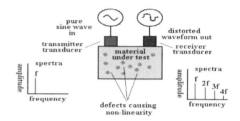

Figure 4. Harmonic generation.

of the sum of two sine waves. Any non linearity will act to produce a multitude of frequency components in the spectra, called inter-modulation products. Considerable practical advantages can be made if only one transducer is used to transmit these complex waveforms. To achieve this, wideband transducers that are acoustically matched to the test material were developed that do not generate non-linear effects internally or at the point of contact with the concrete, these transducers were used in the experiments detailed in this paper.

2 HARMONIC GENERATION

The simplest method in a practical system that measures non-linear effects in a material using acoustic waves is to measure the harmonics generated when a pure tone (pure sine wave) is transmitted through or over the surface of a material. This is illustrated below in Figure 4.

The harmonics are measured by examining the power spectra of the received signal. The transmitted frequencies (fundamental) magnitude is compared to that of the magnitudes of each of the harmonic frequencies. These harmonics are expressed in terms of decibels (dB) down from the fundamental; that is the number of decibels below the fundamentals magnitude. These values can be converted to a distortion factor that is expressed as a percentage.

Other non-harmonically related frequencies may also be generated by the sound wave, particularly in the presence of severe defects, these are called overtones and noise: they result from acoustic emissions, hysteresis and other effects. Figure 5 below shows

Cracked concrete

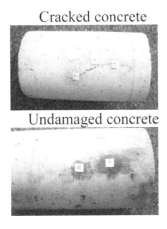

Undamaged concrete

Figure 5.   Concrete test cylinders.

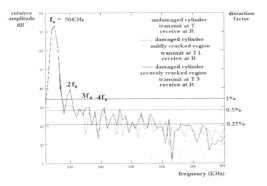

Figure 6.   Harmonics generation over surface of concrete test cylinders.

Figure 7.   Transducers connected to a drilled core test sample.

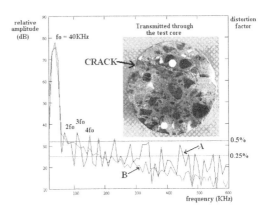

Figure 8.   Harmonic generation-drilled core.

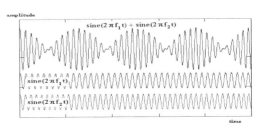

Figure 9.   The power spectra of the sum of two sine waves

the photographs of two concrete test cylinders (size 300 mm long, 150 mm diameter).

Figure 6 below shows the results obtained by sending a 50 kHz sine wave over the surface of these two cylinders of concrete. The second harmonic generated in the severely cracked region is clearly visible and has a level of distortion above 1%. The third and fourth harmonics are not so prominent but have values above 0.5%. The undamaged concrete sample does not produce any clear harmonics and consists of noise predominately below 0.25%.

The photograph in Figure 7 shows a micro-damaged drilled test core with the ultrasonic transmitter on the right and the receiver on the left. The trace A of the spectral plot in Figure 8 shows that transmitting and receiving in a line through the concrete close to the crack produces relatively high levels of 2nd, 3rd and 4th, harmonics above 0.5%. Transmitting and receiving in a line away from the crack, shown as trace B produces little harmonic content.

3   INTER-MODULATION

If two sine waves of different frequency are added together the resulting power spectrum is unaltered, this is illustrated in Figure 9 below.

An ultrasonic wave composed of the sum of two sine waveforms of different frequencies, f1 and f2 with equal amplitude, can be represented by

[sine(a) + sine(b)], where a = $2\pi f_1 t$ and b = $2\pi f_2 t$.

409

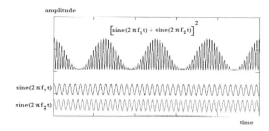

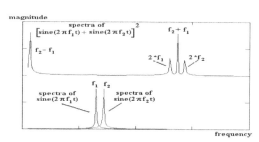

Figure 10. Inter-modulation products resulting from a sine sum waveform being subject to a square law distortion.

If this waveform is passed through a material that exhibits a square law stress-strain relationship. The resultant wave forms can be expressed as:-

$$A(t) = [\sin(a) + \sin(b)]^2$$

by expansion this gives:-

$$A(t) = \sin^2(a) + 2\sin(a)\sin(b) + \sin^2(b)$$

using the standard trigonometric identity formulae

$$\sin(a).\sin(b) = \tfrac{1}{2}[\cos(a-b) - \cos(a+b)]$$

and noting that
$$\sin(a).\sin(a) = \tfrac{1}{2}[\cos(a-a) - \cos(a+a)] = \tfrac{1}{2}[\cos(0) - \cos(2a)]$$

which becomes $= \tfrac{1}{2}[1 - \cos(2a)]$, since $\cos(0) = 1$, then the expression for $A(t)$ becomes:-

$$A(t) = \tfrac{1}{2}[1 - \cos(2a)] + [\cos(a - b) - \cos(a + b)] + \tfrac{1}{2}[1 - \cos(2b)]$$

re-arranging

$$A(t) = 1 + \cos(a - b) - \cos(a + b) - \tfrac{1}{2}\cos(2a) - \tfrac{1}{2}\cos(2b)$$

Figure 10 below shows a graphical representation of this process. Four distinct frequencies and one constant term are generated by this process, the frequencies are; the second harmonics of f_1 and f_2 that is $(2*f_1)$ and $(2*f_2)$. The sum and difference frequencies of f_1 and f_2, that is $(f_1 + f_2)$ and $(f_1 - f_2)$. The second harmonics are half the amplitude of the sum and difference

undamaged cube   damaged cube

Figure 11. Test Arrangement.

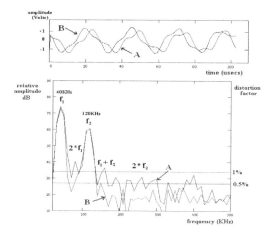

Figure 12. Dual frequency (40/120 KHz) spectra and time data plot of a good and damaged test cube.

frequencies. As there is a larger variation in the generation of the sum and difference frequencies these should provide greater sensitivity in the indication of non-linearity. If the sine wave sum is subject to non-linearity that is of a higher order than a square law stress-strain relationship then many other multiples, sum and difference combinations result, these will all appear in the spectra.

Figure 11 shows a photograph of the transmitter and receiver placed against a test sample cube of concrete (size $50 \times 50 \times 50$ mm). Two concrete test samples were selected and are shown in this figure, one has a crack running through its entire length, the other is undamaged. The transmitter comprises a single piezoelectric wide band actuator that is continuously sending the sum of two sine waveforms at pre-programmed frequencies.

Figure 12 shows the time and spectrum plots for the received waveform having passed through each of the concrete test cubes. The difference in the magnitude of

410

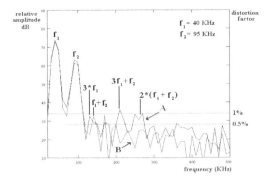

Figure 13. Dual frequency (40/95 KHz).

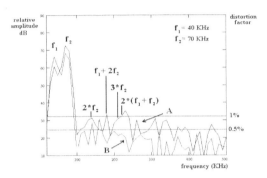

Figure 14. Dual frequency (40/70 KHz).

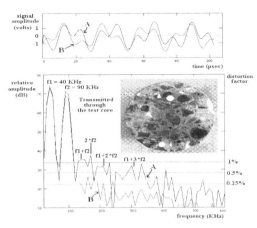

Figure 15. Through test core.

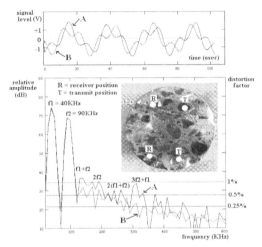

Figure 16. Over surface of test core.

the two frequencies results from the ultrasonic attenuation of concrete being frequency dependent, losses arc greater at higher frequencies. The data for the damaged and undamaged samples are labeled in this figure as A and B respectively.

The damaged sample shows clearly that harmonics and inter-modulation products have been generated by the crack. The upper side band (f1 + f2) at 160 kHz is below 0.5% for the good sample and rises above 1% in the damaged sample. The second harmonic of f2 at 240 kHz changes from, −56 dB (0.16%) in the good sample and rises above 0.5% in the damaged sample. The effect of the combinations of the harmonics and inter-modulation products are very noticeable in the frequency range 200 to 350 kHz. For example, 2f2 (240 kHz), 2f1 + f2 (200 kHz), f1 + 2f2 (280 kHz) and 2f1 + 2f2 (320 kHz). The result is the formation of peaks and troughs within this range, corresponding to the interaction of their frequencies and phases, this effect can mask the changes between the good (trace B) and bad sample (trace A). The correct choice of the two frequencies f2 and f1 is an important factor. Figure 13 illustrates this by showing the spectra resulting from two different frequency combinations, $f_1 + f_2$ is reduced by the effect of the third harmonic of f_1 $(3*f_1)$.

Figure 14 shows a spectral plot taken over the surface of a mildly damaged region of the concrete test cylinders, shown previously in Figure 4. This test was performed at the two frequencies 40 and 70 kHz. The two fundamental frequencies are not sufficiently

separated to form clear spectral peaks, however the inter-modulation products and in particular the second multiple of $f_1 + f_2$ that is $2*(f_1 + f_2)$ shows a very clear pcak above 1% distortion in the damaged region (trace A).

Figure 15 shows a dual frequency being applied to the micro-damaged drilled core. The dual frequency ultrasonic waveform was transmitted through the sample at two locations, one along the crack, shown in and the other away from the crack shown as the two white circles. The difference between the two locations is very clear. The cracked region produces harmonics and inter-modulation products well above 0.5% distortion factor and $f_1 + f_2$ is above 1%. The less cracked region has all levels below 0.5% and for frequencies above 250 kHz is below 0.25%.

Figure 16 shows the same core but this time tested on one side only, the receiver positions are indicated by the letter R and the transmitter positions by letter T. Trace B corresponds to a position away from the crack and trace A near to the crack. There is less difference between the two positions at low frequency however at higher frequencies, above 300 kHz, the cracked region does produce a significantly higher levels of inter-modulation produces particularly at $3f_2 + f_1$ (310 kHz).

4 CONCLUSIONS

Utilizing just two transducers, measurement of harmonic generation together with the production of inter-modulation products resulting from the transmission of an ultrasonic wave or the sum of two ultrasonic waves of different frequency, through or over the surface of a concrete test sample has shown to have the ability to detect cracks and micro-cracks. The method indicates that it can provide a quantitative measurement of the non-linearity of the concrete and thereby giving a measure of the degree of damage.

ACKNOWLEDGEMENTS

The authors would like to thank Les Randle of the University of Exeter for providing the concrete cylinder test samples, Tony Gomez of CNS Farnell for the test cubes and Michael Grantham of Concrete solutions for the micro-cracked drilled core samples.

The excellent photographs of the sectioned micro-cracked concrete samples were provided by Mike Edon of Geomaterials Research Services Ltd.

The instrumentation and transducers used to perform these experiments were originally developed under EC sixth framework research program. ASTJ-CT-2003-502927. (Aeronautics and space) and has been adapted for application to concrete structures and materials.

REFERENCES

Lacouture J.C., Johnson, P.A Cohen-Tenoudji, F. "Study of critical behavior in concrete during curing by application of dynamic linear and non linear means." *J. Acoust. Soc. Am.*, Vol. 113, No. 3, March 2003.

Van Den Abeele, K. Sutin, A, Cameliet, J. Johnson, P.A "Micro-damage diagnostics using nonlinear elastic wave spectroscopy." *NDT&E International* 34 (2001) pp 239–248.

Concrete Solutions – Grantham, Majorana & Salomoni (Eds)
© 2009 Taylor & Francis Group, London, ISBN 978-0-415-55082-6

Permeability testing of site concrete: A review of methods and experience

R.A. Barnes
The Concrete Society, UK

ABSTRACT: The permeability of concrete is of direct relevance to both durable concrete and to leak-resistant concrete for containment. Some specifications for projects in extreme environments or requiring very long lives have specified permeability criteria, to be verified by testing the supplied concrete or precast concrete elements.

However, there does not appear to be any agreement on the test methods that should be adopted. This paper summarises the findings of the recent Concrete Society Technical Report 31, which reviews how permeability can be measured and what typical results are achieved.

1 INTRODUCTION

The permeability of concrete to liquids, ions and gases is of direct relevance to both durable concrete and to leak-resistant concrete for containment. Analytical models used to predict the age at which corrosion of the reinforcement will be initiated require a detailed knowledge of the transport mechanisms involved and the permeability of the concrete (Concrete Society TR61, 2004). Some specifications for projects in extreme environments (e.g. the Middle East) or requiring very long lives (e.g. railway tunnels) have specified permeability criteria, based on some form of analytical durability model, to be verified by testing the supplied concrete or precast concrete elements. A simple permeability test could form part of the quality assurance (QA) scheme for any precast concrete plant, to check on the variability of standard units.

However, there does not appear to be any agreement on the test methods that should be adopted.

To establish values for the permeability of site concrete, measurements can be made either in situ on site, or in the laboratory on samples removed from site. The various test methods available have been categorised under these two headings and within these categories the following are the main processes, although some test methods involve more than one mechanism:

- absorption and capillary effects;
- pressure differential permeability;
- ionic and gas diffusion.

Many of the tests described do not measure permeability directly, but produce a 'permeability index' which is related closely to the method of measurement. In general the test method used should be selected as appropriate for the permeation mechanism relevant to the performance requirements of the concrete being studied.

As indicated earlier, it would be desirable to be able to relate measured permeability values to the durability of reinforced or prestressed concrete under various environmental conditions. However, it would appear that the number of results available for site concrete is relatively limited. As a result, the development of a relationship between measured permeability and predicted durability of concrete structures was beyond the scope of the Report (Concrete Society TR31, 2008).

2 MECHANISMS, DEFINITIONS AND UNITS

'Permeability' is a flow property and is defined as 'that property of a porous medium which characterises the ease with which a fluid will pass through it, under the action of a pressure differential.'

Figure 1 illustrates the stages by which water moves within the concrete where the connected voidage has been idealised as a single pore with a neck at each end.

It will be seen from Figure 1 that the moisture condition of the concrete at the time of the test will have a major influence on the values of absorption, flow or diffusion being measured, as in many cases the concrete is at some intermediate stage of pore filling.

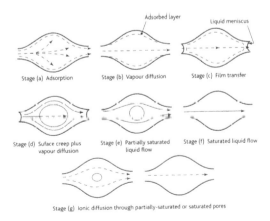

Figure 1. Idealised model of movement of water and ions within concrete.

Samples removed from site for laboratory testing can be conditioned to a controlled moisture condition. For in situ tests this cannot be achieved and pragmatic steps may have to be adopted such as not doing absorption testing within 48 hours of rain.

3 IN SITU TESTS

These tests are ones which can be carried out on concrete structures in situ. They can in addition be carried out on samples taken from structures and tested in a laboratory or on laboratory–produced samples. This requirement for site testing normally results in the tests being carried out on the surface of the concrete or in small diameter holes drilled a short way into the surface.

The drawback to site tests is that the results are inevitably affected by variable ambient factors including the initial moisture content of the concrete, its surface condition and the weather conditions during the test.

Adjustments to the results from varying moisture content may have to be considered and some tests include moisture measurements. The description of the ISAT method requires that 'The surface shall be tested after a period of at least 48 hours during which no water has fallen onto the test surface'. Indeed Neville (Neville, 1995) quotes 'A low value of initial surface absorption may be due either to the inherent low absorption characteristics of the concrete tested or else to the fact that the pores in poor–quality concrete are already full of water'

An advantage of the site tests described in the report is that they do not disfigure the concrete surfaces to any large extent, in contrast to the large core holes to be repaired when samples are removed for laboratory examination.

Most of the tests described are 'permeability index' tests which do not give a measure of true permeability.

Some of the methods measure absorption from the cement–rich surface layer whereas some measure from a drilled cut surface. In concrete, to obtain a true indication of the bulk material properties, the absorbing surface should be sufficiently large to give an average representation of the aggregate and cement matrix. (For a concrete having a maximum aggregate size of 22 mm this means a hole of at least 25 mm in diameter and 100 mm deep).

The tests reviewed in the report include:

- Water Absorption (Standpipe test; Initial Surface Absorption Test (ISAT – see Figure 2); Autoclam; Figg Method; Steinert)
- Gas permeability (Schonlin air permeability test; Surface airflow test; Vacuum test; Autoclam; Torrent Permeability tester; Hong–Parott; Figg air test.)
- Depth of carbonation
- Ionic diffusion by electrical measurement
- Drill hole permeability tests
- Leak testing of completed structures

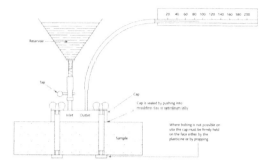

Figure 2. Initial Surface Absorption Test (ISAT) equipment.

The effectiveness of these tests can be illustrated by the work of Romer (Romer, 2005). He reported on an experiment to determine whether the NDT methods designed to measure on site the 'penetrability' of the concrete cover are capable of detecting differences in the w/c ratio and curing conditions of the concretes. Seven concrete slabs with a variety of w/c ratios, different cements, different curing regimes, different temperatures and different moisture conditions were tested with the NDT methods. In five or six out of seven cases the test methods were capable of detecting correctly the expected differences in 'penetrability' at a significant or highly significant level but none of the tests assessed could do it for all 7 cases.

4 LABORATORY TESTS ON SAMPLES FROM SITE

Although a number of the in situ tests listed above can also be carried out in the laboratory, this section deals with those tests designed specifically to measure permeability of concrete on samples in the laboratory. In most instances a sample of concrete taken from a site structure lends itself to a true measurement of permeability better than the in situ test, since the sample can be cut or cored to the precise size required for the particular laboratory test, and the moisture content can be brought to a standard condition.

The various test techniques can be categorised on the basis of the form of transport mechanism involved, as follows:

4.1 Absorption of water

In the absorption tests the volume of voids in the concrete that have been filled or partially filled with water is measured for specific conditions of immersion and time intervals. As such, the tests do not measure a true permeability, but they are relatively simple and quick, and can provide a convenient and important means of obtaining a 'permeability index' for concrete. The absorption tests reviewed in the report included: Shallow immersion and Capillary rise

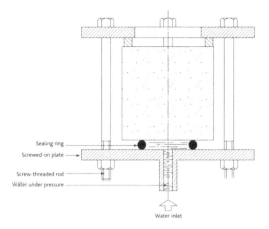

Figure 3. Test cell for depth of penetration measurements.

4.2 *Permeation of liquids or gases, due to a pressure gradient*

The permeation tests provide the means for measuring the true permeability. The basic requirement is that a specimen, usually a core or a cube, should be sealed on all sides other than two opposite parallel faces between which the flow of a liquid or gas can be promoted by an applied pressure. Under steady state conditions a value of permeability can be calculated from the knowledge of sample geometry and fluid characteristics, and the measurement of flow rate and applied pressure. Such tests can prove impractical, especially for liquid flow, on low permeability concretes where flow rates for typical sample dimensions and reasonable pressures will be very low. In such circumstances a test based on the measurement of the depth of penetration of liquid, as observed on subsequent splitting of the test sample, has been found to be more appropriate (see Figure 3). The permeation tests reviewed in the report included: Pressure induced flow – liquids (Measurement by flow, Measurement by penetration) and Pressure–induced flow – gases.

4.3 *Diffusion of gases, water vapour or ions, due to a concentration gradient*

The diffusion of ions, particularly chlorides, and of gases, including water vapour, is a particularly important mechanism which can influence the service life of concrete. Not surprisingly, therefore, a number of laboratory test methods have been developed to measure a diffusion coefficient, which provides another form of 'permeability index' for concrete. The diffusion tests reviewed in the report included: Gas diffusion; Water vapour diffusion; and Ionic diffusion (Drilling and sample analysis, Concentration difference, Electrical Current Flow (Rapid Chloride Permeability test), Rapid migration test, Chloride conductivity Test, Immersion Testing and the Ponding test, Chloride diffusion by electrochemistry). This included the transport of chloride ions into concrete, which is a complicated, multi–mechanistic process. A multitude of tests has been proposed and used to test the resistance of concrete to chloride diffusion and each test has advantages and disadvantages.

5 PERMEABILITY RESULTS FROM SITE CAST CONCRETE – CASE HISTORIES

In general very few results are available for permeabilities measured on in situ concretes. From the limited data reviewed in the report it was not possible to establish any clear relationships between mix design, workability, use of admixtures etc on the permeability achieved in practice. No on-site measurements relating surface absorption and permeability to degree of curing were discovered.

6 USE OF PERMEABILITY RESULTS FOR SPECIFICATION COMPLIANCE

With the increasing importance placed on producing durable concrete it would seem that some form of permeability test could be a useful compliance requirement. However, work reviewed in the report (Pocock, D. & Chadbourn, G., 2006; Harries, N., 2001 and Geiker, M. et al, 1985) has shown that the variability of the test results suggest that this concept is not currently a valid approach.

7 CONCLUSIONS

When testing the permeability of concrete it is not the average 'quality' of the concrete that is being assessed for permeability (unlike the standard compressive cube test) but the easiest passage for the water, or gas, to flow through. A good quality concrete, with a paste structure of minimal permeability, will be very permeable if a crack is present. If no cracks, or other pathways develop, the pore structure of the cement paste will be the governing factor, along with the interfacial zones of aggregate particles and paste. Hence, the importance of initial water/cement ratio, degree and rate of hydration and degree of saturation at the time of test. Not only does drying empty the pore spaces, but it may also lead to cracking.

It is therefore, not surprising that permeability testing appears to be associated with high variability; the more sensitive the test, the greater the variability. It has been suggested that apparent consistency of measured values probably indicates either malfunction or inappropriateness of test technique.

The report reviewed a variety of 'permeability' tests for concrete. The critical factor in measuring 'permeability' values is the moisture condition of the section of structure or sample being tested. This is a major limitation for tests done 'in situ' and for more accurate results it is therefore necessary to remove samples to a laboratory where they can be conditioned to standard moisture state. This removal of samples is often not practicable and measurements 'in situ' can still provide useful and informative data.

There is still little quantitative correlation data available, between permeability results and long term performance.

In the report, four contracts where the ISAT permeability index was successfully used for specification purposes to ensure that the concrete had good weathering properties were identified. However, the variability of test results from subsequent projects suggests that the concept of using permeability testing for specification purposes is not currently a valid approach.

There are still many limitations to measuring the permeability of site concrete and at this stage it is still, not yet possible to arrive at a firm conclusion on methods of test. The report, as the title suggests, provides a review of the methods currently available and of typical results obtained.

REFERENCES

The Concrete Society. *Technical report No. 31. Permeability testing of site concrete: A review of methods and experience (3rd Edition).* The Concrete Society, Camberley, 2008.

The Concrete Society. *Technical report No. 61. Enhancing reinforced concrete durability.* The Concrete Society, Camberley, 2004.

Neville, AM. *Properties of concrete.* Longman, Harlow, 1995.

Romer, M. Comparative test – Part 1 – Comparative test of 'penetrability' methods. *Materials and Structures.* No. 38. Dec. 2005 pp895–906.

Pocock, D. and Chadbourn, G.A. Implementing durability–based design for infrastructure projects. *Proc. 8th Int. Concrete Conf, Bahrain,* 2006.

Harries, N. Compliance testing concrete for durability. *Dubai Municipality Conf. on quality assurance and control.* Dubai, Oct. 2001.

Geiker, M. et al. *Performance criteria for concrete durability.* E and FN Spon, London. 1985.

Concrete Solutions – Grantham, Majorana & Salomoni (Eds)
© 2009 Taylor & Francis Group, London, ISBN 978-0-415-55082-6

A low frequency electromagnetic probe for detection of corrosion in steel-reinforced concrete

J.B. Butcher, M. Lion, C.R. Day & P.W. Haycock
Institute for the Environment, Physical Sciences and Applied Mathematics, Keele University, Staffordshire, UK

M.J. Hocking
SciSite Limited, Keele University Science Park, Keele, Staffordshire, UK

S. Bladon
CRL Surveys, Mitcham, Surrey, UK

ABSTRACT: With the cost to the economy of corrosion in highways, pipelines, and water services alone running at over 1% of GDP in the USA and at similar levels elsewhere in the West, the definitive non-invasive detection of corrosion on steel reinforcing bars embedded in reinforced concrete structures is desirable, albeit difficult. A low frequency electromagnetic probe has been developed which, with suitable data processing, can yield a variety of information about the state of embedded reinforcement without the need to remove or process the overlying material or to make electrical contact to the steel. Data interpretation can be maximized by iteratively altering the physical state of the steel and collecting data on the resulting multiple virtual surfaces. This paper describes the combination of multidimensional perturbed surfaces and data mining techniques as an approach for using the probe to survey large and/or complex reinforced concrete structures. Also explained are means for analyzing electromagnetic signatures to determine the nature of corrosion induced defects and to display their multidimensional nature in a visually accessible manner, together with a bespoke calibration facility. Finally, case studies from a range of civil engineering structures are presented to illustrate the power of the technique.

1 INTRODUCTION

Reinforced concrete is one of the most widely used building materials today. Its uses range from car parks and motorways, to swimming pools and power stations. The repair costs associated with reinforced concrete are vast, with costs in the US at an estimated $1.4 billion (Virmani et al 2001). Therefore, the accurate assessment of steel-reinforced concrete is of huge importance but achieving this is often difficult and costly. Numerous testing methods exist which have varying advantages and disadvantages. A new type of electromagnetic probe has been developed which doesn't require the removal of the surface material above the steel or contact to be made with the steel, therefore saving time and reducing required resources (Hocking et al 2006, Haycock & Lambert 2008). This new probe can, however, produce complex, multidimensional datasets which are difficult and time consuming to interpret. Computational Intelligent (CI) techniques such as Artificial Neural Networks, Support Vector Machines and Fuzzy Logic can be used to overcome this problem.

This paper describes the development of this technique, both in the laboratory and in the field. It concludes with recommendations for future work.

2 TESTING FOR CORROSION

2.1 Structural assessment

As mentioned in Section 1, the need for the accurate assessment of steel-reinforced concrete is paramount. In some cases corrosion can be seen from the outside of the structure through spalling and cracking of the concrete. Corrosion can also be seen from the staining of the structure which occurs as a result of the corrosion process. Despite these tell-tale signs however, corrosion can occur without giving any external indication of what is happening inside. Undetected corrosion leads to structural weakening and, in some cases, partial or total collapse of the structure. It is, therefore, of major importance to have a reliable and accurate technique for corrosion assessment. The techniques used to conduct structural assessment fall into two categories: destructive testing and non-destructive testing.

2.2 Destructive testing

Destructive testing, as the name implies, involves partially destroying part of a structure in order to gain access to the enclosed reinforcing steel. Not only is this approach costly, it is also damaging to the structure which may have been of sound health. The removal of the encasing concrete can leave the steel exposed, causing further problems once it has been re-covered with concrete. One of the most widely used destructive testing methods is coring. This involves drilling into the concrete and extracting a core of concrete and reinforcing steel from the structure to be tested. This approach allows an engineer to see the reinforcing steel in its surrounding environment. However, coring not only has the disadvantages mentioned above, but it is also a very localised testing technique. Cutting the steel reinforcement must be done with caution and with the approval of the supervising structural engineer. As corrosion is also localised, healthy extracted cores can be non-representative of a structure which contains multiple defects. The reliability of coring is also affected by the extraction process as some of the physical properties of the core are destroyed during drilling and extraction. This can also give a misrepresentation of the condition of the structure.

2.3 Non-destructive testing

Non-destructive testing (NDT) involves using non-invasive techniques for the assessment of a structure's reinforced concrete. NDT techniques do not require the removal of concrete and are, therefore, less time consuming, don't damage the structure and can give an accurate assessment of a structure's condition. NDT techniques used for corrosion detection include half-cell potential and resistivity. These only provide an estimate of the likelihood that corrosion maybe occurring, but cannot locate the rust itself and will not provide a positive reading when corrosion processes have run to completion. Half-cell potential requires electrical contact with the reinforcement bars (rebars), which can require removal of concrete around certain areas of the reinforcing cage. This technique can, therefore, become a destructive technique, which leads to the disadvantages mentioned in Section 2.2. Half-cell readings are also affected by environmental factors such as the water content of the concrete, making absolute calibration impossible in many cases.

Electromagnetic NDT approaches include Ground Penetrating Radar (GPR), Eddy Current testing and flux leakage (Hillemeier & Scheel 2003) although it is difficult to detect corrosion using these techniques, which are generally used for locating relatively large scale structural faults.

2.4 New technique

A new technique has been developed as part of collaboration between Keele University and SciSite Limited. A lightweight probe can be wheeled over an area with minimal effort and does not require contact with the steel. In addition, the properties of the concrete, such as water concentration have no effect on the probe. This new technique extracts very small signals from corrosion using an electromagnetic probe to detect changes in the state of the steel and corrosion products under an external influence. To achieve this the reinforcement cage must be energised by some means and a suitable probe used to detect the changes produced. One simple way of doing this is to magnetise the steel and measure the flux leakage. With a sensitive enough probe and measurements taken at a series of magnetic states, a multidimensional dataset can be obtained which contains signatures not just for breaks, but also small amounts of corrosion. More sophisticated high frequency probes have a higher degree of accuracy. However, this paper concentrates on the simple case of DC flux leakage to explain the principle.

Flux leakage has been used for some time to detect defects such as breaks, but its use for corrosion detection has been somewhat limited. Flux leakage is based upon the principles of electromagnetism, where a defect, such as a break within the steel, leads to the creation of poles within the cagecage In a cagecage which is of sound health, the magnetic fields seek to form the fields of lowest energy and form closed loops. When a defect is present, poles are formed which create a magnetic field which is no longer enclosed in the cage. It is this fringing field which is detected by the probe. As some corrosion products, such as magnetite, are magnetic, corrosion can also be detected by looking for small magnetic fields produced by corrosive products within the reinforcing cage. The procedure for using the new technique is as follows.

1. A pre-scan is taken of the structure. This gives an indication of the magnetic state of the reinforcing cage when it has been sitting in the Earth's magnetic field.
2. The structure is energized using a magnet. This puts the cage into a desirable state where defects can be detected.
3. The structure is then scanned again using the probe to detect any differences between the magnetically induced state and the previous, non-magnetized state.
4. Both scans are then compared with the difference between the scans taken. Any areas which are significantly different are areas for further investigation.

A structure can be energised and scanned in multiple magnetic strengths and directions to give more information on the state of the steel cage inside it. This is essential for anything other than the simplest of structures. Figures 1 and 2 show an image of the probe and a sample data set produced from scanning a reinforcing cage.

While this approach appears in principle to be very simple, in practice this is often not the case. When testing large areas such as motorways and car parks, the amount of data which is produced makes manual data processing very intensive and time consuming. In order to overcome this problem, CI techniques are being applied to the data to automatically classify the different types of defects found within a structure. This can reduce the time required to process the data and increase the accuracy of the classification of defects.

Figure 1. The probe.

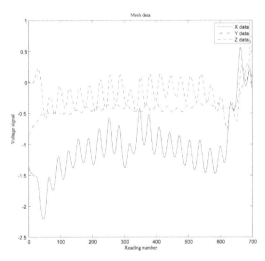

Figure 2. An example of data collected from scanning a steel-reinforcing cage.

As this technique is still in the prototype stage, it requires further calibration and testing. Details of this process can be found in Section 4.

3 COMPUTATIONAL INTELLIGENT TECHNIQUES

3.1 Introduction

Computational Intelligence (CI) techniques are attractive for data mining due to their tolerance of error and noise (Palade et al 2006). This tolerance is often useful for data sets collected from real world examples such as NDT domains where noise is often present. Some of the most widely used CI techniques include Artificial Neural Networks (Haykin 1999), Support Vector Machines (Vapnik 1998 & Shaw-Taylor & Cristianini 2000) and Fuzzy Logic (Cox 1994).

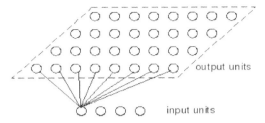

Figure 3. A basic SOM architecture. Only the connections between the first input unit and the first eight output units are shown for simplicity. Similar connections between all other input units and output units exist.

3.2 Artificial neural networks

Artificial Neural Networks (ANNs) are based on an abstract model of neurons found in the human brain. ANNs are able to perform a mapping of numerical input patterns to either an output classification or cluster. ANNs can be categorised as either supervised or unsupervised depending upon the training regime. The most widely used supervised ANN is the multilayer perceptron (MLP). The MLP is adept at performing classification tasks which it learns to solve using the back-propagation algorithm (Rumelhart et al 1986). For example, the classification of bridge vibrational signatures (Pandey & Barai 1995), the conditional assessment of concrete based on age and strength (Lorenzi et al 2004) and the diagnosis of insulation for power transformers (Chen & Chen 2007) are just a few uses in the NDT domain.

The unsupervised ANNs are trained without using output target patterns. Instead, the data is clustered according to regularities in its underlying characteristics. The Self Organising Map (SOM) (Kohonen 1990) is a well known example of an unsupervised ANN. SOMs are particularly suited to problem domains where the data is noisy and the output clusters are hard to determine. Their topological ordering properties mean that the network is readily able to place an input pattern between two (or more) clusters when its characteristics suggest it may be a mixture of those classes. This property is not so readily achieved using the MLP ANNs. A SOM's architecture consists of two layers of neurons, an input and output layer. Each input unit is connected to every unit in the output layer. This basic architecture is shown in Figure 3, with only the connections of one input unit to the first eight output units shown, for simplicity.

As mentioned earlier, SOMs learn by clustering data input patterns based on their distance from the output units weights. The process of training a simple SOM is as follows.

1 An input pattern is presented to the input layer of the SOM.
2 Each output unit then calculates its activation, o, as:

$$o = \sum_{i=1}^{n} w_i \times a_i \qquad (1)$$

419

where n is the number of input units in the input layer, w_i is the weight on the connection between an input unit and the cluster unit, and a_i is the activation of the input unit.

3 The distance, d, between every node's incoming connection weights and the current input is calculated as:

$$d = \sum_{i=1}^{n} \left(w_i - a_i \right)^2 \qquad (2)$$

4 The node with the smallest distance is declared as the winning node and updates its weighted connections to bring it closer still to the input pattern:

$$w_i = w_i + \eta \left(a_i - w_i \right) \qquad (3)$$

where η is the learning rate.

It is also possible to update the cluster nodes around the winning node that lie within a specified distance. This process is repeated for each input patern of a dataset until well defined clusters are formed.

4 CALIBRATION OF TECHNIQUE

An experiment using a 2.1 m × 3.5 m reinforcing steel cage has been created. This provides the typical signatures for 12 different types of defects for a CI technique to distinguish between. The cage has been cast in a tank of concrete and is surrounded by acidic water which permeates through the concrete, causing the cage to corrode. Areas of the cage were purposely corroded and/or broken before the cage was set in concrete to give a set of defects whose type and location is known. As the reinforced concrete is in an acidic environment other defects occur later on in the experiment.

Defects near the ends of the cage are treated differently from those in the middle of the cage, due to the magnetic properties of the cage towards its ends, which will alter the defect signatures. This is due to the cage having unclosed magnetic fields at its ends, which are detected by the probe some several readings away. In order to calculate the area over which end effects could be detected and, therefore, rule them out as defects, early tests on the cage prior to casting were performed. This involved scanning and energising the cage in different directions to see how the data the probe collected differed.

The reinforcing cage was first energised longitudinally, then scanned longitudinally, transversely, and at 30°, 45° and 60° angles. The same scans were then performed after the cage had been energised transversely. These initial experiments gave some interesting results. For instance, the directions of energisation had a major influence on the data. The end of the cage which was energised last had a bigger signal than the other end where the energisation had started. This was as expected from magnetic theory principles. This result also agrees with results from an earlier experiment (Sherratt 2006) in which small steel plates were energised and then scanned. The areas of the plates

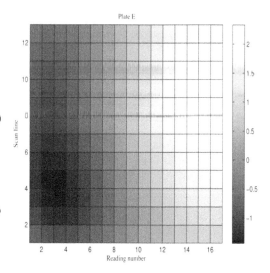

Figure 4. The fringing field from an energised steel plate. Notice the higher magnetic field towards the end of the energisation process (top right) compared to the beginning of the process (bottom left).

where energisation had finished were found to have larger, more concentrated magnetic poles than where it had started. Figure 4 shows a greyscale plot of a steel plate magnetised from left to right, bottom to top. The darkest areas in Figure 4 represent small negative magnetic fields, while the lighter colours represent stronger positive magnetic fields. Figure 5 shows the data collected from scanning the cage. Energisation was applied from left to right.

As Figures 4 and 5 show, a large magnetic pole at the end of the scan due to the energisation process is present. In a blind survey on an embedded cage this could easily be wrongly interpreted as a defect. This phenomenon could also, therefore, be wrongly diagnosed by a CI technique as a defect when it is actually the end of a cage, having a major influence on classification error. An understanding of magnetic fields from edges of cagees would avoid miss-classification; however, the degree to which a CI technique can discriminate between an end and a defect remains the subject of research.

5 CASE STUDY: CAR PARK DATA ANALYSIS

A deck of a multi-storey car park was surveyed. The area scanned consisted of a 6 m × 6 m area of a first floor deck as shown in Figure 6. Eleven scan lines were taken, each 0.5 m apart. Figure 7 shows the data collected from the first scan line. The graphs show the pre-energisation, post-energisation and the pre-post difference respectively. As Figure 7 shows, one would expect to find a defect around the 270th reading and a smaller defect around the area of the 550th reading. This was later verified as correct when the concrete was removed for repair. Figure 8 shows a greyscale plot

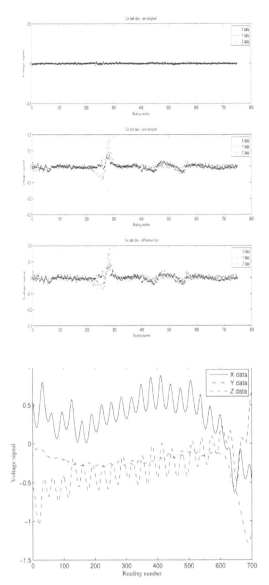

Figure 6.   The deck of the car park surveyed.

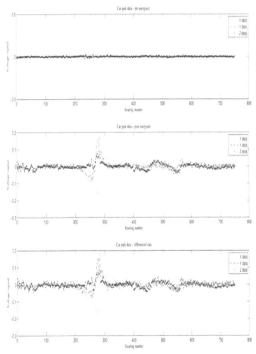

Figure 7.   The data collected for scan line 1 showing pre and post energisation, and the difference between the two.

Figure 5.   A graph showing data collected after energising steel cage from left to right. Notice the large change in the data at the right hand end of the graph.

of the overall scanned area. As Figure 8 shows, the feature identified around the 270th (2 m along the deck) reading is prominent throughout the scanned area. This figure also shows that the smaller feature around the 550th (4 m along the deck) reading is only found in the first four scan lines. Figure 9 shows the deck after the concrete had been removed; notice the corroded steel bar which disappears toward the top end of the picture: this is the defect that was detected around the 270th reading.

Figure 10 shows the underside of the deck area around 550th reading (4 m along) of the first four scan lines which are shown in the greyscale plot shown in Figure 8. The corroded rebars are clearly visible where

the concrete has spalled and broken off, leaving the rebars exposed.

A SOM was used to cluster the data. The topological ordering properties of SOMs made them particularly suited to this kind of data where varying degrees of defects can be found, rather than clear cut defects such as either rust or a break, for example.

Initially, training of the SOM was attempted using each scan line of the dataset.

This, however, was found to be unsuccessful as the SOM was unable to cluster any of the data items. It later became clear that this approach would only learn the position of the peaks of a graph, not the types of peaks within the data. In order to learn the curves or features, the scan line data was split into sub-datasets containing 100 data points each. This could then act

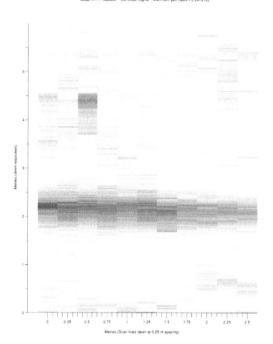

Figure 8. A greyscale plot of the data collected. The light areas indicate weaker signals, while the darker areas indicate strong signals which relate to defects.

Figure 9. The reinforcing cage after removal of the concrete.

as a moving window through the data set. It was also decided to perform some pre-processing on the data to improve the chances of successful clustering. First, every data point was normalised in the range 0–1. This was calculated by:

$$v_n = \frac{v - min(a)}{max(a) - min(a)} \tag{4}$$

where v_n is the new normalised value, v is the original value and $min(a)$ and $max(a)$ are the minimum and maximum values of the dataset respectively. The data was then smoothed using a moving average of 10, and finally sampled to reduce dimensionality, taking

Figure 10. The underside of the deck showing the corroded rebars exposed as a result of the spalling of the concrete. This defect can be seen as the smaller defect in Figure 8, 4 m in, around the left section of the scanned area.

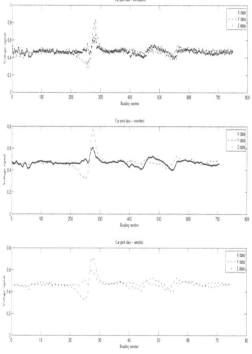

Figure 11. Data processing showing normalisation, smoothing and sampling of the data collected along scan 1.

every 10th data item and putting it into a new dataset. Figure 11 shows these 3 stages respectively.

Once the data had been pre-processed, the SOM could be presented with data for training. By expert inspection, it was decided that the data contained 3 prominent features which could potentially be clustered. Figure 12 shows the 3 features used for training the SOM.

As Figure 12 shows, three different types of features were to be classified by the SOM: 'No Defect', 'Defect A' and 'Defect B'. Two patterns of each feature

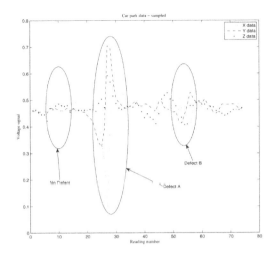

Figure 12. The three types of features used in training and testing of the SOM.

Table 1. Training regime for SOM

Epochs elapsed	Learning rate (η)	Output layer neighbourhood size
0	0.9	5
1	0.8	5
3	0.7	4
7	0.6	4
15	0.5	3
31	0.4	3
63	0.3	2
127	0.2	2
255	0.1	1
511	0.01	1
1023	n/a	n/a

		Feature patterns tested		
		No defect	Defect A	Defect B
SOM classification	No defect	2		
	Defect A		1	
	Defect B			1
	?		1	1
Correct (%)		100	50	50
		66.67		

Figure 13. Confusion matrix of SOM test patterns after 1023 epochs of training.

were used for testing data, with the remaining 9 patterns for No Defect and Defect A and two patterns for Defect B used to train the SOM. The number of training and test patterns were limited due to the small size of the area scanned. Although this will negatively impact the performance of the SOM, it is presented here to show proof of concept that a CI technique can be used to cluster data collected from the probe. After some initial testing, a network architecture consisting of 5 × 5 output units gave the best performance. The training regime used is given in Table 1. Training was halted after 1023 epochs and the clusters were then labelled. The 6 unseen test patterns were then presented to the SOM.

Figure 13 shows the results as a confusion matrix. Each row of the matrix represents each of the classes. Each column of the matrix represents how many patterns the SOM placed into each class. Perfect classification for each pattern would give a diagonal line from the top left to the bottom right of the matrix which would equal the sum of the number of patterns. An addition row was added to the matrix, labelled '?'. This row indicates test patterns which map to unlabelled nodes.

At first glance, the confusion matrix indicates that the performance of the SOM could be better. However, the two patterns which were classified as '?' were investigated further. It was found that the output nodes which both patterns mapped to were next to a labelled node of their class.

One would therefore conclude that the test pattern does in fact, belong to that class, hence giving 100% accuracy. As only two patterns for Defect B were used for training, the SOM's output layer did contain two nodes which were ambiguously labelled as No Defect and Defect B. This is due to the similarity between the two patterns, as Defect B only gives a small oscillation in what is otherwise a flat surface. To remove ambiguity from the SOM and improve performance, a bigger dataset is required for both training and testing.

As the new technique uses a lightweight and easy to use electromagnetic probe, the types of structures which have been assessed have varied greatly. Such examples include swimming pools, bridge joints, cable suspended bridges, flyovers, motorways (Haycock & Lambert 2008) and a Marina. Further details of the surveys conducted using the new technique can be found at http://www.scisite.co.uk.

6 CONCLUSIONS

A new non-destructive technique has been presented which combines an easy to use probe with CI techniques to provide accurate detection of corrosion within steel reinforced concrete. A proprietary magnetic probe has been used which requires no contact with the reinforcing steel and is not effected by environmental conditions such as the water concentration within the concrete. As the probe can produce high dimensional datasets which are difficult to interpret, CI techniques are being applied to automatically classify defects detected by the probe. Early results are promising as a SOM has been successfully applied to data collected from a car park deck. Even though some patterns mapped to unlabelled nodes, further analysis of the formed clusters confirmed that the SOM classified all test patterns correctly. Larger datasets are now needed to validate its performance further.

Further work involves collecting and processing calibration data from the cage experiment outlined earlier. This will help to gather a dataset of signatures for different types of defects. This can then be processed using CI techniques such as SOMs, SVMs or Echo State Networks (Jaeger 2001). These techniques will be evaluated and their performance compared. The distinction between end effects and defects will also be investigated further as this is an area which could influence performance greatly. Using the low frequency magnetic probe generally leads to very large datasets and heavy reliance on CI for complete analysis. Use of a more sophisticated higher frequency probe along the same lines leads to significantly decreased ambiguity in interpretation and potentially simpler CI methods. This has not been discussed since the form of the data is more complicated which would obscure the underlying technique: collection of multidimensional data from energetically perturbed reinforcement and extraction of corrosion signatures. However, the method employed by SciSite makes use of a more proprietary detection system.

REFERENCES

Chen, H.C. & Chen, P.H., 2007. Application of back-propagation neural network to power transformer insulation diagnosis. *Advances in Neural Networks ISNN 2007*, 4493.

Cox, E., 1994. *The fuzzy systems handbook: a practitioner's guide to building, using, maintaining fuzzy systems*. Boston: AP Professional.

Haycock, P.W. & Lambert, R., 2008. Rust detection – The Scisite Method. *Concrete* 42. p.42–43.

Hillemeier, B. & Scheel, H. 2003. Fast Location of Prestressing Steel Fractures in Bridge Decks and Parking Lots. *International Symposium of Non-Destructive Testing in Civil Engineering*.

Haykin, S., 1999, *Neural Networks A Comprehensive Foundation*, Second Edition, New-Jersey: Prentice-Hall

Hocking, M.J., North, L., Wright, A.K., Haycock, P.W., Bladon, S., 2006. In: Grantham, M.G., Jaubertie, R., Lanos, C., (eds). *Proceedings 2nd International Conference on Concrete Repair* (BRE press). p.235.

Jaeger, H., 2001. The "echo state" approach to analysing and training recurrent neural networks. Technical Report 148. German National Research Centre for Information Technology.

Kohenen, T., 1990. The Self Organising Map. *Proceedings of the IEEE* 78(9). p.1464–1480.

Lorenzi, A., Silva Filho L.C.P., Campagnolo J.L., 2004. Using a back-propagation algorithm to create a neural network for interpreting ultrasonic readings of concrete. *In WCNDT 2004*.

Palade, V. & Bocaniala, C.D., 2006. In Palade, V., Jain, L., Bocaniala, C.D. (eds), *Computational Intelligence Methodologies in Fault Diagnosis: Review and State of the Art*, London: Springer.

Pandey P.C. & Barai, S.V., 1995. Vibrational signature analysis using artificial neural networks. *Journal of Computing in Civil Engineering* 9(4). p. 259–265.

Rumelhart, D.E., Hinto, G.E., Williams R.J., 1986. Learning internal representations by error propagation. *Parallel Distributed Processing, Explorations in the Microstructure of Cognition*, 1.

Shawe-Taylor J. & Cristianini, N., 2000. *An introduction to Support Vector Machines and other kernel-based learning methods*. Cambridge: Cambridge University Press.

Sherratt, S., 2006. Doctoral Progression Report. Unpublished Report. Keele University.

Vapnik, V.N., 1998. *Statistical Learning Theory*. New York: Wiley.

Virmani, Y.P., Koch, G.H., Brongers, M.P.H., Thompson, N.G., Payer, J.H., 2001. Corrosion Costs and Preventative Strategies in the United States, *Report by CC Technologies Laboratories, Inc. to Federal Highway Administration (FHWA), Office of Infrastructure Research and Development*. Publication No. FHWA-RD-01-156.

Concrete Solutions – Grantham, Majorana & Salomoni (Eds)
© 2009 Taylor & Francis Group, London, ISBN 978-0-415-55082-6

Investigation of cracking in foundation bases affected by DEF using underwater ultrasonic probes

Michael G. Grantham

MG Associates/CET Safehouse Group and Queen's University Belfast, UK

ABSTRACT: Work previously carried out by ourselves at a structure in Plymouth in Devon, UK, in 1998 showed that cracking in foundation bases in a void area underneath one of the buildings was due to secondary or delayed ettringite formation (DEF). The structure was built in the hot summer of 1976 and used a local high alkali cement, both of which were factors in causing the DEF.

Petrographic examination of similar bases had shown significant microcracking. A query arose as to whether a zone of reduced cracking might exist in a conical shaped zone beneath the columns, where loading may have reduced the cracking. In this paper, we report the results of an ultrasonic survey using downhole probes (known as "sonic coring") and also confirmatory work using petrographic methods.

1 INTRODUCTION

Delayed ettringite formation is a rare phenomenon in concrete. It occurs when the concrete temperature at the time of placing or soon afterwards exceeds about 70°C. This suppresses the formation of ettringite in the concrete, which instead appears later, after hardening, in an amorphous form. The sulfate concentration in the pore liquid is high for an unusually long period of time in the hardened concrete. Eventually, in the presence of moisture, the sulfate reacts with calcium- and aluminium-containing phases of the cement paste and the cement paste expands. Due to this expansion empty cracks (gaps) are formed around aggregates. The cracks may remain empty or later be partly or even completely filled with ettringite. (Concrete Experts, 2006). In its early stages, crystalline ettringite is not seen, though some darkening of the paste may be observed (Grantham et al, 1999). Once exposed to moisture, significant cracking can result in the concrete with expansions of 5 mm/m or more being observed.

In this work, remedial action was being taken by the owner of the structure and their technical advisors, Scott Wilson Kirkpatrick, to inhibit further expansion of the concrete, using a ring beam cast around the bases. A query arose as to whether, under load from the columns bearing onto the foundation bases, expansion due to DEF may have been suppressed. In particular, the investigation aimed to show whether a conical shaped crack-suppressed zone in the base beneath the column might exist.

The investigation was carried out used a combination of ultrasonic investigation using downhole probes in twin core holes drilled into the top of one of the bases and also petrographic examination of the removed

concrete cores by plane and polarised light microscopy of thin sections of the concrete.

1.1 Earlier work

An earlier investigation into the bases in question was carried out in 1998 and resulted in a paper in 1999 (Grantham et al, 1999). This concluded that the cracking in the bases was the result of the combined effects of thermal cracking due to heat of hydration and also DEF.

2 THE INVESTIGATION

2.1 Ultrasonic investigation

Two pairs of 100 mm diameter core samples were removed by diamond drilling close to the column in the base examined. Figure 1 shows the arrangement.

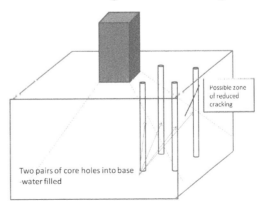

Figure 1a. Test arrangement.

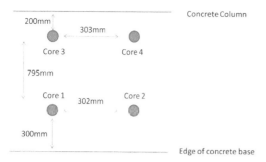

Ultrasound Tests – Base E5

Figure 1b.   Test arrangement top of base.

Initially, it was feared that filling the holes with water for the sonic coring would be unsuccessful. In that case it was planned to use wallpaper paste as the couplant; in the event water worked perfectly well without significant leakage.

Once water filled, an ultrasonic survey was carried out in the holes using a PUNDIT 7® manufactured by CNS Farnell and using their underwater ultrasonic probes. Readings were taken between pairs of holes

with the probes immersed in water and level with each other, at 25 mm intervals down the holes Typical test results are presented in Table 1.

2.2   Petrographic examination

- The samples were examined as received and photographed
- A longitudinal polished plate was prepared from each core. Two polished plates were prepared from one sample: one from the outer end and one from the inner end.
- The polished plates were examined with a binocular microscope and by the method of point counting.
- Fluorescent resin impregnated thin sections were prepared from each core. The thin sections were taken from the inner ends of three Samples and from the outer ends of the remaining two samples. Each thin section measured approximately 56 × 70 mm.
- The thin sections were examined with a Zeiss petrological photomicroscope.

2.2.1   Electron microscopy/microprobe analysis

- Two samples were selected for EMP analysis. The samples were prepared as high quality polished surfaces using diamond abrasives.

Table 1.   Typical Results of Ultrasonic Downhole Tests.

Core Holes 3-4				Core Holes 1-2			
Depth (mm)	Transit Time (us)	Path Length (mm)	Pulse Velocity (km/s)	Depth (mm)	Transit Time (us)	Path Length (mm)	Pulse Velocity (km/s)
12.5	67.0	303	4.5	12.5	66.0	303	4.6
37.5	67.1	303	4.5	37.5	66.1	303	4.6
62.5	66.8	303	4.5	62.5	66.2	303	4.6
87.5	67.6	303	4.5	87.5	66.6	303	4.5
112.5	66.9	303	4.5	112.5	66.5	303	4.6
137.5	67.5	303	4.5	137.5	66.0	303	4.6
162.5	67.9	303	4.5	162.5	66.9	303	4.5
187.5	68.6	303	4.4	187.5	69.7	303	4.3
212.5	68.3	303	4.4	212.5	70.0	303	4.3
237.5	69.3	303	4.4	237.5	68.0	303	4.5
262.5	69.0	303	4.4	262.5	68.0	303	4.5
287.5	69.4	303	4.4				
312.5	69.5	303	4.4				
337.5	69.2	303	4.4				
362.5	70.1	303	4.3				
387.5	70.0	303	4.3				
412.5	69.6	303	4.4				
437.5	70.5	303	4.3				
462.5	69.0	303	4.4				
Core Holes 2-4				Core Holes 1-3			
12.5	160.1	795	5.0	12.5	160.2	795	5.0
37.5	158.0	795	5.0	37.5	158.8	795	5.0
62.5	157.6	795	5.0	62.5	158.2	795	5.0
87.5	155.4	795	5.1	87.5	158.2	795	5.0
112.5	154.7	795	5.1	112.5	158.6	795	5.0
137.5	154.4	795	5.1	137.5	157.4	795	5.1
162.5	155.7	795	5.1	162.5	157.4	795	5.1
187.5	155.5	795	5.1	187.5	156.4	795	5.1
212.5	156.9	795	5.1	212.5	156.4	795	5.1
237.5	155.0	795	5.1	237.5	155.2	795	5.1
262.5	156.0	795	5.1	262.5	156.9	795	5.1
				287.5	157.1	795	5.1

- The polished surfaces were coated with carbon and examined with a scanning electron microscope.
- Chemical analyses were made of the materials present in the binder of each sample using an energy dispersive X-ray microanalysis system attached to the electron microscope. X-ray mapping techniques were used to examine the distribution of ettringite in the cement paste and around aggregate surfaces.

2.3 Categorising the Level of DEF

It is generally possible by petrographic examination to obtain an indication of the severity of DEF from the relative abundance of ettringite and the distribution of ettringite as follows:

- *Low severity DEF:* Ettringite-filled peripheral microcracks occur in small quantities around aggregate surfaces. This is usually visible even where the degree of expansion resulting from DEF is very low or non-existent.
- *Moderate severity DEF:* Ettringite-filled peripheral microcracks occur around a small proportion of the fine aggregate surfaces and are moderately abundant around coarse aggregate surfaces. Ettringite-filled cracks in the paste are rare.
- *High severity DEF:* Ettringite-filled microcracks in the paste are abundant. This is most commonly encountered only where the degree of expansion resulting from DEF is relatively high. Ettringite-filled cracks in the paste are commonly associated with voids in pores in the cement hydrates that are completely infilled with ettringite.

On the basis of the above outlined classification of severity, the samples examined from the inner ends of the cores showed high severity DEF. No unambiguous evidence was found for the occurrence of DEF in the thin sections prepared from the outer 70 mm of the cores.

The lack of DEF in the outer 70 mm of the cores and the abundance of DEF at the inner ends of the cores is consistent with heat buildup during curing being greater in the concrete at depth than at the surfaces and is characteristic of DEF distribution in large volume concrete.

The estimated composition of the samples by point counting is given in Table 2.

The cement contents of all samples examined petrographically were unusually high and would be expected to contribute to the potential heat build up during the curing of the concrete and the amount of sulfate available for DEF.

2.4 Electron microscopy and microprobe examination of Samples 7350/3 and 8

The microprobe confirmed the abundance of ettringite around aggregate surfaces in these samples and also showed the presence of patchy concentrations of ettringite within pores in the cement hydrates. The X-ray map in Figure 2 illustrates the distribution of ettringite around aggregate surfaces and in the paste in these samples.

The chemical analyses made of the cement paste (Table 3 shows a typical example) show the sulfate contents of the paste without obvious ettringite filled cracks or voids to be on average a little lower than would be expected for typical hydrated Portland cement and this is regarded as being indicative of some migration of sulfate from within the cement hydrates to centres of ettringite formation on aggregate surfaces and in pores in the cement hydrates. The unusual development of ettringite in voids in some of the aggregate particles (Figure 35 in Appendix B) indicates that pore fluids in the cement paste have carried the sulfur, calcium and aluminium into pores and voids in the aggregate particles as well as onto the surfaces of the aggregate particles.

Table 2. Compositions by point counting.

Laboratory Reference	7350-1 inner	7350-3 inner & outer	7350-9 inner	7350-10 outer
Sample Reference	E18 C1	E18 C3,a,b,c	L15	N10
VOLUME PROPORTIONS:				
Paste %	37.6	37.3	40.0	36.8
Aggregate %	59.4	60.5	59.4	61.0
Void %	3.0	2.2	0.6	2.2
Normalised volume proportions (excluding void)				
Paste%	38,8	38,1	40,2	37.6
Aggregate%	61.2	61.9	59.8	62.4
Water/cement ratio (assessed petrographically)	0.50	0.50	0.50	0.50
WEIGHT PROPORTIONS:				
Aggregate (kg/m^3)	1653	1670	1613	1684
Cement (kg/m^3)	474	466	492	460
Water (kg/m^3)	237	233	246	230
Total (kg/m^3)	2364	2369	2351	2374
Aggregate/cement ratio	3.5	3.6	3.3	3.7
Cement content (Wt. %)*	21.2	20.8	22.2	20.4

2.5   Ultrasound Testing

The ultrasound tests showed that, despite the deterioration indicated from the petrography, the test results suggested that the concrete was still quite competent. Levels of UPV in excess of 4 km/s are normally regarded as indicating reasonable concrete strength and integrity. Both aggregate type and moisture content influence the results, however, and in this case, the limestone aggregate used and the probable high moisture contents will both have tended to increase the values obtained.

The ultrasound method correlates well with dynamic modulus (and so indirectly with strength, too). Studying the results from the test holes suggested that there was not a great change in UPV with depth, though a possible trend for a slight reduction in UPV was observed, which would be consistent with the petrographic findings of increased levels of ettringite with depth.

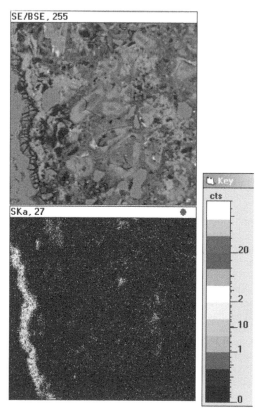

Figure 2.   X-ray map for sulfur showing the contact between the surface of a limestone coarse aggregate particle and the surrounding cement paste. The aggregate particle has an ettringite-filled peripheral crack around its surface.
Scale: Each image represents an area measuring about $200 \times 200\,\mu m$

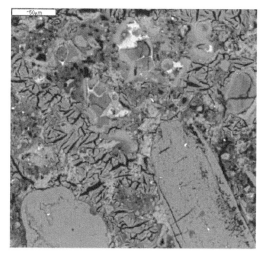

Figure 3.   *Polished surface, backscattered electron image:* View showing an area of paste with abundant ettringite. Some of the ettringite occurs within cracks and some ettringite occupies irregular voids in the paste.

Table 3.   Sample 7350/3: Microprobe analyses of the paste.

	0.1 mm paste area without ettringite filled cracks or voids	0.1 mm paste area without ettringite filled cracks or voids	0.1 mm paste area with ettringite filled cracks or voids	0.1 mm paste area without ettringite filled cracks or voids	0.1 mm area of paste with ettringite-filled pores and voids	0.1 mm area of paste close to ettringite-filled crack
SiO_2	17.9	19.2	17.4	19.3	14.3	20.8
TiO_2	0.3	0.4	0.2	0.3	0.2	0.2
Al_2O_3	5.1	6.1	5.4	4.9	5.6	5.1
Fe_2O_3	2.3	2.7	2.1	2.4	1.5	2.0
MnO_3	−0.1	0.3	0.2	0.1	0.1	0.0
MgO	0.7	0.9	0.6	1.0	0.7	0.9
CaO^{Note1}	71.0	67.9	69.6	69.2	73.8	68.6
Na_2O	0.3	0.3	0.3	0.4	0.1	0.1
K_2O	0.9	1.2	0.8	0.8	0.8	0.9
SO_3	1.6	1.1	3.4	1.6	2.8	1.5

Note: The analyses are in weight %, are normalised to 100% and exclude CO_2 and H_2O.
Note 1: The high calcium contents reflect the unavoidable inclusion of limestone dust in the areas analysed with the electron microprobe.

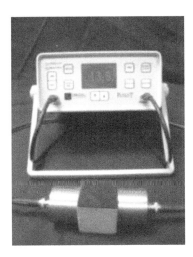

Figure 4. PUNDIT 7 UPV Tester used in investigation.

The specific reason for using ultrasound in this case was to attempt to establish whether the level of microcracking in the concrete decreased with depth, adjacent to the column, which may have indicated that expansion was being inhibited in the compression zone around the concrete columns. There seemed little evidence to support this, in fact, as mentioned above, perhaps the reverse was true, though the changes in UPV were so small that they are probably not very significant.

3 SUMMARY

- The petrographic examination confirmed that the cracks occurring to the bases had resulted from secondary or delayed ettringite formation.
- The ultrasound data suggested that the concrete remained reasonably competent, despite the microcracking and formation of DEF
- There seemed little evidence that the level of microcracking decreased with depth, suggesting that a zone of reduced expansion in the compression zone in the foundation base under the column probably did

not exist. The evidence, if anything, suggested the reverse – that cracking increased slightly with depth, which was confirmed by the petrographic work.

4 CONCLUSIONS

- The investigation confirmed that a zone of reduced microcracking in the concrete in the compression zone of the foundation under the column did not exist.
- The concept of using sonic coring by ultrasonic downhole probes worked well and is a potentially useful method for examining buried defects in structures.
- Remedial measures have been undertaken to the bases in question, using a ring beam cast around the base to contain further expansion and to maintain the integrity of the bases. At the same time, measures to exclude moisture in the void area beneath the building have been taken, to minimize future expansion.

ACKNOWLEDGMENTS

Grateful thanks are extended to Mike Eden of Geomaterials Research Services (Now Sandberg) for providing the excellent petrographic work and to CNS Farnell for the loan of their underwater probes. Finally, grateful thanks are extended to Babcock Marine Division for their permission to publish this paper.

REFERENCES

Bungey, J. Millard, S and Grantham, M. *Testing of Concrete in Structures*, 4th Ed 2006, Taylor & Francis
Collepardi, M. A state-of-the-art review on delayed ettringite attack on concrete, Elsevier Science, 2003
Concrete Experts , 2006, Accessed 10 April 2009 http://www.concrete-experts.com/pages/def.htm
Grantham M, Gray M. and Eden, M. "Delayed Ettringite Formation in Foundation Bases – A Case Study. *Proceedings of the 8th International Structural Faults and Repair Conference* 1999. Engineering Technics press.

Concrete Solutions – Grantham, Majorana & Salomoni (Eds)
© 2009 Taylor & Francis Group, London, ISBN 978-0-415-55082-6

A ferromagnetic resonance probe

S.L. Sherratt, L.J. North,* N.J. Cassidy & P.W. Haycock

Institute for the Environment, Physical Sciences and Applied Mathematics, Keele University, Staffordshire, UK

S.R. Hoon

School of Environmental and Geographical Sciences, Manchester Metropolitan University, Manchester, UK

ABSTRACT: Ferromagnetic resonance is a means of probing the nature of magnetic materials. The technique relies on the resonant absorption of microwaves at a frequency which matches the energy of spinwaves in the material. The observed resonant signal from steel and its corrosion products is different, in principle enabling discrimination between steel reinforcement that is in good condition from corroded. However, FMR has generally been used as a laboratory based technique for characterizing small samples of materials. This paper describes an extension of the method through the use of ultra-wideband antennae to allow the characterization of the surface of steel reinforcements within concrete. Use of data collected at a series of resonant conditions allows for the extraction of information about the steel-cement interface by means of perturbed parameter auto-deconvolution. The paper explains both the free-space system for use in the field and a transmission line based analogue for calibration purposes.

1   INTRODUCTION

When a ferrimagnetic or ferromagnetic material is placed in a static magnetic bias field it absorbs electromagnetic radiation. The frequency at which this absorption occurs depends on the material and the bias field strength. This phenomenon is known as ferromagnetic resonance (FMR) and is the electron analogue of nuclear magnetic resonance (NMR), the principle behind MRI machines (refer to Blakemore (1985) and Soohoo (1985) for a more detailed description). In the early stages of corrosion, the initial iron oxides formed are lepidocrocite (γ-FeOOH) and goethite (α-FeOOH) which at room temperature are paramagnetic and antiferromagnetic, respectively (Cornell & Schwertmann 2003). Over time, ferrimagnetic magnetite (Fe_3O_4), and to a lesser extent maghemite (γ-Fe_2O_3) and weakly paramagnetic haematite (α-Fe_2O_3) are formed. All the corrosion products exhibit, to a greater or lesser extent, FMR behavior.

Ground penetrating radar is a widely used tool for sub-surface imaging in civil engineering (McCann & Forde 2001; Arunachalam et al. 2006; Sbartai et al. 2006; Shaw et al. 2005). It is a non-invasive, non-destructive method that employs high frequency electromagnetic radiation to detect buried objects, surfaces and defects. The frequencies used in high resolution civil engineering GPR applications are, coincidentally, those at which FMR may be observed by magnetic corrosion products.

By applying a magnetic field to a section of corroded steel reinforced concrete while probing it using GPR FMR, absorption in the magnetic corrosion products can be observed. The amount of absorption is very small; however, by comparing GPR results from the same area but acquired with no magnetic bias field it is possible to differentiate these small changes from the background noise.

The amount of electromagnetic radiation that passes through a material is described by its complex transmission coefficient, z^*. It is a function of frequency, ω, length, L, complex relative magnetic permeability, μ_r^*, and complex permittivity, ε_r^* and is given by.

$$z^* = \exp\left(iL(\omega/c)\mu_r^*\varepsilon_r^*\right) \qquad (1)$$

The amount of energy absorbed as it is transmitted through a material depends on the imaginary components of the permeability, μ_r'', and permittivity ε_r''.

The amount of energy reflected at the interface between two materials, for example corrosion and concrete, is described by the reflection coefficient, Γ. It is a function of corrosion permeability, μ_{rC}^*, and permittivity, ε_{rC}^*, and the permittivity of the surrounding concrete ε_{rS}^*. It is given by:

$$\Gamma = \left(\sqrt{\mu_{rC}^*/\varepsilon_{rC}^*} - \sqrt{1/\varepsilon_{rS}^*}\right)\Big/\left(\sqrt{\mu_{rC}^*/\varepsilon_{rC}^*} + \sqrt{1/\varepsilon_{rS}^*}\right) \qquad (2)$$

Applying a bias field causes FMR which alters the magnetic permeability of the corrosion, μ_{rC}^* and,

* Present address: School of Applied Sciences, Cranfield University, Cranfield, UK

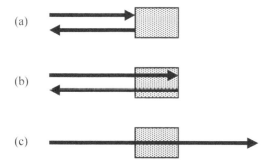

Figure 1. Signals acquired from the sample: reflections from the front face (a), back face (b) and transmitted (c).

Figure 2. A scanning electron micrograph of micro-particulate magnetite. The micrograph shows individual crystals of varying sizes from less than 50 nm to greater than 1 μm.

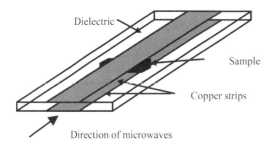

Figure 3. Parallel strip transmission line geometry.

therefore, the amount of signal that is reflected, transmitted and absorbed.

Therefore, the FMR phenomenon may be observed directly and used to locate and identify magnetic corrosion products. Despite steel being ferromagnetic its permittivity is so high that the energy is totally reflected and, therefore, not absorbed. This means that the steel GPR signal will not vary when a bias field is applied.

This technique has advantages compared with other currently used methods as it directly detects the presence of corrosion and is completely non-destructive.

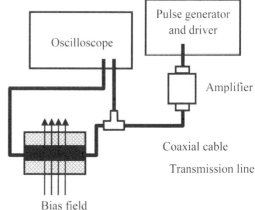

Figure 4. Experimental setup for the transmission line system.

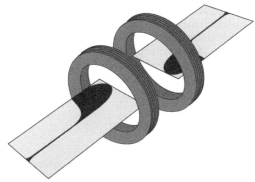

Figure 5. Vivaldi antennae and magnets used in the free space system.

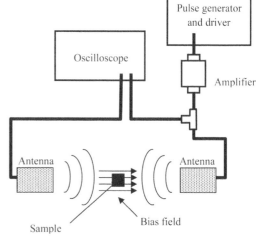

Figure 6. Experimental setup for the free space system.

In a laboratory setup, both the reflected and transmitted GPR signals can be collected; however, when out in the field it is likely to only be the reflected signal that is recorded. The reflected signal will contain

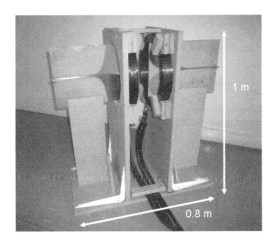

Figure 7. Experimental setup for the free-space system.

reflections from the front and back faces of the sample as shown in Figure 1.

Before this technique can be used in the field, it is necessary to characterize different corrosion products over a range of magnetic bias fields and frequencies. A transmission line has been designed to represent the free space system in a controlled laboratory environment. Using a method developed by Nicolson & Ross (1970) and Weir (1974) it is possible to directly calculate μ_r^* and ε_r^* of the sample. However, this involves taking some calibration measurements, which is not possible on site.

2 EXPERIMENTAL PROCEDURES

2.1 Sample preparation

Nano-particulate and micro-particulate magnetite and haematite samples were prepared for the transmission

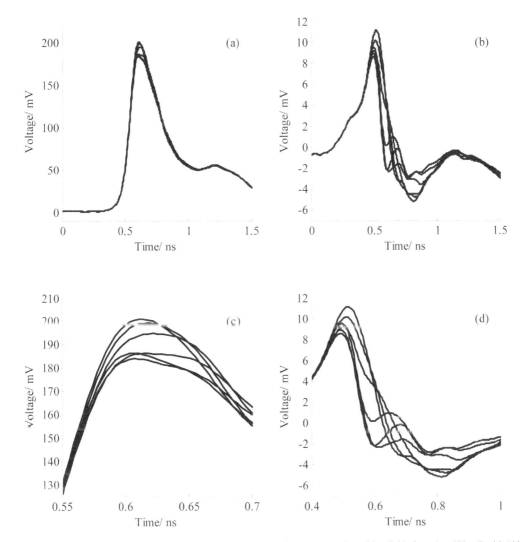

Figure 8. Transmitted (a) and reflected (b) signals from nano-particulate magnetite at bias fields from 0 to 500 mT with 100 mT increments. Parts (c) and (d) show the same data as (a) and (b) respectively but in more detail.

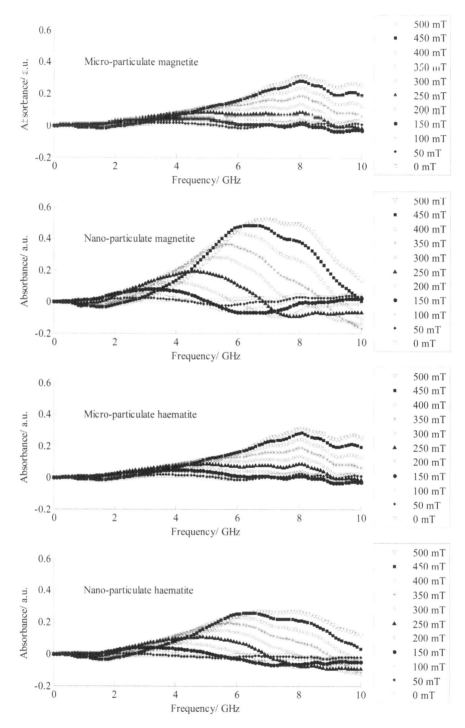

Figure 9. Absorbance from micro-particulate and nano-particulate magnetite and haematite.

line by mixing the iron oxide with epoxy resin at a mass ratio of 60:40. These were then placed into a Teflon mould and left to harden. The samples contained 57 mg of iron oxide and had a volume of 53.5 mm³.

A $10 \times 10 \times 10$ cm micro-particulate sample with a mass ratio of 50:50 was prepared for testing in the free space system. The micro-particulate magnetite crystals are shown in Figure 2.

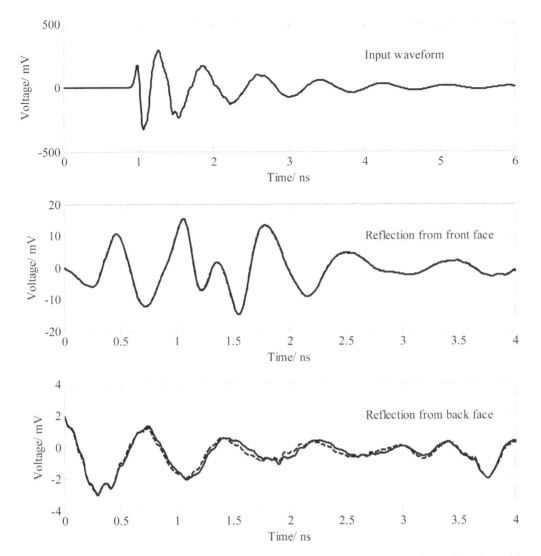

Figure 10. GPR signals from the free-space system: The input waveform and reflections from the front and back faces of the sample. Acquired with the magnet off (solid) and on (dashed).

2.2 Transmission line procedure

The transmission line was made from copper-etched FR4 PCB and consisted of a 4 mm parallel strip. The sample and parallel strips are shown in Figure 3.

A TD1110C tunnel diode pulse driver was combined with a Picosecond Pulse Labs TD1107 tunnel diode step generator to give a 230 mV step with 30 ps rise time. This was amplified with a MA COM MAAMGM0002 to give a −600 mV step with a 45 ps rise time. The signal was split and an Agilent 86100b oscilloscope was used to collect the input, reflected and transmitted signals. LabVIEW™ 8.0 was used to automate data collection with closed loop control of the magnetic field from 0 to 500 mT in 50 mT increments. The experimental setup is shown in Figure 4.

2.3 Free-space system procedure

A Vivaldi antenna was designed and manufactured at Keele Univeristy using the same method as for the transmission line. The shape was optimized to work over a broad frequency range with a plane wavefront. A Helmholtz pair electromagnet was made to produce a uniform field of 50 mT across the sample as shown in Figure 5. The experimental setup is shown in Figures 6 and 7.

2.4 Data analysis

The time-domain data collected by the transmission line system was processed and fast Fourier transformed using code written in Matlab. The frequency-domain data for each bias field was then

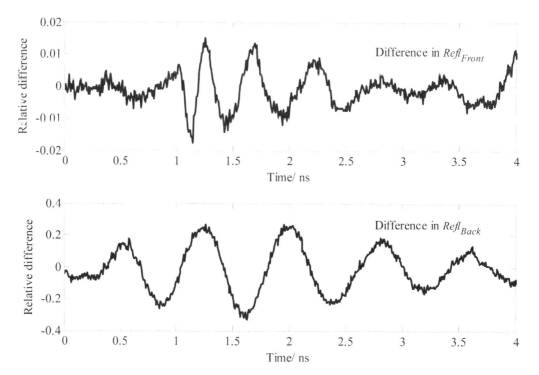

Figure 11. Relative difference between the reflected GPR signals with applied bias fields of 0 and 50 mT.

divided by the signal collected with a zero field to deconvolve the shape of the microwave impulse.

3 RESULTS AND DISCUSSION

3.1 Transmission line system

The raw time-domain data in Figure 8 shows that, as the bias field strength is varied, the amount of microwave energy both reflected and transmitted also varies. The Fourier transformed data for micro- and nano-particulate magnetite and haematite in Figure 9 show the imaginary part of the transmitted GPR signal, which in this case, can be thought of as absorbance. Clearly, as the field is increased from 50 to 500 mT the maximum absorbance frequency also increases from around 2–7 GHz. The relationship between resonance frequency and bias field is linear and also depends on the particle size and the type of iron oxide. The data in Figure 9 depends on the sample length and μ_r^* and ε_r^* of the transmission line dielectric.

Theoretically it is possible to retrieve the μ_r^* and ε_r^* of the sample from this data but this requires further improvements to the transmission line.

3.2 Free space system

The GPR signals collected from the free space system were:

- the input waveform
- the reflection from the front of sample (Ref_{Front})

- the reflection from the back of sample (Ref_{Back})
- the transmission through the sample

The total reflection signal was easily separated into Ref_{Front} and Ref_{Back} by time-domain gating.

Figure 10 shows the raw reflected data collected with magnetic bias fields of 0 and 50 mT. The GPR signals are clearly very similar. However, with a good signal-to-noise ratio, slight differences can be extracted, particularly for Ref_{Back}. The difference between the two GPR signals is shown in Figure 11 where it has been normalized to the maximum amplitude; $Ref_{Front} = 15$, $Ref_{Back} = 1.2$. The maximum difference compared to the amplitude is $Ref_{Front} \sim 0.02$, $Ref_{Back} \sim 0.26$. The small variation in Ref_{Front} is due to the change in the reflection coefficient from the air to the sample. Ref_{Back} has a larger difference as it is reflected from the back face of the sample as well as being absorbed due to passing through the sample twice.

4 CONCLUSIONS AND FURTHER WORK

The results presented in this paper make it clear that certain corrosion products absorb microwave frequency energy when placed in a static magnetic field. The absorbance line is broad and depends on the particle size and type of iron oxide. The transmission line system is a useful tool for characterizing the iron oxides.

Further work will include making improvements to the transmission line so that the relative permeability

μ_r^*, and relative permittivity, ε_r^* can be retrieved. The imaginary part of the permeability is a more accurate measure of the absorbance than the imaginary component of the transmitted signal, as it is a materials parameter and has no dependence on the system. Permeability spectra will be collected for different pure phase corrosion products as well as actual corrosion samples.

Using the free space system, it has been shown that applying a magnetic bias field of 50 mT does alter the GPR signal of the magnetite sample. Despite having a large sample size, a poor signal-to-noise ratio was achieved; however, the quality of the data can be improved by using higher microwave power. Currently 600 mV (15 VA) is spread over a broad range of frequencies giving a very low energy at each frequency, particularly with the free space system where not all the microwaves are received by the antenna. Using a narrower range of frequencies or, possibly, a set of frequency ranges and analysing the ratios to focus in on a specific corrosion 'signature' will also improve the sensitivity of the technique. Results from the transmission line will enable the appropriate choice of frequencies and bias field strengths, as well as providing a calibration with regard to the type of corrosion product present.

REFERENCES

Arunachalam, K., Melapudi, V.R., Udpa, L., Udpa, S.S. 2006. Microwave NDT of Cement-Based Materials Using Far-Field Reflection Coefficients. *NDT and E International* 39: 585–593.

Blakemore, J.S. 1985. *Solid State Physics. 2nd ed.* Cambridge: Cambridge University Press.

Cornell, R.M., Schwertmann, U. 2003. *The Iron Oxides.* Weinheim: Wiley-VCH.

McCann, D.M., Forde, M.C. 2001. Review of NDT Methods in the Assessment of Concrete and Masonry Structures. *NDT and E International* 34 (2): 71–84.

Nicholson, A.M., Ross, G.F. 1970. Measurement of the Intrinsic Properties of Materials by Time-Domain Techniques. *IEEE Transactions on Instrument and Measurement* 19: 377–382.

Sbartai, Z.M., Laurens, S., Balayssac, J.P., Arliguie, G., Ballivy, G. 2006. Ability of the Direct direct Wave of Radar Ground-Coupled Antenna for NDT of Concrete Structures. *NDT and E International* 39. 100 107

Shaw, M.R., Millard, S.G., Molyneaux, T.C.K., Taylor, M.J., Bungey, J.H. 2005. Location of Steel Reinforcement in Concrete Using Ground Penetrating Radar and Neural Networks, *NDT and E International* 38 (3): 203–212

Soohoo, R.F. 1985. *Microwave Magnetics.* New York: Harper and Row Publishers.

Weir, W.B. 1974. Automatic Measurement of Complex Dielectric Constant and Permeability at Microwave Frequencies, *Proceedings of the IEEE* 62(1): 33–36

Concrete Solutions – Grantham, Majorana & Salomoni (Eds)
© 2009 Taylor & Francis Group, London, ISBN 978-0-415-55082-6

Predicting service life of reinforced concrete structures based on corrosion rate measurements

D.W. Law
School of the Built Environment, Heriot-Watt University, Riccarton Campus, Edinburgh, UK

T.C.K. Molyneaux
Department of Civil, Environmental and Chemical Engineering, RMIT University, City Campus, Melbourne, Australia

F. Blin & K. Wilson
Advanced Materials Group, Maunsell Australia, Melbourne, Australia

ABSTRACT: The corrosion of reinforcing steel in concrete structures is one of the primary causes of concern for asset owners worldwide. When signs of distress are visually observed or anticipated a range of tests are commonly used to assess the condition of the structure and in particular the state of the reinforcing steel. These include the measurement of the instantaneous corrosion rate of steel bars. However, because corrosion is a dynamic process it fluctuates with time as changes occur in the ambient environmental conditions. In order to accurately predict the residual service life of a structure, a correlation must be made between the instantaneous corrosion rate measurement and the mean annual corrosion rate, as well as the level of corrosion that has already occurred and any subsequent changes in the corrosion rate in the future. This paper describes a model that accounts for these parameters in estimating the future service life of the structure.

1 INTRODUCTION

The corrosion of reinforcing steel in concrete structures is one of the primary causes of concern for asset owners worldwide. It has been estimated that the annual cost of the deterioration of reinforced concrete structures in Australia is in the order of $200 million (AUD) annually (Homayouni, 2004). At present many inspections and remediation works are conducted when there is visual evidence of corrosion, such as rust staining and/or cracking/spalling. When signs of corrosion damage are observed a number of techniques are commonly used to assess the condition of the structure and in particular the state of the reinforcing steel. These may include: chloride analysis, carbonation depth evaluation, resistivity testing, cover depth survey, half cell potential mapping and corrosion rate measurements (Concrete Society/Institute of Corrosion, 2004).

The Non Destructive Testing (NDT) techniques currently available provide information upon the condition of the structure, and that of the reinforcing steel. As most inspection surveys are undertaken once visual evidence of deterioration has been observed, it is essential to ascertain both the present condition of the structure and the existing level of corrosion. In some cases it may be possible to break the concrete out to the reinforcement to determine the actual level of corrosion present. However, the destructive nature of this test makes it limited in application.

The environmental conditions also influence the results obtained not only at the time of the measurement but also in the period before testing (Millard et al, 2001) In order for corrosion to occur, moisture and oxygen need to be present; as such, the amount of rainfall over the period leading up to the measurements can have a significant effect. Rainfall prior to the readings can lead to a measured corrosion rate that is higher than the average annual value. On the other hand, a dry period of several days prior to the reading can yield a low or negligible corrosion rate, which again may not reflect the mean value. Temperature has also been shown to have an effect on the corrosion rate and hence the time of day or time of year can affect the result (Millard et al, 2006).

While electrochemical half-cell surveys provide a quick method of locating probable corrosion activity and indicating the likely severity of the problem, this method is both indirect and only qualitative. As such this technique does not provide a measurement of the ongoing rate of corrosion of the steel reinforcement. The chloride profile on the other hand, in conjunction with a cover survey, can provide information on the likelihood of as well as a time estimate of corrosion initiation. In order to obtain information on the rate of corrosion, a Linear Polarization Resistance (LPR) measurement can be used to provide an instantaneous measure of the corrosion rate under the prevailing environmental conditions at the time of testing.

In order to estimate the residual service life of the structure it is necessary to use the NDT data to both establish the current condition of the reinforcing steel (i.e. level of existing corrosion) and to then extrapolate the data to provide a model for the corrosion rate in the future, the associated loss of steel and subsequently the structural impact of this loss. This assessment may include the time to the initiation of corrosion, the time to cracking, the time to spalling, or the time until structural failure becomes a possibility.

2 TEST METHODS

Prior to any detailed testing a visual survey should be undertaken, in which the location, size and orientation of any defects observed is recorded. Where possible, the survey should include any breakout of the rebar to measure the loss of section in the reinforcing steel. The environmental conditions at the time of the test, as well as for the preceding days, should be recorded. A visual survey can be most effective when an agreed system is used to assign a condition rating to the level of deterioration observed (Blin et al, 2008). The visual inspection should also be used to select the location(s) of any detailed surveys. These location may be selected to;

1. Establish a baseline of corrosion level measured against the observed visual deterioration. This baseline may be used to estimate the deterioration expected in other similar areas of the structure, once corrosion has reached this level in the future
2. Investigate the condition of areas identified as at risk of active corrosion, though no visual deterioration has been observed to date. These may be based upon experience of areas where corrosion generally occurs in this type of structure or on analysis of data available for the structure.
3. Survey general areas of the structure to identify the corrosion risk

At the conclusion of the cover survey, a half cell survey should be undertaken, as this will provide an indication of the likelihood of corrosion within the area tested (C876-91, 1999) This data should be used as a guide as the potential is dependent upon a range of factors. For instance, a very low potential of $-500\,mV$ or less (Ag/AgCl) may not be due to corrosion, but to saturation of the concrete and oxygen starvation (e.g. concrete in lower tidal zone). The half cell results may also be used to select the location for the LPR measurements: typically the areas displaying the most negative potentials are expected to be those with the highest level of corrosion. However, if the potential is very low, as mentioned above, LPR measurements are not necessary as they are likely to produce very low corrosion rates (the rate in such instances would be controlled by the diffusion rate of oxygen through the concrete).

From the LPR measurement the polarisation resistance, R_p can be determined by taking the Stern-Geary constant as $26\,mV$ for active steel and $52\,mV$ for passive steel (Andrade et al, 1996) The solution resistance, R_s, can be found by applying a high frequency input signal between the auxiliary electrode and reinforcing bar and measuring the resulting current. The total corrosion current measured, I_{corr} is then related to the area of reinforcing steel being polarised, A, to give the corrosion current density.

$$i_{corr} = \frac{I_{corr}}{A} \qquad \text{Equation 1}$$

As the auxiliary electrode is placed on the surface of the concrete it has been suggested that only the top half of the bar is polarised (Gonzalez, J.A. et al, 1985). Another consideration is the length of bar polarised. It is often assumed that A is the "shadow area" beneath the auxiliary electrode. However, there is a high degree of uncertainty as to whether the perturbation current only affects the steel directly below the auxiliary electrode, or spreads outwards over a larger unknown area. Efforts have been made to confine the polarising current using an annular guard ring, but even this is not always entirely successful, (Law et al, 2000).

To be able to maximise the accuracy of the assessment of the area of steel beneath the auxiliary electrode the cover survey should be used to position the auxiliary electrode directly above the rebar. The area of bar polarised may be estimated by equation 2, where h is the diameter of the auxiliary electrode and r the radius of the reinforcing steel.

$$A = 2\pi hr \text{ (single bar)} \qquad \text{Equation 2}$$

In order to model the residual service life, the chloride content at the rebar and the chloride diffusion coefficient must be known. A chloride profile can be obtained by taking a core from the structure. The apparent chloride diffusion coefficient can then be obtained using Crank's solution of Fick's 2nd Law of Diffusion (DuraCrete, EU-Project (Brite Euram III) No. BE95-1347, 1999).

3 SERVICE LIFE MODEL

The estimated corrosion rates, coupled with the other NDT data may be used to predict the residual service life of a structure. This model can both provide an estimate of whether the structure will achieve the required design life and facilitate the development of a cost effective management plan.

As stated above, the environmental conditions can influence the corrosion rate and in order to model the future performance of the structure the instantaneous corrosion rate, i_{corr}, must be related to an annual mean corrosion rate. This can be achieved in a number of ways, though a degree of uncertainty exists with each. The preferred approach is to take a number of instantaneous corrosion rate measurements over the course of a year to allow a mean value to be determined. The statistical parameters and particularly the standard deviation can then be used to provide a level of confidence to be assigned to the predictive model. The larger the number of data points the greater the degree of confidence.

As it is generally not possible to take a large number, or often more than one measurement due to commercial restraints, the value of i_{corr} measured must be used to estimate the annual mean corrosion rate, i_m. This can be achieved, in order of simplicity by;

a) Assuming that i_{corr} is the mean corrosion rate. This is acceptable if the environmental conditions at the time of the measurement are typical of those found throughout the year,

b) Assume that i_{corr} is the active corrosion rate, providing that measurements were undertaken in wet conditions, and then determine the number of wet days per year from meteorological records. Assume zero corrosion in dry conditions and i_{corr} in wet conditions and take an average over the year,

c) Compensate for i_{corr} by comparing the environmental conditions during measurement with conditions over the year (i.e. making an assessment of how representative the day of measurement is compared to the average weather condition).

The initiation of corrosion can be estimated using the estimated chloride diffusion coefficient. The calculation of the diffusion coefficient should account for changes in the diffusion coefficient with time[10,11] and the build-up of chlorides on the surface of the concrete. In marine structures the exposure period until a constant surface value is reached may be taken as being instantaneous[12]. However, for those structures where deposition comes from the application of de-icing salt or atmospheric particles a rate of deposition should be applied (DuraCrete, EU-Project(Brite Euram III) No. BE95-1347, 1999)

Thus, at the time of the inspection the time-adjusted diffusion coefficient may be determined and hence the time of corrosion initiation may then be estimated. This latter is defined as the time at which the chloride concentration at the rebar is calculated as exceeding the threshold level required for corrosion to initiated (DuraCrete, EU-Project(Brite Euram III) No. BE95-1347, 1999). The section loss to the current point in time may be estimated by assuming the mean annual corrosion rate measured from the time of the initiation of corrosion. However, the rate of chloride induced corrosion has been shown to increase with time (Bamforth, 1997, Andrade, et al, 1996). As such, while the application of the current mean annual rate will give an upper bound for section loss, a more accurate estimate may be achieved by predicting the mean annual corrosion rate each year from the time of initiation, t_i, to the current time, t.

An exponential relationship between chloride concentration and corrosion rate (Equ. 3) has been proposed (Bamforth, 1997). Where b is a constant, C_x is the chloride content by weight of sample at the

Table 1. Chloride content at the rebar and measured corrosion rate.

Corrosion Rate (μm/year)	Chloride Content (% by weight of sample)	Corrosion Rate (μm/year)	Chloride Content (% by weight of sample)
38	0.06	16	0.05
81	0.09	4	0.03
11	0.14	31	0.07
1	0.20	8	0.08
2	0.62	0.01	0.02
36	0.15	1	0.03
22	0.08	3	0.01
3	0.08	0.2	0.03
2	0.05		

bar and CR is the corrosion rate in micron/year. When C_x is less than the threshold value, CR = 0. An upper limit on the corrosion rate is also proposed of 80 micron/year. This is based upon actual site measurements, Table 1, where the maximum corrosion rate measured on a number of sites throughout the Australasian region has been 81 micron/year. This data has been obtained on coastal structures and it may be that a lower value should be applied on other structures, where continual wet/dry cycling does not occur. Other authors have suggested limits of 120 micron/year in the tidal zone, 60 microns/year for airborne seawater and cyclic wet/dry and 6 microns/year for wet, rarely dry conditions (Andrade et al, 1996).

$$CR = 0.55.e^{b.Cx} \qquad \text{Equation 3}$$

The data indicates that each structure, and indeed, each location on a structure should be treated differently. While an average b value may be used, based on either all chloride and corrosion rate data available or on all data from a single structure, a higher degree of accuracy may be achieved by fitting the b value for each location. Using the annual CR value the total penetration, P(t) estimated by equation 4 is:-

$$P(t) = \sum_{t_i}^{t} CR \qquad \text{Equation 4}$$

These equations assume a single pit on the steel surface at any one point. Because it may be possible for two, or more, pits to be located on the same section of circumference of the bar, a conservative estimate for section loss may be achieved by notally using a value of 2P(t) for section loss.

A number of models have been proposed for the time to cracking (CONTECTVET, 2001, Vidal et al, 2004). Based upon the CONTECTVET model, the time to cracking can be estimated by

$$p_{cr} = (83.8 + 7.4c/d_b - 22.6f_{ct.sp})/1000 \qquad \text{Equation 5}$$

Where p_{cr} is the corrosion penetration to crack initiation, c is the cover, d_b the bar diameter and $f_{ct.sp}$ the tensile splitting strength. Thus, when $P(t_f) = p_{cr}$ that will be the time to cracking of the concrete, where $P(t_f)$ is the future penetration with time.

$$P(t_f) = P(t) + \sum^{t_i} CR \qquad\qquad \text{Equation 6}$$

Hence, the time to failure can be estimated by calculating the residual cross section, A_{res},

$$A_{res} = (\pi(d_b - 2P(t_f))^2)/4 \qquad\qquad \text{Equation 7}$$

The residual cross section can then be used to determine when structural failure will occur. This can be done by setting a safety margin for the loss of section such as 10%. At the point where $A_{res}/A < 90\%$ the structure is considered unsafe and the end of useful service life has been reached.

4 CONCLUSIONS

The use of on site corrosion rate measurements can be used, together with other non destructive tests to predict the residual service life of a structure. This involves a number of steps;

- Measuring the corrosion rate and noting the environmental conditions at the time;
- Converting the instantaneous corrosion rate to a mean annual rate;
- Establishing the chloride diffusion coefficient;
- Determining the time at which corrosion was initiated;
- Estimating the corrosion penetration (section loss) at the current time;
- Predicting the time to cracking of the cover concrete; and
- Predicting the time to structural failure by determining the residual reinforcement cross section.

Based upon these predictions, possible future action can be recommended. Such recommendations may include the need for further modelling, additional testing, remedial action, or in extreme cases the immediate closure of the structure.

REFERENCES

Andrade C. & Alonso C., 1996 Corrosion Rate Monitoring in the Laboratory and on Site, *Constr. & Build. Mat.*, 10, No.5, pp. 315–328

ASTM C876-91, Standard Test Method for Half-Cell Potentials for Uncoated Reinforcing Steel in Concrete, 1999. (Withdrawn, 2008)

Bamforth P. B., 1997, Guide for Prevention of Corrosion in Reinforced Concrete Exposed to Soil, Partners in Technology Programme Contract CI39/3/231, A Predictive Model for Chloride Induced Corrosion and It's Use in Defining Service Life

Bamforth P. & Pocock D., 2000, Design for Durability of Reinforced Concrete Exposed to Chlorides, *Workshop on Structures with Service Life of 100 Years or More*, Bahrain

Blin F, Law D. W., Dacre M. C., op't Hoog C. J., Gray B. & Newcombe R., 2008, Extension of Design Life of Existing Maritime Infrastructure – A Durability Perspective, Proceedings Corrosion Control 2008, Australian Corrosion Association, Wellington, New Zealand, 16–19th November 2008

Concrete Society/Institute of Corrosion, 2004, Electrochemical tests for reinforcement corrosion, *Technical Report No. 60*, Joint Concrete Society/Institute of Corrosion report, August

CONTECTVET, 2001, A Validated User Manual for Assessing the Residual Service Life of Concrete Structures, British Cement Association

DuraCrete, 1999, Modelling of Degradation, EU-Project (Brite Euram III) No. BE95-1347, Probabilistic Performance Based Durability Design of Concrete Structures Report

Gonzalez J.A., Molina A., Escudero M.L. & Andrade C., 1985, Errors in the electrochemical evaluation of very small corrosion rates – I. Polarisation resistance method applied to corrosion of steel in concrete, *Corrosion Science*, Vol. 25, No. 10, pp. 917–930

Homayouni F., 2004, Research eats at concrete cancer, *Campus Review*, 1004, pp. 10

Law D. W., Millard S. G. & Bungey J. H., 2000, Linear Polarisation resistance measurements using a potentiostatically controlled guard ring, *NDT & E International*, Vol 1, pp. 15–21

Millard S. G., Law D. W., Bungey J. H. & Cairns J. J., 2001, Environmental influences on linear polarisation corrosion rate measurements in reinforced concrete, *NDT and E International*, Vol. 34, pp. 409–417

Millard S. G., Ho H. S. & Law D. W., 2006, Field measurements of corrosion rate of steel in reinforced concrete, *Concrete Solutions* 2006, St-Malo, Brittany, France, June 27–29

Vidal T., Castel A., & Francois R., 2004, Analysing crack width to predict corrosion in reinforced concrete, *Cement & Concrete Research*, Vol. 34, pp. 165–174

Concrete Solutions – Grantham, Majorana & Salomoni (Eds)
© *2009 Taylor & Francis Group, London, ISBN 978-0-415-55082-6*

Rehabilitation planning and corrosion surveys of bridge decks from the soffit: Enhanced opportunities for corrosion assessment under traffic

Ulrich Schneck

CITec Concrete Improvement Technologies GmbH, Dresden OT Cossebaude, Germany

ABSTRACT: Experience from various corrosion surveys shows that potential mapping does no only give notice about possible corrosion activity on the reinforcement layer next to the measurement surface, but can indicate also about corrosion problems on rebar layers in greater concrete depth. With a careful approach for measurement and data interpretation it is possible to distinct reliably between corrosion activity on several reinforcement layers or various concrete depths. Furthermore, a corrosion survey can be done on the opposite side of a corrosion suspect concrete slab, e.g. on the soffit of a bridge deck slab, to access corrosion activity on the top side. This article describes the general approach and results of some practical cases.

1 INTRODUCTION

In most cases of corrosion surveys on bridge decks potential mapping is done when the site works have started and both asphalt layer and sealing have been removed from the concrete. Single trial openings of the asphalt which are made during the stage of planning for visual inspection – mainly of tendons and tendon ducts – are guided only by visual effects and may cause wrong conclusions about the extent of repair, and sound advice about the necessary concrete repair can usually given only after launching the site works.

Experience from various corrosion surveys (Sodiekat et al, 2002, Schneck, 2003) shows that potential mapping not only gives notice about possible corrosion activity on the reinforcement layer next to the measurement surface, but can indicate also corrosion problems on rebar layers at greater concrete depths or in the opposite side of the investigated concrete slab. With a careful approach for measurement and data interpretation it is possible to reliably distinguish between corrosion activity on several reinforcement layers or various concrete depths.

Furthermore, a corrosion survey can be done on the opposite side of a corrosion suspected concrete slab, e.g. on the soffit of a bridge deck slab, to assess corrosion activity on the deck side. This opens a new approach for rehabilitation planning: Condition surveys can be done well in advance of the site works without traffic interruptions, they can result in sound knowledge about the necessary extent of repair, and there is an extended choice of repair strategy using performance-based planning that is based not only on the stated damage but on understanding of the individual situation of a structure.

2 GENERAL APPROACH

Potential surveys are widely used to investigate reinforced concrete structures rapidly and non-destructively for (chloride induced) corrosion activity. The various factors that have great influence on the measurement result, including concrete humidity, do not allow a direct correlation of the measurement values to corrosion activity in many cases, and it is not possible to conclude directly from the measurement data about the state of corrosion on the reinforcement.

All corrosion surveys are based on indirect measurements, because none of the known measurement methods for larger areas

- potential mapping
- surface resistivity measurement
- concrete cover measurement
- chloride, concrete humidity and pH-analysis
- visual inspection of the concrete surface

can be evaluated due to a general, corrosion related pre-specification. The systematic combination of potential mapping data with other measurement results provides a much better basis for corrosion assessment by overlaying different data and eliminating the interpretation uncertainties of any measurement. Basic considerations for that approach have been given (Schneck,2003).

A more discriminating attempt at a corrosion survey is to distinguish between the corrosion active reinforcement layers of a concrete slab or massive reinforced concrete section. The common measurements can also be used in this case, but have to be interpreted rather differently. Fig 1 shows a general approach which has to be adopted for special investigation cases, but outlines the idea:

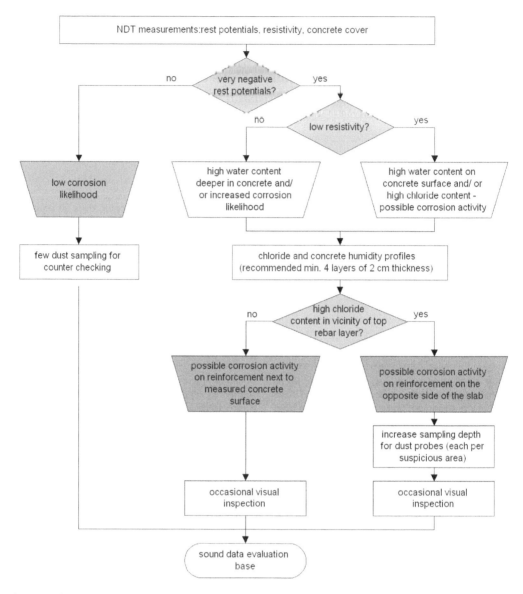

Figure 1. Flowchart for assessing corrosion problems in different depths of concrete slabs.

With an initial mapping of potentials and resistivity all areas suspected of possible corrosion activity can be found. It is important to know that basically only chloride induced macro cell corrosion can be detected by the origin of the potential measurement. Now all spots with negative potentials and/ or high potential gradients may have corrosion activity, but may have also a high water content in the concrete, causing low oxygen content and thus have no corrosion activity in this case, despite the high potentials.

The surface resistivity readings are influenced by both chloride and humidity content on the concrete surface – if the resistivity values are low, it can indicate possible corrosion activity on the reinforcement layer next to the measurement surface, if the potentials are negative in the same area. If the areas of negative potentials are within the areas of low resistivity, but smaller, the chloride or water front may have only partly reached the rebar level. Areas of special interest are those, where a spot of negative potential readings is found, but no correlation with the resistivity readings can be seen. There, a corrosion problem may be found on the opposite reinforcement layer of a concrete slab.

All these assumptions have to be verified by concrete sampling, chloride and water analysis and the visual inspection of the reinforcement and, as well, these investigations should be generally accompanied by conventional testing on delamination and spalling.

Figure 2. View on the Carola Bridge in Dresden.

It is advised to refer to a plot of the cumulative frequency of the potential readings in order to get an idea about potential values that separate the measured data into passive, intermediate and corrosion active values (Elsener, 2003)

If the corrosion problem is assumed on the opposite side of the measurement surface of the concrete slab anyway (e.g. the survey of a bridge deck from the soffit), the whole survey refers mainly to the potential mapping. Concrete cover information is less important, and the resistivity readings may indicate that in case of low measurement values a wet and/ or chloride containing spot has spread across the whole cross-section.

Regarding post-tensioned pre-stressed concrete, it is not possible to get information about possible corrosion activity on the tendon strands, because they are shielded by the ducts. But usually corrosion problems start at the surrounding reinforcement, and with the help of the corrosion survey at least areas can be found, where possibly tendons are corrosion affected as well. The verification is matter of a visual inspection. Cases, where chloride containing water has entered the tendon inside and has moved then to deeper situated areas along the tendon, where no outside corrosion activity takes place, are more difficult and have to be taken into account according to local circumstances.

3 PROJECT CASES

Three different monitoring cases are presented: a 16 cm high box girder floor slab, a 30 cm high bridge deck and an 80 cm thick bridge deck – the latter in a comparative survey on the soffit and on the top side:

3.1 Carola Bridge in Dresden

The Carola Bridge in Dresden is an important link across the river Elbe. It consists of 3 separate structures. The dimensions and reinforcement layout follow the forces and moments, thus it has a very dynamic appearance.

On the pre-stressed, 40 year old structure some concrete replacement was done in 1994 in order to repair damage caused by a collapsed inner drainage pipe for surface water and chloride ingress in the 16 cm high box girder floor slab. From 2000, an increasing number of hollow zones had been ob-served under the mineral coating that was attached after the repair to the floor slab surface. The monitoring project has been introduced in a paper by Schneck (Schneck, 2003) and one special aspect of the corrosion survey will be discussed here:

The potential and resistivity measurement on the top side of the box girder floor slab showed, in one of the sections, an area of ca $10 \, m^2$ with a high likelihood of corrosion activity according to the rest potentials (ringed in figure 3 – between -300 and $-400 \, mV$ vs. CSE). There was no obvious correlation with either low concrete cover readings or low surface resistivity.

Dust samples were taken up to 8 cm of concrete depth, and the reinforcement did not show signs of corrosion when inspected visually. The chloride values were all below 0.5%, related to the cement mass, and at a concrete humidity of ca 3% there was no explanation for the negative potential readings.

After consulting older inspection reports the situation could be clarified: There were chloride analysis data available from the soffit of the box girder floor slab in exactly these coordinates, showing chloride values between 1.5 and 2% of cement mass. So the potential mapping gave notice about corrosion activity on the reinforcement layer on the soffit of the floor slab. The high chloride content was caused in the concrete by chloride containing water left inside the box girder via emergency outlets in the floor slab before penetrating the top side of the slab. The sampling depth of 8 cm was not deep enough to trace the chloride entering from the soffit.

Combining all available information, the corrosion active area could be identified (and get repaired later). This kind of damage case was unexpected during the survey, but could be traced since the data taken on the top side did not fit and required further action.

3.2 Rohrbach Bridge close to Stuttgart

This ca. 35 year old highway bridge should get a higher traffic capacity from 3 to 4 lanes by changing the emergency lane into a regular traffic lane. An additional slab attached on top of the old bridge deck was planned to take and to distribute the new live loads. It was essential to know the extent of possibly damaged transversal tendons; the structural analysis allowed every second tendon to fail, but never two neighboring tendons. An extended report can be read in a paper by the author (Schneck, 2005).

The corrosion survey was done from the soffit, covered almost the whole soffit surface ($3,650 \, m^2$) and resulted in ca $700 \, m^2$ zones with negative rest potentials and the opportunity of chloride induced corrosion activity ($=20\%$ of the total area, see also fig 7 at the end of the article). The potential- and resistivity mapping was made in a grid of 60×60 cm using the CITec Survey apparatus, available from CITec. Within

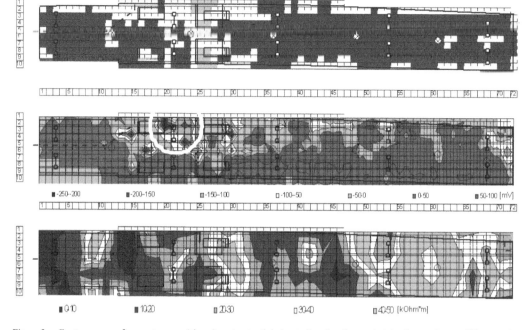

Figure 3. Contour maps of concrete cover (above), rest potentials (center) and surface resistivity (bottom) – possible corrosion activity within ringed area.

Figure 4. View under the Rohrbach Bridge (right structure), photo courtesy A. Bechert.

these suspect areas dust samples were taken in 5 layers of 4 cm thickness (20 cm sampling depth), which got quite close under the top surface of the bridge deck slab. The survey gave the following results:

- Clearly the sealing under the asphalt was leaking in several areas, and large amounts of water could concentrate in the concrete – up to 8%, related to the concrete mass. The high water content was mainly causing negative rest potentials (ca 650 m² see fig 6)
- Only within 30 m², distributed on 10 sub areas, an increased chloride content was found and had caused corrosion activity – see verification spots in fig 6. Water entering the concrete is only partly chloride containing (in Winter), and chloride diffusion in water saturated concrete is much slower than capillary suction in moderately wet or dry concrete.

- The remaining 3,600 m² of surveyed area could be stated as free from corrosion activity

With that knowledge, the proposed construction principle could be verified already: Even without the visual inspection of the tendons, from all obtained data, the possible corrosion active areas were small enough to launch the project without expecting bad surprises.

Immediately after removing the asphalt layer and seal from the bridge deck some months later, all 10 suspect areas were visually inspected, and all tendons were in excellent shape. Only the covering reinforcement was partly heavily corroded, and the corroded areas were exactly shaped as the areas indicated by the potential survey from the soffit.

The following data were measured:

- Rest potentials and surface resistivity on the upper side and soffit in a grid system of 60 × 60 cm (see pictures in fig 5; the measurement on the soffit was partly limited by the heavy current of the river)
- The concrete cover on the top side
- The chloride and humidity profiles in sampling depths of 20/40/60/80 mm in locations that were suspected of corrosion due to the results of the non-destructive measurements
- concrete damage, rebound values and other corrosion relevant data

3.3 Bridge across the Weisse Elster river

With a non-destructive corrosion survey, corrosion problems on both reinforcement and post-tensioned

446

Figure 5.   Cross-section of the bridge deck, measurements on top side and on the soffit.

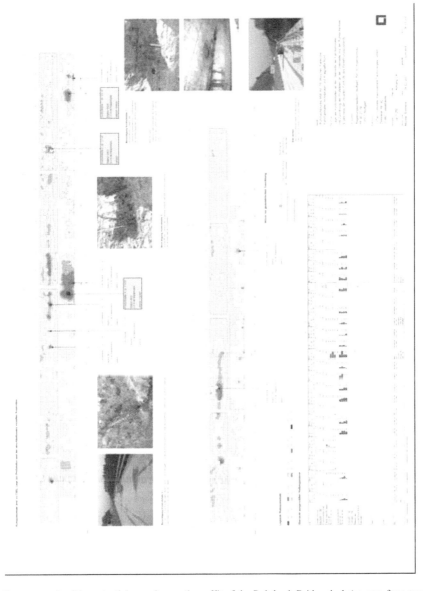

Figure 6.   Contour graph of the potential mapping on the soffit of the Rohrbach Bridge deck (as seen from top side) with corrosion suspect areas, results of chloride and humidity analysis and plan of coordinated for the visual inspection.

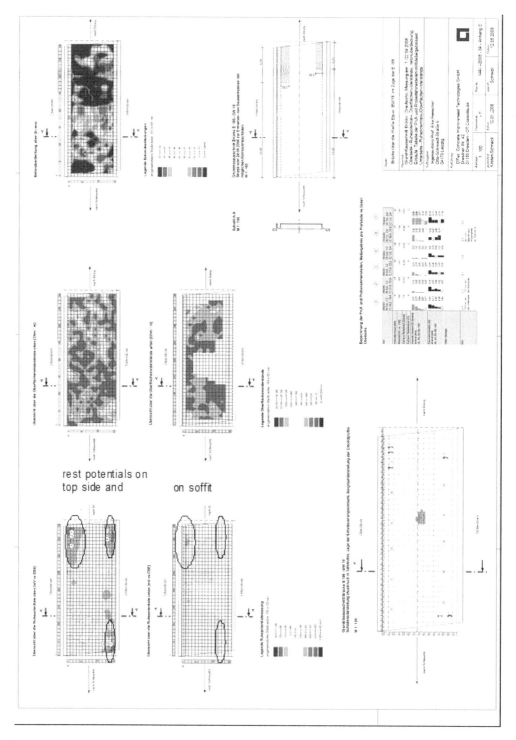

rest potentials on top side and on soffit

Figure 7. Bridge across the Weisse Elster river – potential mapping results from top side and from the soffit, corrosion suspect areas marked on both measurement surfaces.

steel were evaluated on the superstructure of a bridge along the B186 road across the Weisse Elster river, close to Leipzig. As a special subject, a comparative measurement of rest potentials and surface resistivity on the deck surface and on the soffit of the 80 cm thick bridge deck was used to establish how reliable corrosion measurements can be done from the soffit and under traffic for identifying corrosion problems on the deck surface side.

The drawing with the measurement results (see fig 7) shows that the rest potentials (top left) have a similar profile on both the upper side and soffit and are ca. 150 mV more positive on the soffit. This corresponds with experience from other projects, but this is the first time this has been done on such a thick cross section.

Potential differences of more than 400 mV suggest corrosion elements, but in the suspect locations, only low or negligible chloride contents were found (ranging between 0.1 and 0.4% in the reinforcement vicinity, related to the cement mass). Due to the very high water content in the concrete (between 5 and 10%) it is likely that the oxygen content at higher concrete cover was lower than on the surface (top right in fig 7), and this absence of oxygen caused more negative potentials in these areas and large potential differences across the surface without indicating corrosion activity. This is not uncommon in damp concrete.

It could be stated that no chloride induced corrosion activity on the reinforcement and the pre-stressing steel should take place and that no corrosion products should be found.

The corrosion activity of post-tensioned steel had to be assessed here with an indirect approach, because corrosion problems related to chloride ingress are likely to be seen first on the tendon ducts. Locations which are suspected of corrosion activity had to be opened and inspected visually. For this case, as forecast by the survey results, both reinforcement, tendon ducts and pre-stressing steel were found in perfect condition.

4 CONCLUSIONS

In summary, following conclusions can be made from the corrosion survey cases:

- It has been shown that non-destructive, full scale corrosion surveys are possible without traffic disruption. According to the measurement results, locations of possible damage can be inspected precisely for safe, object related data evaluation.
- Even at thick cross sections, rest potentials measured at the soffit are similar to the respective values measured at the deck surface – only shifted by an offset value. Hence it is also possible to assess corrosion activity on the top side by measurements on the soffit.
- The decision whether possible damaged areas are exhibiting real corrosion problems must be made by supporting investigations. However, areas with no corrosion activity can be clearly identified.
- The extent and appropriate timing of repairs can be determined exactly at the stage of design and planning and can be concentrated on necessary areas. This raises the cost efficiency and durability of repairs to a significantly improved level.

REFERENCES

Elsener, B.: Half cell potential measurements – potential mapping on reinforced concrete structures. RILEM recommendations from TC 154 – EMC, Materials and Structures 36, Aug – Sept 2003, pp 461–471

Sodeikat, Chr., Gehlen, Chr., Schießl, P.: Auffinden von Bewehrungskorrosion mit Hilfe der Potentialfeldmessung. Ein ungewöhnlicher Praxisfall. Beton- und Stahlbetonbau 97 (2002), no. 9 pp 437–444

Schneck, U.: Zerstörungsfreie Ortung von Bewehrungskorrosion an Stahlbeton-platten "von der anderen Seite" – ein Praxisbericht, Dresden 2003

Schneck, U., Grünzig, H., Mucke, S., Winkler, T.: Making Informed Decisions on a High Traffic Bearing Bridge with the Support of a Non-Destructive Condition Survey – a Case Study. Proc. 1st Concrete Solutions Conference, St Malo, 2003

Schneck, U., Bechert, A.: Erweiterung der Rohrbachbrücke im Zuge der BAB A8 auf 4 Fahrspuren – Absicherung des Projekts durch zerstörungsfreie Korrosions-untersuchung unter Verkehr, Proc. 18. Dresdner Brückenbausymposium, TU Dresden, 2008

Concrete Solutions – Grantham, Majorana & Salomoni (Eds)
© 2009 Taylor & Francis Group, London, ISBN 978-0-415-55082-6

Delamination detection of FRP sheet reinforced concrete specimens using a microstrip patch antenna

S.K. Woo & Y.C. Song
Korea Electric Power Research Institute, Daejeon, Korea

H.C. Rhim
Yonsei University, Seoul, Korea

ABSTRACT: A series of experimental studies have been conducted to evaluate the capability of a microstrip patch antenna system in detecting delamination in Fibre Reinforced Polymer (FRP) sheet reinforced concrete. For that purpose, a prototype microstrip patch antenna was developed with 15 GHz center frequency and 1 GHz bandwidth. For the comparison, a horn antenna with 15 GHz center frequency and 10 GHz bandwidth was used for the measurements of the same specimens. The laboratory size specimens had dimensions of 600 mm (length) × 600 mm (width) × 50 mm (thickness) with a series of delaminations of 300 mm (length) × 300 mm (width) × 5, 10, 15 mm (thickness). FRP of 1.5 mm thickness and epoxy of 3 mm thickness were placed on the top of artificially created delamination to represent an actual FRP reinforced concrete condition. In all cases, the delamination has been successfully identified. Also, it was shown that the imaging results in the microstrip patch antenna were improved by signal processing.

1 INTRODUCTION

Fibre Reinforced Polymer sheet (below FRP sheet) is widely used as a structural reinforcement for construction or engineering work structures. It is used to strengthen the internal force capacity of the bottom of beams and slabs for construction structures (An et al, 2005 and Chen & Li, 2005) and to repair the bottom part of the plates at the edge of bridges or reinforce inside tunnels in engineering work structures (Choi et al, 2001 and Reed et al, 2005). The usage of FRP sheet is widely spreading as reinforcement since it is lighter than steel, does not rust, and is relatively easier to install. On the other hand, strengthening using FRP sheets can induce the destruction of the structure due to the delamination of the concrete and the bonded FRP sheet (Ko, 2006). The delamination of the FRP sheet and concrete is difficult to observe with the eye and as the range of the delamination is small, difficulties follow in distinguishing whether they have bonded after construction. Furthermore, as time passes and delamination at an early stage occurs, exploring this progress beforehand is needed to secure the safety of the structure.

The current delamination detection methods can be largely divided into ultrasonic wave inspection methods and electromagnetic wave measurement methods. For the ultrasonic wave inspection, the instrument is easy to use as it is simple and the price is lower compared to the measuring equipment for electromagnetic waves, but the measuring time can become longer since the part that is measured must be measured

in fixed lattices of about 5~10 cm (Bastianini et al, 2001). For the electromagnetic wave measurement method, the time required for the measurement is shorter than the ultrasonic wave method and a wide area can be measured at once, but it is expensive and technical difficulties still exist in analyzing the measured data (Bakhtiari et al, 1994, Kim et al, 2003).

In this paper, an experiment was carried out to find the delamination of FRP sheets and concrete by using the electromagnetic method (Feng et al, 2002), and this paper aims to develop such measurement system. Measurement using electromagnetic waves requires the bandwidth and frequency of antennae that can perceive the width of the concrete and FRP sheet delamination. Antennae that are generally utilized have a frequency of 900 MHz~1.5 GHz, which has a low resolution to find delaminations of on the mm scale. Thus, a high-frequency antenna with a frequency range of 10~15 GHz was used and a signal treating method needed for this was developed.

Adequate antennae for this paper can be the horn antennae and microstrip patch antennae, and the horn antenna is more superior to the microstrip patch antenna for the antenna gain and frequency range. Therefore, for the microstrip patch antenna, such weaknesses have been overcome by designing through arranging the antenna devices. This paper conducted an experiment using a commercial horn antenna for a comparison test with the microstrip patch antenna. The microstrip patch antenna is an antenna with a center frequency of 15 GHz and a bandwidth of 1 GHz and for the comparison test, a commercial waveguide antenna

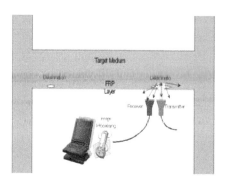

Figure 1. Schematic diagram of microstrip patch antenna measurement.

with a center frequency of 15 GHz and a bandwidth of 10 GHz was used for the horn antenna.

In this experiment, three FRP sheet strengthened specimens were measured that had 5, 10, 15 mm delaminations. As a result of analyzing the experimental result values by using the signal treating program, the existence and non-existence of delamination on the strengthened part of the specimen was checked.

2 MICROSTRIP PATCH ANTENNA

2.1 Characteristics of the microstrip patch antenna

The antenna that is generally used currently is the Ground Penetrating Radar (GPR) antenna of the horn antenna series, but the delamination measurement was carried out in this paper by using a microstrip patch antenna. The horn antenna has a wider frequency range compared to the 10 GHz microstrip patch and as the gain rate is high, the reliability of the measurements is high. However, the high price is a weakness compared to the patch antenna. The microstrip patch antenna used has a strength in which the price is 1/2~1/3 of that of the horn antenna, but since it has the weakness of a low frequency range of 1 GHz and a low gain rate in measurement, this was improved by designing through arranging each patch antenna.

The electromagnetic waves generated from the electric network are supplied to the antenna through electrical wires. Here, the electromagnetic waves transformed in the microstrip patch antenna reach the medium and transmissions and reflections occur in each level of the medium. Then, the existence/non-existence of delamination can be predicted as the difference in the time it takes for the antenna to absorb the reflected electromagnetic waves. The schematic diagram of the measurement system is shown in Fig. 1.

2.2 Design of the microstrip patch antenna

The determination of the size of the microstrip patch antenna is determined by width × length, and the size of the antenna should be made to be smaller than the size of the specimen which is subject to measurement. In addition, when considering the area and form of

arrangement of the sorted patch to increase resolution, the size is formed of 4 × 8 arrays of 150 mm × 80 mm.

The second characteristic that must be decided when designing an antenna is that without any time differences between each patch, the feeder of the electromagnetic waves must simultaneously synchronize the waves, when measuring the antenna. The synchronization of electromagnetic waves is achieved through feeders and synchronization is needed for each patch to emit the electromagnetic waves that were transmitted simultaneously. To have high directivity and a low side lobe, the distance between the arrays of antennae must be fixed appropriately and in the case where dielectric constant is from 2.0 to 2.5, which is a method generally used, antennae were designed to have an emitted wavelength with a separated distance of 0.75 to 0.9.

3 ELECTROMAGNETIC CHARACTERISTICS OF TEST SPECIMENS

Non-destructive tests of structures using radar are accomplished by analyzing phenomena such as reflections that are generated when electromagnetic waves strike the structures. Furthermore, although electromagnetic waves travel at the velocity of light c in a vacuum state or in the air, in a dielectric substance, the velocity (v) becomes slower as the medium has greater dielectric constant as in Eq. (1).

$$v = \left(\frac{C}{\sqrt{\varepsilon}} \right) \qquad (1)$$

where,

v = electromagnetic wave transmitting velocity;
C = velocity of light; and ε = dielectric constant.

Such a phenomenon has a direct influence on measurement with electromagnetic waves. Thus, setting electromagnetic properties in the process of a non-destructive test of concrete using radar is imperative in raising the accuracy and reliability of the test

As a fundamental stage for this, the electromagnetic characteristics of concrete, FRP sheets, and epoxy were measured. The dielectric constant and the loss coefficient exist as representative electromagnetic material properties of matter and in this paper, the dielectric constant that is a coefficient used directly in the experiment was measured over a high-frequency range of 1~25 GHz.

3.1 Test results of the dielectric constant of concrete

Since the electromagnetic characteristics of concrete are affected by the moisture content, they were measured by using an open-ended coaxial probe and a network analyzer on specimens with different moisture content. The standard of moisture content for the electromagnetic characteristics of concrete is presented in Table 1.

The concrete specimens were constructed to have a diameter of 100 mm and a thickness of 15 mm. When

Table 1. Moisture content of concrete specimens.

Specimen condition	Moisture content (%)	Moisture control
Dry 1 (absolute dry condition)	0.0	Condition in which furnace drying was conducted at 105° for 48 hours
Dry 2 (air dry condition)	2.6	Dried in air (Relative humidity 57%, temperature 22°)
Wet 1 (wet condition)	6.3	Covered in wet towel and preserved for 24 hours
Wet 2 (saturated condition)	8.6	Taking it out after immersing in a water tank for 24 hours

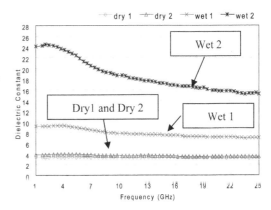

Figure 3. Dielectric constant of 21 MPa concrete specimen.

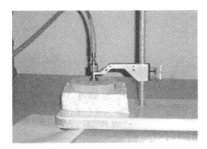

Figure 2. Measurement of concrete dielectric constant.

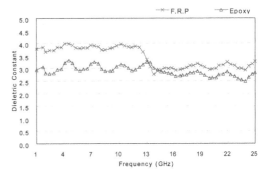

Figure 4. Dielectric constant of FRP and epoxy.

the thickness of the concrete specimen is greater than 5 mm at minimum, no influence of the material behind the specimen exists, but the thickness was made to be 15 mm considering the moist control of the specimen and the error from the fabrication process.

Four specimens were created for each condition in this paper, measured five times, and the mean values were compared. The frequency range was indicated as the high-frequency range (10~20 GHz). Concrete specimens were measured by dividing according to the results of the moisture content and the results according to the specimen strength (Fig. 2).

The dielectric constant according to the four types (absolute dry, air dry, wet, saturated condition) of moisture content is shown in Fig. 3. Even though the dielectric constant was greater as the moisture content was higher, almost no difference existed for the absolute dry and air dry conditions. The strength of specimens used in the experiment was 21 MPa and a dielectric constant of about 4.0 was derived in the air dry condition. Thus, 4.0 was used for the dielectric constant utilized in the experiment for concrete. Furthermore, for the air dry condition specimens, to be used in the experiment, almost no difference existed in dielectric constant even though the frequency range changed.

3.2 Test results of the dielectric constant of FRP sheet and epoxy

The dimensions of the FRP sheet specimens were constructed to be 120 mm × 120 mm in the form of

squares and as an element of error exists due to the thinness (about 1.5 mm), measurements were carried out by changing the plies from 1 ply to 5 ply. Moreover, epoxy was constructed to be a circle with a diameter of 100 mm and thickness of 30 mm and was measured.

As the results of measuring FRP sheets and epoxy, FRP sheets showed a dielectric constant value within the range of 3.5~4.0 and for epoxy, within the range of 2.8~3.3. The loss coefficient showed up as a value close to 0 that indicating that almost no electromagnetic loss occurred (Fig. 4).

Dielectric constant and the loss coefficient of FRP sheets show significant changes according the changes in frequency, but for epoxy, almost none exists. That is, it was identified that FRP sheets respond sensitively to frequency changes.

4 EXPERIMENTAL RESULTS OF ANTENNA

4.1 Test specimens

After dividing the conditions of the specimens into four sections as in Fig. 5, the experiment was conducted according to each section by inserting delaminations of three thicknesses of 5, 10, 15 mm. The samples used in the experiment were composed of four matters of FRP sheets, epoxy, Styrofoam, and concrete (21 MPa) as shown in Fig. 5, and here, since Styrofoam was used

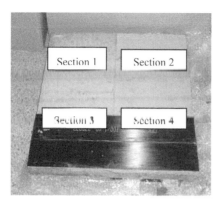

Figure 5. View of specimen.

Table 2. Composition of each measurement zone.

Section 1	Styroform + Concrete
Section 2	Concrete
Section 3	FRP + Concrete
Section 4	FRP + Styroform + Concrete

for the purpose of generating delamination and has the same dielectric constant as air, it signifies a zone where delamination has been formed. The structural diagram of each zone of specimens is shown in Table 2.

4.2 Microstrip patch antenna

The experiment was carried out in the order of small delaminations from 15 mm by utilizing a 4×8 microstrip patch antenna with a center frequency of 15 GHz and a frequency bandwidth of 1 GHz. Furthermore, after measuring the differences in the sections with and without delamination out of the specimens in the non-strengthened zone, the differences in the sections with and without delamination in the zone strengthened with FRP sheets were examined.

The measurement of delamination in the strengthened zone of all three types of specimens was possible out of the specimens with 5, 10, 15 mm delaminations, and as the thickness of delamination increased, the delamination measurement results became more distinct. Furthermore, the measurement of delamination was possible in both the strengthened and non-strengthened zones and it was identified that the strengthened zone reduced the differences of the measured results compared to the non-strengthened zone. Fig. 6 and 7 are the experimental results of measuring a 5 mm delamination which had the smallest differences in measurement in the experiment. On the existence and non-existence of delamination, differences were confirmed in both strengthened zones and non-strengthened zones.

The measurement results of the microstrip patch antenna in the non-strengthened zone of the specimen with 5 mm delamination are shown in Fig. 6. In Fig. 6, the measurement result of zone 1 where the concrete

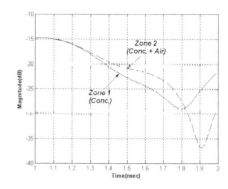

Figure 6. Microstrip patch antenna measurement result of zone 1 and zone 2 (5 mm delamination specimen).

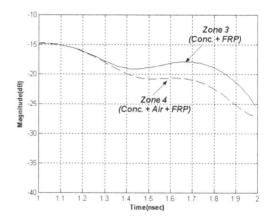

Figure 7. Microstrip patch antenna measurement result of zone 3 and zone 4 (5 mm delamination specimen).

was 50 mm thick is indicated as a solid line and the measurement result of zone 2 that had 5 mm delamination is presented as a dotted line. The differences are evident in all areas as checked in Fig. 6.

The measurement result of the microstrip patch antenna in the strengthened zone of the specimen with 5 mm delamination also proved that a precise measurement was possible (Fig. 7). In Fig. 7, the measurement result of zone 3 where the concrete was 50 mm thick is indicated as a solid line and the measurement result of zone 4 that had 5 mm delamination is presented as a dotted line. As shown in Fig.7, definite differences were identified in the zones from 1.3 ns to 2.1 ns.

4.3 Horn Antenna

To increase the reliability of the measurement for the microstrip patch antenna, the experiment was conducted by utilizing the waveguide antenna which currently is commercialized. The horn antenna has an equivalent center frequency of 15 GHz compared to the microstrip patch antenna, but as it had a wide frequency range of 10 GHz that is ten times that of the microstrip patch antenna, an accurate measurement result was expected. Similar to the microstrip patch

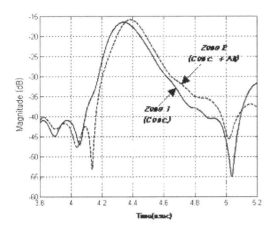

Figure 8. Horn antenna measurement result of zone 1 and zone 2 (5 mm delamination specimen).

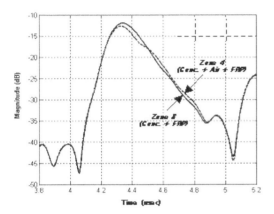

Figure 9. Horn antenna measurement result of zone 3 and zone 4 (5 mm delamination specimen).

antenna, the results showed that measurement was possible in all three types of specimens with 5, 10, 15 mm delamination and as the experiment was carried out on a specimen with thin delamination, it was confirmed that the differences decreased. Moreover, compared to the microstrip patch antenna, definite differences in the numeric values were observed.

Fig. 8 and 9 show the measurement results of using a horn antenna for the specimen with 5 mm delamination. Fig. 8 presents the specimen with a delamination thickness of 5 mm and the measured result in zone 2 where the thickness of concrete is 50 mm is shown as a solid line and the measured value of zone 1 that has a delamination of 5 mm is shown as a dotted line. As confirmed in Fig. 8, a distinct difference can be observed.

Fig. 9 is a figure showing the FRP sheet strengthening part of the 5 mm delamination specimen. Zone 3 is presented as a solid line and zone 4 as a dotted line. An evident difference can be identified in the 4.4 ns section and the 4.7 ns section.

5 CONCLUSIONS

The research results of this experiment are as follows.

(1) An antenna was developed for the delamination detection of Fibre Reinforced Polymer (FRP) sheet of reinforced concrete. For usability, an integrated antenna was developed, the delamination generated on the surface was measured, and to obtain a definite resolution, a high-frequency, wide bandwidth antenna with a center frequency of 15 GHz and a frequency bandwidth of 1 GHz was developed.

(2) The results of the capacity assessment of the microstrip patch antenna (center frequency 15 GHz, frequency bandwidth 1 GHz) showed that in both the strengthened zone and non-strengthened zone, delamination detection is possible.

(3) As a result of measuring by using a horn antenna (center frequency 15 GHz, frequency bandwidth 10 GHz) for the comparison test, a precise measurement of all three types of specimens used in the experiment was possible which is similar to the results of the microstrip patch antenna. In addition, since the horn antenna has a wide frequency bandwidth, an even more precise measurement result was obtained compared to that of the microstrip patch antenna. However, since an accurate measurement is also possible for the microstrip patch antenna, it is judged that the microstrip patch antenna, which has a relatively superior efficiency, is more suitable for the delamination detection of the FRP sheets.

ACKNOWLEDGEMENTS

This research program was conducted under the financial support of the Electric Power Industry Technology Evaluation & Planning (ETEP) of Korea and its kind support is gratefully acknowledged.

REFERENCES

An, S.H. & Lee, S.H. 2005. Flexural Strengthening Design of Reinforced Concrete Beams Strengthened with FRP. *Journal for the Architectural Institute of Korea Structure and Construction* 21(5): 51–58.

Bakhtiari, S., Qaddoumi, N., Ganchev, S. & Zoughi, R. 1994. Microwave Noncontact Examination of Debond and Thickness Variation in Stratified Composite Media. *IEEE Transactions on Microwave Theory and Techniques*. 42(3): 389–395.

Bastianini, F., Tommaso, A. & Pascale, G. 2001. Ultrasonic Non-Destructive Assessment of Bonding Defects in Composite Structural Strengthening. *Composite Structures* 53: 463–467.

Chen, C. & Li, C. 2005. Punching Shear Strength of Reinforced Concrete Slabs Strengthened with Glass Fibre-Reinforced Polymer Laminates. *ACI Structural Journal* 102(4): 535–542.

Choi, Y.G., Kwon, O.Y., Bae, G.J. & Cho, M.S. 2001. Study of Application of FRP Materials as a Tunnel Reinforcement. *Journal for the Korean Tunneling Association* 3(1): 11–19.

Feng, M., Flavis, F. & Kim, Y.J. 2002. Use of Microwaves for Damage Detection of Fibre Reinforced Polymer-Wrapped Concrete Structures. *Journal of Engineering Mechanics* 128(2): 172–183.

Feng, M., Liu, C., He, X. & Shinozuka, M. 2000. Electromagnetic Image Reconstruction for Damage Detection. *Journal of Engineering Mechanics* 126(7): 725–729.

Kim, Y., Jofre, L., Flaviis, F. & Feng, M. 2003. Microwave Reflection Tomographic Array for Damage Detection of Civil Structures. *IEEE Transactions on Antennae and Propagation* 51(11): 3022–3032.

Ko, H.B. 2006. Basic Research on the Bonding Capacity of FRP Sheets and Concrete. *Journal for the Architectural Institute of Korea Structure and Construction* 22(8): 69–76.

Reed, M., Barnes, R., Schindler, A. & Lee H. 2005. Fibre-Reinforced Polymer Strengthening of Concrete Bridges that Remain Open to Traffic. *ACI Structural Journal* 102(6): 823–831.

Rhim, H. & Buyukozturk, O. 2000. Wideband Microwave Imaging of Concrete for Nondestructive Testing. ASCE-Journal of Structural Engineering 126(12): 1451–1457.

CITec Concrete Improvement Technologies GmbH

Your partner for the extensive, performance-based assessment and the non-destructive rehabilitation of corrosion issues in reinforced concrete structures

Services

Qualified, non-destructive condition survey on corrosion damage, performance-based damage analysis and sound advice for appropriate maintenance actions

Non-destructive, careful treatment of corrosion damage after chloride attack and carbonation without traffic limitations according to CEN TS 14038

New: electrochemical desalination of sandstone and masonry - non-destructively, fast, with high efficiency also in deeper zones of the material, applicable under aspects of the preservation of historic monuments

Products

CITec Survey®: Integrated system for the condition survey and assessment of corrosion damages with online input of rest potential, surface resistivity, temperature, interface to HILTI Ferroscan and powerful tools for visualizing and evaluating of measurement data

Reference electrodes: MnO_2 miniature reference electrodes for potential measurement in conrete for briefly applications (e.g. electrochemical chloride extraction) and for permanent installation (cathodic protection)

CITec CeControl®, CombinationElectrodes: Worldwide patented procedure and tools for the non-destructive, efficient and individual rehabilitation of corrosion-damaged reinforced concrete structures

CITec Concrete Improvement Technologies GmbH
Dresdner Str. 42
D-01156 Dresden OT Cossebaude
Germany

Phone: +49 351 436 0130
Fax: +49 351 436 0134
e-Mail: citec@citec-online.com

www.citec-online.com

Test Equipment... you can rely on

Ultrasonic Pulse Velocity Test Equipment

The PUNDIT family of instruments is the leading range of portable test instruments for non-destructive integrity testing of concrete. Using any of the instruments, it is possible to detect the presence of cracks, voids and other defects. A wide range of accessories enables the PUNDIT family to be used in a variety of applications for testing many different materials.

Suitable for use in the research lab, commercial testing lab or on site, any of the instruments can be used for testing concrete, timber, ceramics, geological specimens, refractory materials, carbon electrodes and cast iron.

PUNDIT*plus*

The PUNDIT*plus* is the premier instrument of its type. With advanced functionality, it can provide users with values for transit time, path length, velocity and elastic modulus. With variable transmitter output settings and control over the pulse rate, it is the ideal choice for the advanced user. The instrument has a data-logging facility, a built-in RS232 output and is supplied complete with PC download software.

PUNDIT7

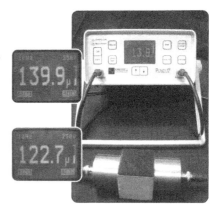

This instrument is a state-of-the-art replacement for both the PUNDIT 6 and the PUNDITpc. It is packed full of new features which include a novel signal strength indicator bar on the display. Perfect for use on site, simple and easy to use yet powerful enough for the expert user.

It can be controlled directly from a PC or factory control system giving the user full remote control; ideal for extended tests in the research lab or high volume production testing applications.

Ultrasonic Accessories

There is a whole range of ultrasonic transducers, exponential probes, wheel probes, attenuators, amplifiers and other accessories to make your testing easier. With the widest range of frequencies (24kHz to 1MHz) available to cover many applications from concrete, ceramics and timber through to cast iron. Waterproof, deep sea, load bearing and combined compression and shear wave transducers are also available.

www.cnsfarnell.com

CNSFARNELL

World Class Ultrasonic
Test Equipment

Concrete Test Equipment

Concrete Test Hammers

Rebound hammers are used to obtain an estimate of strength and quality of hardened concrete at the surface. These are widely used throughout the world to obtain a very quick evaluation of a concrete structure. There is a range of hammers to suit all applications including a standard unit and either paper or electronic recording hammers.

Resistivity Meter

The Resistivity Meter is a portable instrument designed specifically to measure the resistivity of concrete whose steel reinforcement has the potential to corrode. Resistivity measurements can be made in areas identified as high risk (using the SCHOLAR) to enable an estimate to be made of the probable rate of corrosion.

A test method has recently been developed enabling it to assess the chloride penetration resistance of concrete.

SCHOLAR

The SCHOLAR is a unique, combined logging cover meter and half-cell potential meter. With a host of features making it easy to use on site, and a huge memory capability, it is the ideal tool for anyone carrying out serious site investigation work, particularly for reinforced concrete structures. Standard configuration is as a cover meter but accessories include deep search and mini heads for cover measurement, a hand-held Half Cell and the Potential Wheel (a rolling half-cell).

Also often used with a Resistivity Meter to get an in-depth picture of the corrosion state of reinforced concrete.

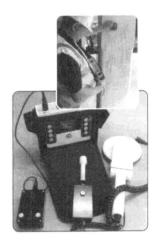

ERUDITE

The ERUDITE is an electronic resonant frequency test system used by quality control and research departments to determine the dynamic moduli of elasticity and rigidity of sample blocks of concrete or other materials. The ERUDITE automatically measures the fundamental longitudinal, torsional and transverse resonant frequencies. It also calculates the damping constant, Q. Primarily used for testing concrete, ceramics, refractories, stone, rock, carbon and graphite.

AUTOCLAM

Concrete permeability testing is made easy with the AUTOCLAM. It measures air and water permeability and water absorption into the surface zone of concrete and other building materials, including surface repair materials, without damage. It is an exceptionally good tool for assessing the durability of these surfaces when exposed to normal or aggressive environments.

MG Associates was formed in 1992 to provide an independent service in the testing and inspection of structures. In May 2008, we merged with CET Safehouse and are now part of the CET Safehouse group of companies. (see www.cetsafehouse.com). We specialise in durability surveys of large concrete structures such as concrete bridges, reinforced concrete car parks, and multi storey and high rise concrete buildings, but we also deal with domestic housing, offices, culverts - in fact all kinds of structures. In addition to our expertise in concrete, we also have expertise in the following areas:-

Plasters, Renders, Screeds, Mortars, Brickwork and Blockwork, Paints, Metals, including Steel and Cast-Iron, Ceramic Tiling

We also offer laboratory testing services ranging from chemical analysis, to physical testing and also petrographic (microscopic) examination of samples. This can be on samples taken by the Client or by our own site teams.

We also provide training courses in concrete as a material and in concrete inspection, testing and repair in the UK and abroad. Training courses are run in an air-conditioned training facility in Borehamwood and can also be run in-house for a Client directly

If you are unsure whether we can help please Email us.

Company Profile

The Principals have spent many years in the construction industry and have dealt with failures in structures of all types, from Structural Problems to Reinforcement Corrosion, ASR, Carbonation, High Alumina Cement, Chlorides, Sulfate attack, Concrete Cube Failures and many more. We also deal with most building materials, plasters, renders, mortars, screeds, sealants, mastics, plastics, paint, metals, timber, ceramic tiling - in fact just about anything related to buildings

Senior staff comprises Chartered Chemists and Engineers. We have a considerable amount of in-house engineering knowledge, but additional expertise is provided through external consultants, as rePrincipals lecture regularly on concrete inspection and repair. We also undertake expert witness work.

Contact Information

We can be contacted by telephone, post, fax or E mail.

Telephone

+44 (0)1332 813740

FAX

+44 (0)1332 813723

Postal address

MG Associates, Northdown House, Ashford Road, Harrietsham, ME17 1QW, United Kingdom

Email: info@concrete-testing.com

www.concretebookshop.com

Phone: +44(0)700 460 7777
Email: enquiries@concretebookshop.com

www.concretebookshop.com
The Concrete Bookshop is wholly owned by The Concrete Society

The Institute of
Concrete Technology

Delivering
experience and
professionalism to
the construction
industry

Fellows

Members

Associates

Students &
Graduates

Technicians

The Institute's aim is to promote concrete technology as a recognised
engineering discipline and consolidate the professional status of practising
concrete technologists worldwide.

For more information on membership grades and professional affiliation see:
http://ict.concrete.org.uk

Professional affiliate of the
Engineering Council UK

The Institute of Concrete
Technology - the professional
development and education wing
of the Concrete Society

CRL
CONCRETE REPAIRS LIMITED

multi-disciplined *structural* renovation

BS EN 1504 CRL COMPLIANT

Concrete inspection & testing
Concrete repair
Sprayed concrete/Guniting
Composite strengthening
Cathodic protection
Crack injection grouting
Protective Coatings

Cathite House, 23a Willow Lane, Mitcham, Surrey CR4 4TU
Tel: 020 8288 4848   Fax: 020 8288 4847

Email: mail@concrete-repairs.co.uk   Website: www.concrete-repairs.co.uk

OFFICES IN: LONDON   CHESTERFIELD   BRISTOL   FALKIRK   WARRINGTON

463

Author index